Linear Algebra with Applications

Linear Algebra with Applications

Charles G. Cullen
The University of Pittsburgh

Scott, Foresman/Little, Brown College Division
Scott, Foresman and Company   Glenview, Illinois • Boston • London

An interactive software package, for the IBM-PC and compatibles, is available in the *Instructor's Guide and Software Toolkit*. This software is designed to help teach the subject as well as to ease the computational effort required to do meaningful examples.

Library of Congress Cataloging-in-Publication Data

Cullen, Charles G.
  Linear algebra with applications.

  Includes indexes.
  1. Algebras, Linear.   I. Title.
QA184.C85   1988   512.5   87–26417
ISBN 0–673–18570–2

  2 3 4 5 6 - KPF - 93 92 91 90 89 88

Preface

Linear Algebra with Applications provides a carefully motivated introduction to linear algebra and some of its many applications. The text is suitable for introductory courses that include a significant number of students who are not majoring in mathematics. The students in this group are usually mathematically unsophisticated, possibly not having completed the full calculus sequence. Some applications to calculus are included in the text, but they are easily omitted so that the text is understandable to students who have had little or no calculus background.

The emphasis is on those topics from linear algebra that the students will encounter in their later work; extraneous material is carefully avoided. Interesting and accessible applications are included, in a nonintrusive way, throughout the text. The last two chapters are devoted to more-involved applications. The importance of the computer to the development of linear algebra and to its practical applications is acknowledged and incorporated. Applications to computer science are consistently emphasized, with many pointers for implementing the techniques of linear algebra on a computer. Interactive software (for the IBM-PC and compatibles or the Apple Macintosh), which actually helps to teach the subject, is included with the *Instructor's Guide and Software Toolkit*. This software can be most helpful in focusing attention on the interpretation of calculations, rather than on the details of the calculations themselves.

The treatment is strongly computational, with many examples. Proofs of theorems have been included when I felt that they added either to understanding or provided the basis for computational techniques; difficult, less relevant proofs have been omitted. The proof is frequently presented first, the formal statement of the related theorem acting as a summary statement. Where proofs are omitted, the material is presented in a logical manner so that all the details can be added by the interested student. Exercises provide the opportunity to complete or extend the theory; hints are often given. Because it is reasonable to expect some increase in mathematical maturity as the course evolves, the level of mathematical sophistication gradually increases as the book progresses. There is enough general vector-space theory to give students a feel for the generality of the subject without overwhelming them with abstract theory. There is an emphasis on reducing general vector-space questions to equivalent questions about systems of linear equations. In most cases the emphasis is on methods that involve first principles.

Because students learn best by doing, the 750 exercises form a very important part of the text. There are many computational exercises, plus exercises designed to reinforce, supplement, and extend the development of the text. The use of motivating exercises (experimental evidence in support of general results to follow) is a key feature of this text. All exercises that are referred to in the main body of the text are flagged with a dagger. Complete or partial answers to about half of the exercises are included in the text; those exercises have green numbers. The *Instructor's Guide and Software Toolkit* contains answers to the remainder of the computational problems and outlines of solutions for the theoretical problems. There is a set of review exercises at the end of each of the first five chapters.

Text

Chapter 1 begins with the linear equations problem and the elimination methods of solution. Matrix algebra is then developed and related to the elimination methods via row equivalence. There is a thorough discussion of systems with multiple right-hand sides that is used frequently in later chapters. An optional section discusses machine arithmetic, partial pivoting, and the *LU* factorization. By using these ideas it is possible to describe a reasonably robust linear equations solver that is similar to most production codes for solving this problem.

Chapter 2 develops the essential properties of the determinant function from a recursive point of view. Cramer's rule and the adjoint formula are developed and placed in their proper computational perspective. Eigenvalues and eigenvectors are introduced in Section 2.4 as an application of determinants; this section can be postponed until the beginning of Chapter 5 if desired.

Chapter 3 deals with vector spaces. There is a brief geometric introduction for $\mathcal{R}^2$ and $\mathcal{R}^3$ and then a discussion of $\mathcal{R}^n$. Most important vector space concepts are introduced first for $\mathcal{R}^n$ and then examined more fully in the discussion of general real vector spaces that follows. Vector norms, orthogonality, and the Gram-Schmidt process are treated only for $\mathcal{R}^n$. The chapter ends with a discussion of the change-of-basis problem.

Chapter 4 introduces linear transformations, their fundamental subspaces, and their matrix representatives. The algebra of linear transformations is developed, compared with the algebra of matrices, and used to interpret linear transformations on $\mathcal{R}^n$ geometrically. The representation theory for linear operators, developed in Sections 3.9, 4.5, and 5.1, is the hardest theoretical material in the text; this material can be omitted if the instructor is willing to give up the geometric motivation for the eigenvalue problem.

Chapter 5 uses the change-of-basis problem to introduce similarity and completes the discussion of eigenvalues, eigenvectors, and diagonalizability. Orthogonal similarity and reduction to triangular form are discussed in Section 5.4; this important and useful topic is frequently omitted from books at this level. There is an optional section on the power method.

Chapter 6 on linear programming requires only Chapter 1 as background. The problem is treated geometrically in the plane and then the tableau version of the simplex method is introduced and used in several interesting applications. The two-phase method is introduced so that one is not restricted to problems in standard form. The software is essential for this section.

Chapter 7 contains four independent applications: graph theory, least squares approximations, quadratic forms, and linear economic models. These sections can be covered any time after the indicated prerequisite sections have been covered.

Chapters 1–5 form the core of the text; the prerequisites for the remainder of the material are indicated in the following diagram.

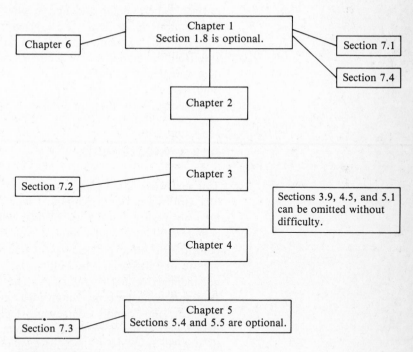

Software

The software provided to users of this text (IBM-PC version included with the *Instructor's Guide and Software Toolkit*) is very user-friendly and can be a valuable aid in teaching and learning the material. It consists of two main programs, MATMAN and MATALG. These programs are designed primarily to help teach linear algebra; they are not designed as production codes for large-scale problems. The programs are menu driven and require no programming knowledge; all that is required is to know where to insert the program disk and how to turn on the machine.

MATMAN (The MATrix MANipulator) is an interactive program that does row reductions; it can be used to solve linear systems, invert matrices, evaluate determinants, and solve linear programming problems using the tableaux implementation of the simplex algorithm. In the primary mode the user specifies the row operations to be performed and the machine does the arithmetic quickly and accurately. The arithmetic can be done using either rational or real numbers. An automatic switch can be unlocked by the instructor after the student has progressed to the desired point.

MATALG (The MATrix ALGebraist) is an interactive program for automating basic matrix operations. It can be used to save computation time with moderate-sized problems, to illustrate round-off error, and to provide experimental evidence in support of many of the theoretical observations in the text. These programs will run on most IBM-PC compatibles. A version that will run on the Apple Macintosh is available from the author. These programs were written in FORTRAN; the disk contains only the compiled versions of the programs. The program disk and the *Instructor's Guide and Software Toolkit* contain full documentation for these programs.

Acknowledgments

This text was started while I was on sabbatical leave at the University of California at Santa Barbara; I would like to thank the faculty for their hospitality and the use of their word-processing facilities during a most delightful winter term. I would also like to acknowledge the contribution of several classes of linear algebra students at the University of Pittsburgh who helped me class test and debug the original manuscript and its revisions. The software was written by John Burkardt, who took my rough ideas and turned them into accurate, efficient, and friendly courseware. I greatly admire his skill and his patience. The development of the courseware was partially supported by a grant from the College of General Studies of the University of Pittsburgh.

I wish to thank the staff at Scott, Foresman and Company for their excellent work in producing this text. In particular, Project Editor Elizabeth Brennen has been a pleasure to work with. I would also like to thank my colleagues at other schools who reviewed the manuscript and made many valuable suggestions. They include T. L. Herdman, Virginia Polytechnic and State University; Eugene W. Johnson, The University of Iowa; William J. Keane, Boston College; David Kraines, Duke University; John Petro, Western Michigan University; and Steven Pruess, Colorado School of Mines.

Charles G. Cullen

Overview

Contents

Linear Algebra with Applications

1

Linear Systems and Matrices

In the first chapter of this text we will consider techniques for the efficient solution of systems of linear algebraic equations. Such systems arise frequently in all areas of investigation in which mathematical modeling is a useful tool. Frequently the number of unknowns and the number of equations is very large, and computational assistance, in the form of a computer, is needed to solve the system. In this chapter we will introduce matrices and matrix algebra in order to present efficient methods for solving linear systems.

1.1   *Geometric View of Linear Systems*

Let us begin by considering the following scheduling problem, which is a simplified version of a typical business application.

A manufacturer of stuffed animals makes three products: Yogi Bear, Peter Rabbit, and Donald Duck. Each toy must pass through three separate departments: sewing, stuffing, and decorating. Yogi spends 24 minutes in the sewing department, 18 minutes in the stuffing department, and 9 minutes in decorating. Peter spends 16 minutes in sewing, 12 minutes in stuffing, and 8 minutes in decorating, while Donald requires 18 minutes of sewing, 9 minutes of stuffing, and 4 minutes of decorating. The sewing department can supply 50 hours of labor per day, the stuffing department 33 hours, and the decorating department 18 hours. How many of each animal should be produced in order to make full use of the available labor?

To analyze this situation we let

x = the number of bears produced;

y = the number of rabbits produced; and

z = the number of ducks produced.

The total usage of the sewing department is thus $24x + 16y + 18z$ minutes. Since the sewers are available for 50 hours or 3000 minutes we must have

$$24x + 16y + 18z = 3000.$$

The total usage of the stuffing department is thus $18x + 12y + 9z$ minutes and this department can supply 33 hours or 1980 minutes per day. Thus,

$$18x + 12y + 9z = 1980.$$

Similar analysis of the use of the decorators leads to the equation

$$9x + 8y + 4z = 1080.$$

Thus, the unknowns x, y, and z must satisfy each of the three equations

$$24x + 16y + 18z = 3000$$
$$18x + 12y + 9z = 1980$$
$$9x + 8y + 4z = 1080.$$

This is an example of a system of linear equations. One major concern of this text will be with techniques for solving such systems; much of what we do will be motivated by the desire to deal with such systems efficiently and intelligently.

The general linear equation in three variables is an equation of the form $ax + by + cz = k$, where a, b, c, and k are constants. If the problem has more than three variables, we normally use subscripted variables like x_i, y_j, or z_k and denote the constants in the equation by a_i, b_j, or c_k. Thus, the GENERAL LINEAR EQUATION IN N UNKNOWNS is an equation of the form

$$a_1x_1 + a_2x_2 + \ldots + a_nx_n = b,$$

where b and the a_i are constants. The SOLUTION SET of this equation consists of all n-tuples of values of the variables that satisfy the equation. To find a particular solution of a linear equation you may randomly pick values for all but one of the unknowns and then use the equation to compute the appropriate value of the remaining unknown. For the linear equation $4x + 5y = 3$ we could choose x and then compute y, or we could first choose y and then compute x. Thus, if we choose $x = 1$ it follows that $5y = 3 - 4 = -1$ or $y = -1/5$; if we choose $y = 4$, then $4x = 3 - 20$ or $x = -17/4$. Thus, $(1, -1/5)$ and $(-17/4, 4)$ are two members of the solution set of $4x + 5y = 3$. Can you find a solution for which $x = 5$?

For the general case we make the following definition.

Definition 1.1

An m by n LINEAR SYSTEM is a set of m linear equations in n unknowns. A SOLUTION of this linear system is an n-tuple of values for the variables that satisfy each equation in the system.

Given such a system, we need to ask the following questions.

(1) Do solutions exist?
(2) How many solutions are there?
(3) How can we find all solutions?

We begin with a very simple case in which there are only two variables and only two equations.

$$3x - 4y = 12$$
$$4x + 3y = 0 \qquad\qquad (1.1)$$

It is easy to check that $(4, 0)$, that is, $x = 4$ and $y = 0$, is a solution of the first equation but not of the second, while $(3, -4)$ is a solution of the second equation but not of the first. You should check that $(36/25, -48/25)$ is a solution of both equations and hence a solution of System 1.1. We will see one way to obtain this solution shortly.

In the case of two variables there is a familiar and useful geometric interpretation of linear equations:

> The equation $ax + by = c$ represents a straight line in the xy-plane.

The solutions of the linear equation $ax + by = c$ are just the co-ordinates of the points that lie on the line. The solution set is the set of all points that lie on the line.

The solution set of the first equation of system 1.1 is easy to obtain. Since

$$3x - 4y = 12 \quad \text{is equivalent to} \quad y = -3 + \frac{3x}{4},$$

we can choose for x any real number, say $x = c$, and then compute $y = -3 + (3/4)c$. Thus, $x = c$, $y = -3 + (3/4)c$, is a solution of the equation for any choice of c ($c = 4$ gives the first solution). We can describe the solution set for this equation as

$$\left\{ \left(c, -3 + \frac{3c}{4} \right) \,\middle|\, c \text{ real} \right\} \quad \text{or} \quad \left\{ \left(x, -3 + \frac{3x}{4} \right), \,\middle|\, x \text{ real} \right\}. \tag{1.2}$$

Note that this amounts to giving a parametric description of the straight line whose equation is $3x - 4y = 12$; $x = c$ is the parameter.

The problem of solving System 1.1 amounts to finding the point of intersection of the two lines; see Figure 1.1.

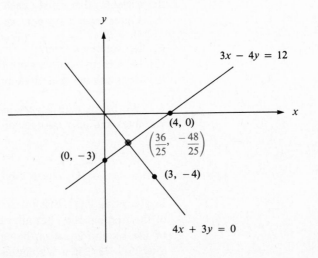

FIGURE 1.1

Finding the point of intersection of the two lines amounts to finding a value of x for which the two y values are the same. Thus we solve each equation for y, obtaining

$$y = -3 + \frac{3x}{4} \quad \text{and} \quad y = \frac{-4x}{3}.$$

These y values will be the same if

$$-3 + \frac{3x}{4} = \frac{-4x}{3} \quad \text{or, equivalently,} \quad \frac{3x}{4} + \frac{4x}{3} = \frac{25x}{12} = 3.$$

It follows that $x = 36/25$ and that $y = -4x/3 = (-4/3)(36/25) = -48/25$. Thus, the point of intersection of the two lines has coordinates $(36/25, -48/25)$.

Since solving two equations in two unknowns is equivalent to finding the point of intersection of two straight lines, it is clear that such a system may have either:

(1) no solutions, if the lines are parallel;
(2) exactly one solution, if the lines intersect; or
(3) an infinite number of solutions, if the lines are the same.

Figure 1.2 illustrates these three cases.

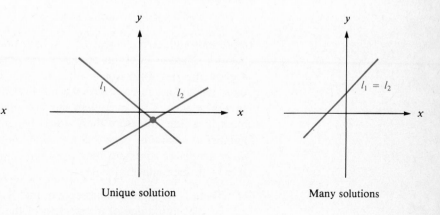

No solutions   Unique solution   Many solutions

FIGURE 1.2

For n equations in two unknowns $(n > 2)$, the same three possibilities exist, as illustrated for the case $n = 3$ in Figure 1.3.
Note that in the first case none of the lines need be parallel. For $n > 2$ the first possibility (no solutions) is clearly the most likely, at least if the lines are chosen at random. If the equations describe some physical system, then it is more likely that a solution exists.

FIGURE 1.3

In general, *if there are more equations than unknowns, then it is a very real possibility that no solutions exist.*

If you have studied three-dimensional analytic geometry, perhaps in the third term of calculus, then you know that a linear equation in three variables such as

$$ax + by + cz = d, \tag{1.3}$$

represents a plane in 3-space (If you have not seen this before, be sure to work Exercise 18 of this section.) This observation allows us to give a geometric interpretation of the problem of solving n equations in three unknowns; in each case we are looking for a point common to several planes. If $n = 2$, then the two planes are either parallel or meet in a line; therefore there are either no solutions or an infinite number of solutions. Note that *we cannot have a unique solution when there are fewer equations than unknowns.*

If $n = 3$, then either:

(1) There is no point of intersection; that is, the third plane is parallel to the line of intersection of the first two planes (like the ceiling, floor, and one side wall of your room);
(2) The three planes meet in a unique point (where the third plane meets the line of intersection of the first two); or
(3) There are an infinite number of solutions; that is, the three planes have at least one line in common (like the pages of a book have the binding in common).

Figure 1.4 illustrates these three cases. If n > 3, then the same three possibilities occur, the most likely (if the planes are chosen at random) being that there are no solutions.

No solutions

Unique solution

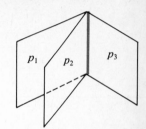

Many solutions

FIGURE 1.4

Note that in both cases considered there are always three possibilities: no solutions (in which case we say that the system is INCONSISTENT), a unique solution, or an infinite number of solutions. From our examples it appears that the seemingly desirable case of a unique solution is most likely when the number of unknowns is equal to the number of equations. We will establish results later that will tell us precisely when a linear system has a unique solution.

Example 1

Find the solution set of $3x - 5y + z = 6$.

In this case we will treat two of the variables as parameters; if we set $x = c_1$ and $y = c_2$, then we must have

$$z = 6 - 3x + 5y = 6 - 3c_1 + 5c_2.$$

Thus, we can describe the SOLUTION SET of $3x - 5y + z = 6$ as

$$\{(c_1, c_2, 6 - 3c_1 + 5c_2) \,|\, c_1, c_2 \text{ real}\}$$

or, equivalently, as

$$\{(x, y, 6 - 3x + 5y) \,|\, x, y \text{ real}\}. \quad \blacksquare$$

Let us now consider a general system of m equations in the n unknowns $x_1, x_2, \ldots, x_n$:

$$
\begin{aligned}
a_{11}x_1 + a_{12}x_2 + \cdots + a_{1n}x_n &= b_1 \\
a_{21}x_1 + a_{22}x_2 + \cdots + a_{2n}x_n &= b_2 \\
\vdots \qquad \vdots \qquad\qquad \vdots \qquad\ \ \vdots \\
a_{m1}x_1 + a_{m2}x_2 + \cdots + a_{mn}x_n &= b_m.
\end{aligned}
\qquad (1.4)
$$

Note carefully the notation used to describe the general system; *the coefficients are doubly subscripted, with the first subscript indicating the equation and the second subscript indicating the variable; for example, a_{ij} is the coefficient of x_j in equation i.*

In the case $m = 4$, $n = 3$, System 1.4 becomes

$$a_{11}x_1 + a_{12}x_2 + a_{13}x_3 = b_1$$

$$a_{21}x_1 + a_{22}x_2 + a_{23}x_3 = b_2$$

$$a_{31}x_1 + a_{32}x_2 + a_{33}x_3 = b_3$$

$$a_{41}x_1 + a_{42}x_2 + a_{43}x_3 = b_4.$$

A MATRIX is a rectangular array of numbers; we will associate two important matrices with System 1.4.

The COEFFICIENT MATRIX of System 1.4 is the array

$$A = \begin{bmatrix} a_{11} & a_{12} & \cdots & a_{1n} \\ a_{21} & a_{22} & \cdots & a_{2n} \\ \cdot & \cdot & & \cdot \\ \cdot & \cdot & & \cdot \\ \cdot & \cdot & & \cdot \\ a_{m1} & a_{m2} & \cdots & a_{mn} \end{bmatrix}. \tag{1.5}$$

Note that the first subscript indicates the row in which the entry occurs while the second subscript determines the column. Thus, a_{21} is in row 2 and column 1, while a_{47} is in row 4 and column 7, and a_{rc} is the entry in row r and column c of the matrix A. We will say that the matrix A is m by n $(m \times n)$, with the first number indicating the number of rows of A and the second indicating the number of columns of A.

The AUGMENTED MATRIX is the coefficient matrix with one extra column, that column being the matrix B whose entries are the constants from the right-hand side of the system. Thus, the augmented matrix of System 1.4 is

$$[A \,|\, B] = \begin{bmatrix} a_{11} & a_{12} & \cdots & a_{1n} & b_1 \\ a_{21} & a_{22} & \cdots & a_{2n} & b_2 \\ \cdot & \cdot & & \cdot & \cdot \\ \cdot & \cdot & & \cdot & \cdot \\ \cdot & \cdot & & \cdot & \cdot \\ a_{m1} & a_{m2} & \cdots & a_{mn} & b_m \end{bmatrix}, \tag{1.6}$$

$$\text{where } B = \begin{bmatrix} b_1 \\ b_2 \\ \cdot \\ \cdot \\ \cdot \\ b_m \end{bmatrix}.$$

As a general rule, if A is $m \times n$, then the augmented matrix is $m \times (n + 1)$.

There is an obvious one-to-one correspondence between systems and augmented matrices. For example, the augmented matrix of the system

$$\begin{array}{rcl} 3x - 4y &=& 12 \\ 4x + 3y &=& 0 \end{array} \quad \text{is} \quad \begin{bmatrix} 3 & -4 & 12 \\ 4 & 3 & 0 \end{bmatrix},$$

while the system with augmented matrix

$$\begin{bmatrix} 3 & 0 & 4 & 2 \\ 5 & 8 & -2 & 4 \\ 4 & 2 & 8 & 11 \end{bmatrix} \quad \text{is} \quad \begin{array}{rcl} 3x_1 + 4x_3 &=& 2 \\ 5x_1 + 8x_2 - 2x_3 &=& 4 \\ 4x_1 + 2x_2 + 8x_3 &=& 11 \,. \end{array}$$

Be careful to supply zeros in the appropriate places when going from the system to the augmented matrix. For example, the coefficient and augmented matrices of the system

$$\begin{array}{rcl} 2x_1 - x_2 &=& 5 \\ -x_1 + 2x_2 - x_3 &=& 0 \\ -x_2 + 2x_3 - x_4 &=& 0 \\ -x_3 + 2x_4 &=& 9 \end{array}$$

are, respectively,

$$\begin{bmatrix} 2 & -1 & 0 & 0 \\ -1 & 2 & -1 & 0 \\ 0 & -1 & 2 & -1 \\ 0 & 0 & -1 & 2 \end{bmatrix} \quad \text{and} \quad \begin{bmatrix} 2 & -1 & 0 & 0 & 5 \\ -1 & 2 & -1 & 0 & 0 \\ 0 & -1 & 2 & -1 & 0 \\ 0 & 0 & -1 & 2 & 9 \end{bmatrix}.$$

We conclude this section with an example that shows how linear systems arise in curve-fitting problems.

Example 2

Identify the constants a, b, and c so that the parabola $y = ax^2 + bx + c$ will pass through the points $(1, 3)$, $(0, 5)$, and $(3, 11)$.

Each point on the curve leads to a linear equation in a, b, and c. Thus, because the point $(1, 3)$ is on the curve we must have

$$a + b + c = 3 \,.$$

Similarly, because $(0, 5)$ and $(3, 11)$ are to be on the curve, we must also have

$$0a + 0b + c = 5 \quad \text{and} \quad 9a + 3b + c = 11 \,.$$

Thus $a, b,$ and c must satisfy the 3 by 3 linear system

$$a + b + c = 3$$
$$c = 5$$
$$9a + 3b + c = 11.$$

This system is easy to solve because the second equation tells us that $c = 5$. The other two equations then reduce to

$$a + b = -2$$
$$9a + 3b = 6.$$

From the first equation we have $b = -2 - a$, which, when substituted into the second equation, yields $9a + 3(-2 - a) = 6$ or $6a = 12$. Thus, $a = 2,$ $b = -4,$ and the desired parabola is $y = 2x^2 - 4x + 5.$ ■

Exercise Set 1.1

1. For each of the following systems, graph the lines represented by the equations and, from the graphs, find the solution(s) of the systems if any exist.

   (a) $x + y = 4$ (b) $x + y = 4$ (c) $x + y = 4$
   　　$2x - 2y = 8$ 　　$2x + 2y = 6$ 　　$x - y = 8$
   　　　　　　　　　　　　　　　　　　　　　　　　　　$2x + 3y = 6$

*2. Find the solution sets of each of the following linear equations.

   (a) $3x - 5y = 15$ (b) $4x_1 + 3x_2 = 9$
   (c) $3x + 5y - 7z = 10$ (d) $4x_1 + 2x_2 - 6x_3 = 12$

3. Which of the following equations are linear?

   (a) $x + 2y + z = 2$ (b) $x + y + z = \sin 3$
   (c) $x + \frac{1}{y} + 3z^2 = 11$ (d) $x + y + \sqrt{2}z = 4$
   (e) $3x + \sin y + z = 0$ (f) $3x^2 + 5xy = 11$

4. Find solutions of the systems in Exercise 1 using the technique illustrated in the solution of System 1.1.

5. Find the coefficient matrix and the augmented matrix for the following systems.

   (a) $2x - 3z = 11$ (b) $4x_1 - 3x_2 + 15x_3 = 0$
   　　$4x + 5y + z = 6$ 　　$5x_1 + 17x_2 = 14$
   　　$4y + 12z = 0$ 　　$3x_1 + 5x_2 + x_3 = 4$

   (c) $7x + 8y + 8z = -x$ (d) $x_1 - x_3 = 0$
   　　$9x - 16y - 18z = -y$ 　　$x_2 + 3x_3 = 1$
   　　$-5x + 11y + 13z = -z$ 　　$x_1 + x_2 + x_3 = 5$

---

*The answers to exercises with green numbers are given in the back of the text.

(e) $4x_{i-1} + 6x_i - 5x_{i+1} = 10 + i, \quad i = 1, 2, \ldots, 5; x_0 = x_6 = 0$

(f) $\sum_{i=1}^{4} (i + j)x_i = j, \quad j = 1, 2, 3, 4$

6. Solve the systems in exercises 5(a) and 5(d) using the substitution technique used to solve System 1.1.

†7. Find the system whose augmented matrix is the following matrix.

(a) $\begin{bmatrix} 2 & 1 & 5 & 1 & 5 \\ 1 & 1 & -3 & -4 & -1 \\ 3 & 6 & -2 & 1 & 8 \\ 2 & 2 & 2 & -3 & 2 \end{bmatrix}$
(b) $\begin{bmatrix} 3 & 0 & 0 & 0 & 9 \\ 2 & 1 & 0 & 0 & 0 \\ -2 & 3 & 1 & 0 & 4 \\ 0 & 3 & 2 & 1 & 7 \end{bmatrix}$

(c) $\begin{bmatrix} 4 & 1 & 0 & 0 & 5 \\ 1 & 4 & 1 & 0 & 0 \\ 0 & 1 & 4 & 1 & 0 \\ 0 & 0 & 1 & 4 & 7 \end{bmatrix}$
(d) $\begin{bmatrix} 1 & -3 & 0 & 3 \\ 0 & 1 & 1 & 1 \\ 0 & 0 & -6 & -7 \end{bmatrix}$ sub back

(e) $\begin{bmatrix} 2 & 0 & 0 & 5 \\ 0 & 1 & 0 & 3 \\ 0 & 0 & 7 & 2 \end{bmatrix}$
(f) $\begin{bmatrix} 1 & 2 & 4 & 8 \\ 0 & 0 & 3 & 9 \\ 0 & 0 & 0 & 5 \end{bmatrix}$ - imposs

Which, if any, of the above systems are easy to solve?

†8. Show that if (x_0, y_0) is a solution of the system

$$ax + by = c$$
$$dx + ey = g,$$

then (x_0, y_0) is also a solution of the equation

$$(d + ka)x + (e + kb)y = (g + kc),$$

obtained by adding k times the first equation to the second.

9. Show that if (a, b) and (c, d) are both in the solution set of $rx + sy = k$, then so is $[ta + (1 - t)c, tb + (1 - t)d]$ for any real t. Interpret this result geometrically. What point corresponds to $t = 1/2$?

10. Sketch the following sets in the xy-plane.
    (a) $\{(x, y) \mid 2x + 3y = 6\}$
    (b) $\{(x, y) \mid 2x + 3y < 6\}$
    (c) $\{(x, y) \mid 2x + 3y > 6\}$

11. Most computers do not understand subscript notation, so one uses $A(I, J)$ or $A[I, J]$ and $X(I)$ or $X[I]$ in place of a_{ij} and x_i. Rewrite System 1.4 and Matrices 1.5 and 1.6 using this notation.

12. A certain farm has only cows and chickens. It is observed that altogether there are 60 heads and 200 legs on the farm. How many of each kind of animal are on the farm?

---

†These exercises are particularly important. They either provide motivation for results to follow or they have been referred to in the text.

13. A candy store wishes to make 100 lbs. of mixed nuts, including peanuts, cashews, almonds, and sunflower seeds. Peanuts cost $1.25/lb., cashews are $3.00/lb., almonds are $2.50/lb., and sunflower seeds are $1.75/lb. The mix is to contain twice as many peanuts as cashews, 3 times as many cashews as almonds and is to be worth $1.80/lb. How much of each raw ingredient is needed?

14. Show that it is *not* possible to choose 17 coins, with a total value of $1.85, from a pile of quarters, dimes, and nickels in such a way that you have twice as many dimes as quarters.

15. Find the equation of the parabola $y = ax^2 + bx + c$ that passes through the points $(1, 1)$, $(2, 12)$, and $(-1, -9)$.

16. Find the equation of the circle $x^2 + y^2 + ax + by = c$ that passes through the points $(5, 5)$, $(4, 6)$, and $(5, -1)$.

17. In order to control a certain crop pest it is recommended that the crop be sprayed with a mixture of 6 units of chemical A, 10 units of chemical B and 8 units of chemical C. These ingredients are available in three ready-mixed proportions: one barrel of spray S1 contains 1, 3, and 4 units of chemicals A, B and C, respectively; one barrel of spray S2 contains 3 units of each chemical, and one barrel of spray S3 contains 2 units of A and 5 units of B. How much of each commercial spray should be used to obtain the exact amounts of chemical needed?

†18. Study the diagram below and then sketch the plane represented by each of the following linear equations:

   (a) $2x + 3y + 4z = 12$ (b) $3x - 2y + 3z = 6$ (c) $2x + 5y = 10$.

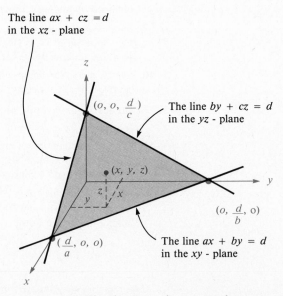

The line $ax + cz = d$ in the xz - plane

$(o, o, \frac{d}{c})$ The line $by + cz = d$ in the yz - plane

(x, y, z)

$(o, \frac{d}{b}, o)$

$(\frac{d}{a}, o, o)$ The line $ax + by = d$ in the xy - plane

The plane $ax + by + cz = d$

1.2   *Gaussian and Gauss-Jordan Elimination*

We will say that two linear systems are EQUIVALENT if they have pre-cisely the same solution set. The best method of solving systems of linear equations is to replace the given system by an equivalent system that is easier to solve, as shown in Figure 1.5.

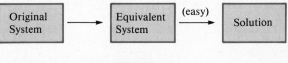

FIGURE 1.5

In order to implement this solution technique we need to answer the following two questions.

(1) When is a linear system easy to solve?
(2) What can we do to a linear system to change it into an equivalent system?

The exercises in Section 1.1 contained some hints to the answers to these questions. Certainly a system is easy to solve if the coefficient matrix is a DIAGONAL MATRIX; that is, if $a_{ij} = 0$ for $i \neq j$. In the case of a system of four equations in four unknowns, the augmented matrix would look like

$$\left[\begin{array}{cccc|c} a_{11} & 0 & 0 & 0 & b_1 \\ 0 & a_{22} & 0 & 0 & b_2 \\ 0 & 0 & a_{33} & 0 & b_3 \\ 0 & 0 & 0 & a_{44} & b_4 \end{array}\right].$$

The corresponding system consists of equations such as

$$a_{ii}x_i = b_i,$$

which are very easy to solve. Note that if any $a_{ii} = 0$ while $b_i \neq 0$, then this system has no solution.

A system is also easy to solve when its coefficient matrix is an UPPER TRIANGULAR MATRIX, that is, when $a_{ij} = 0$ for $j < i$. In this case, with four equations in four unknowns the augmented matrix would have the following form:

$$\left[\begin{array}{cccc|c} a_{11} & a_{12} & a_{13} & a_{14} & b_1 \\ 0 & a_{22} & a_{23} & a_{24} & b_2 \\ 0 & 0 & a_{33} & a_{34} & b_3 \\ 0 & 0 & 0 & a_{44} & b_4 \end{array}\right].$$

A system like this can be solved by a technique called BACK SUBSTITUTION. We illustrate this technique for the 3-by-3 system

$$x + 2y + 3z = 5$$
$$y - 2z = 6$$
$$2z = 4,$$

whose augmented matrix is

$$\begin{bmatrix} 1 & 2 & 3 & 5 \\ 0 & 1 & -2 & 6 \\ 0 & 0 & 2 & 4 \end{bmatrix}.$$

The last equation is easy to solve for z; the solution is $z = 2$. We then substitute this value of z into the next-to-last equation, obtaining $y - 2(2) = 6$, or $y = 10$. Finally, we substitute the values of y and z into the first equation to obtain $x = -21$.

In either of the preceding cases the solution process amounts to solving a series of equations of the form $ax = b$ which can fail to have a solution *only* if $a = 0$ when $b \neq 0$.

We will generate new systems, equivalent to our original system, by manipulating the equations. Note that the following operations will lead to equivalent systems:

(1) multiply an equation by a nonzero constant;
(2) interchange the order of two equations; and
(3) replace a given equation by the sum of itself and a multiple of another equation (see Exercise 8, Section 1.1).

In Example 1 we show how to use these operations to solve a system of three equations in three unknowns. Each time we change the system we will list the new augmented matrix. Note that we could go from one augmented matrix to the next by operating on the rows of the augmented matrix.

Example 1

Solve the following system for x, y, and z.

$$\begin{array}{ll} \textit{System} & \textit{Augmented Matrix} \end{array}$$

$$\begin{array}{ll} \begin{aligned} x + 2y + 3z &= 1 \\ 2x + 5y + 5z &= -3 \\ 3x + 5y + 11z &= 2 \end{aligned} & \begin{bmatrix} 1 & 2 & 3 & 1 \\ 2 & 5 & 5 & -3 \\ 3 & 5 & 11 & 2 \end{bmatrix} \end{array}$$

We will eliminate x from the second equation by replacing the second equation by the sum of itself plus (-2) times the first equation. This operation yields

$$x + 2y + 3z = \boxed{1}$$
$$y - z = \boxed{-5} \quad \text{and} \quad \begin{bmatrix} 1 & 2 & 3 & \boxed{1} \\ 0 & 1 & -1 & \boxed{-5} \\ 3 & 5 & 11 & \boxed{2} \end{bmatrix}.$$
$$3x + 5y + 11z = \boxed{2}$$

In order to eliminate x from the third equation we replace the third equation by itself plus (-3) times the first equation to obtain

$$x + 2y + 3z = \boxed{1}$$
$$y - z = \boxed{-5} \quad \text{and} \quad \begin{bmatrix} 1 & 2 & 3 & \boxed{1} \\ 0 & 1 & -1 & \boxed{-5} \\ 0 & -1 & 2 & \boxed{-1} \end{bmatrix}.$$
$$-y + 2z = \boxed{-1}$$

In the next step we eliminate y from the third equation by replacing the third equation by the sum of itself plus the second equation. This step yields

$$x + 2y + 3z = \boxed{1}$$
$$y - z = \boxed{-5} \quad \text{and} \quad \begin{bmatrix} 1 & 2 & 3 & \boxed{1} \\ 0 & 1 & -1 & \boxed{-5} \\ 0 & 0 & 1 & \boxed{-6} \end{bmatrix}.$$
$$z = \boxed{-6}$$

At this point the system can be solved by the back-substitution technique, but we continue the reduction by adding the third equation to the second and then adding (-3) times the third equation to the first. These operations yield

$$x + 2y = \boxed{19}$$
$$y = \boxed{-11} \quad \text{and} \quad \begin{bmatrix} 1 & 2 & 0 & \boxed{19} \\ 0 & 1 & 0 & \boxed{-11} \\ 0 & 0 & 1 & \boxed{-6} \end{bmatrix}.$$
$$z = \boxed{-6}$$

Finally we add (-2) times the second equation to the first to obtain the solution of the original system.

$$x = \boxed{41}$$
$$y = \boxed{-11} \quad \text{and} \quad \begin{bmatrix} 1 & 0 & 0 & \boxed{41} \\ 0 & 1 & 0 & \boxed{-11} \\ 0 & 0 & 1 & \boxed{-6} \end{bmatrix}. \quad \blacksquare$$
$$z = \boxed{-6}$$

It is clearly possible, and somewhat more convenient, to work just with the augmented matrices; certainly this is the case with modern machine calculations.* The preceding operations on the equations of a system are equivalent to operations on the rows of the augmented matrix; we now define these operations formally.

---

*There now exist machines that do symbolic algebra, but it is still more efficient to deal with just the augmented matrix.

Definition 1.2

An ELEMENTARY ROW OPERATION on a matrix A is one of the following.

Type I: Multiply row i by $k \neq 0$ $(R_i \leftarrow kR_i)$

Type II: Interchange row i and row j $(R_i \leftrightarrow R_j)$

Type III: Replace row j by itself plus k times row i $(R_j \leftarrow kR_i + R_j)$

The abbreviations in parentheses will frequently be used to describe the details of a specific row reduction; the arrow indicates replacement, R_i represents the ith row of the matrix being reduced, and R_j indicates the jth row of the same matrix. The process of changing one matrix to another by elementary row operations is called ROW REDUCTION; it is the most basic computational procedure of linear algebra.

The schematic diagram of our solution process, given in Figure 1.5, can now be more accurately described in Figure 1.6. This solution process, or any of a number of minor variants, is called GAUSSIAN ELIMINATION if the reduction is stopped when the coefficient matrix first becomes upper triangular. In machine calculations we do not normally change the diagonal elements to 1, as is convenient with hand calculations. If the reduction is carried as far as possible toward a diagonal coefficient matrix, then the process is called GAUSS-JORDAN ELIMINATION.

FIGURE 1.6

Example 2

Solve the following system by Gaussian elimination:

$$x + y + 2z = 2$$
$$x + y - 3z = 2$$
$$2x + y + 5z = 5 .$$

The augmented matrix is

$$\begin{bmatrix} 1 & 1 & 2 & 2 \\ 1 & 1 & -3 & 2 \\ 2 & 1 & 5 & 5 \end{bmatrix},$$

and the operations $(R_2 \leftarrow -R_1 + R_2)$ and $(R_3 \leftarrow -2R_1 + R_3)$ yield

$$\begin{bmatrix} 1 & 1 & 2 & 2 \\ 0 & 0 & -5 & 0 \\ 0 & -1 & 1 & 1 \end{bmatrix}.$$

Multiplying the third row by -1 $(R_3 \leftarrow -R_3)$, and then interchanging rows 2 and 3 $(R_2 \leftrightarrow R_3)$, leads to

$$\begin{bmatrix} 1 & 1 & 2 & 2 \\ 0 & 1 & -1 & -1 \\ 0 & 0 & -5 & 0 \end{bmatrix},$$

which is upper triangular. The corresponding system can now be solved by back substitution. Since the third equation of the system is $-5z = 0$ we have $z = 0$, and it follows easily that $y = -1$ and $x = 3$. ∎

Example 3

Solve the following system by Gauss-Jordan elimination.

$$\begin{aligned} x_1 + 2x_2 - x_3 &= 4 \\ 2x_1 + 4x_2 + 3x_3 &= 5 \\ -x_1 - 2x_2 + 6x_3 &= -7 \end{aligned}$$

The augmented matrix is

$$\begin{bmatrix} 1 & 2 & -1 & 4 \\ 2 & 4 & 3 & 5 \\ -1 & -2 & 6 & -7 \end{bmatrix},$$

and the operations $(R_2 \leftarrow -2R_1 + R_2)$ and $(R_3 \leftarrow R_1 + R_3)$ yield

$$\begin{bmatrix} 1 & 2 & -1 & 4 \\ 0 & 0 & 5 & -3 \\ 0 & 0 & 5 & -3 \end{bmatrix}.$$

The row operations $(R_3 \leftarrow -R_2 + R_3)$, $(R_2 \leftarrow 1/5R_2)$, and $(R_1 \leftarrow R_2 + R_1)$ now yield

$$\begin{bmatrix} 1 & 2 & 0 & 17/5 \\ 0 & 0 & 1 & -3/5 \\ 0 & 0 & 0 & 0 \end{bmatrix}.$$

The system corresponding to this augmented matrix is

$$x_1 + 2x_2 = \frac{17}{5}$$

$$x_3 = \frac{-3}{5}.$$

A different solution of this system can be obtained for each value of x_2. If we arbitrarily let $x_2 = c$, then the most general solution of the system is

$$x_1 = \frac{17}{5} - 2c$$

$$x_2 = c$$

$$x_3 = \frac{-3}{5}.$$

We fail to obtain a unique solution because the final form of the system has fewer equations than unknowns.

Note carefully that it was absolutely critical that the entry in the $(3, 4)$ position of the final matrix was zero. If that entry had been $h \neq 0$, then the last equation would have been

$$0x_1 + 0x_2 + 0x_3 = h \neq 0.$$

Since this equation has no solutions at all, neither would the original system; that is, the system would have been inconsistent.   ∎

Example 4

Consider the insulated metal rod in Figure 1.7; the numbers represent the temperatures at the indicated points.

FIGURE 1.7

Find the three interior temperatures t_1, t_2, and t_3 if it is assumed that the temperature at each interior point is the average of the temperatures at the two adjacent points.

The given assumption about the interior temperatures leads to the three following linear equations.

$$t_1 = \frac{50 + t_2}{2}$$

$$t_2 = \frac{t_1 + t_3}{2}$$

$$t_3 = \frac{t_2 + 100}{2}$$

If we rewrite these equations in the usual form we have the 3-by-3 linear system

$$2t_1 - t_2 \qquad = 50$$
$$-t_1 + 2t_2 - t_3 = 0$$
$$- t_2 + 2t_3 = 100 .$$

The augmented matrix of this system is

$$\begin{bmatrix} 2 & -1 & 0 & 50 \\ -1 & 2 & -1 & 0 \\ 0 & -1 & 2 & 100 \end{bmatrix},$$

which easily row reduces (verify this yourself) to

$$\begin{bmatrix} 1 & -2 & 1 & 0 \\ 0 & 1 & -2 & -100 \\ 0 & 0 & 4 & 350 \end{bmatrix}.$$

Thus, the original system is equivalent to the system

$$t_1 - 2t_2 + t_3 = 0$$
$$t_2 - 2t_3 = -100$$
$$4t_3 = 350 .$$

Solving this system by back substitution yields

$$t_3 = \frac{350}{4} = 87.5$$

$$t_2 = -100 + 2t_3 = -100 + 175 = 75$$
$$t_1 = 2t_2 - t_3 = 150 - 87.5 = 62.5 . \quad \blacksquare$$

It is important to note that in each of the above reductions we have simplified the augmented matrix in a very orderly fashion — one column at a time, from left to right. Please do the same in the exercises that follow.

Exercise Set 1.2

1. Find the most general solutions of the systems whose augmented matrices are as follows.

(a) $\begin{bmatrix} 1 & 3 & -2 & 0 \\ 0 & 1 & 5 & 2 \\ 0 & 0 & 1 & -11 \end{bmatrix}$ (b) $\begin{bmatrix} 1 & -3 & 4 & 2 & 0 \\ 0 & 1 & 2 & 0 & 0 \\ 0 & 0 & 0 & 1 & 5 \end{bmatrix}$

(c) $\begin{bmatrix} 1 & 2 & 0 & 0 & 5 \\ 2 & 0 & 1 & 0 & 3 \\ -3 & 0 & 2 & 1 & -2 \end{bmatrix}$ (d) $\begin{bmatrix} 1 & -2 & 0 & 5 & 0 & 0 & 0 & 3 \\ 0 & 0 & 1 & 3 & -1 & 0 & 0 & 2 \\ 0 & 0 & 0 & 0 & 0 & 1 & 0 & 3 \\ 0 & 0 & 0 & 0 & 0 & 0 & 1 & 0 \\ 0 & 0 & 0 & 0 & 0 & 0 & 0 & 0 \end{bmatrix}$

(e) $\begin{bmatrix} 1 & 0 & 0 & 4 & 3 \\ 0 & 1 & 0 & -7 & 5 \\ 0 & 0 & 1 & 8 & 2 \\ 0 & 0 & 0 & 0 & 5 \end{bmatrix}$ (f) $\begin{bmatrix} 2 & 0 & 0 & 10 \\ 5 & 3 & 0 & 15 \\ 6 & 1 & -5 & 0 \end{bmatrix}$

Use Gaussian elimination to find all solutions of the following systems.

2. $\begin{aligned} x_1 + 2x_2 \quad\quad &= 7 \\ x_2 + 3x_3 &= 1 \\ x_1 \quad\quad - x_3 &= 2 \end{aligned}$

3. $\begin{aligned} 3x_1 - x_2 + 2x_3 &= -4 \\ 2x_1 + x_2 + x_3 &= -1 \\ x_1 + 3x_2 \quad\quad &= 2 \end{aligned}$

4. $\begin{aligned} x_1 - x_2 - x_3 - x_4 &= 5 \\ x_1 + 2x_2 + 3x_3 + x_4 &= -2 \\ 3x_1 + x_2 \quad\quad + 2x_4 &= 1 \\ 2x_1 + \quad\quad 2x_3 + 3x_4 &= 3 \end{aligned}$

5. $\begin{aligned} 2x_1 + 4x_2 + 3x_3 + 2x_4 &= 2 \\ 3x_1 + 6x_2 + 5x_3 + 2x_4 &= 2 \\ 2x_1 + 5x_2 + 2x_3 - 3x_4 &= 3 \\ 4x_1 + 5x_2 + 14x_3 + 14x_4 &= 11 \end{aligned}$

6. $\begin{aligned} 3x_1 - 7x_2 + 4x_3 &= 10 \\ -x_1 - 2x_2 + 3x_3 &= 1 \\ x_1 + x_2 + 2x_3 &= 8 \end{aligned}$

7. $\begin{aligned} x_1 + 3x_2 + 7x_3 + 2x_4 &= 2 \\ x_1 - 2x_2 + x_3 - 4x_4 &= 1 \\ x_1 - 12x_2 - 11x_3 - 16x_4 &= 5 \end{aligned}$

8. $\begin{aligned} x_1 + x_2 + 2x_3 &= 8 \\ - x_2 + 5x_3 &= 9 \\ 3x_1 - 7x_2 + 4x_3 &= 10 \end{aligned}$

9. $\begin{aligned} x_1 - x_2 + 2x_3 - x_4 &= -1 \\ 2x_1 + x_2 - 2x_3 - 2x_4 &= -2 \\ -x_1 + 2x_2 - 4x_3 + x_4 &= 1 \\ 2x_1 + 2x_2 - 4x_3 - 2x_4 &= -2 \end{aligned}$

10. $\begin{aligned} 3x + 2y - z &= -15 \\ 5x + 3y + 2z &= 0 \\ 3x + y + 13z &= 11 \\ 11x + 7y \quad\quad &= -30 \end{aligned}$

11. Solve each of the systems in Exercises 2 through 9 by Gauss-Jordan elimination.

12. Find all values of h for which the following system is solvable.

$$\begin{aligned} x_1 + 2x_2 + 3x_3 + x_4 &= 3 \\ 3x_1 + 2x_2 + x_3 + 4x_4 &= 7 \\ 2x_2 + 4x_3 + x_4 &= 1 \\ x_1 + x_2 + x_3 + x_4 &= h \end{aligned}$$

13. Solve the following system by Gauss-Jordan elimination. After each row operation graph the equations of the new system.

$3x + 4y = 12$

$2x + 3y = 10$

14. For which values of t will the following system have a unique solution? No solutions? Many solutions?

$x_1 + 2x_2 - \qquad 3x_3 = \qquad 4$

$4x_1 + \quad x_2 + \qquad 2x_3 = \qquad 2$

$4x_1 + \quad x_2 + (t^2 - 14)x_3 = t - 2$

15. Count the number of multiplications which would be required to solve a 3×3 system in the following ways:

(a) by a Gaussian elimination algorithm which does not change the diagonal elements to 1;

(b) by a Gaussian elimination algorithm which changes the diagonal elements to 1 before the column is reduced;

(c) by a Gauss-Jordan elimination algorithm in which the coefficient matrix is first reduced to traingular form, and then the columns are simplified from right to left; and

(d) by a Gauss-Jordan elimination in which each column is completely reduced before moving to the next column.

Note that exact 1s and exact 0s should be set, not computed; thus they require no computational effort.

16. Try to solve the general 2×2 system

$$a_{11}x_1 + a_{12}x_2 = b_1$$
$$a_{21}x_1 + a_{22}x_2 = b_2$$

by Gaussian elimination. What condition on the coefficients will ensure that a unique solution exists?

17. A certain company makes three models of its basic product: the regular, deluxe, and limited versions. Each unit requires processing by departments A, B, and C. The regular model requires 3 time units in department A, 7 in department B, and 5 in department C. The deluxe model requires 4 time units in department A, 8 in department B, and 7 in department C. The limited model requires 4 time units in department A, 9 in department B, and 6 in department C. The company has a special order which must be filled by overtime work. Department A can supply 129 units of overtime, B can supply 285 units, and C can supply 210 units. Management decides to use all the available overtime labor. How many of each model can be supplied?

18. A certain farm has 3 kinds of livestock: sheep, goats, and pigs. There are three food types available: A, B, and C. Each sheep consumes 1 unit of food A, 1 unit of B, and 2 units of C per month. Each pig consumes 3 units of A, 4 units of B, and 5 units of C per month, while each goat eats 2 units of A, 1 unit of B, and 5 units of C. How many animals of each type can the farm support if each month there are available 44 units of food A, 49 units of food B, and 83 units of food C, and all the available food is consumed?

19. Consider the insulated metal rod pictured below. The numbers represent the temperatures at the indicated points. Find the four interior temperatures. Make the same assumption as in Example 4 of this section.

20. Consider the insulated metal plate pictured below. The numbers represent temperatures at the indicated points. Find the four interior temperatures t_1, t_2, t_3, and t_4 if it is assumed that each interior temperature is the average of the temperatures at the four neighboring grid points.

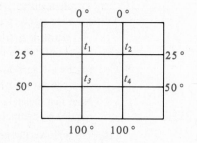

1.3   *Row Equivalence and Echelon Matrices*

In the last section we saw how to solve linear systems by row reducing the augmented matrix. In each case we reduced the augmented matrix to the point where the system was easy to solve. In this section we will address the following question:

> Given the matrix A, what is the simplest matrix R that we can get from A using elementary row operations?

By *simple* we mean that the linear system corresponding to R is easy to solve; thus we want R to be as nearly diagonal as possible.

The experience obtained in the last two sections should make it clear that we can reduce any matrix A to a matrix such as

$$R = \begin{bmatrix} 1 & 2 & 0 & 4 & 0 & 0 & 7 \\ 0 & 0 & 1 & 2 & 0 & 0 & -2 \\ 0 & 0 & 0 & 0 & 1 & 0 & 4 \\ 0 & 0 & 0 & 0 & 0 & 1 & 3 \\ 0 & 0 & 0 & 0 & 0 & 0 & 0 \end{bmatrix},$$

which has the following properties.

(1) If a row of R contains at least one nonzero entry, then the first nonzero entry in that row (the ROW LEADER) is 1.
(2) The zero rows, if any, come last.
(3) In any two consecutive nonzero rows, the leading 1 in the lower row occurs farther to the right than the leading 1 in the upper row.
(4) Each column that contains a row leader has zeros everywhere else.

The bulk of the work in the reduction is aimed at satisfying Properties 3 and 4, and is accomplished mainly by Type III operations, with some minor use of Type II operations to get the rows in the proper order. Any zero rows which occur are moved to the bottom using Type II operations. The row leaders are changed to 1 by using Type I operations.

Definition 1.3

A REDUCED ROW ECHELON MATRIX is any matrix satisfying the four properties just listed. Such a matrix is said to be in REDUCED ROW ECHELON FORM. A matrix satisfying the first three properties is said to be a ROW ECHELON MATRIX, or to be in ROW ECHELON FORM.

The terminology in Definition 1.3 is suggested by the nontechnical meaning of the word *echelon* which is "step-like."
The matrices

$$\begin{bmatrix} 1 & 2 & 3 & 1 \\ 0 & 1 & 3 & 5 \\ 0 & 0 & 0 & 1 \end{bmatrix}, \begin{bmatrix} 1 & 5 & 2 & 3 \\ 0 & 2 & 0 & 2 \\ 0 & 0 & 1 & 1 \end{bmatrix}, \text{ and } \begin{bmatrix} 1 & 3 & 9 & 6 \\ 0 & 0 & 0 & 1 \\ 0 & 0 & 1 & 0 \end{bmatrix}$$

are close to being in reduced row echelon form. Can you explain why they are not? Which are in row echelon form?
The matrices

$$\begin{bmatrix} 1 & 0 & 0 & 3 \\ 0 & 1 & 0 & 5 \\ 0 & 0 & 1 & 7 \end{bmatrix}, \begin{bmatrix} 0 & 1 & 3 & 0 \\ 0 & 0 & 0 & 1 \\ 0 & 0 & 0 & 0 \end{bmatrix}, \text{ and } \begin{bmatrix} 1 & 2 & 0 & 2 & 4 \\ 0 & 0 & 1 & 3 & 5 \\ 0 & 0 & 0 & 0 & 0 \\ 0 & 0 & 0 & 0 & 0 \end{bmatrix}$$

are all in reduced row echelon form.
Note that the Gaussian elimination process for solving a linear system amounts to reducing the augmented matrix to row echelon form while the Gauss-Jordan elimination method amounts to reducing

the augmented matrix to reduced row echelon form. The Gauss-Jordan method replaces the back substitution of the Gaussian elimination method with the further reduction of the coefficient matrix. A careful accounting of the arithmetic required (one normally counts only the number of multiplications of two numbers required) shows that the Gaussian elimination and back-substitution combination, when properly implemented, is slightly more efficient than the Gauss-Jordan reduction; it is the method of choice for use on a digital computer. When working small problems by hand most students prefer Gauss-Jordan elimination.

Both of these methods are quite easy to program for use on a modern digital computer. The Gauss-Jordan algorithm is described here in a way that is independent of machine or programming language. You can obtain the Gaussian elimination algorithm by omitting steps 3 and 6.

THE GAUSS-JORDAN ELIMINATION ALGORITHM

(1) Locate the leftmost nonzero column.

(2) By interchanging rows, if necessary, move a nonzero entry to the top position of the column in step 1.

(3) Change the top entry in step 2 to 1 by multiplying the first row by a nonzero constant.

(4) Add suitable multiples of the top row to the rows below so that all entries below the row leader become 0.

(5) Remove the top row and the column just reduced from consideration and repeat steps 1–4 until the matrix is in row echelon form (see Figure 1.8).

(6) Beginning with the last row and working upward, add suitable multiples of each row to the rows above in order to introduce zeros above the row leaders.

(7) Solve the reduced system.

The Gauss-Jordan elimination algorithm will fail to produce a solution of the original linear system only if there is a nonzero row, in the reduced matrix, whose row leader is in the last column. Such a row would be associated with the equation

$$0x_1 + 0x_2 + \ldots + 0x_n = 1 ,$$

which clearly is inconsistent.

| Already reduced | * * · * | * | These rows are already reduced and will not be changed again until step 6. |
| All zeros here | * * * · · * | The unreduced part | This is the part of the matrix currently under consideration. |

↑ *This* column is to be reduced next

| Already in REF | * * · * | * | These rows are already reduced and will not be changed again until step 6. |
| All zeros here | 1 0 0 · · 0 | * This is changed but still unreduced | * This is the part of the matrix to be considered at the next stage of the reduction. |

FIGURE 1.8

Any computer program that is designed to implement this elimination algorithm should make provisions to stop the computation and alert the user if the system is found to be inconsistent. (In practice, round-off error makes it unlikely that exact zeros will occur, but one should nevertheless provide for this case.) In Section 1.8 we will describe a better way of implementing this algorithm. If the reduced system is found to be consistent, then it has a unique solution if the number of nonzero equations is equal to the number of unknowns. If there are fewer equations than unknowns, then the solution is not unique. We summarize this discussion in the following theorem.

Theorem 1.1

Consider a consistent system of m linear equations in n unknowns whose augmented matrix has been reduced to a reduced row echelon matrix R that has r nonzero rows. The solution of the system is unique if $r = n$. If $r < n$, then the solution is not unique; the general solution involves expressing r of the unknowns in terms of the remaining $n - r$ unknowns whose values are arbitrary.

Example 1

Find all solutions of the system whose augmented matrix has been reduced to

$$\begin{bmatrix} 1 & 2 & 0 & 4 & 0 & 0 & 7 \\ 0 & 0 & 1 & 2 & 0 & 0 & -2 \\ 0 & 0 & 0 & 0 & 1 & 0 & 4 \\ 0 & 0 & 0 & 0 & 0 & 1 & 3 \\ 0 & 0 & 0 & 0 & 0 & 0 & k \end{bmatrix}.$$

If $k \neq 0$, then the system fails the consistency test and has no solutions. If $k = 0$, then the associated consistent system is

$$\begin{aligned} x_1 + 2x_2 \quad + 4x_4 \qquad\qquad &= 7 \\ x_3 + 2x_4 \qquad\qquad &= -2 \\ x_5 \quad &= 4 \\ x_6 &= 3. \end{aligned}$$

These four equations in six unknowns determine four of the unknowns (the ones associated with the row leaders) in terms of the other two. If we set $x_2 = c_1$ and $x_4 = c_2$, then the most general solution of the system is

$$\begin{aligned} x_1 &= 7 - 2c_1 - 4c_2 \\ x_2 &= c_1 \\ x_3 &= -2 - 2c_2 \\ x_4 &= c_2 \\ x_5 &= 4 \\ x_6 &= 3. \end{aligned}$$

This system has two arbitrary constants in its general solution; such a system is said to have two *degrees of freedom*. ◼

A system whose augmented matrix or coefficient matrix is in echelon form is certainly easy to solve, as we have seen in Example 1 and the exercises and examples of the last section. In fact, the system may still be easy to solve even if properties 1 and 2 (on page 23) are not satisfied.

Example 2

Solve the system whose augmented matrix has been reduced to

$$\begin{bmatrix} 2 & -1 & 3 & 0 \\ 0 & 0 & 3 & 6 \\ 0 & 5 & 1 & 17 \end{bmatrix}.$$

Although this matrix fails to be in row echelon form because the rows are out of order and the row leaders are not 1, the corresponding system is still easy to solve. The associated system is

$$2x_1 - x_2 + 3x_3 = 0$$
$$3x_3 = 6$$
$$5x_2 + x_3 = 17,$$

and from the second equation we obtain $x_3 = 2$. Substituting this value into the third equation yields $5x_2 + 2 = 17$, so $x_2 = 3$. Finally, substituting these values into the first equation yields $x_1 = -3/2$.   ■

In our next example we consider a curve-fitting problem which comes up in the computer-graphics problem of fitting a function with a cubic-spline approximation.

Example 3

Find a cubic polynomial $y = ax^3 + bx^2 + cx + d$ which passes through $(1, 3)$ with slope -5 and through $(3, -7)$ with slope -1.

Since $(1, 3)$ and $(3, -7)$ are to be on the curve, we must have

$$a + b + c + d = 3 \quad \text{and} \quad 27a + 9b + 3c + d = -7.$$

Recall, from elementary calculus, that the slope at a point (x_1, y_1) on the curve is the value $y'(x_1)$ of the derivative at $x = x_1$. For $y = ax^3 + bx^2 + cx + d$ the derivative is $y' = 3ax^2 + 2bx + c$. Since the slope at $(1, 3)$ is to be -5 we must have

$$3a + 2b + c = -5,$$

and since the slope at $(3, -7)$ is to be -1 we must also have

$$27a + 6b + c = -1.$$

Thus a, b, c, and d must satisfy the linear system

$$
\begin{aligned}
a + b + c + d &= 3 \\
27a + 9b + 3c + d &= -7 \\
3a + 2b + c &= -5 \\
27a + 6b + c &= -1.
\end{aligned}
$$

The augmented matrix and the details of the row reduction are as follows.

$$
\begin{bmatrix}
1 & 1 & 1 & 1 & 3 \\
27 & 9 & 3 & 1 & -7 \\
3 & 2 & 1 & 0 & -5 \\
27 & 6 & 1 & 0 & -1
\end{bmatrix}
\xrightarrow[\substack{R_2 \leftarrow -27R_1 + R_2 \\ R_3 \leftarrow -3R_1 + R_3 \\ R_4 \leftarrow -27R_1 + R_4}]{}
\begin{bmatrix}
1 & 1 & 1 & 1 & 3 \\
0 & -18 & -24 & -26 & -88 \\
0 & -1 & -2 & -3 & -14 \\
0 & -21 & -26 & -27 & -82
\end{bmatrix}
$$

$$
\xrightarrow[\substack{R_2 \leftrightarrow R_3 \\ R_3 \leftarrow -18R_2 + R_3 \\ R_4 \leftarrow -21R_2 + R_4}]{}
\begin{bmatrix}
1 & 1 & 1 & 1 & 3 \\
0 & -1 & -2 & -3 & -14 \\
0 & 0 & 12 & 28 & 164 \\
0 & 0 & 16 & 36 & 212
\end{bmatrix}
$$

$$R_2 \leftarrow -R_2$$
$$R_4 \leftarrow -4/3 R_3 + R_4 \quad \begin{bmatrix} 1 & 1 & 1 & 1 & 3 \\ 0 & 1 & 2 & 3 & 14 \\ 0 & 0 & 12 & 28 & 164 \\ 0 & 0 & 0 & -4/3 & -20/3 \end{bmatrix}$$

$$R_4 \leftarrow -3/4 R_4$$
$$R_3 \leftarrow 1/12 R_3 \quad \begin{bmatrix} 1 & 1 & 1 & 1 & 3 \\ 0 & 1 & 2 & 3 & 14 \\ 0 & 0 & 1 & 7/3 & 41/3 \\ 0 & 0 & 0 & 1 & 5 \end{bmatrix}$$

The back substitution now yields $d = 5$, $c = 41/3 - 35/3 = 2$, $b = -5$, and $a = 1$. Thus the desired cubic polynomial is

$$y = x^3 - 5x^2 + 2x + 5. \quad \blacksquare$$

The key to success in using the elimination methods to solve linear systems is to carry out the reduction in an orderly way.

Once you have mastered the basic technique you may wish to experiment with some modifications of the basic algorithm. In particular, if the problem involves only integers and you are working by hand, it is desirable to minimize work with fractions; we all are faster and more accurate with integer arithmetic than with fractions.

For example, to produce a 0 in the $(1, 3)$ position of

$$\begin{bmatrix} 1 & 5 & 6 & 3 & 8 \\ 0 & 0 & 7 & 4 & 11 \end{bmatrix}$$

the basic algorithm would use $(R_2 \leftarrow 1/7 R_2)$ and $(R_1 \leftarrow -6R_2 + R_1)$, yielding

$$\begin{bmatrix} 1 & 5 & 0 & -3/7 & -10/7 \\ 0 & 0 & 1 & 4/7 & 11/7 \end{bmatrix}.$$

Subsequent calculations using these fractions could become a burden. An alternate approach would be to sacrifice the leading 1 in the first row and use the operations $(R_1 \leftarrow 7R_1)$ and $(R_1 \leftarrow -6R_2 + R_1)$ to produce first

$$\begin{bmatrix} 7 & 35 & 42 & 21 & 56 \\ 0 & 0 & 7 & 4 & 11 \end{bmatrix} \quad \text{and then} \quad \begin{bmatrix} 7 & 35 & 0 & -3 & -10 \\ 0 & 0 & 7 & 4 & 11 \end{bmatrix}.$$

If such modifications are made to the basic process, and the scaling of the rows is delayed until the final step, then arithmetic with fractions can be avoided completely, provided that there are no fractional elements in the original matrix. We illustrate this procedure in our next example.

Example 4

Consider the system whose augmented matrix is

$$\begin{bmatrix} 3 & -1 & 2 & 1 \\ 2 & 1 & 1 & 1 \\ 1 & -3 & 0 & 2 \end{bmatrix}.$$

The normal first step would be to divide the first row by 3 to obtain a 1 in the $(1, 1)$ position. However, two alternatives are available; we could interchange the first and third rows, or we could add (-1) times row 2 to row 1. We choose the first option and follow it with the operations $(R_2 \leftarrow -2R_1 + R_2)$ and $(R_3 \leftarrow -3R_1 + R_3)$ to yield

$$\begin{bmatrix} 1 & -3 & 0 & 2 \\ 0 & 7 & 1 & -3 \\ 0 & 8 & 2 & -5 \end{bmatrix}.$$

In order to obtain a 1 in the $(2, 2)$ position we could divide the second row by 7; to avoid creating fractions, we choose instead to use the operations $(R_2 \leftarrow -R_3 + R_2)$ and $(R_2 \leftarrow -R_2)$ to produce

$$\begin{bmatrix} 1 & -3 & 0 & 2 \\ 0 & 1 & 1 & -2 \\ 0 & 8 & 2 & -5 \end{bmatrix}.$$

The operations $(R_1 \leftarrow 3R_2 + R_1)$ and $(R_3 \leftarrow -8R_2 + R_3)$ now yield

$$\begin{bmatrix} 1 & 0 & 3 & -4 \\ 0 & 1 & 1 & -2 \\ 0 & 0 & -6 & 11 \end{bmatrix}.$$

To clear the third column without introducing fractions we will first make the elements in the $(1, 3)$ and $(2, 3)$ positions multiples of 6 by the operations $(R_1 \leftarrow 6R_1)$ and $(R_2 \leftarrow 6R_2)$. These operations, followed by $(R_3 \leftarrow -R_3)$, yield

$$\begin{bmatrix} 6 & 0 & 18 & -24 \\ 0 & 6 & 6 & -12 \\ 0 & 0 & 6 & -11 \end{bmatrix}.$$

The operations $(R_2 \leftarrow -R_3 + R_2)$ and $(R_1 \leftarrow -3R_3 + R_1)$ now lead to

$$\begin{bmatrix} 6 & 0 & 0 & 9 \\ 0 & 6 & 0 & 9 \\ 0 & 0 & 6 & -11 \end{bmatrix},$$

from which we obtain, by multiplying each row by $1/6$,

$$\begin{bmatrix} 1 & 0 & 0 & 9/6 \\ 0 & 1 & 0 & -1/6 \\ 0 & 0 & 1 & -11/6 \end{bmatrix}.$$

This final matrix is in reduced row echelon form, and we have obtained it without doing any arithmetic with fractions. ■

In order to study elementary row operations more efficiently we make the following definition.

Definition 1.4

The matrices A and B are said to be ROW EQUIVALENT ($A \underset{R}{\sim} B$) if B can be obtained from A by a finite sequence of elementary row operations.

Although row equivalence is a less restrictive relation on matrices than equality, it nevertheless satisfies three very important properties that are also satisfied by equality:

(1) REFLEXIVE ($A \underset{R}{\sim} A$ for every matrix A);
(2) SYMMETRIC (If $A \underset{R}{\sim} B$, then $B \underset{R}{\sim} A$); and
(3) TRANSITIVE (If $A \underset{R}{\sim} B$ and $B \underset{R}{\sim} C$, then $A \underset{R}{\sim} C$).

The first and third of these properties are essentially obvious; to verify the second we observe that *every elementary row operation can be reversed by an elementary row operation of the same type.*

(1) $(R_i \leftarrow 1/kR_i)$ reverses $(R_i \leftarrow kR_i)$.
(2) $(R_i \leftrightarrow R_j)$ reverses itself.
(3) $(R_j \leftarrow -kR_i + R_j)$ reverses $(R_j \leftarrow kR_i + R_j)$.

For example, $(R_2 \leftarrow 1/2R_2)$ reverses $(R_2 \leftarrow 2R_2)$, $(R_4 \leftarrow 3R_2 + R_4)$ reverses $(R_4 \leftarrow -3R_2 + R_4)$, and $(R_2 \leftrightarrow R_3)$ reverses itself. It follows that if a sequence of elementary row operations reduces A to B, then we can get from B back to A by using the sequence of operations that reverse the operations in the original sequence. The operations must be undone in the reverse order; that is, the last operation must be reversed first, the next to last operation must be reversed second, and so on.

If A is square, it is quite possible that the reduced row echelon matrix associated with A is the diagonal matrix

$$I_n = \begin{bmatrix} 1 & 0 & 0 & \cdots & 0 \\ 0 & 1 & 0 & \cdots & 0 \\ \vdots & \vdots & \vdots & & \vdots \\ 0 & 0 & 0 & \cdots & 1 \end{bmatrix}.$$

This particular diagonal matrix is called an IDENTITY MATRIX for reasons that will become clear in later sections of this chapter. The subscript indicates that this matrix is $n \times n$; the subscript is usually omitted when the size of the matrix is clear from the context.

The augmented matrix of a system of n equations in n unknowns is $n \times (n + 1)$. The reduced row echelon matrix of such a system may very well be of the form

$$[I \,|\, B] = \begin{bmatrix} 1 & 0 & 0 & \cdots & 0 & b_1 \\ 0 & 1 & 0 & \cdots & 0 & b_2 \\ \vdots & \vdots & \vdots & & \vdots & \vdots \\ 0 & 0 & 0 & \cdots & 1 & b_n \end{bmatrix}$$

This is the case if the solution of the original system is unique. The only way that the reduced row echelon matrix for such a system could fail to be of this form is if a zero row appeared in the augmented matrix during the reduction process.

We conclude this discussion with an important theorem whose proof will be omitted.

Theorem 1.2

Every matrix is row equivalent to a unique matrix in reduced row echelon form.

Theorem 1.2 assures us that we will always arrive at the same reduced row echelon form even if we use different sequences of operations to do the reduction. Thus, the final solution to the original linear system will not be changed by the choice of the operations used for the reduction. The important thing is to be orderly and to do the arithmetic correctly; machines have a distinct advantage at this point.

We next consider a very simple example of a transportation problem. Such problems are of great importance in the military and in business.

Example 5

A manufacturer of small computers receives orders for 150 units from each of three stores located in Cleveland, St. Louis, and Dallas. There are 150 units in a Pittsburgh warehouse and 300 units in a Houston warehouse. The shipping costs from Pittsburgh to the three stores are $5, $15, and $20 per unit, while the shipping costs from Houston to the three stores are $20, $10, and $5 per unit. How should the order be filled in order to minimize shipping costs?

Let x, y, and z be the number of units shipped from Pittsburgh to Cleveland, St. Louis, and Dallas respectively, and let u, v, and w be the number of units shipped from Houston to Cleveland, St. Louis, and Dallas respectively. We describe the data in the problem with the data graph in Figure 1.9.

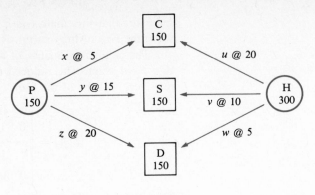

FIGURE 1.9

Since each store needs 150 units we must satisfy the demand conditions

$$x + u = 150 \quad \text{(Cleveland)}$$

$$y + v = 150 \quad \text{(St. Louis)}$$

$$z + w = 150 \quad \text{(Dallas)}.$$

Since each warehouse has limited stock we must also satisfy the supply conditions

$$x + y + z = 150 \quad \text{(Pittsburgh)}$$

$$u + v + w = 300 \quad \text{(Houston)}.$$

Thus, our six unknowns must satisfy a system of five linear equations. Because it has more unknowns than equations, this system, if it is consistent, will have many solutions. Of these solutions, we wish to choose a solution that will minimize

$$C = 5x + 15y + 20z + 20u + 10v + 5w,$$

where C is the cost of filling the orders.

The augmented matrix of the system and its echelon form are

$$\begin{bmatrix} 1 & 0 & 0 & 1 & 0 & 0 & 150 \\ 0 & 1 & 0 & 0 & 1 & 0 & 150 \\ 0 & 0 & 1 & 0 & 0 & 1 & 150 \\ 1 & 1 & 1 & 0 & 0 & 0 & 150 \\ 0 & 0 & 0 & 1 & 1 & 1 & 300 \end{bmatrix} \rightarrow \begin{bmatrix} 1 & 0 & 0 & 0 & -1 & -1 & -150 \\ 0 & 1 & 0 & 0 & 1 & 0 & 150 \\ 0 & 0 & 1 & 0 & 0 & 1 & 150 \\ 0 & 0 & 0 & 1 & 1 & 1 & 300 \\ 0 & 0 & 0 & 0 & 0 & 0 & 0 \end{bmatrix}.$$

The general solution of the constraint system is thus

$$x = -150 + v + w$$

$$y = 150 - v$$

$$z = 150 - w$$

$$u = 300 - v - w,$$

where v and w are arbitrary, except that $0 \leq v \leq 150$, $0 \leq w \leq 150$, and, because $x \geq 0$, $150 \leq (v + w) \leq 300$. The cost function now becomes

$$C = 5(-150 + v + w) + 15(150 - v) + 20(150 - w)$$
$$+ 20(300 - v - w) + 10v + 5w$$
$$= 10500 - 20v - 30w.$$

Clearly, C will be smallest when v and w are as large as possible, that is, when $v = w = 150$. Then $x = 150$, $y = z = u = 0$, and the minimum cost is \$3000. ■■

Exercise Set 1.3

1. Find a row echelon form for each of the following matrices.

(a) $\begin{bmatrix} 3 & -1 & 2 & 1 & 0 & 0 \\ 2 & 1 & 1 & 0 & 1 & 0 \\ 1 & -3 & 0 & 0 & 0 & 1 \end{bmatrix}$ (b) $\begin{bmatrix} 1 & 2 & 3 & 1 & 1 & 1 \\ 2 & 4 & 5 & 1 & -3 & 2 \\ 3 & 5 & 6 & 1 & 2 & -2 \end{bmatrix}$

(c) $\begin{bmatrix} 1 & 1 & 2 & 6 \\ 1 & -1 & -4 & -8 \\ 3 & -2 & 5 & 11 \\ 2 & 5 & -2 & 3 \end{bmatrix}$ (d) $\begin{bmatrix} 1 & -1 & 2 & 3 & 1 & 0 & 0 & 0 \\ 2 & -1 & 0 & 2 & 0 & 1 & 0 & 0 \\ 4 & 1 & -11 & -1 & 0 & 0 & 1 & 0 \\ 1 & 2 & 3 & 83 & 0 & 0 & 0 & 1 \end{bmatrix}$

2. Find the reduced row echelon form for each of the matrices in Exercise 1.

3. Find the general solution of the systems whose augmented matrices are the following.

(a) $\begin{bmatrix} 3 & 0 & 5 & 15 \\ 0 & 0 & 7 & 21 \\ 0 & 1 & 2 & 7 \end{bmatrix}$ (b) $\begin{bmatrix} 1 & 5 & 0 & 0 & 5 & -1 \\ 0 & 0 & 1 & 0 & 3 & 1 \\ 0 & 0 & 0 & 1 & 4 & 2 \\ 0 & 0 & 0 & 0 & 0 & 0 \end{bmatrix}$

(c) $\begin{bmatrix} 3 & 0 & 5 & 15 \\ 0 & 0 & 0 & 21 \\ 0 & 1 & 2 & 7 \end{bmatrix}$ (d) $\begin{bmatrix} 1 & 0 & 2 & 0 \\ 0 & 1 & 3 & 0 \\ 0 & 0 & 0 & 0 \end{bmatrix}$

4. From your work in either 1(a) or 2(a) find the solution of each of the following systems.

(a) $3x - y + 2z = 1$
$\quad 2x + y + z = 0$
$\quad \; x - 3y \quad\;\; = 0$

(b) $3x - y + 2z = 0$
$\quad 2x + y + z = 1$
$\quad \; x - 3y \quad\;\; = 0$

(c) $3x - y + 2z = 0$
$\quad 2x + y + z = 0$
$\quad \; x - 3y \quad\;\; = 1$

5. From your work in either 1(b) or 2(b) find the solution of each of the three following systems.

(a) $x + 2y + 3z = 1$
$\quad 2x + 4y + 5z = 1$
$\quad 3x + 5y + 6z = 1$

(b) $x + 2y + 3z = \;\;\; 1$
$\quad 2x + 4y + 5z = -3$
$\quad 3x + 5y + 6z = \;\;\; 2$

(c) $x + 2y + 3z = 1$
$2x + 4y + 5z = 2$
$3x + 5y + 6z = -2$

6. Theorem 1.2 states the reduced row echelon form of any matrix A is unique. Is the row echelon form of A also unique? Give an example of a matrix with a nonunique row echelon form.

7. Reduce $A = \begin{bmatrix} 1 & 2 & 3 & 1 \\ 2 & 5 & 5 & -3 \\ 5 & 10 & 16 & -1 \end{bmatrix}$ to a reduced row echelon matrix R,

and then find a sequence of elementary row operations which will change R back to A.

8. What is the inverse of the composite row operation $(R_i \leftarrow aR_i + bR_j)$?

9. The following algorithm will reduce an $m \times n$ $(m < n)$ matrix A to a matrix which is nearly in row echelon form.

> For $i = 1$ to $m - 1$
>   For $k = i + 1$ to m
> $t = a_{ki}/a_{ii}$
>   For $j = i$ to n
> $a_{kj} = a_{kj} - ta_{ij}$

(a) Describe carefully what these instructions will do to A provided that no $a_{ii} = 0$.

(b) Write a segment of computer code, in your favorite programming language, that will implement the above algorithm. Assume that the matrix A is already in the computer.

(c) Modify the code in part (b) so that the resulting matrix will be in row echelon form.

(d) Further modify the code so that the resulting matrix will be in reduced row echelon form.

(e) How would the code need to be changed to handle the case $a_{ii} = 0$?

(f) How would the code need to be changed if you wanted to save the original matrix?

(g) Provide proper input, output, and dimension statements for your code. Test your code on the matrix of Example 3.

†10. In this problem you will need the following algebraic identities:

$$1 + 2 + 3 + \ldots + (k - 1) + k = \frac{k(k + 1)}{2}$$

$$1 + 2^2 + 3^2 + \ldots + (k - 1)^2 + k^2 = \frac{k(k + 1)(2k + 1)}{6}.$$

Count the number of multiplications required for

(a) the solution of n equations in n unknowns if the coefficient matrix is in row echelon form with no zero rows;

(b) the reduction of an $n \times (n + 1)$ matrix to row echelon form; and

(c) the reduction of an $n \times (n + 1)$ matrix to reduced row echelon form.

11. Solve the system of equations (found on page 2) associated with the stuffed-animal-manufacturing problem at the beginning of Section 1.1.

12. Find a cubic polynomial $y = ax^3 + bx^2 + cx + d$ which passes through $(1, 2)$ with slope -1 and through $(2, 5)$ with slope 8.

13. Find a cubic polynomial $y = ax^3 + bx^2 + cx + d$ which passes through $(1, 1)$ with slope 3 and which also passes through $(2, 11)$ and $(-1, -1)$.

14. Solve the transportation problem of Example 5 if each of the two warehouses contains 225 units.

15. Solve the transporation problem whose data graph is shown below.

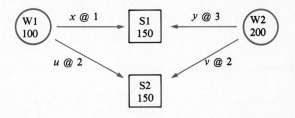

16. Solve the transportation problem whose data graph is shown below.

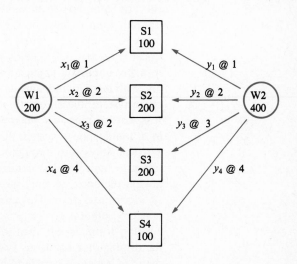

1.4 Homogeneous Systems

In this section we will treat two special cases of the linear equations problem which are important enough to deserve special emphasis: homogeneous systems and systems with multiple right-hand sides.

The general system of linear equations

$$a_{11}x_1 + a_{12}x_2 + \cdots + a_{1n}x_n = b_1$$
$$a_{21}x_1 + a_{22}x_2 + \cdots + a_{2n}x_n = b_2$$
$$\vdots \qquad \vdots \qquad \qquad \vdots \qquad \vdots \qquad (1.7)$$
$$a_{m1}x_1 + a_{m2}x_2 + \cdots + a_{mn}x_n = b_m$$

is said to be HOMOGENEOUS if all the constants b_i on the right-hand side of the equations are zero. Thus, the general homogeneous system looks like

$$a_{11}x_1 + a_{12}x_2 + \cdots + a_{1n}x_n = 0$$
$$a_{21}x_1 + a_{22}x_2 + \cdots + a_{2n}x_n = 0$$
$$\vdots \qquad \vdots \qquad \qquad \vdots \qquad \vdots \qquad (1.8)$$
$$a_{m1}x_1 + a_{m2}x_2 + \cdots + a_{mn}x_n = 0.$$

For linear systems in general we know that there are three possibilities: no solutions, a unique solution, or an infinite number of solutions. For a homogeneous system the first case can be ruled out by observing that a solution is obtained by setting all $x_i = 0$ (We will call this the ZERO SOLUTION or the TRIVIAL SOLUTION). If the solution is unique, then it must be the zero solution. The only time there are nonzero solutions is when there are an infinite number of such solutions; homogeneous systems with an infinite number of solutions occur in many important applications.

Solving a homogeneous system is no different from solving any other system; the elimination methods are still the methods of choice. It is worth noting that when we apply the Gaussian elimination method to a homogeneous system the last column of the augmented matrix is zero and stays zero throughout the reduction process. There is a temptation to omit this last column in the homogeneous case, but we will always include it to emphasize the homogeneous nature of the system.

It follows from Theorem 1.1 that a homogeneous system has a nonzero solution precisely when the number of nonzero rows, in the echelon form, is less than the number of unknowns. In particular, if the original homogeneous system has fewer equations than unknowns, then the system is certain to have nonzero solutions—an infinite number of them. We summarize these important facts about the solutions of homogeneous systems in Theorem 1.3.

Theorem 1.3

The homogeneous system 1.8 has an infinite number of nonzero solutions whenever there are fewer equations than unknowns, that is, if $m < n$. In general, System 1.8 will have an infinite number of nonzero solutions whenever the number of nonzero rows, in a row echelon form for the augmented matrix of System 1.8, is less than the number of unknowns.

Example 1

Find all solutions of the homogeneous system

$$x + 2y + 3z = 0$$
$$4x + 5y + 6z = 0$$
$$7x + 8y \quad\ = 0.$$

The augmented matrix and the details of the row reduction are given below; the operations used are listed beneath the arrows.

$$\begin{bmatrix} 1 & 2 & 3 & 0 \\ 4 & 5 & 6 & 0 \\ 7 & 8 & 0 & 0 \end{bmatrix} \xrightarrow[\begin{array}{l} R_2 \leftarrow -4R_1 + R_2 \\ R_3 \leftarrow -7R_1 + R_3 \end{array}]{}$$

$$\begin{bmatrix} 1 & 2 & 3 & 0 \\ 0 & -3 & -6 & 0 \\ 0 & -6 & -21 & 0 \end{bmatrix} \xrightarrow[\begin{array}{l} R_2 \leftarrow -1/3R_2 \\ R_3 \leftarrow 6R_2 + R_3 \end{array}]{}$$

$$\begin{bmatrix} 1 & 2 & 3 & 0 \\ 0 & 1 & 2 & 0 \\ 0 & 0 & -9 & 0 \end{bmatrix} \xrightarrow[R_3 \leftarrow -1/9R_3]{} \begin{bmatrix} 1 & 2 & 3 & 0 \\ 0 & 1 & 2 & 0 \\ 0 & 0 & 1 & 0 \end{bmatrix}.$$

The final matrix is a row echelon form for the augmented matrix of the original system. Since the number of nonzero rows (three) in the row echelon matrix is equal to the number of unknowns, this system has no nonzero solution. The only solution of the system is $x = y = z = 0$. ■

Example 2

Find all solutions of the homogeneous system

$$x_1 + x_2 + x_3 - 3x_4 = 0$$
$$3x_1 - 2x_2 - 17x_3 + 16x_4 = 0$$
$$3x_1 + 2x_2 - x_3 - 4x_4 = 0.$$

Since this homogeneous system has fewer equations (three) than unknowns (four), there are certain to be an infinite number of solutions. To find the solutions we will use Gauss-Jordan elimination:

$$\begin{bmatrix} 1 & 1 & 1 & -3 & 0 \\ 3 & -2 & -17 & 16 & 0 \\ 3 & 2 & -1 & -4 & 0 \end{bmatrix} \xrightarrow[\begin{array}{l} R_2 \leftarrow -3R_1 + R_2 \\ R_3 \leftarrow -3R_1 + R_3 \end{array}]{}$$

$$\begin{bmatrix} 1 & 1 & 1 & -3 & 0 \\ 0 & -5 & -20 & 25 & 0 \\ 0 & -1 & -4 & 5 & 0 \end{bmatrix} \xrightarrow[\begin{array}{l} R_3 \leftarrow -R_3 \\ R_2 \leftarrow 5R_3 + R_2 \\ R_2 \leftrightarrow R_3 \\ R_1 \leftarrow -R_2 + R_1 \end{array}]{} \begin{bmatrix} 1 & 0 & -3 & 2 & 0 \\ 0 & 1 & 4 & -5 & 0 \\ 0 & 0 & 0 & 0 & 0 \end{bmatrix}.$$

The reduced system is

$$x_1 - 3x_3 + 2x_4 = 0 \qquad x_1 = 3x_3 - 2x_4$$
$$\text{or}$$
$$x_2 + 4x_3 - 5x_4 = 0 \qquad x_2 = -4x_3 + 5x_4$$

and the general solution is obtained by assigning arbitrary values to x_3 and x_4. If we set $x_3 = c_1$ and $x_4 = c_2$ we obtain, as the general solution,

$$x_1 = 3c_1 - 2c_2, \quad x_2 = -4c_1 + 5c_2, \quad x_3 = c_1, \quad x_4 = c_2. \quad \blacksquare$$

Example 3

Find all values of λ such that the following system has nonzero solutions

$$2x + y = \lambda x$$
$$4x - y = \lambda y.$$

At first glance this system does not appear to be homogeneous, but when the terms on the right are moved to the left side we obtain the homogeneous system

$$(2 - \lambda)x + \qquad y = 0$$
$$4x - (1 + \lambda)y = 0.$$

The elementary row operations $(R_2 \leftarrow 1/4R_2)$, $(R_1 \leftrightarrow R_2)$, and $(R_2 \leftarrow -(2 - \lambda)R_1 + R_2)$ reduce the augmented matrix from

$$\begin{bmatrix} (2 - \lambda) & 1 & 0 \\ 4 & -(1 + \lambda) & 0 \end{bmatrix} \text{ to } \begin{bmatrix} 1 & -(1 + \lambda)/4 & 0 \\ 0 & 1 + (2 - \lambda)(1 + \lambda)/4 & 0 \end{bmatrix}$$

$$= \begin{bmatrix} 1 & -(1 + \lambda)/4 & 0 \\ 0 & -(\lambda^2 - \lambda - 6)/4 & 0 \end{bmatrix}.$$

This system will have a nonzero solution precisely when the second row of the last matrix is zero. This will occur when $\lambda^2 - \lambda - 6 = 0$. The solutions of this quadratic equation are easily seen to be $\lambda = 3$ and $\lambda = -2$. $\quad \blacksquare$

The second topic we deal with in this section is the frequently occurring case in which there are several systems with the same coefficient matrix but different right-hand sides. Consider the case of three systems like 1.7 on page 36, each with the same coefficient matrix A, but with right-hand sides

$$B = \begin{bmatrix} b_1 \\ b_2 \\ \vdots \\ b_m \end{bmatrix}, \quad C = \begin{bmatrix} c_1 \\ c_2 \\ \vdots \\ c_m \end{bmatrix}, \quad \text{and} \quad D = \begin{bmatrix} d_1 \\ d_2 \\ \vdots \\ d_m \end{bmatrix}$$

This situation occurs frequently in applications where the algebraic system models a physical system in which the constants on the right-hand sides of the equations represent inputs to the system and the unknowns represent the response of the system to those inputs. In this case our problem asks for the response of the system to several different inputs.

We could, of course, solve each of the systems separately using the techniques of this chapter. The bulk of the work in each case would be the reduction of the coefficient matrix A to echelon form. In each case the same operations would be used, so it would be inefficient to do this work three times. A better approach to this set of problems was suggested in Exercises 4 and 5 in Section 1.3, where we saw that if

$$[A \mid B \mid C \mid D] \xrightarrow[\text{reduce}]{\text{row}} [R \mid H \mid K \mid J],$$

then the system with right-hand side B is equivalent to the system with coefficient matrix R and right-hand side H, the system with right-hand side C is equivalent to the system with coefficient matrix R and right-hand side K, and so on. In particular, if $R = I$, then H, K, and J are the unique solutions of the three systems.

Example 4

Solve the following three systems with a single row reduction.

$$
\begin{array}{lll}
x + 2y = 4 & x + 2y = 1 & x + 2y = 2 \\
2x + 3y = 7 & 2x + 3y = 1 & 2x + 3y = 9 \\
x + 4y = 6 & x + 4y = 3 & x + 4y = 5
\end{array}
$$

The 3-fold augmented matrix and the routine row reduction are given below

$$
\begin{bmatrix}
1 & 2 & 4 & 1 & 2 \\
2 & 3 & 7 & 1 & 9 \\
1 & 4 & 6 & 3 & 5
\end{bmatrix}
\rightarrow
\begin{bmatrix}
1 & 2 & 4 & 1 & 2 \\
0 & -1 & -1 & -1 & 5 \\
0 & 2 & 2 & 2 & 3
\end{bmatrix}
$$

$$
\rightarrow
\begin{bmatrix}
1 & 0 & 2 & -1 & 12 \\
0 & 1 & 1 & 1 & -5 \\
0 & 0 & 0 & 0 & 13
\end{bmatrix}
$$

From this calculation we see, using the third columns, that the first system has $x = 2$, $y = 1$ as its unique solution. From the fourth columns we see that the unique solution of the second system is $x = -1$, $y = 1$. From the fifth columns, we see that the third system is inconsistent, because $13 \neq 0$. ■

Our next example shows how homogeneous systems can be used in balancing the equations which describe chemical reactions.

Example 5

When sodium carbonate (Na_2CO_3) is combined with bromine vapor (Br_2) the by-products are sodium bromide (NaBr), sodium bromate ($NaBrO_3$) and carbon dioxide (CO_2). Balance the following equation that describes the reaction:

$$Na_2CO_3 + Br_2 \longrightarrow NaBr + NaBrO_3 + CO_2 .$$

The above description of the reaction does not indicate the relative amounts of each chemical involved. In order to provide that information we need to balance the chemical equation. Thus we seek positive integers x_1, x_2, x_3, x_4, and x_5 so that

$$x_1Na_2CO_3 + x_2Br_2 = x_3NaBr + x_4NaBrO_3 + x_5CO_2 .$$

We can identify linear relations between the x_i by comparing the number of atoms of each of the chemical elements involved in the above equation. Thus, for sodium (Na) we must have

$$2x_1 = x_3 + x_4 .$$

Similarly, we must have

$$x_1 = x_5 \text{ for the carbon (C),}$$

$$3x_1 = 3x_4 + 2x_5 \text{ for the oxygen (O), and}$$

$$2x_2 = x_3 + x_4 \text{ for the bromine (Br)} .$$

Thus we see that the five x_i must satisfy a system of four homogeneous equations. Since there are fewer equations than unknowns we know that nonzero solutions exist. We will find these solutions by Gauss-Jordan elimination. The augmented matrix of the system and the reduction to echelon form are given below.

$$
\begin{bmatrix}
2 & 0 & -1 & -1 & 0 & 0 \\
1 & 0 & 0 & 0 & -1 & 0 \\
3 & 0 & 0 & -3 & -2 & 0 \\
0 & 2 & -1 & -1 & 0 & 0
\end{bmatrix}
\rightarrow
\begin{bmatrix}
1 & 0 & 0 & 0 & -1 & 0 \\
0 & 0 & -1 & -1 & 2 & 0 \\
0 & 0 & 0 & -3 & 1 & 0 \\
0 & 1 & -1/2 & -1/2 & 0 & 0
\end{bmatrix}
$$

$$
\rightarrow
\begin{bmatrix}
1 & 0 & 0 & 0 & -1 & 0 \\
0 & 1 & -1/2 & -1/2 & 0 & 0 \\
0 & 0 & 1 & 1 & -2 & 0 \\
0 & 0 & 0 & 1 & -1/3 & 0
\end{bmatrix}
\rightarrow
$$

$$
\begin{bmatrix}
1 & 0 & 0 & 0 & -1 & 0 \\
0 & 1 & -1/2 & 0 & -1/6 & 0 \\
0 & 0 & 1 & 0 & -5/3 & 0 \\
0 & 0 & 0 & 1 & -1/3 & 0
\end{bmatrix}
\rightarrow
\begin{bmatrix}
1 & 0 & 0 & 0 & -1 & 0 \\
0 & 1 & 0 & 0 & -1 & 0 \\
0 & 0 & 1 & 0 & -5/3 & 0 \\
0 & 0 & 0 & 1 & -1/3 & 0
\end{bmatrix}
$$

We obtain the general solution of this homogeneous system by letting x_5 be arbitrary, say $x_5 = c$. Then we have $x_1 = c$, $x_2 = c$, $x_3 = 5c/3$, $x_4 = c/3$, and $x_5 = c$. Since we want the x_i to be positive integers we choose c to be some multiple of 3. If we choose $c = 3$, then the balanced chemical equation is

$$3Na_2CO_3 + 3Br_2 = 5NaBr + NaBrO_3 + 3CO_2. \quad \blacksquare$$

Exercise Set 1.4

1. Find all solutions of the following homogeneous systems.

   (a) $\begin{aligned} 3x_1 + 2x_2 + 16x_3 + 5x_4 &= 0 \\ 2x_2 + 10x_3 + 8x_4 &= 0 \\ x_1 + x_2 + 7x_3 + 3x_4 &= 0 \end{aligned}$

   (b) $\begin{aligned} 2x_1 + x_2 + 3x_3 &= 0 \\ x_1 + 2x_2 \quad\quad\quad &= 0 \\ x_2 + x_3 &= 0 \end{aligned}$

   (c) $\begin{aligned} x_1 + 3x_2 + 5x_3 + x_4 &= 0 \\ 4x_1 - 7x_2 - 3x_3 - x_4 &= 0 \\ 3x_1 + 2x_2 + 7x_3 + 3x_4 &= 0 \end{aligned}$

   (d) $\begin{aligned} 7x_1 - 8x_2 - 8x_3 &= -x_1 \\ 9x_1 - 16x_2 - 18x_3 &= -x_2 \\ -5x_1 + 11x_2 + 13x_3 &= -x_3 \end{aligned}$

   (e) $\begin{aligned} x_1 \quad\quad - 4x_3 &= 3x_1 \\ 5x_2 + 4x_3 &= 3x_2 \\ -4x_1 + 4x_2 + 3x_3 &= 3x_3 \end{aligned}$

   (f) $\begin{aligned} x_1 + x_2 - 3x_3 + x_4 &= 0 \\ 2x_1 \quad\quad + x_3 - x_4 &= 0 \\ x_1 + 3x_2 - 10x_3 + 4x_4 &= 0 \end{aligned}$

2. Consider the general homogeneous linear system 1.8, on page 36, with $m = n = 3$. Suppose we know that

   $$x_1 = a, \; x_2 = b, \; x_3 = c \quad \text{and} \quad x_1 = a', \; x_2 = b', \; x_3 = c'$$

   are two solutions of the system. Let h and k be arbitrary real constants and show that

   $$x_1 = ha + ka', \; x_2 = hb + kb', \; x_3 = hc + kc'$$

   is also a solution of the system. Is this true if the system is *not* homogeneous? What condition on h and k would make the statement true in the nonhomogeneous case?

3. Find all values of λ such that the following systems have nontrivial solutions.

   (a) $\begin{aligned} 5x + y &= \lambda x \\ 4x + 8y &= \lambda y \end{aligned}$

   (b) $\begin{aligned} 3x - 5y &= \lambda x \\ 5x + 3y &= \lambda y \end{aligned}$

   (c) $\begin{aligned} 2x + y + z &= \lambda x \\ 2x + 3y + 2z &= \lambda y \\ x + y + 2z &= \lambda z \end{aligned}$

   (d) $\begin{aligned} y &= \lambda x \\ z &= \lambda y \\ -2x + 3y &= \lambda z \end{aligned}$

4. From your solutions of Exercises 2–10 in Section 1.2 find the general solutions of the associated homogeneous systems (change the right-hand side to zero in each case). No additional calculations should be necessary.

5. Solve the following set of systems of linear equations, each of which has coefficient matrix A and respective right-hand sides B_1, B_2, and B_3.

$$A = \begin{bmatrix} 1 & 0 & 1 \\ 2 & 1 & 0 \\ 0 & 0 & 1 \\ 1 & -2 & 0 \end{bmatrix} \quad B_1 = \begin{bmatrix} 0 \\ 1 \\ -1 \\ 3 \end{bmatrix} \quad B_2 = \begin{bmatrix} 0 \\ 0 \\ 1 \\ -5 \end{bmatrix} \quad B_3 = \begin{bmatrix} 0 \\ 0 \\ -3 \\ 5 \end{bmatrix}$$

6. Solve the following set of systems of linear equations, each of which has coefficient matrix A and respective right-hand sides B_1, B_2, and B_3.

$$A = \begin{bmatrix} 3 & -1 & 2 \\ 2 & 2 & 1 \\ 1 & -3 & 0 \end{bmatrix} \quad B_1 = \begin{bmatrix} 1 \\ 0 \\ 0 \end{bmatrix} \quad B_2 = \begin{bmatrix} 0 \\ 1 \\ 0 \end{bmatrix} \quad B_3 = \begin{bmatrix} 0 \\ 0 \\ 1 \end{bmatrix}$$

†**7.** What condition on the coefficients of the general 2×2 homogeneous system

$$a_{11}x_1 + a_{12}x_2 = 0$$
$$a_{21}x_1 + a_{22}x_2 = 0$$

will ensure that a nonzero solution exists?

8. Obtain balanced chemical equations for the following reactions.

(a) $N_2 + H_2 \longrightarrow NH_3$

(b) $Cl_2 + KOH \longrightarrow KCl + KClO_3 + H_2O$

(c) $C_3H_5(NO_3)_3 \longrightarrow H_2O + CO + N_2 + O_2$

9. Consider the insulated rod pictured below. Recall Example 4 of Section 1.2 and then use the techniques of this section to estimate the internal temperatures t_1, t_2, t_3, and t_4 under the following circumstances.

(a) $t_0 = 0°$, $t_5 = 75°$ (b) $t_0 = 32°$, $t_5 = 212°$

(c) $t_0 = -20°$, $t_5 = 100°$ (d) $t_0 = 212°$, $t_5 = 0°$

10. Consider the insulated metal plate pictured below. The numbers represent temperatures at the indicated points. Assume that each interior temperature is the average of the temperatures at the four neighboring grid points.

Find the four interior temperatures t_1, t_2, t_3, and t_4 if the other temperatures are as listed below.

(a) $b_1 = b_2 = 50°$, $b_3 = b_4 = 100°$, $b_5 = b_6 = 200°$.

(b) $b_1 = b_3 = 0°$, $b_2 = b_4 = b_5 = b_6 = 100°$

(c) $b_1 = 220°$, $b_2 = b_3 = b_4 = b_5 = b_6 = 0°$

11. Describe how the application described in Exercise 17 in Section 1.1 could lead to a set of systems with the same coefficient matrix.

12. Describe how the curve-fitting problem of Example 3 in Section 1.3 could lead to a set of systems with the same coefficient matrix.

1.5   *Matrix Algebra*

Matrices have already been encountered in our discussion of systems of linear equations. In this section we will develop an algebra for matrices that has many applications. In particular, the results of this section will make it much easier to deal with systems of linear equations.

We begin by formally repeating the basic definitions and notational conventions used in dealing with matrices.

Definition 1.5

A MATRIX is a rectangular array of numbers. The normal way of describing a matrix with m rows and n columns is

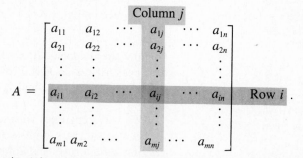

The matrix A is said to be $m \times n$ (the row size is always listed first) and the entry a_{ij} occurs in row i and column j. Two matrices are EQUAL if they are identical in every way; that is, they are the same size and have the same entry in every position.

The DIAGONAL ENTRIES of A are a_{11}, a_{22}, a_{33}, . . . , that is, those entries with equal row and column indices. The UPPER TRIANGULAR ENTRIES of A are those a_{ij} with $i < j$ while the LOWER TRIANGULAR ENTRIES of A are those a_{ij} with $j < i$. See Figure 1.10.

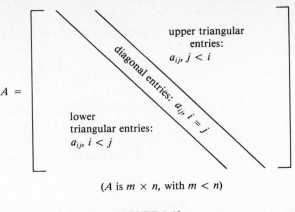

$$(A \text{ is } m \times n, \text{ with } m < n)$$

FIGURE 1.10

A matrix is said to be UPPER TRIANGULAR if all its lower triangular entries are zero; that is, all the nonzero entries are diagonal or upper triangular entries. A matrix is said to be LOWER TRIANGULAR if all its upper triangular entries are zero. A matrix is DIAGONAL if it is square ($m = n$) and all the nondiagonal entries are zero, that is, $a_{ij} = 0$ if $i \neq j$. Of the matrices

$$D = \begin{bmatrix} 1 & 0 & 0 \\ 0 & 3 & 0 \\ 0 & 0 & 2 \end{bmatrix}, \quad L = \begin{bmatrix} 1 & 0 & 0 \\ 2 & 0 & 0 \\ 3 & 3 & 5 \end{bmatrix}, \quad U = \begin{bmatrix} 1 & 2 & 1 & 1 \\ 0 & 3 & 2 & 3 \\ 0 & 0 & 4 & 5 \end{bmatrix}, \quad R = \begin{bmatrix} 1 & 2 & 3 \end{bmatrix},$$

$$C = \begin{bmatrix} 5 \\ 6 \\ 3 \end{bmatrix}, \quad A = \begin{bmatrix} 0 & 3 & 2 \\ 3 & 0 & 5 \\ 2 & 6 & 0 \end{bmatrix}, \quad \text{and} \quad Z = \begin{bmatrix} 0 & 0 & 0 \\ 0 & 0 & 0 \\ 0 & 0 & 0 \end{bmatrix},$$

D and Z are diagonal, L, D, and Z are lower triangular, U, D, and Z are upper triangular, R is a row matrix ($m = 1$), C is a column matrix ($n = 1$), and D, L, A, and Z are square matrices while U, R, and C are not.

We will sometimes wish to consider submatrices of a given matrix A. By a SUBMATRIX of A we mean any matrix obtained from A by deleting certain rows and columns of A. Of particular interest are submatrices obtained by partitioning the matrix into other submatrices. For example, the 5×5 matrix

$$A = \begin{bmatrix} 1 & 3 & 7 & 4 & 2 \\ 3 & 4 & 6 & 5 & 0 \\ 8 & 0 & 2 & 1 & 2 \\ 1 & 3 & 2 & 4 & 1 \\ 4 & 6 & 7 & 2 & 2 \end{bmatrix} = \begin{bmatrix} A_{11} & A_{12} \\ A_{21} & A_{22} \end{bmatrix},$$

where $A_{11} = \begin{bmatrix} 1 & 3 & 7 \\ 3 & 4 & 6 \\ 8 & 0 & 2 \end{bmatrix}$ is a 3 × 3 submatrix of A

and $A_{22} = \begin{bmatrix} 4 & 1 \\ 2 & 2 \end{bmatrix}$ is a 2 × 2 submatrix of A.

We already utilized this notation when we wrote the augmented matrix of the linear system 1.4 on page 7 as $[A \,|\, B]$ in the last section. Of particular importance are the rows and columns of the matrix A; we will use the following notation for these submatrices:

$$\text{Row}_i(A) = [a_{i1} \quad a_{i2} \quad \cdots \quad a_{in}] \text{ is the } i\text{th row of } A$$

and

$$\text{Col}_j(A) = \begin{bmatrix} a_{1j} \\ a_{2j} \\ \vdots \\ a_{mj} \end{bmatrix} \text{ is the } j\text{th column of } A.$$

We will, on occasion, wish to partition A into its rows

$$A = \begin{bmatrix} \text{Row}_1(A) \\ \text{Row}_2(A) \\ \vdots \\ \text{Row}_m(A) \end{bmatrix} = \begin{bmatrix} a_{11} & a_{12} & \cdots & a_{1n} \\ a_{21} & a_{22} & \cdots & a_{2n} \\ \vdots & \vdots & & \vdots \\ a_{m1} & a_{m2} & \cdots & a_{mn} \end{bmatrix}$$

or into its columns

$$A = [\; \text{Col}_1(A) \; \cdots \; \text{Col}_n(A) \;] = \begin{bmatrix} a_{11} & a_{12} & \cdots & a_{1n} \\ a_{21} & a_{22} & \cdots & a_{2n} \\ \vdots & \vdots & & \vdots \\ a_{m1} & a_{m2} & \cdots & a_{mn} \end{bmatrix}.$$

We begin our discussion of the algebra of matrices by defining the sum of two matrices.

Definition 1.6

If A and B are matrices of the same size, then the SUM $A + B$ is the matrix obtained by adding the corresponding entries of A and B; that is, if A and B are both $m \times n$ then $C = A + B$ is the $m \times n$ matrix whose entries satisfy:

$$c_{ij} = a_{ij} + b_{ij}; \qquad i = 1, 2, \ldots, m, \qquad j = 1, 2, \ldots, n.$$

If A and B are not the same size, then their sum is not defined.

Example 1

For the matrices

$$A = \begin{bmatrix} 2 & 0 \\ 4 & 3 \\ 6 & 7 \end{bmatrix}, \quad B = \begin{bmatrix} 1 & 5 \\ 0 & 3 \\ 2 & 6 \end{bmatrix}, \quad C = \begin{bmatrix} 1 & 2 & 3 \\ 4 & 5 & 6 \\ 7 & 8 & 9 \end{bmatrix},$$

$$\text{and} \quad D = \begin{bmatrix} 0 & 1 & 1 \\ 2 & 0 & 2 \\ 3 & 3 & 0 \end{bmatrix}$$

it follows from Definition 1.5 that

$$A + B = \begin{bmatrix} 3 & 5 \\ 4 & 6 \\ 8 & 13 \end{bmatrix} \quad \text{and} \quad C + D = \begin{bmatrix} 1 & 3 & 4 \\ 6 & 5 & 8 \\ 10 & 11 & 9 \end{bmatrix}.$$

Note that $A + C$, $A + D$, $B + C$, and $B + D$ are not defined. ▬

We next define the product of a scalar (real number) and a matrix. This definition allows us to write $A + A + A = 3A$, and so on.

Definition 1.7

If A is any matrix and k is any number, then the SCALAR MULTIPLE kA is the matrix obtained by multiplying each entry of A by k. We define the ADDITIVE INVERSE (negative) of A by $-A = (-1)A$.

Example 2

If A, B, C, and D are the matrices of Example 1, then

$$2A + 3B = 2\begin{bmatrix} 2 & 0 \\ 4 & 3 \\ 6 & 7 \end{bmatrix} + 3\begin{bmatrix} 1 & 5 \\ 0 & 3 \\ 2 & 6 \end{bmatrix}$$

$$= \begin{bmatrix} 4 & 0 \\ 8 & 6 \\ 12 & 14 \end{bmatrix} + \begin{bmatrix} 3 & 15 \\ 0 & 9 \\ 6 & 18 \end{bmatrix} = \begin{bmatrix} 7 & 15 \\ 8 & 15 \\ 18 & 32 \end{bmatrix},$$

$$5C - 4D = 5C + (-4)D = \begin{bmatrix} 5 & 10 & 15 \\ 20 & 25 & 30 \\ 35 & 40 & 45 \end{bmatrix} + \begin{bmatrix} 0 & -4 & -4 \\ -8 & 0 & -8 \\ -12 & -12 & 0 \end{bmatrix}$$

$$= \begin{bmatrix} 5 & 6 & 11 \\ 12 & 25 & 22 \\ 23 & 28 & 45 \end{bmatrix}, \quad \text{and}$$

$$C - C = C + (-1)C = \begin{bmatrix} 1 & 2 & 3 \\ 4 & 5 & 6 \\ 7 & 8 & 9 \end{bmatrix} + \begin{bmatrix} -1 & -2 & -3 \\ -4 & -5 & -6 \\ -7 & -8 & -9 \end{bmatrix}$$

$$= \begin{bmatrix} 0 & 0 & 0 \\ 0 & 0 & 0 \\ 0 & 0 & 0 \end{bmatrix} = Z. \quad \blacksquare$$

The matrix Z, all of whose entries are zero, is called the ZERO MATRIX. We will denote the zero matrix either by Z or 0; no confusion will result from using the same notation for the zero scalar and the zero matrix.

Another important algebraic operation on matrices is matrix multiplication. We first give the definition for an important special case.

Definition 1.8

The PRODUCT of a $1 \times n$ row matrix R and an $n \times 1$ column matrix C is defined to be

$$RC = [r_1 \quad r_2 \quad \cdots \quad r_n] \begin{bmatrix} c_1 \\ c_2 \\ \vdots \\ c_n \end{bmatrix}$$

$$= r_1 c_1 + r_2 c_2 + \ldots + r_n c_n = \sum_{i=1}^{n} r_i c_i.$$

Note that we have defined the product of a $1 \times n$ matrix and an $n \times 1$ matrix to be a scalar. (We will not distinguish between scalars and 1×1 matrices.) This definition allows us to write the general linear equation

$$a_1 x_1 + a_2 x_2 + \ldots + a_n x_n = b$$

as the matrix equation

$$[a_1 \quad a_2 \quad \cdots \quad a_n] \begin{bmatrix} x_1 \\ x_2 \\ \vdots \\ x_n \end{bmatrix} = b.$$

We now use Definition 1.8 to give the general definition of matrix multiplication.

Definition 1.9

If A is an $m \times r$ matrix and B is an $r \times n$ matrix, then the PRODUCT AB is the $m \times n$ matrix C whose entries are given by

$$c_{ij} = \text{Row}_i(A)\text{Col}_j(B)$$

$$= a_{i1}b_{1j} + a_{i2}b_{2j} + \ldots + a_{in}b_{nj} = \sum_{k=1}^{n} a_{ik}b_{kj}.$$

The row-by-column characterization of matrix multiplication is clearly the way to remember Definition 1.9. Note that, in the expression for

c_{ij}, each term has an a with first subscript i and a b with second subscript j. Moreover, in every term the second subscript on a and the first subscript on b are the same, as shown:

$$AB = \begin{bmatrix} a_{11} & a_{12} & \cdots & a_{1r} \\ \vdots & \vdots & & \vdots \\ a_{i1} & a_{i2} & \cdots & a_{ir} \\ \vdots & \vdots & & \vdots \\ a_{m1} & a_{m2} & \cdots & a_{mr} \end{bmatrix} \begin{bmatrix} b_{11} & \cdots & b_{1j} & \cdots & b_{1n} \\ \vdots & & \vdots & & \vdots \\ b_{i1} & \cdots & b_{ij} & \cdots & b_{in} \\ \vdots & & \vdots & & \vdots \\ b_{r1} & \cdots & b_{rj} & \cdots & b_{rn} \end{bmatrix}$$

$$= \begin{bmatrix} c_{11} & \cdots & c_{1j} & \cdots & c_{1n} \\ \vdots & & \vdots & & \vdots \\ c_{i1} & \cdots & c_{ij} & \cdots & c_{in} \\ \vdots & & \vdots & & \vdots \\ c_{m1} & \cdots & c_{mj} & \cdots & c_{mn} \end{bmatrix} = C.$$

Note that the matrix product is defined only when the number of columns of the first factor is equal to the number of rows of the second factor. The following diagram is useful in remembering how the sizes of the factors affect the size of the product:

$$(A) \quad (B) = AB$$
$$(m \times r)(r \times n) = m \times n.$$

Example 3

Compute AB and BA if

$$A = \begin{bmatrix} 1 & 4 & 5 \\ 2 & 1 & -3 \end{bmatrix} \quad \text{and} \quad B = \begin{bmatrix} 2 & 0 \\ -1 & 2 \\ 3 & 7 \end{bmatrix}.$$

$$AB =$$
$$\begin{bmatrix} 1 \times 2 + 4 \times (-1) + 5 \times 3 & 1 \times 0 + 4 \times 2 + 5 \times 7 \\ 2 \times 2 + 1 \times (-1) + (-3) \times 3 & 2 \times 0 + 1 \times 2 + (-3) \times 7 \end{bmatrix}$$

$$= \begin{bmatrix} 13 & 43 \\ -6 & -19 \end{bmatrix}$$

$$BA = \begin{bmatrix} 2 & 0 \\ -1 & 2 \\ 3 & 7 \end{bmatrix} \begin{bmatrix} 1 & 4 & 5 \\ 2 & 1 & -3 \end{bmatrix} = \begin{bmatrix} 2 & 8 & 10 \\ 3 & -2 & -11 \\ 17 & 19 & -6 \end{bmatrix}$$

Note that the 2×2 matrix AB and the 3×3 matrix BA are not equal.
In fact, they are not even the same size!

Example 4

Compute CD and DC if

$$C = \begin{bmatrix} 1 & 2 \\ 2 & 4 \end{bmatrix} \quad \text{and} \quad D = \begin{bmatrix} 2 & -4 \\ -1 & 2 \end{bmatrix}.$$

Directly from Definition 1.9 we obtain

$$CD = \begin{bmatrix} 0 & 0 \\ 0 & 0 \end{bmatrix} \quad \text{while} \quad DC = \begin{bmatrix} -6 & -12 \\ 3 & 6 \end{bmatrix}. \quad ■$$

Again, note that CD and DC are not equal even though in this case they are the same size. Note also that for matrices A and C of Examples 3 and 4, AC is not defined although CA is a well-defined 2×3 matrix.

These last two examples point out some major differences between matrix multiplication and the multiplication of scalars.

---

1. The product of two matrices is not always defined.
2. Matrix multiplication is NONCOMMUTATIVE (in general $AB \neq BA$).
3. The product of two nonzero matrices may be the zero matrix.

We will investigate the algebraic properties of matrix algebra in more detail in the next section.

It is very important to note that the general linear system

$$a_{11}x_1 + a_{12}x_2 + \ldots + a_{1n}x_n = b_1$$
$$a_{21}x_1 + a_{22}x_2 + \ldots + a_{2n}x_n = b_2$$
$$\vdots \qquad \vdots \qquad \qquad \vdots \qquad \vdots$$
$$a_{m1}x_1 + a_{m2}x_2 + \ldots + a_{mn}x_n = b_m$$

can be written as the single matrix equation

$$AX = B.$$

Here A is the coefficient matrix, X is the matrix of unknowns and B is the matrix of constants from the right-hand sides of the equations:

$$X = \begin{bmatrix} x_1 \\ x_2 \\ \vdots \\ x_n \end{bmatrix} \quad \text{and} \quad B = \begin{bmatrix} b_1 \\ b_2 \\ \vdots \\ b_m \end{bmatrix}.$$

A careful examination of the definition of matrix multiplication shows that each of the elements in the jth column of the product matrix AB is computed by taking the product of a row of A and the jth column of B; that is, only the jth column of B is involved in computing the jth column of AB. In fact, if A is $m \times n$, we have

$$\text{Col}_j(AB) = \begin{bmatrix} \text{Row}_1(A)\text{Col}_j(B) \\ \text{Row}_2(A)\text{Col}_j(B) \\ \vdots \\ \text{Row}_m(A)\text{Col}_j(B) \end{bmatrix} = \begin{bmatrix} a_{11}b_{1j} + a_{12}b_{2j} + \ldots + a_{1n}b_{nj} \\ a_{21}b_{1j} + a_{22}b_{2j} + \ldots + a_{2n}b_{nj} \\ \vdots \qquad \vdots \qquad \qquad \vdots \\ a_{m1}b_{1j} + a_{m2}b_{2j} + \ldots + a_{mn}b_{nj} \end{bmatrix}$$

$$= A\text{Col}_j(B)$$

$$= \text{Col}_1(A)b_{1j} + \ldots + \text{Col}_n(A)b_{nj}. \tag{1.9}$$

A useful expression equivalent to Equation (1.9) is

$$AB = A[\text{Col}_1(B) \quad \text{Col}_2(B) \quad \cdots \quad \text{Col}_n(B)]$$

$$= [A\text{Col}_1(B) \quad A\text{Col}_2(B) \quad \cdots \quad A\text{Col}_n(B)].$$

Similarly, one sees that

$$AB = \begin{bmatrix} \text{Row}_1(A) \\ \text{Row}_2(A) \\ \vdots \\ \text{Row}_m(A) \end{bmatrix} B = \begin{bmatrix} \text{Row}_1(A)B \\ \text{Row}_2(A)B \\ \vdots \\ \text{Row}_m(A)B \end{bmatrix},$$

where

$$\text{Row}_i(AB) = \text{Row}_i(A)B = a_{i1}\text{Row}_1(B) + \ldots + a_{in}\text{Row}_n(B). \tag{1.10}$$

For the matrices A and B of Example 3 we see that

$$\text{Col}_2(AB) = \begin{bmatrix} 43 \\ -19 \end{bmatrix} = A\begin{bmatrix} 0 \\ 2 \\ 7 \end{bmatrix} = 0\,\text{Col}_1(A) + 2\,\text{Col}_2(A) + 7\,\text{Col}_3(A)$$

and

$$\text{Row}_1(AB) = [13 \quad 43] = [1 \quad 4 \quad 5]B$$

$$= \text{Row}_1(B) + 4\text{Row}_2(B) + 5\text{Row}_3(B).$$

These two results will be important in the development in Section 1.7. Note in particular that solving the linear system $AX = B$ amounts to finding scalars x_i such that

$$\text{Col}_1(A)x_1 + \text{Col}_2(A)x_2 + \ldots + \text{Col}_n(A)x_n = B;$$

that is, B is expressed as a linear combination of the columns of A. In general, a LINEAR COMBINATION of a set of matrices $C_1, C_2, \ldots, C_n$ is any sum of scalar multiples of the C_i; that is, any sum of the form

$$a_1 C_1 + a_2 C_2 + \ldots + a_n C_n.$$

Our next example shows how matrix multiplication is useful in studying population growth.

Example 5

A new species of game fish is to be introduced into a small lake. The initial stocking consists of 1000 one-year-old fish, 100 two-year-old fish and 10 three-year-old fish. For this species it is estimated that 25 percent of the one-year-old fish and 50 percent of the two-year-old fish will survive the first year. The three-year-old fish will spawn and then die. The average number of eggs that hatch per three-year-old fish is 1000, only 2 percent of which will survive the first year. What will be the population of the species after four years? If the lake can support 2900 of this species, how long will it take to become overcrowded?

Let $f_1(k)$ be the population of 1-year-old fish, k years later,
    $f_2(k)$ be the population of 2-year-old fish, k years later, and
    $f_3(k)$ be the population of 3-year-old fish, k years later.
The above assumptions lead to the following equations:

$$f_2(k + 1) = .25 f_1(k)$$

$$f_3(k + 1) = .50 f_2(k)$$

$$f_1(k + 1) = 1000 f_3(k) \times .02 = 20 f_3(k).$$

In matrix form these equations are

$$F(k + 1) = \begin{bmatrix} f_1(k + 1) \\ f_2(k + 1) \\ f_3(k + 1) \end{bmatrix} = \begin{bmatrix} 0 & 0 & 20 \\ .25 & 0 & 0 \\ 0 & .5 & 0 \end{bmatrix} F(k) = AF(k),$$

where the matrix A is called the LESLIE MATRIX of this population. The population-distribution matrices $F(k)$ for this population are then

$$F(1) = \begin{bmatrix} 1000 \\ 100 \\ 10 \end{bmatrix}, \quad \text{which is the given initial distribution, and}$$

$$F(2) = AF(1) = A \begin{bmatrix} 1000 \\ 100 \\ 10 \end{bmatrix} = \begin{bmatrix} 200 \\ 250 \\ 50 \end{bmatrix},$$

$$F(3) = AF(2) = \begin{bmatrix} 1000 \\ 50 \\ 125 \end{bmatrix}, \quad F(4) = AF(3) = \begin{bmatrix} 2500 \\ 250 \\ 25 \end{bmatrix},$$

$$F(5) = AF(4) = \begin{bmatrix} 500 \\ 625 \\ 125 \end{bmatrix}, \quad \text{and} \quad F(6) = AF(5) = \begin{bmatrix} 2500 \\ 125 \\ 312.5 \end{bmatrix}.$$

From these calculations we see that the population at the end of the fourth year is 2775 and that the population at the end of the sixth year is 2937. Thus overcrowding becomes a problem after six years. ■

The problems in the following exercise set are designed not only to help you understand the definitions given above, but also to provide laboratory evidence for some of the general discussion of the next section. They require your diligent attention.

Exercise Set 1.5

1. Compute AB and BA if

$$A = \begin{bmatrix} 1 & 2 & 3 \\ 4 & 5 & 6 \end{bmatrix} \quad \text{and} \quad B = \begin{bmatrix} 1 & 0 \\ 2 & 3 \\ 0 & 4 \end{bmatrix}.$$

2. (a) Compute AB and BA if

$$A = \begin{bmatrix} 3 & -1 & 2 & 2 \\ 2 & 1 & 1 & 2 \\ 1 & -3 & 0 & -3 \end{bmatrix} \quad \text{and} \quad B = \begin{bmatrix} -3/6 & 6/6 & 3/6 \\ -1/6 & 2/6 & -1/6 \\ 7/6 & -8/6 & -5/6 \\ 6/6 & -6/6 & -6/6 \end{bmatrix}.$$

   (b) Compute $I_3 A$, $I_4 B$, and BI_3, where I_n is the $n \times n$ identity matrix.

3. (a) Compute AB, BA, IA, and BI if

$$A = \begin{bmatrix} 0 & 1 & 2 \\ -1 & 3 & 0 \\ 1 & -2 & 1 \end{bmatrix} \quad \text{and} \quad B = \begin{bmatrix} -3 & 5 & 6 \\ -1 & 2 & 2 \\ 1 & -1 & -1 \end{bmatrix}.$$

   (b) Verify that $\text{Col}_2(AB) = A\text{Col}_2(B)$ and that $\text{Row}_3(AB) = \text{Row}_3(A)B$.

   (c) Compute $2A + 3B$.

4. Compute AB and BA if

$$A = \begin{bmatrix} 2 & -1 & 1 \\ -1 & 2 & -1 \\ 1 & -1 & 2 \end{bmatrix} \quad \text{and} \quad B = \begin{bmatrix} 6 & -5 & 5 \\ -5 & 6 & -5 \\ 5 & -5 & 6 \end{bmatrix}.$$

5. (a) Compute $(AB)C$ and $A(BC)$ for

$$A = \begin{bmatrix} 2 & -1 & 5 \\ 3 & 2 & 4 \\ 8 & 0 & -2 \end{bmatrix}, \quad B = \begin{bmatrix} -3 & 5 & 6 \\ -1 & 2 & 2 \\ 1 & -1 & -1 \end{bmatrix},$$

$$\text{and} \quad C = \begin{bmatrix} 0 & 1 & 2 \\ -1 & 3 & 0 \\ 1 & -2 & 1 \end{bmatrix}.$$

(b) Compute $A(B + C)$ and $AB + AC$ for these matrices.

6. For the matrices

$$A = \begin{bmatrix} 1 & 2 & 0 \\ 5 & -3 & 4 \end{bmatrix}, \quad B = \begin{bmatrix} -2 & 5 & 6 \\ 2 & 0 & -1 \end{bmatrix},$$

$$\text{and} \quad C = \begin{bmatrix} 1 & 0 & 1 & 7 \\ 5 & 3 & 5 & -2 \\ -2 & 4 & 3 & 2 \end{bmatrix},$$

verify the right-handed distributive law; that is, show directly that
$$(A + B)C = AC + BC.$$

7. (a) Find the most general matrix B which commutes with
$$A = \begin{bmatrix} 0 & 1 \\ -1 & 0 \end{bmatrix}.$$

(*Hint:* Assume $B = \begin{bmatrix} a & b \\ c & d \end{bmatrix}$, compute AB and BA, set $AB = BA$, and then solve for the entries of B.)

(b) Show that any two matrices that commute with A commute with each other.

†8. Compute LU and UL for

$$L = \begin{bmatrix} 1 & 0 & 0 \\ 3 & 2 & 0 \\ 0 & 1 & -1 \end{bmatrix} \quad \text{and} \quad U = \begin{bmatrix} 2 & 6 & 2 \\ 0 & -2 & 1 \\ 0 & 0 & 0 \end{bmatrix}.$$

†9. Suppose $A = \begin{bmatrix} a & b \\ c & d \end{bmatrix}$ with $ad - bc \neq 0$.

Let $B = \dfrac{1}{ad - bc} \begin{bmatrix} d & -b \\ -c & a \end{bmatrix}$ and compute AB and BA.

†10. Compute the following matrix products.

(a) $\begin{bmatrix} 1 & 0 & 0 \\ 0 & 0 & 1 \\ 0 & 1 & 0 \end{bmatrix} \begin{bmatrix} a & b & c \\ d & e & f \\ g & h & i \end{bmatrix}$ **(b)** $\begin{bmatrix} 1 & 0 & 0 \\ 0 & k & 0 \\ 0 & 0 & 1 \end{bmatrix} \begin{bmatrix} a & b & c \\ d & e & f \\ g & h & i \end{bmatrix}$

(c) $\begin{bmatrix} 1 & 0 & 0 \\ 0 & 1 & 0 \\ k & 0 & 1 \end{bmatrix} \begin{bmatrix} a & b & c \\ d & e & f \\ g & h & i \end{bmatrix}$ **(d)** $\begin{bmatrix} a & b & c \\ d & e & f \\ g & h & i \end{bmatrix} \begin{bmatrix} 1 & 0 & 0 \\ 0 & 0 & 1 \\ 0 & 1 & 0 \end{bmatrix}$

(e) $\begin{bmatrix} a & b & c \\ d & e & f \\ g & h & i \end{bmatrix} \begin{bmatrix} 1 & 0 & 0 \\ 0 & k & 0 \\ 0 & 0 & 1 \end{bmatrix}$ **(f)** $\begin{bmatrix} a & b & c \\ d & e & f \\ g & h & i \end{bmatrix} \begin{bmatrix} 1 & 0 & 0 \\ 0 & 1 & 0 \\ k & 0 & 1 \end{bmatrix}$

11. Suppose A is $m \times r$ and B is $s \times t$.

(a) When are AB and BA both defined?

(b) When can $AB = BA$?

(c) When are AB and $A + B$ both defined?

12. Find the most general matrix B such that $AB = BA$ if $A = \begin{bmatrix} 2 & 1 \\ 0 & 2 \end{bmatrix}$

13. Find the most general matrix B that commutes with $A = \begin{bmatrix} 2 & 1 \\ -1 & 2 \end{bmatrix}$.

Show that any two such matrices commute.

14. Write the following systems of linear equations as matrix equations of the form $AX = B$.

(a)
$$\begin{aligned} 3x + 5y + 8z &= 12 \\ 4x - 2y - 4z &= 0 \\ 5x + 3y + 12z &= 13 \end{aligned}$$

(b)
$$\begin{aligned} 2x_1 + x_2 + 5x_3 + x_4 &= 5 \\ x_1 + x_2 - 3x_3 - 4x_4 &= -1 \\ 3x_1 + 6x_2 - 2x_3 + x_4 &= 8 \\ 2x_1 + 2x_2 + 2x_3 - 3x_4 &= 2 \end{aligned}$$

(c)
$$\begin{aligned} 2x_1 - x_2 + x_3 &= 4x_1 \\ -x_1 + 2x_2 - x_3 &= 4x_2 \\ x_1 - 2x_2 - x_3 &= 4x_3 \end{aligned}$$

15. Write the following system as a single matrix equation and then use matrix multiplication to make the substitutions
$$\begin{aligned} x &= 5u + 7v - 2w \\ y &= 2u + w \\ z &= 3v + 7w \end{aligned}$$
in the linear system of Exercise 14(a).

16. Show that if A has a zero row, then so does AB, and if A has two equal rows, then so does AB.

†17. Let A, B, and C be arbitrary $n \times n$ matrices and h and k be real numbers. Use summation notation to write a formula for the element in the i, j position of

(a) $A + (B + C)$ (b) $(A + B) + C$ (c) $(h + k)A$

(d) $hA + kA$ (e) $A(B + C)$ (f) $AB + AC$

(g) $A(BC)$ (h) $(AB)C$.

†18. Referring to your answers to Exercise 17,

(a) write out in full and compare the answers to parts (e) and (f) for $n = 3$; and

(b) write out in full and compare the answers to parts (g) and (h) for $n = 3$.

19. (a) Compute the matrix product $\begin{bmatrix} x & y & z \end{bmatrix} \begin{bmatrix} 4 & -1 & 2 \\ -1 & 0 & 1 \\ 2 & 1 & 0 \end{bmatrix} \begin{bmatrix} x \\ y \\ z \end{bmatrix}$.

(b) Write the quadratic polynomial
$$3x^2 + 8xy + 2y^2 + 6xz - 3z^2$$
as a matrix product like the one in part (a).

†20. Show that it is not possible to find a, b, c, and d so that
$$A = \begin{bmatrix} 0 & 1 \\ 2 & 3 \end{bmatrix} = LU, \quad \text{where} \quad L = \begin{bmatrix} 1 & 0 \\ a & 1 \end{bmatrix} \quad \text{and} \quad U = \begin{bmatrix} b & d \\ 0 & c \end{bmatrix}.$$

21. Count the number of multiplications required to compute the product AB if A is n by r and B is r by m. What does this count reduce to if A and B are both n by n?

22. **(a)** Use a calculator or MATALG to compute the matrix product

$$AB = \begin{bmatrix} 3.12 & 4.00 & -5.07 \\ 5.23 & 7.00 & 3.89 \\ 11.40 & 0.00 & 5.27 \end{bmatrix} \begin{bmatrix} 0.35 & 4.01 & -1.00 \\ 1.00 & 3.00 & 4.30 \\ 1.87 & 6.37 & -9.09 \end{bmatrix}.$$

   (b) Compute $3A^2B + 5B^3 - 7BA$ using MATALG.

23. Show that for $A = \begin{bmatrix} 1 & 2 \\ 2 & 4 \end{bmatrix}$ it is not possible to find a 2×2 matrix B such that $BA = I$.

24. **(a)** Show that the following algorithm computes $C = AB$ if A is $m \times r$ and B is $r \times n$:

   For $i = 1$ to m
      For $j = 1$ to n
         $c_{ij} = 0$
         For $k = 1$ to r
            $c_{ij} = c_{ij} + a_{ik}b_{kj}$.

   (b) Write a computer program in your favorite programming language to compute $C = AB$.

25. Write a computer program to compute $C = aA + bB$ if A and B are $m \times n$.

26. In a certain small town there are 3000 families and three supermarkets. Let $N_i(k)$ be the number of families which shop at market i in the kth week and let p_{ij} be the fraction of those who shop in market i one week but will shop in market j the next week. Given the matrix

$$P = \begin{bmatrix} .5 & .2 & .4 \\ .3 & .6 & .1 \\ .2 & .2 & .5 \end{bmatrix},$$

   and the fact that in the first week 1000 families shopped in each market, determine how many customers each market had in the fourth week.

27. Consider a population in which no one lives longer than 100 years. Let $n_1(k)$ be the number of people between 0 and 10 years old during the kth decade. Let $n_2(k)$ be the population in the 11–20 age bracket, and so on for each ten-year bracket. Let s_i be the survival rate of the ith group over a 10-year period and let b_i be the birthrate of the ith group over a 10-year period. Given that the birthrate matrix is

$$B = [0 \quad .05 \quad .50 \quad .40 \quad .05 \quad 0 \quad 0 \quad 0 \quad 0 \quad 0]$$

   and that the survival-rate matrix is

$$S = [.95 \quad .95 \quad .90 \quad .90 \quad .85 \quad .75 \quad .60 \quad .40 \quad .30 \quad 0],$$

   find the Leslie model for this population; that is, find a matrix A such that $N(k + 1) = AN(k)$.

1.6

Properties of Matrix Operations

In the last section we introduced algebraic operations for matrices and showed how a system of linear equations can be expressed as a single matrix equation. In order to exploit this point of view we need to explore some of the manipulative properties of matrix algebra. It would be particularly useful if these properties were like the properties of addition and multiplication of numbers, that is, the 1×1 case. In this section we will explore those properties, trying to point out carefully the areas of similarity and dissimilarity. A start has already been made on this exploration in the examples and exercises of the last section. It is very important to have worked those exercises before reading this section.

Since matrix addition and scalar multiplication are so like the addition and multiplication of numbers, it is not surprising that they behave very similarly. In the next theorem we list the important properties of matrix addition and scalar multiplication. All of these properties are familiar in the case of ordinary algebra (the 1×1 case). In fact, careful proofs of these properties rely heavily on the corresponding properties of ordinary algebra.

Theorem 1.4

Let A and B be any pair of matrices of the same size and let h and k be any scalars. The following properties are then true.

(1) $A + B = B + A$ Matrix addition is commutative.
(2) $A + (B + C) = (A + B) + C$ Matrix addition is associative.
(3) $h(A + B) = hA + hB$ Scalar multiplication distributes over matrix addition.
(4) $(h + k)A = hA + kA$ Scalar multiplication distributes over scalar addition.
(5) $(hk)A = h(kA)$ Scalar multiplication is associative.
(6) There is a unique matrix Z such that for all matrices A,
   $A + Z = A$ (Z is called the ADDITIVE-IDENTITY ELEMENT).
(7) For each matrix A there is a unique matrix C such that
   $A + C = Z$ (C is called the ADDITIVE INVERSE OF A).

Proofs of several of these properties were outlined in Exercises 17 and 18 in Section 1.5. The proof of any one of the properties requires only an examination of a typical element of each of the two matrices. For example, the entry in the (i, j) position of $h(A + B)$ is $h(a_{ij} + b_{ij})$, while the entry in the (i, j) position of $hA + hB$ is $ha_{ij} + hb_{ij}$. Clearly these two numbers are the same, which establishes part 4 of the theorem. The additive-identity matrix of property 6 is the ZERO MATRIX of the same size as A:

$$Z = 0 = \begin{bmatrix} 0 & \cdots & 0 \\ \vdots & & \vdots \\ 0 & \cdots & 0 \end{bmatrix}.$$

The additive inverse of property 7 is the negative of the matrix A introduced in Definition 1.7; that is, $C = -A = (-1)A$. ☐

The only real difference between matrix addition and scalar addition is that matrix addition is not always defined. Matrix multiplication is quite another story. We have already seen that matrix multiplication is noncommutative (in general $AB \neq BA$) and that the product of two nonzero matrices can be the zero matrix.

In the next theorem we summarize the important properties of matrix multiplication. All of these properties have familiar analogs in the case of ordinary algebra.

Theorem 1.5

Let A, B, and C be any three matrices. Then, provided only that the indicated operations are defined, we have the following properties.

(1) $A(BC) = (AB)C$ Matrix multiplication is associative;
(2) (a) $A(B + C) = AB + AC$ Matrix multiplication distributes
    (b) $(B + C)A = BA + CA$ over matrix addition;
(3) There is a unique $n \times n$ matrix I such that $IA = AI = A$ for every $n \times n$ matrix A; and
(4) $(kA)B = k(AB) = A(kB)$ for any scalar k.

The general proofs of these properties are not particularly instructive, only messy, so we will not write them all down. The proof of the associative property was outlined in Exercises 17 and 18 of Section 1.5, and some experimental evidence for the other properties was given in the last exercise set. To verify the right-handed distributive law (property 2(b)) we assume that B and C are $m \times n$ and that A is $n \times t$. Then the entry in the (i, j) position of $(B + C)A$ is

$$\sum_{k=1}^{n} (b_{ik} + c_{ik})a_{kj} = \sum_{k=1}^{n} (b_{ik}a_{kj} + c_{ik}a_{kj}) = \sum_{k=1}^{n} b_{ik}a_{kj} + \sum_{k=1}^{n} c_{ik}a_{kj},$$

while the entry in the (i, j) position of $BA + CA$ is

$$\sum_{k=1}^{n} b_{ik}a_{kj} + \sum_{k=1}^{n} c_{ik}a_{kj}.$$

Since these two expressions are the same, and since $(B + C)A$ and $BA + CA$ are both $m \times t$, it follows that $(B + C)A = BA + CA$ as claimed.

The matrix I in the third property is called the IDENTITY matrix. The $n \times n$ matrix

$$I_n = \begin{bmatrix} 1 & 0 & \cdots & 0 \\ 0 & 1 & \cdots & 0 \\ \vdots & \vdots & & \vdots \\ 0 & 0 & \cdots & 1 \end{bmatrix},$$

first encountered in Section 1.3, is clearly what we need here; it is the matrix analog of the number 1. Note that if A is not square, say $m \times n$ with $m \neq n$, then $I_n A$ is undefined. It is, however, true that $I_m A = A I_n = A$. $\square$

To complete our comparison of matrix multiplication and multiplication of numbers we need a matrix analog of the reciprocal of a nonzero number. The RECIPROCAL, or multiplicative inverse, of the number a, denoted by $1/a$ or a^{-1}, is the unique number x such that $ax = 1$.

Definition 1.10

If A is a square matrix and B satisfies

$$AB = BA = I,$$

then B is a MULTIPLICATIVE INVERSE of A.

Matrices that have inverses are said to be NONSINGULAR or INVERTIBLE. Only square matrices can have inverses. If a square matrix A does not have an inverse, it is said to be SINGULAR or NONINVERTIBLE. Exercise 23 in Section 1.5 shows that not every nonzero square matrix has a multiplicative inverse.

Example 1

The matrix $A = \begin{bmatrix} 1 & 3 \\ 1 & 2 \end{bmatrix}$ has $B = \begin{bmatrix} -2 & 3 \\ 1 & -1 \end{bmatrix}$ as a multiplicative inverse. To show this we compute

$$AB = \begin{bmatrix} 1 & 3 \\ 1 & 2 \end{bmatrix} \begin{bmatrix} -2 & 3 \\ 1 & -1 \end{bmatrix} = \begin{bmatrix} 1 & 0 \\ 0 & 1 \end{bmatrix}$$

$$\text{and} \quad BA = \begin{bmatrix} -2 & 3 \\ 1 & -1 \end{bmatrix} \begin{bmatrix} 1 & 3 \\ 1 & 2 \end{bmatrix} = \begin{bmatrix} 1 & 0 \\ 0 & 1 \end{bmatrix}. \quad \blacksquare$$

It is easy to establish, from the associative property of matrix multiplication, that a matrix can have no more than one inverse. To see that this is so, suppose that B and C are both inverses of A; that is,

$$AB = BA = I \quad \text{and} \quad AC = CA = I.$$

It follows that

$$C(AB) = CI = C$$

and, from the associativity of matrix multiplication, that

$$C(AB) = (CA)B = IB = B.$$

Therefore $C = B$, as claimed.

Definition 1.11

If the matrix A is invertible, then the unique multiplicative inverse of A will be denoted by A^{-1}.

We avoid using the reciprocal notation I/A because of the possible ambiguity of expressions like B/A. (Does it mean BA^{-1} or $A^{-1}B$? In general these two expressions will be unequal because of the noncommutativity of matrix multiplication.) There is a simple formula for the inverse of a 2×2 matrix which is worth noting:

$$\begin{bmatrix} a & b \\ c & d \end{bmatrix}^{-1} = \frac{1}{ad - bc} \begin{bmatrix} d & -b \\ -c & a \end{bmatrix}, \quad \text{if } ad - bc \neq 0. \qquad (1.11)$$

You should do the multiplication necessary to check this last assertion, (recall Exercise 9 in Section 1.5) and verify that this formula gives the correct answer for the matrix A of Example 1. For the matrix

$$C = \begin{bmatrix} 5 & -4 \\ 2 & 3 \end{bmatrix}, \text{ Equation 1.11 yields } C^{-1} = \frac{1}{23} \begin{bmatrix} 3 & 4 \\ -2 & 5 \end{bmatrix}.$$

In general it will not be practical to give such general formulas for A^{-1}; we will instead provide an algorithm for computing A^{-1}. Given a candidate for the inverse, one can always check to see if it really is the inverse.

Example 2

If $T = \begin{bmatrix} 1 & 0 & 0 \\ 1 & 2 & 0 \\ 2 & 1 & 3 \end{bmatrix}$, does $T^{-1} = \begin{bmatrix} 1 & 0 & 0 \\ -1/2 & 1/2 & 0 \\ -1/2 & -1/6 & 1 \end{bmatrix}$?

To answer this question we compute the product of T and the proposed inverse:

$$\begin{bmatrix} 1 & 0 & 0 \\ 1 & 2 & 0 \\ 2 & 1 & 3 \end{bmatrix} \begin{bmatrix} 1 & 0 & 0 \\ -1/2 & 1/2 & 0 \\ -1/2 & -1/6 & 1 \end{bmatrix} = \begin{bmatrix} 1 & 0 & 0 \\ 0 & 1 & 0 \\ 0 & 0 & 3 \end{bmatrix} \neq I.$$

It follows that the given matrix is not T^{-1}. It appears to be close, but not quite right. We leave it to you to verify that replacing the $(3,3)$ element by $1/3$ will yield the correct T^{-1}. ■

Some elementary properties of the matrix inverse are easy to establish by direct multiplication.

Theorem 1.6

If A and B are invertible n by n matrices, then the following assertions are true.

(1) A^{-1} is also invertible and $(A^{-1})^{-1} = A$.
(2) If $k \neq 0$, then kA is invertible and $(kA)^{-1} = (1/k)A^{-1}$.
(3) The product AB is invertible and $(AB)^{-1} = B^{-1}A^{-1}$, that is, the inverse of the product is the product of the inverses in the reverse order.

In order to verify the third assertion of the theorem we compute

$$(AB)(B^{-1}A^{-1}) = A(BB^{-1})A^{-1} = AIA^{-1} = I$$

and

$$(B^{-1}A^{-1})(AB) = B^{-1}(A^{-1}A)B = B^{-1}IB = I.$$

Therefore, because of the uniqueness of matrix inverses, $(AB)^{-1} = B^{-1}A^{-1}$ as claimed. The other proofs are left to the exercises. □

In the next example we will illustrate a brute-force method for finding the inverse of a matrix A. A more complete discussion of techniques for computing matrix inverses will be given in the next section.

Example 3

Find the inverse of $A = \begin{bmatrix} 1 & 2 \\ 3 & 5 \end{bmatrix}$.

Let us assume that $A^{-1} = B = \begin{bmatrix} b_{11} & b_{12} \\ b_{21} & b_{22} \end{bmatrix}$ and try to choose the b_{ij} so that $AB = I$. By direct multiplication

$$\begin{bmatrix} 1 & 2 \\ 3 & 5 \end{bmatrix}\begin{bmatrix} b_{11} & b_{12} \\ b_{21} & b_{22} \end{bmatrix} = \begin{bmatrix} b_{11} + 2b_{21} & b_{12} + 2b_{22} \\ 3b_{11} + 5b_{21} & 3b_{12} + 5b_{22} \end{bmatrix} = \begin{bmatrix} 1 & 0 \\ 0 & 1 \end{bmatrix}.$$

Thus we need to satisfy the four equations

$$\begin{matrix} b_{11} + 2b_{21} = 1 \\ 3b_{11} + 5b_{21} = 0 \end{matrix} \quad \text{or} \quad A\begin{bmatrix} b_{11} \\ b_{21} \end{bmatrix} = \begin{bmatrix} 1 \\ 0 \end{bmatrix}$$

$$\begin{matrix} b_{12} + 2b_{22} = 0 \\ 3b_{12} + 5b_{22} = 1 \end{matrix} \quad \text{or} \quad A\begin{bmatrix} b_{12} \\ b_{22} \end{bmatrix} = \begin{bmatrix} 0 \\ 1 \end{bmatrix}.$$

Note that these four equations really amount to two 2×2 systems with the same coefficient matrix. We can use the method suggested in Section 1.4 to efficiently find a solution of this pair of systems:

$$\left[A \begin{array}{c|c} 1 & 0 \\ 0 & 1 \end{array}\right] = [A\,|\,I] = \left[\begin{array}{cc|cc} 1 & 2 & 1 & 0 \\ 3 & 5 & 0 & 1 \end{array}\right] \rightarrow \left[\begin{array}{cc|cc} 1 & 2 & 1 & 0 \\ 0 & -1 & -3 & 1 \end{array}\right]$$

$$\rightarrow \left[\begin{array}{cc|cc} 1 & 0 & -5 & 2 \\ 0 & 1 & 3 & -1 \end{array}\right].$$

From the above row reduction we see that the solutions of the two systems are

$$\begin{bmatrix} b_{11} \\ b_{21} \end{bmatrix} = \begin{bmatrix} -5 \\ 3 \end{bmatrix} \quad \text{and} \quad \begin{bmatrix} b_{12} \\ b_{22} \end{bmatrix} = \begin{bmatrix} 2 \\ -1 \end{bmatrix}$$

so that $B = \begin{bmatrix} -5 & 2 \\ 3 & -1 \end{bmatrix}$ satisfies $AB = I$.

A simple calculation shows that $BA = I$ also; hence $B = A^{-1}$. ■

Exercise Set 1.6

1. Verify that

$$A^{-1} = \begin{bmatrix} 0 & 1 & 2 \\ -1 & 3 & 0 \\ 1 & -2 & 1 \end{bmatrix} \quad \text{if} \quad A = \begin{bmatrix} -3 & 5 & 6 \\ -1 & 2 & 2 \\ 1 & -1 & -1 \end{bmatrix}$$

   and that

$$B^{-1} = \tfrac{1}{6}\begin{bmatrix} -3 & 6 & 3 \\ -1 & 2 & -1 \\ 7 & -8 & -5 \end{bmatrix} \quad \text{for} \quad B = \begin{bmatrix} 3 & -1 & 2 \\ 2 & 1 & 1 \\ 1 & -3 & 0 \end{bmatrix}.$$

   Find $(AB)^{-1}$, $(BA)^{-1}$, $(A^2)^{-1}$, and $(ABA)^{-1}$.

2. Let A and B be square matrices of the same size. Which of the following are valid equations?

   (a) $(AB)^2 = A^2B^2$

   (b) $(AB)^2 = A(BA)B$

   (c) $(A + B)^2 = A^2 + 2AB + B^2$

   (d) $(A - B)^2 = A^2 - AB - BA + B^2$

3. Find a formula for $(A + B)^3$. What does this reduce to if $AB = BA$?

4. For $A = \begin{bmatrix} 1 & 1 \\ 0 & 1 \end{bmatrix}$ and $B = \begin{bmatrix} 2 & 1 \\ 5 & 3 \end{bmatrix}$ use Formula 1.11 to compute A^{-1}, B^{-1}, $A^{-1} + B^{-1}$, and $(A + B)^{-1}$.

5. Let B be a nonsingular matrix for which $B^{-1} = \begin{bmatrix} 3 & 4 \\ 5 & 6 \end{bmatrix}$. Find B.

6. If $(5A)^{-1} = \begin{bmatrix} 4 & 5 \\ 2 & 3 \end{bmatrix}$, find A.

7. Establish the following cancellation laws.

   (a) If A^{-1} exists and $AB = AC$, then $B = C$.

   (b) If A^{-1} exists and $BA = CA$, then $B = C$.

8. Let A and B be the matrices of Exercise 1. Show that
$$\text{Row}_2(AB) = -\text{Row}_1(B) + 2\text{Row}_2(B) + 2\text{Row}_3(B)$$
and that
$$\text{Col}_3(AB) = 2\text{Col}_1(A) + \text{Col}_2(A).$$

9. Let B be a square matrix such that $B^3 + 5B^2 - 7B + I = 0$. Show that $B^{-1} = 7I - 5B - B^2$.

10. For $J = \begin{bmatrix} 3 & 1 & 0 \\ 0 & 3 & 0 \\ 0 & 0 & -1 \end{bmatrix}$ show that $J^3 - 5J^2 + 3J + 9I = 0$

Show that $J^{-1} = (1/9)(-J^2 + 5J - 3I)$ and use this result to compute J^{-1}.

11. Let A and B be $n \times n$. Show that if A has a zero row, then so does AB. Show also that if A has a zero row, then A must be singular.

12. Let A and B be $n \times n$. Show that if A has two equal rows, then so does AB and hence A must be singular.

13. Show that the product of two lower triangular $n \times n$ matrices must be lower triangular.

14. Let $AX = B$ be any consistent system of linear equations, and let X_p be a fixed solution. Show that every solution to the system can be written in the form $X = X_p + X_h$ where X_h is a solution of the homogeneous system $AX = 0$. Show also that every matrix of this form is a solution.

15. Let A, B, and C be $n \times n$ nonsingular matrices. Show that $(ABC)^{-1} = C^{-1}B^{-1}A^{-1}$. Can you generalize this result? If so, how?

16. Let $M = \left[\begin{array}{ccc|c} 1 & -1 & 0 & 2 \\ 3 & 2 & 1 & 5 \\ 4 & -2 & 1 & 2 \\ 3 & 5 & 0 & 1 \end{array} \right] = \begin{bmatrix} M_{11} & M_{12} \\ M_{21} & M_{22} \end{bmatrix}$

and $N = \left[\begin{array}{cc|cc} 5 & 0 & -3 & 1 \\ 2 & 2 & 0 & 1 \\ 3 & 0 & 2 & 1 \\ 5 & 0 & 0 & 2 \end{array} \right] = \begin{bmatrix} N_{11} & N_{12} \\ N_{21} & N_{22} \end{bmatrix}.$

Verify by direct calculation that
$$MN = \begin{bmatrix} M_{11}N_{11} + M_{12}N_{21} & M_{11}N_{12} + M_{12}N_{22} \\ M_{21}N_{11} + M_{22}N_{21} & M_{21}N_{12} + M_{22}N_{22} \end{bmatrix}.$$

17. What is the unique solution of the linear system $AX = K$ if the coefficient matrix A is invertible?

18. Show that if A^{-1} exists and $AB = I$, then $BA = I$.

19. Complete the proof of Theorem 1.6.

20. In the next chapter we will formally define A^T, the TRANSPOSE of A by the statement,

"The entry in the (i, j) position of A^T is a_{ji},"

and think of A^T as the matrix obtained from A by interchanging rows and columns. For

$$\begin{bmatrix} 1 & 0 & 0 & -4 & -13 & -73/3 & -4 \\ 0 & 1 & 0 & 7 & 22 & 122/3 & 8 \\ 0 & 0 & 1 & -3 & -10 & -53/3 & -2 \end{bmatrix} = [I \,|\, X] = [I \,|\, A^{-1}B].$$

Since the coefficient matrix is row equivalent to I, the system has the unique solution

$$X = A^{-1}B = \begin{bmatrix} -4 & -13 & -73/3 & -4 \\ 7 & 22 & 122/3 & 8 \\ -3 & -10 & -53/3 & -2 \end{bmatrix}. \quad \blacksquare$$

The particular case $B = I$ of the above problem is worthy of special attention. If A is invertible, then the unique solution of the equation $AX = I$ is $X = A^{-1}$ and it can be computed by Equation 1.12, that is,

$$[A \,|\, I] \xrightarrow[\text{reduce}]{\text{row}} [I \,|\, A^{-1}]. \tag{1.13}$$

Equation 1.13 describes the most efficient general-purpose method for computing the inverse of a nonsingular matrix. The method is nothing more than the Gauss-Jordan elimination method for solving the matrix equation $AX = I$. Note that this method is the brute-force method described in Example 3 of Section 1.6.

Example 2

Find the inverse of $C = \begin{bmatrix} 1 & -3 & 0 & -2 \\ 3 & -12 & -2 & -6 \\ -2 & 10 & 2 & 5 \\ -1 & 6 & 1 & 3 \end{bmatrix}.$

The augmented matrix of the system $CX = I$ is

$$[C \,|\, I] = \begin{bmatrix} 1 & -3 & 0 & -2 & 1 & 0 & 0 & 0 \\ 3 & -12 & -2 & -6 & 0 & 1 & 0 & 0 \\ -2 & 10 & 2 & 5 & 0 & 0 & 1 & 0 \\ -1 & 6 & 1 & 3 & 0 & 0 & 0 & 1 \end{bmatrix}$$

$$\rightarrow \begin{bmatrix} 1 & -3 & 0 & -2 & 1 & 0 & 0 & 0 \\ 0 & -3 & -2 & 0 & -3 & 1 & 0 & 0 \\ 0 & 4 & 2 & 1 & 2 & 0 & 1 & 0 \\ 0 & 3 & 1 & 1 & 1 & 0 & 0 & 1 \end{bmatrix},$$

where the first column has been reduced by the row operations $(R_2 \leftarrow -3R_1 + R_2)$, $(R_3 \leftarrow 2R_1 + R_3)$, and $(R_4 \leftarrow R_1 + R_4)$. The operation $(R_2 \leftarrow R_3 + R_2)$ produces a 1 in the $(2, 2)$ position without introducing fractions, and then the row operations $(R_1 \leftarrow 3R_2 + R_1)$, $(R_3 \leftarrow -4R_2 + R_3)$, and $(R_4 \leftarrow -3R_2 + R_4)$ lead to

$$\begin{bmatrix} 1 & 0 & 0 & 1 & -2 & 3 & 3 & 0 \\ 0 & 1 & 0 & 1 & -1 & 1 & 1 & 0 \\ 0 & 0 & 2 & -3 & 6 & -4 & -3 & 0 \\ 0 & 0 & 1 & -2 & 4 & -3 & -3 & 1 \end{bmatrix}$$

$$\rightarrow \begin{bmatrix} 1 & 0 & 0 & 1 & -2 & 3 & 3 & 0 \\ 0 & 1 & 0 & 1 & -1 & 1 & 1 & 0 \\ 0 & 0 & 1 & -2 & 4 & -3 & -3 & 1 \\ 0 & 0 & 0 & 1 & -2 & 2 & 3 & -2 \end{bmatrix},$$

where the second matrix above comes from the first by the operations $(R_3 \leftarrow -2R_4 + R_3)$ and $(R_3 \leftrightarrow R_4)$.

Finally, operations $(R_3 \leftarrow 2R_4 + R_3)$, $(R_2 \leftarrow -R_4 + R_2)$, and $(R_1 \leftarrow -R_4 + R_1)$ yield

$$\begin{bmatrix} 1 & 0 & 0 & 0 & 0 & 1 & 0 & 2 \\ 0 & 1 & 0 & 0 & 1 & -1 & -2 & 2 \\ 0 & 0 & 1 & 0 & 0 & 1 & 3 & -3 \\ 0 & 0 & 0 & 1 & -2 & 2 & 3 & -2 \end{bmatrix} = [I \,|\, C^{-1}]. \quad \blacksquare$$

It should be clear that the inversion algorithm of Equation 1.13 involves considerably more work than the solution of a single linear system with coefficient matrix A; thus one should not try to solve a single linear system $AX = K$ by finding A^{-1} and then computing $X = A^{-1}K$. If one wishes to solve a set of linear systems $AX = B_i$; $i = 1, 2, \ldots, s$, then the inverse may be useful, especially if not all the B_i are available initially. In Section 1.8 we will describe a still better way of dealing with the case of multiple right-hand sides.

In order to give more insight into why some matrices have inverses and others do not, we need to investigate the relationship between matrix multiplication and the elementary row operations used to solve linear systems. The key tool in the development to follow was given in Equation 1.10, which gives a relation between the rows of the product AB and the rows of B; that is,

$$\text{Row}_i(AB) = \text{Row}_i(A)B.$$

If we write this last equation in terms of the rows of B it becomes

$$\text{Row}_i(AB) = a_{i1}\text{Row}_1(B) + a_{i2}\text{Row}_2(B) + \ldots + a_{in}\text{Row}_n(B),$$
$$(1.14)$$

which can be described as saying that the rows of the product are linear combinations of the rows of the second factor with coefficients from the corresponding row of the first factor. This last observation can be used to obtain the following theorem which gives a connection between row equivalence and matrix multiplication. Recall Exercise 10 in Section 1.5.

Theorem 1.7

An elementary row operation on a matrix A can be accomplished by multiplying on the left by an identity matrix on which the elementary row operation has been performed. Symbolically, if E is an elementary row operation and $E(A)$ is the result of performing E on A, then

$$E(A) = [E(I)]A .$$

A formal proof of this theorem divides naturally into three parts, one for each type of elementary row operation. We give the details for a Type III operation $(R_j \leftarrow kR_i + R_j)$. In this case the matrix $E = E(A)$ differs from I only in the (j, i) position, where there is a k. If $i < j$,

$$EA = \begin{array}{c} \\ \text{Row } i \rightarrow \\ \\ \text{Row } j \rightarrow \\ \\ \end{array} \begin{bmatrix} 1 & \cdot & 0 & \cdot & \cdot & 0 & \cdot & 0 \\ \cdot & & \cdot & & & \cdot & & \cdot \\ 0 & \cdot & 1 & \cdot & \cdot & 0 & \cdot & 0 \\ \cdot & & \cdot & & & \cdot & & \cdot \\ 0 & \cdot & k & \cdot & \cdot & 1 & \cdot & 0 \\ \cdot & & \cdot & & & \cdot & & \cdot \\ 0 & \cdot & 0 & \cdot & \cdot & 0 & \cdot & 1 \end{bmatrix} \begin{bmatrix} \text{Row}_1(A) \\ \cdot \\ \text{Row}_i(A) \\ \cdot \\ \text{Row}_j(A) \\ \cdot \\ \text{Row}_n(A) \end{bmatrix} .$$

From Equation (1.14) it follows that

$$\text{Row}_j(EA) = \text{Row}_j(E)A = k\text{Row}_i(A) + \text{Row}_j(A)$$

and that all other rows of EA and A agree. This establishes the assertion of the theorem for Type III operations. The other proofs are similar and will be left to the exercises. □

Definition 1.12

An ELEMENTARY MATRIX is a matrix that has been obtained from the identity matrix by applying one elementary row operation.

Theorem 1.7 says that elementary row operations can be accomplished by multiplying on the left by elementary matrices.

We list below examples of 4 by 4 elementary matrices corresponding respectively to the operations $(R_3 \leftarrow 5R_3)$, $(R_2 \leftrightarrow R_3)$, and $(R_4 \leftarrow 2R_3 + R_4)$.

$$\begin{array}{ccc} R_3 \leftarrow 5R_3 & R_2 \leftrightarrow R_3 & R_4 \leftarrow 2R_3 + R_4 \\ \begin{bmatrix} 1 & 0 & 0 & 0 \\ 0 & 1 & 0 & 0 \\ 0 & 0 & 5 & 0 \\ 0 & 0 & 0 & 1 \end{bmatrix} & \begin{bmatrix} 1 & 0 & 0 & 0 \\ 0 & 0 & 1 & 0 \\ 0 & 1 & 0 & 0 \\ 0 & 0 & 0 & 1 \end{bmatrix} & \begin{bmatrix} 1 & 0 & 0 & 0 \\ 0 & 1 & 0 & 0 \\ 0 & 0 & 1 & 0 \\ 0 & 0 & 2 & 1 \end{bmatrix} \end{array}$$

For our purposes the most important fact about elementary matrices is that they are always invertible and that their inverses are also elementary matrices. This follows easily from the observation in Section 1.3 that every elementary row operation can be reversed by an elementary row operation of the same type.

The inverse of the elementary matrix for $(R_i \leftarrow kR_i)$ is the elementary matrix for $(R_i \leftarrow 1/kR_i)$, for example,

$$\begin{bmatrix} 1 & 0 & 0 \\ 0 & k & 0 \\ 0 & 0 & 1 \end{bmatrix}^{-1} = \begin{bmatrix} 1 & 0 & 0 \\ 0 & 1/k & 0 \\ 0 & 0 & 1 \end{bmatrix}.$$

The inverse of the elementary matrix for $(R_i \leftrightarrow R_j)$ is itself; for example,

$$\begin{bmatrix} 1 & 0 & 0 \\ 0 & 0 & 1 \\ 0 & 1 & 0 \end{bmatrix}^{-1} = \begin{bmatrix} 1 & 0 & 0 \\ 0 & 0 & 1 \\ 0 & 1 & 0 \end{bmatrix}.$$

The inverse of the elementary matrix for $(R_j \leftarrow kR_i + R_j)$ is the elementary matrix for $(R_j \leftarrow -kR_i + R_j)$, for example

$$\begin{bmatrix} 1 & 0 & 0 \\ k & 1 & 0 \\ 0 & 0 & 1 \end{bmatrix}^{-1} = \begin{bmatrix} 1 & 0 & 0 \\ -k & 1 & 0 \\ 0 & 0 & 1 \end{bmatrix}.$$

Suppose now that a sequence of elementary row operations reduces the matrix A to a matrix B; that is, A is row equivalent to B. Let $E_1, E_2, \ldots, E_r$ be the elementary matrices corresponding to the elementary row operations used to reduce A to B. From Theorem 1.7 we have

$$E_r E_{r-1} \ldots E_2 E_1 A = B.$$

The matrix

$$P = E_r E_{r-1} \ldots E_2 E_1 = E_r E_{r-1} \ldots E_2 E_1 I$$

is nonsingular, because it is a product of elementary matrices that are each nonsingular, and satisfies the equation $PA = B$. If we write $P = PI$ we see that P can be obtained by performing on I exactly the same operations which changed A to B. Thus the matrix P constitutes a "record" of the elementary row operations used to reduce A to B.

Theorem 1.8

If A and B are row equivalent, then there exists a nonsingular matrix P such that $PA = B$. The matrix P can be computed, as A is being reduced to B, via

$$[A \mid I] \xrightarrow[\text{reduce}]{\text{row}} [B \mid P]. \tag{1.15}$$

If $B = I$, then $PA = I$ and $P = A^{-1}$.

Example 3

Consider the following row reduction:

$$\begin{bmatrix} 1 & 2 & 3 & 1 & 0 & 0 \\ 3 & 1 & -2 & 0 & 1 & 0 \\ 4 & 5 & 6 & 0 & 0 & 1 \end{bmatrix} \begin{array}{l} \xrightarrow{\hspace{2cm}} \\ R_2 \leftarrow -3R_1 + R_2 \\ R_3 \leftarrow -4R_1 + R_3 \end{array}$$

$$\begin{bmatrix} 1 & 2 & 3 & 1 & 0 & 0 \\ 0 & -5 & -11 & -3 & 1 & 0 \\ 0 & -3 & -6 & -4 & 0 & 1 \end{bmatrix}$$

$$\xrightarrow[R_3 \leftarrow -3/5R_2 + R_3]{} \begin{bmatrix} 1 & 2 & 3 & 1 & 0 & 0 \\ 0 & -5 & -11 & -3 & 1 & 0 \\ 0 & 0 & 3/5 & -11/5 & -3/5 & 1 \end{bmatrix}.$$

The information about the operations used is recorded in the last three columns and, from Theorem 1.8, it follows that

$$\begin{bmatrix} 1 & 0 & 0 \\ -3 & 1 & 0 \\ -11/5 & -3/5 & 1 \end{bmatrix} \begin{bmatrix} 1 & 2 & 3 \\ 3 & 1 & -2 \\ 4 & 5 & 6 \end{bmatrix} = \begin{bmatrix} 1 & 2 & 3 \\ 0 & -5 & -11 \\ 0 & 0 & 3/5 \end{bmatrix},$$

which you should verify by direct calculation. ∎

The final theorem of this section provides several conditions equivalent to the nonsingularity of an $n \times n$ matrix. Several other conditions will be discovered later and a more general theorem will then be stated (see Theorem 3.14 on page 185).

Theorem 1.9

For an $n \times n$ matrix A the following statements are equivalent:

(1) A is nonsingular.
(2) A has a left inverse (there exists B such that $BA = I$).
(3) $AX = 0$ only if $X = 0$.
(4) A is row equivalent to I.
(5) A is a product of elementary matrices.

We will prove the equivalence by establishing the chain of implications $1 \rightarrow 2 \rightarrow 3 \rightarrow 4 \rightarrow 5 \rightarrow 1$.

$(1 \rightarrow 2)$ Obvious.

$(2 \rightarrow 3)$ If there exists B such that $BA = I$ and $AX = 0$, then

$$X = B(AX) = B0 = 0.$$

$(3 \rightarrow 4)$ If A is not row equivalent to I, then the solution process of Section 1.3 will result in a zero row and consequently will leave one of the unknowns arbitrary; hence nonzero solutions will exist. It follows that A must be row equivalent to I.

$(4 \rightarrow 5)$ Since A is row equivalent to I, there exist elementary matrices $E_1, E_2 \ldots E_k$ such that $E_k \ldots E_2 E_1 A = I$. Then

$$A = (E_k \ldots E_2 E_1)^{-1} I = E_1^{-1} E_2^{-1} \ldots E_k^{-1}$$

and, since each elementary matrix has an elementary inverse, we have succeeded in expressing A as a product of elementary matrices.

$(5 \rightarrow 1)$ This is obvious since each elementary matrix is invertible and the product of invertible matrices is invertible. $\square$

Example 4

Express $A = \begin{bmatrix} 1 & 3 \\ 2 & 8 \end{bmatrix}$ as a product of elementary matrices.

Consider the following reduction of A to I:

$$\begin{bmatrix} 1 & 3 \\ 2 & 8 \end{bmatrix} \rightarrow \begin{bmatrix} 1 & 3 \\ 0 & 2 \end{bmatrix} \rightarrow \begin{bmatrix} 1 & 3 \\ 0 & 1 \end{bmatrix} \rightarrow \begin{bmatrix} 1 & 0 \\ 0 & 1 \end{bmatrix}.$$

The elementary matrices

$$E_1 = \begin{bmatrix} 1 & 0 \\ -2 & 1 \end{bmatrix}, \quad E_2 = \begin{bmatrix} 1 & 0 \\ 0 & 1/2 \end{bmatrix}, \quad \text{and } E_3 = \begin{bmatrix} 1 & -3 \\ 0 & 1 \end{bmatrix}$$

correspond respectively to the three elementary row operations $(R_2 \leftarrow -2R_1 + R_2)$, $(R_2 \leftarrow 1/2 R_2)$, and $(R_1 \leftarrow -3R_2 + R_1)$ used in the above reduction. It follows, as in the proof of Theorem 1.9, that

$$E_3 E_2 E_1 A = \begin{bmatrix} 1 & -3 \\ 0 & 1 \end{bmatrix} \begin{bmatrix} 1 & 0 \\ 0 & 1/2 \end{bmatrix} \begin{bmatrix} 1 & 0 \\ -2 & 1 \end{bmatrix} \begin{bmatrix} 1 & 3 \\ 2 & 8 \end{bmatrix} = \begin{bmatrix} 1 & 0 \\ 0 & 1 \end{bmatrix}$$

and then that

$$\begin{bmatrix} 1 & 3 \\ 2 & 8 \end{bmatrix} = \left(\begin{bmatrix} 1 & -3 \\ 0 & 1 \end{bmatrix} \begin{bmatrix} 1 & 0 \\ 0 & 1/2 \end{bmatrix} \begin{bmatrix} 1 & 0 \\ -2 & 1 \end{bmatrix} \right)^{-1}$$

$$= \begin{bmatrix} 1 & 0 \\ -2 & 1 \end{bmatrix}^{-1} \begin{bmatrix} 1 & 0 \\ 0 & 1/2 \end{bmatrix}^{-1} \begin{bmatrix} 1 & -3 \\ 0 & 1 \end{bmatrix}^{-1}$$

$$= \begin{bmatrix} 1 & 0 \\ 2 & 1 \end{bmatrix} \begin{bmatrix} 1 & 0 \\ 0 & 2 \end{bmatrix} \begin{bmatrix} 1 & 3 \\ 0 & 1 \end{bmatrix} = A.$$

This is the desired factorization of A as a product of elementary matrices; in general such factorizations will not be unique.   ■

In the final example of this section we will show how matrices and their inverses can be used in constructing codes which are very difficult to break.

Example 5

Consider a message-coding process which associates with the letters A through Z the numbers 1 through 26 and associates with a blank space the number 0. After the initial coding the numbers are split into groups of three and then scrambled using multiplication by the matrix

$$A = \begin{bmatrix} -3 & 5 & 6 \\ -1 & 2 & 2 \\ 1 & -1 & -1 \end{bmatrix}, \quad \text{for which } A^{-1} = \begin{bmatrix} 0 & 1 & 2 \\ -1 & 3 & 0 \\ 1 & -2 & 1 \end{bmatrix}.$$

The message "SUPPLIES LOW" would initially be coded as the string of integers

$$19, 21, 16, 16, 12, 9, 5, 19, 0, 12, 15, 23,$$

arranged as a set of 3×1 column matrices, and then scrambled. These steps lead to

$$\begin{bmatrix} -3 & 5 & 6 \\ -1 & 2 & 2 \\ 1 & -1 & -1 \end{bmatrix}\begin{bmatrix} 19 \\ 21 \\ 16 \end{bmatrix} = \begin{bmatrix} 144 \\ 55 \\ -18 \end{bmatrix}, \quad A\begin{bmatrix} 16 \\ 12 \\ 9 \end{bmatrix} = \begin{bmatrix} 66 \\ 26 \\ -5 \end{bmatrix},$$

$$A\begin{bmatrix} 5 \\ 19 \\ 0 \end{bmatrix} = \begin{bmatrix} 80 \\ 33 \\ -14 \end{bmatrix}, \quad \text{and } A\begin{bmatrix} 12 \\ 15 \\ 23 \end{bmatrix} = \begin{bmatrix} 177 \\ 64 \\ -26 \end{bmatrix}.$$

The message would then be transmitted as the string of integers

$$144, 55, -18, 66, 26, -5, 80, 33, -14, 177, 64, -26.$$

Note that this code would be very difficult to break because a letter is not always coded the same way.

The recipient of the message would unscramble the message using A^{-1} and then assign letters to the integers as above. In order to translate the coded message

$$217, 79, -39, 78, 27, -13, 125, 42, -21, 26, 10, -5, 1, 5, 9$$

we arrange the given message into columns and unscramble using A^{-1}. Thus the unscrambled message would be

$$\begin{bmatrix} 0 & 1 & 2 \\ -1 & 3 & 0 \\ 1 & -2 & 1 \end{bmatrix}\begin{bmatrix} 217 \\ 79 \\ -39 \end{bmatrix} = \begin{bmatrix} 1 \\ 20 \\ 20 \end{bmatrix},$$

$$A^{-1}\begin{bmatrix} 78 \\ 27 \\ -13 \end{bmatrix} = \begin{bmatrix} 1 \\ 3 \\ 11 \end{bmatrix}, \quad A^{-1}\begin{bmatrix} 125 \\ 42 \\ -21 \end{bmatrix} = \begin{bmatrix} 0 \\ 1 \\ 20 \end{bmatrix},$$

$$A^{-1}\begin{bmatrix} 26 \\ 10 \\ -5 \end{bmatrix} = \begin{bmatrix} 0 \\ 4 \\ 1 \end{bmatrix}, \quad \text{and } A^{-1}\begin{bmatrix} 1 \\ 5 \\ 9 \end{bmatrix} = \begin{bmatrix} 23 \\ 14 \\ 0 \end{bmatrix}.$$

The string

$$1, 20, 20, \quad 1, \quad 3, 11, 0, 1, 20, 0, 4, \quad 1, 23, 14, 0$$

then yields the message

A T T A C K   A T   D A W N.

The complexity of such a code would depend on the size of the matrix used. There would be advantages to having a coding matrix with integer entries whose inverse also had integer entries. It would be a simple matter to design a computer code to unscramble an incoming message. It would also be easy to change the code by changing the coding matrix. ▪

Exercise Set 1.7

Solve the following matrix equations for X.

1. $\begin{bmatrix} 1 & 2 \\ 3 & 4 \end{bmatrix} X = \begin{bmatrix} 5 & 4 & 3 \\ 1 & 2 & 0 \end{bmatrix}$

2. $\begin{bmatrix} 1 & 2 & 1 \\ -1 & -1 & 1 \\ 0 & 1 & 3 \end{bmatrix} X = \begin{bmatrix} 3 & 0 & 2 \\ 1 & 0 & 3 \\ -2 & 0 & 7 \end{bmatrix}$

3. $\begin{bmatrix} 5 & 5 & -3 \\ 2 & 2 & -1 \\ -1 & -1 & 1 \end{bmatrix} X = \begin{bmatrix} 10 & 2 & 7 \\ 4 & 1 & 3 \\ -2 & 0 & -1 \end{bmatrix}$

4. $\begin{bmatrix} 1 & 2 \\ 3 & 6 \end{bmatrix} X = \begin{bmatrix} 3 & 0 & 1 \\ 9 & 0 & 1 \end{bmatrix}$

Find the inverses of the following matrices, if possible.

5. $A = \begin{bmatrix} 1 & 1 & 1 \\ 0 & 2 & 1 \\ 1 & 0 & 1 \end{bmatrix}$

6. $A = \begin{bmatrix} 3 & 2 & 1 \\ 2 & -1 & -1 \\ 1 & 4 & 0 \end{bmatrix}$

7. $B = \begin{bmatrix} -3 & 5 & 6 \\ -1 & 2 & 2 \\ 1 & -1 & -1 \end{bmatrix}$

8. $J = \begin{bmatrix} 3 & 0 & 0 \\ 1 & 3 & 0 \\ 0 & 1 & 3 \end{bmatrix}$

9. $T = \begin{bmatrix} 1 & 2 & 6 \\ 0 & -1 & 4 \\ 0 & 0 & 3 \end{bmatrix}$

10. $A = \begin{bmatrix} 1 & 3 & 3 \\ 1 & 3 & 4 \\ 1 & 4 & 3 \end{bmatrix}$

11. $B = \begin{bmatrix} 2 & 1 & -1 & 2 \\ 1 & 3 & 2 & -3 \\ -1 & 2 & 1 & -1 \\ 2 & -3 & -1 & 4 \end{bmatrix}$

12. $C = \begin{bmatrix} 2 & 4 & 3 & 2 \\ 3 & 6 & 5 & 2 \\ 2 & 5 & 2 & -3 \\ 4 & 5 & 14 & 14 \end{bmatrix}$

13. $R = \begin{bmatrix} \cos \theta & -\sin \theta \\ \sin \theta & \cos \theta \end{bmatrix}$

14. $S = \dfrac{1}{6} \begin{bmatrix} 3 & 3 & 3 & 3 \\ 3 & -5 & 1 & 1 \\ 3 & 1 & -5 & 1 \\ 3 & 1 & 1 & -5 \end{bmatrix}$ **15.** $L = \begin{bmatrix} 1 & 0 & 0 & 0 \\ 2 & 1 & 0 & 0 \\ -3 & 5 & 1 & 0 \\ 4 & 2 & 3 & 1 \end{bmatrix}$

16. Express each matrix in Exercises 6–11 as a product of elementary matrices.

17. Find a nonsingular matrix P such that $PA = B$, where

$$A = \begin{bmatrix} 2 & 3 & 4 \\ 4 & 3 & 1 \\ 1 & 2 & 4 \end{bmatrix}, \text{ and } B = \begin{bmatrix} 1 & 2 & -1 \\ -1 & 1 & 2 \\ 2 & -1 & 1 \end{bmatrix}.$$

18. Let $A = \begin{bmatrix} 4 & 1 & 2 & 6 \\ 1 & 2 & 4 & 1 \\ 0 & -1 & -2 & 1 \\ 0 & 6 & 12 & 0 \end{bmatrix}$. Find a nonsingular matrix P such that PA

is in row echelon form. Is P unique?

19. Let P be an approximation to A^{-1} (PA differs from I in only a few places). Show how to find A^{-1} from P and PA. (*Hint:* row reduce $[PA \mid P]$.)

†20. Show that the inverse of a lower triangular matrix is also a lower triangular matrix.

21. Define ELEMENTARY COLUMN OPERATIONS by replacing "row" by "column" everywhere it occurs in Definition 1.2. Let $C(A)$ be the result of performing the column operation C on the matrix A.

(a) Show that $C(A) = A[C(I)]$.

(b) Show that every elementary column matrix $C(I)$ is an elementary (row) matrix.

(c) Show that if $\begin{bmatrix} A \\ I \end{bmatrix} \xrightarrow[\text{reduce}]{\text{column}} \begin{bmatrix} B \\ Q \end{bmatrix}$, then $AQ = B$.

(d) Use the result of part (c) to compute the inverse of

$$A = \begin{bmatrix} 1 & 4 & 0 \\ 2 & -1 & -1 \\ 3 & 2 & 1 \end{bmatrix}.$$

22. Use the matrix code of Example 5 to decode the following messages.

(a) $22, 9, 1, 24, 8, 4, 135, 47, -21, 113, 38, -19, 75, 25, -6, 36, 18, 1,$
$56, 21, -6, -60, -20, 20$

(b) $169, 63, -22, 113, 46, -21, 89, 30, -11, 121, 42, -19$

(c) $-37, -12, 16, 57, 19, -4, 78, 26, -6, 1, 2, 3, 7, 4, 2, 121, 42, -19$

(d) $48, 22, -3, -39, -13, 19, 45, 21, -3, 151, 57, -19, 97, 37, -16,$
$-20, -4, 12, 153, 56, -21, 181, 65, -32, 64, 23, 0, -57, -19, 19$

23. Use matrix A of Exercise 10 to code the following messages.

(a) RETREAT SLOWLY

(b) PARTY AT SUNSET

24. Write a computer program which will unscramble messages coded with a given $n \times n$ matrix A. Test your program on the messages in Exercise 22.

1.8   *Partial Pivoting and the LU Factorization**

In this section we will look a little more closely at the Gaussian elimination algorithm. We will also describe some of the refinements which are usually included in a good general-purpose algorithm for solving the linear equation $AX = B$ on a modern digital computer.

In order to understand the reasons for the refinements to be made we need to take a brief look at how a typical computer represents numbers and does arithmetic. Numbers in a computer are represented in normalized floating-point form. Thus a typical machine number is in the form

$$x = \pm 0.d_1 d_2 \ldots d_k \times 10^n$$

where the digits $d_1, d_2, \ldots, d_k$ are integers between 0 and 9, $d_1 \neq 0$, and n is an integer. The number k and the possible values of n are characteristic of a particular machine and will vary from machine to machine. Many machines use number systems with bases other than 10; 2, 8, and 16 are common. We will restrict our attention to the familiar case of base 10. Some examples of numbers and their machine representations, for a machine with $k = 5$, follow.

Number	Machine Representation
312.52	$.31252 \times 10^3$
$-.002371$	$-.23710 \times 10^{-3}$
$\pi = 3.141592654$	$.31416 \times 10^1$
$\frac{40}{3} = 13.33\overline{33}$	$.13333 \times 10^2$
$-.126786$	$-.12679 \times 10^0$

Note that some real numbers cannot be represented exactly in the machine; they can only be approximated correct to k digits. The difference between a number and its machine representation is an example of ROUND-OFF ERROR. Round-off error also occurs when two machine numbers are added or multiplied. Since the true sum or product will usually have more than k digits, some rounding is necessary to change the sum or product into a machine-representable number. For example, on a machine with $k = 5$, the sum

$$(.31252 \times 10^3) + (.23710 \times 10^{-3}) = (.3125202371 \times 10^3)$$

---

*This section contains optional material which may be omitted without loss of continuity.

would be changed to $(.31252 \times 10^3)$ before it could be stored. Similarly, the product

$$(.31252 \times 10^3) \times (.23710 \times 10^{-3}) = (7.4098492 \times 10^{-2})$$

would be changed to $.74098 \times 10^{-3}$. The difference

$$(.31252 \times 10^3) - (.31248 \times 10^3) = (.00004 \times 10^3)$$

would be stored as $(.40000 \times 10^{-4})$. Note that the difference has only one significant figure. This kind of loss of accuracy is a source of many inaccuracies in machine calculations; it is called CATASTROPHIC CANCELLATION and has nothing to do with round-off error.

Because of round-off error the machine will have a difficult time distinguishing between two very small numbers. Thus it is best to avoid testing to see if two machine numbers are equal or testing to see if a machine number is zero.

The first example of this section illustrates what can happen to the Gaussian elimination algorithm when it is implemented on a computer.

Example 1

Solve the system

$$1.00 \times 10^{-4}x + 1.00y = 1.00$$

$$1.00x + 1.00y = 2.00.$$

The solution of the system, correct to the number of digits shown, is $x = 1.00010001$, $y = .99989999$. Suppose that we attempt to solve this system on a machine that carries only 3 significant figures; that is, the results of all calculations are rounded to 3 significant figures.

The operation $R_2 \leftarrow -10^4 R_1 + R_2$ reduces the augmented matrix

$$\begin{bmatrix} 1.00 \times 10^{-4} & 1.00 & 1.00 \\ 1.00 & 1.00 & 2.00 \end{bmatrix} \text{ to}$$

$$\begin{bmatrix} 1.00 \times 10^{-4} & 1.00 & 1.00 \\ 0 & -1.00 \times 10^4 & -1.00 \times 10^4 \end{bmatrix},$$

from which we obtain, by back substitution, $y = 1.00$ and $x = 0.00$. This is a very bad result by any reasonable standard.

If we first interchange rows 1 and 2 and then use the operation $R_2 \leftarrow -1.00 \times 10^{-4} R_1 + R_2$, we obtain the reduction

$$\begin{bmatrix} 1.00 \times 10^{-4} & 1.00 & 1.00 \\ 1.00 & 1.00 & 2.00 \end{bmatrix} \rightarrow \begin{bmatrix} 1.00 & 1.00 & 2.00 \\ 1.00 \times 10^{-4} & 1.00 & 1.00 \end{bmatrix}$$

$$\rightarrow \begin{bmatrix} 1.00 & 1.00 & 2.00 \\ 0 & 1.00 & 1.00 \end{bmatrix},$$

from which we obtain $x = 1, y = 1$. This last result is a very good approximation to the true solution. ∎

We have previously observed that a ZERO PIVOT, that is, a zero appearing on the diagonal during the Gaussian elimination algorithm, forces us to modify the algorithm to obtain a nonzero element in the pivot position. This modification takes the form of an elementary row operation of the form $(R_i \leftrightarrow R_j)$. Example 1 illustrates that small pivots are as troublesome as zero pivots since they lead to multiplications by very large numbers. For this reason it is important to incorporate, into the Gaussian elimination algorithm for solving linear systems, a process called PARTIAL PIVOTING, which protects against small elements appearing on the diagonal. In partial pivoting the element with the largest absolute value in the column to be reduced is moved to the diagonal position *before* the column is reduced. The new diagonal element is called the PIVOT for that stage of the reduction. In this way all the multipliers are in absolute value less than or equal to 1. The net effect of this partial-pivoting procedure is that the numbers will stay small and the growth of round-off error will be controlled. This is clearly a machine procedure, not a hand-calculation procedure.

Example 2

Solve the following system by Gaussian elimination with partial pivoting.

$$2x_1 + 3x_2 - x_3 + 2x_4 = 3$$
$$-4x_1 - 6x_2 + 2x_3 + x_4 = 3$$
$$2x_1 + 4x_2 + 4x_3 - x_4 = 0$$
$$4x_1 + 8x_2 + 2x_3 + 7x_4 = 12$$

The augmented matrix for this system is

$$\begin{bmatrix} 2 & 3 & -1 & 2 & 3 \\ -4 & -6 & 2 & 1 & 4 \\ 2 & 4 & 4 & -1 & 0 \\ 4 & 8 & 2 & 7 & 12 \end{bmatrix}.$$

The element with largest absolute value in the first column is the -4 in the second row. We select this element as the first pivot, move it to the $(1, 1)$ position using the elementary row operation $(R_1 \leftrightarrow R_2)$ and then use it to reduce the first column. This series of steps yields

$$\begin{array}{c} \xrightarrow{\hspace{2cm}} \\ R_2 \leftarrow .5R_1 + R_2 \\ R_3 \leftarrow .5R_1 + R_3 \\ R_4 \leftarrow R_1 + R_4 \end{array} \begin{bmatrix} -4 & -6 & 2 & 1 & 4 \\ 0 & 0 & 0 & 2.5 & 5 \\ 0 & 1 & 5 & -.5 & 2 \\ 0 & 2 & 4 & 8 & 16 \end{bmatrix}.$$

The largest element in the second column, excluding the first row, is the 2 in the $(4, 2)$ position. We select this element as the next pivot and move it to the diagonal position with the operation $(R_2 \leftrightarrow R_4)$. One Type III operation then completes the reduction:

$$\begin{bmatrix} -4 & -6 & 2 & 1 & 4 \\ 0 & 2 & 4 & 8 & 16 \\ 0 & 1 & 5 & -.5 & 2 \\ 0 & 0 & 0 & 2.5 & 5 \end{bmatrix} \xrightarrow{\quad R_3 \leftarrow -.5R_2 + R_3 \quad}$$

$$\begin{bmatrix} -4 & -6 & 2 & 1 & 4 \\ 0 & 2 & 4 & 8 & 16 \\ 0 & 0 & 3 & -4.5 & -6 \\ 0 & 0 & 0 & 2.5 & 5 \end{bmatrix}.$$

The routine back substitution now yields

$$x_4 = 2, \quad x_3 = 1, \quad x_2 = -2, \quad \text{and} \quad x_1 = 3. \quad \blacksquare$$

For a more complete discussion of partial pivoting and other methods for the control of round-off error you are urged to consult a book on numerical analysis.

Several of our earlier examples and exercises have been designed to suggest that it is frequently, but not always, possible to factor a nonsingular square matrix into the product of a lower triangular matrix and an upper triangular matrix (see Exercises 8 and 20 of Section 1.5 and Example 3 of Section 1.7). Before we investigate such factorizations let us see what use we can make of them. Thus consider the linear system $AX = K$ and let us assume that we have a factorization $A = LU$ where L is a lower triangular matrix and U is an upper triangular matrix. If we write

$$AX = L(UX) = K$$

and let $Y = UX$, then we must have $LY = K$. Thus the solution process has been split into three parts:

(1) factor $A = LU$;
(2) solve the lower triangular system $LY = K$ for Y; and
(3) solve the upper triangular system $UX = Y$ for X.

We have already observed that the solution of a triangular system is relatively easy, requiring only about $n^2/2$ multiplications (see Exercise 10 of Section 1.3). It turns out that the method suggested above is exactly equivalent to Gaussian elimination for a single linear system; the advertised gain in efficiency comes when we have a set of systems to solve, perhaps with not all the right-hand sides available at the same time. Once one has the factorization, then the solution of a new system with the same coefficient matrix requires only the solution of two triangular systems and this requires only about n^2 multiplications compared with the $n^3/3$ multiplications required to solve the system from scratch by Gaussian elimination.

Example 3

Given the factorization

$$A = \begin{bmatrix} 2 & 6 & 2 \\ 6 & 14 & 8 \\ 0 & -2 & -1 \end{bmatrix} = LU = \begin{bmatrix} 1 & 0 & 0 \\ 3 & 1 & 0 \\ 0 & 1/2 & 1 \end{bmatrix}\begin{bmatrix} 2 & 6 & 2 \\ 0 & -4 & 2 \\ 0 & 0 & -2 \end{bmatrix}$$

and $K = \begin{bmatrix} 1 \\ 1 \\ 1 \end{bmatrix}$,

solve the linear systems $AX_1 = K$, $AX_2 = X_1$, and $AX_3 = X_2$.

The system $LY = K$ of step 2, for the first system, is

$$\begin{aligned} y_1 &= 1 & y_1 &= 1 \\ 3y_1 + y_2 &= 1, & \text{for which} \quad y_2 &= -2. \\ 1/2y_2 + y_3 &= 1 & y_3 &= 2 \end{aligned}$$

The system $UX = Y$ of step 3, for the first system, is

$$\begin{aligned} 2x_1 + 6x_2 + 2x_3 &= 1 & x_1 &= 3/2 \\ -4x_2 + 2x_3 &= -2, \text{ for which } x_2 &= 0; \text{ therefore } X_1 = \begin{bmatrix} 3/2 \\ 0 \\ -1 \end{bmatrix}. \\ -2x_3 &= 2 & x_3 &= -1 \end{aligned}$$

This is the solution of the first system and the input for the second system. We now repeat steps 2 and 3 for the second system. The solution of the lower triangular system

$$\begin{bmatrix} 1 & 0 & 0 \\ 3 & 1 & 0 \\ 0 & 1/2 & 1 \end{bmatrix}\begin{bmatrix} y_1 \\ y_2 \\ y_3 \end{bmatrix} = \begin{bmatrix} 3/2 \\ 0 \\ -1 \end{bmatrix} \quad \text{is} \quad Y = \begin{bmatrix} 3/2 \\ -9/2 \\ 5/4 \end{bmatrix},$$

and the solution of the upper triangular system

$$\begin{bmatrix} 2 & 6 & 2 \\ 0 & -4 & 2 \\ 0 & 0 & -2 \end{bmatrix}\begin{bmatrix} x_1 \\ x_2 \\ x_3 \end{bmatrix} = \begin{bmatrix} 3/2 \\ -9/2 \\ 5/4 \end{bmatrix} \quad \text{is} \quad X_2 = \begin{bmatrix} -17/16 \\ 13/16 \\ -5/8 \end{bmatrix}.$$

This is the solution of the second system and the input to the third system. Repeating steps 2 and 3 for the third system we have

$$\begin{bmatrix} 1 & 0 & 0 \\ 3 & 1 & 0 \\ 0 & 1/2 & 1 \end{bmatrix}\begin{bmatrix} y_1 \\ y_2 \\ y_3 \end{bmatrix} = \begin{bmatrix} -17/16 \\ 13/16 \\ -5/8 \end{bmatrix}, \quad \text{so } Y = \begin{bmatrix} -17/16 \\ 4 \\ -21/8 \end{bmatrix}$$

and then

$$\begin{bmatrix} 2 & 6 & 2 \\ 0 & -4 & 2 \\ 0 & 0 & -2 \end{bmatrix}\begin{bmatrix} x_1 \\ x_2 \\ x_3 \end{bmatrix} = \begin{bmatrix} -17/16 \\ 4 \\ -21/8 \end{bmatrix},$$

$$\text{from which } X_3 = \begin{bmatrix} -13/16 \\ -11/32 \\ 21/16 \end{bmatrix}. \quad \blacksquare$$

Example 3 shows that a factorization of the coefficient matrix into the product of a lower triangular matrix and an upper triangular matrix would be useful in dealing with the case of multiple right-hand sides. We will now discuss how to find this factorization and the conditions under which it exists.

It is relatively easy to see that the desired factorization exists if and only if no row interchanges are required in the Gaussian reduction of the coefficient matrix. If this is the case, then only Type III elementary row operations are required and all of them correspond to elementary matrices which are lower triangular. Thus we have, from Theorem 1.6, the factorization

$$L_k \ldots L_2 L_1 A = U,$$

where U is an upper triangular matrix and the L_i are lower triangular elementary matrices corresponding to the elementary row operations used in the reduction. Now the product of lower triangular matrices is still lower triangular (see Exercise 13 of Section 1.6) and the inverse of a lower triangular matrix is also lower triangular (see Exercise 20 of Section 1.7). Thus

$$A = LU \text{ where } L = (L_k \ldots L_2 L_1)^{-1} = L_1^{-1} L_2^{-1} \ldots L_k^{-1}$$

$$(1.16)$$

is a lower triangular matrix.

In order to see what the matrix L is like we consider the reduction of a nonsingular 3×3 matrix A.

$$\begin{bmatrix} a_{11} & a_{12} & a_{13} \\ a_{21} & a_{22} & a_{23} \\ a_{31} & a_{32} & a_{33} \end{bmatrix} \rightarrow \begin{bmatrix} a_{11} & a_{12} & a_{13} \\ 0 & b_{22} & b_{23} \\ 0 & b_{32} & b_{33} \end{bmatrix} \rightarrow \begin{bmatrix} a_{11} & a_{12} & a_{13} \\ 0 & b_{22} & b_{23} \\ 0 & 0 & c_{33} \end{bmatrix}$$

In this reduction we have not changed the row leaders to 1, so only three elementary row operations are needed to reduce A to upper triangular form; they are of the form $(R_2 \leftarrow aR_1 + R_2)$, $(R_3 \leftarrow bR_1 + R_3)$, and $(R_3 \leftarrow cR_2 + R_3)$, where $a = -a_{21}/a_{11}$, $b = -a_{31}/a_{11}$, and $c = -b_{32}/b_{22}$. From Equation 1.16 we have

$$L = \left(\begin{bmatrix} 1 & 0 & 0 \\ 0 & 1 & 0 \\ 0 & c & 1 \end{bmatrix} \begin{bmatrix} 1 & 0 & 0 \\ 0 & 1 & 0 \\ b & 0 & 1 \end{bmatrix} \begin{bmatrix} 1 & 0 & 0 \\ a & 1 & 0 \\ 0 & 0 & 1 \end{bmatrix} \right)^{-1}$$

$$= \begin{bmatrix} 1 & 0 & 0 \\ a & 1 & 0 \\ 0 & 0 & 1 \end{bmatrix}^{-1} \begin{bmatrix} 1 & 0 & 0 \\ 0 & 1 & 0 \\ b & 0 & 1 \end{bmatrix}^{-1} \begin{bmatrix} 1 & 0 & 0 \\ 0 & 1 & 0 \\ 0 & c & 1 \end{bmatrix}^{-1}$$

$$= \begin{bmatrix} 1 & 0 & 0 \\ -a & 1 & 0 \\ 0 & 0 & 1 \end{bmatrix} \begin{bmatrix} 1 & 0 & 0 \\ 0 & 1 & 0 \\ -b & 0 & 1 \end{bmatrix} \begin{bmatrix} 1 & 0 & 0 \\ 0 & 1 & 0 \\ 0 & -c & 1 \end{bmatrix}$$

$$= \begin{bmatrix} 1 & 0 & 0 \\ -a & 1 & 0 \\ -b & -c & 1 \end{bmatrix}.$$

Note that the first operation in the reduction was $(R_2 \leftarrow aR_1 + R_2)$, which was designed to produce a zero in the $(2, 1)$ position, and the scalar $-a$ appears in the $(2, 1)$ position of L. Moreover, the second operation used was $(R_3 \leftarrow bR_1 + R_3)$ and the scalar $-b$ appears in the $(3, 1)$ position of L while the third operation used was $(R_3 \leftarrow cR_2 + R_3)$ and the scalar $-c$ appears in the $(3, 2)$ position of L.

The situation in general is just like in the above 3×3 example; L has 1s on the diagonal and the elements below the diagonal are the negatives of the multipliers used in the reduction. In most computer implementations of the Gaussian elimination algorithm the original matrix A is overwritten at each step by the reduced matrix (see Exercise 9 in Section 1.3). In the case of the above 3×3 nonsingular matrix the algorithm changes A to an upper triangular matrix of the form

$$U = \begin{bmatrix} x & x & x \\ 0 & x & x \\ 0 & 0 & x \end{bmatrix}.$$

An efficient way to find the LU factorization of A is to use the above observation about the elements of L to construct L during the reduction, that is, store the negatives of the multipliers in the zeroed positions. For the reduction described above we have

$$A \rightarrow \begin{bmatrix} a_{11} & a_{12} & a_{13} \\ -a & b_{22} & b_{23} \\ -b & b_{32} & b_{33} \end{bmatrix} \rightarrow \begin{bmatrix} a_{11} & a_{12} & a_{13} \\ -a & b_{22} & b_{23} \\ -b & -c & c_{33} \end{bmatrix},$$

from which we can recover both L and U. Note that it is not necessary to store the diagonal elements of L since they are always 1, such a matrix will be called UNIT LOWER TRIANGULAR. We illustrate the procedure in the next example.

Example 4

Find the LU factorization of $A = \begin{bmatrix} 2 & -4 & 4 \\ -2 & 2 & 0 \\ 4 & 1 & -8 \end{bmatrix}$.

According to the procedure suggested above we make the following reduction:

$$\begin{bmatrix} 2 & -4 & 4 \\ -2 & 2 & 0 \\ 4 & 1 & -8 \end{bmatrix} \xrightarrow[\substack{R_2 \leftarrow R_1 + R_2 \\ R_3 \leftarrow -2R_1 + R_3}]{} \begin{bmatrix} 2 & -4 & 4 \\ -1 & -2 & 4 \\ 2 & 9 & -16 \end{bmatrix}$$

$$\xrightarrow[R_3 \leftarrow 9R_2/2 + R_3]{} \begin{bmatrix} 2 & -4 & 4 \\ -1 & -2 & 4 \\ 2 & -9/2 & 2 \end{bmatrix}.$$

From this reduction we recover the factorization

$$A = \begin{bmatrix} 2 & -4 & 4 \\ -2 & 2 & 0 \\ 4 & 1 & -8 \end{bmatrix} = LU$$

$$= \begin{bmatrix} 1 & 0 & 0 \\ -1 & 1 & 0 \\ 2 & -9/2 & 1 \end{bmatrix} \begin{bmatrix} 2 & -4 & 4 \\ 0 & -2 & 4 \\ 0 & 0 & 2 \end{bmatrix}.$$

Note that at the second step of the above reduction we reduced only the 2×2 submatrix in the lower right hand corner; we did not change the multipliers in the first column at the second step. ∎

The above procedure requires some modifications if an untimely zero appears on the diagonal, necessitating a row interchange. The key to understanding how to revise the above calculation, in order to handle the case where row interchanges are either necessary (to avoid zero diagonal elements in hand calculations) or desirable (to include partial pivoting in machine calculations), lies in the following simple calculation:

$$PLP^{-1} = \begin{bmatrix} 1 & 0 & 0 & 0 \\ 0 & 0 & 1 & 0 \\ 0 & 1 & 0 & 0 \\ 0 & 0 & 0 & 1 \end{bmatrix} \begin{bmatrix} 1 & 0 & 0 & 0 \\ a & 1 & 0 & 0 \\ b & 0 & 1 & 0 \\ c & 0 & 0 & 1 \end{bmatrix} \begin{bmatrix} 1 & 0 & 0 & 0 \\ 0 & 0 & 1 & 0 \\ 0 & 1 & 0 & 0 \\ 0 & 0 & 0 & 1 \end{bmatrix}$$

$$= \begin{bmatrix} 1 & 0 & 0 & 0 \\ b & 1 & 0 & 0 \\ a & 0 & 1 & 0 \\ c & 0 & 0 & 1 \end{bmatrix} = L'.$$

Here P is the elementary matrix associated with the row interchange $(R_2 \leftrightarrow R_3)$, and L is the product of the elementary matrices associated with the elementary row operations $(R_2 \leftarrow aR_1 + R_2)$, $(R_3 \leftarrow bR_1 + R_3)$, and $(R_3 \leftarrow cR_1 + R_4)$. Note that the only difference between L and L' is that the second and third entries of the first column have been interchanged.

Suppose that after the first column of a 4×4 matrix A has been reduced we encounter a zero in the $(2, 2)$ position; for example,

$$L_1 A = U_1 = \begin{bmatrix} x & x & x & x \\ 0 & 0 & x & x \\ 0 & x & x & x \\ 0 & x & x & x \end{bmatrix}.$$

Interchanging rows 2 and 3 leads to

$$PL_1 A = PU_1$$

which can be rewritten as

$$PL_1 A = PL_1(P^{-1}P)A = (PL_1 P^{-1})(PA) = PU_1$$

or, using the calculation above,

$$L_1'(PA) = PU_1 = \begin{bmatrix} x & x & x & x \\ 0 & x & x & x \\ 0 & 0 & x & x \\ 0 & x & x & x \end{bmatrix}.$$

The second column can now be reduced to yield

$$L_2 L_1'(PA) = \begin{bmatrix} x & x & x & x \\ 0 & x & x & x \\ 0 & 0 & x & x \\ 0 & 0 & x & x \end{bmatrix} = U_2.$$

If the $(3, 3)$ entry is zero and the $(3, 4)$ entry is not zero, then we would need to interchange rows 3 and 4 which would lead to a matrix of the form

$$P'L_2 L_1'(PA) = P'U_2 = \begin{bmatrix} x & x & x & x \\ 0 & x & x & x \\ 0 & 0 & x & x \\ 0 & 0 & 0 & x \end{bmatrix} = U.$$

Proceeding as above we write

$$P'L_2 L_1'(PA) = [P'(L_2 L_1')P'^{-1}]P'(PA) = L''(P'PA)$$

and, since $P'^{-1} = P'$ because P' is an elementary matrix of Type II, L'' is still a lower triangular matrix. In fact L'' differs from $L_2 L'$ only because the lower triangular elements in the last two rows have been interchanged. For our example,

$$L_2 L_1' = \begin{bmatrix} 1 & 0 & 0 & 0 \\ b & 1 & 0 & 0 \\ a & 0 & 1 & 0 \\ c & d & 0 & 1 \end{bmatrix} \quad \text{and} \quad L'' = \begin{bmatrix} 1 & 0 & 0 & 0 \\ b & 1 & 0 & 0 \\ c & d & 1 & 0 \\ a & 0 & 0 & 1 \end{bmatrix}.$$

We now define a PERMUTATION MATRIX to be any product of elementary matrices of Type II; such a matrix is simply the identity matrix I with its rows rearranged. A permutation matrix is a product of elementary matrices of Type II.

By generalizing the above calculations in a straightforward way we have the following theorem.

Theorem 1.10

If A is an $n \times n$ nonsingular matrix, then there exists a permutation matrix P such that PA can be factored as $PA = LU$ where L is a unit lower triangular matrix and U is upper triangular.

The modified form of the factorization works just as well for solving the linear system $AX = K$ as the original form, since the system $AX = K$ is equivalent to the system $PAX = PK$, which is just the original system with the equations (rows of the augmented matrix) reordered. The solution of the second system involves the following three steps.

(1) Factor $PA = LU$.
(2) Solve the lower triangular system $LY = PK$ for Y.
(3) Solve the upper triangular system $UX = Y$ for X.

Note that one needs to know the permutation matrix P in order to rearrange the right-hand side in the second step of the solution process.

The compact form of the reduction needs only to be slightly modified to accommodate row interchanges; one must interchange rows in the entire matrix while remembering to use Type III operations only on the unreduced part of the matrix.

Example 5

Find the LU factorization for either A or PA for a suitable permutation matrix P.

$$A = \begin{bmatrix} 2 & 3 & -1 & 2 \\ -4 & -6 & 2 & 1 \\ 2 & 4 & 4 & -1 \\ 4 & 8 & 2 & 7 \end{bmatrix} \qquad P = \begin{bmatrix} 1 & 0 & 0 & 0 \\ 0 & 1 & 0 & 0 \\ 0 & 0 & 1 & 0 \\ 0 & 0 & 0 & 1 \end{bmatrix}$$

$$\downarrow$$

$$\begin{bmatrix} 2 & 3 & -1 & 2 \\ -2 & 0 & 0 & 5 \\ 1 & 1 & 5 & -3 \\ 2 & 2 & 4 & 3 \end{bmatrix}$$

$$\downarrow$$

$$\begin{bmatrix} 2 & 3 & -1 & 2 \\ 1 & 1 & 5 & -3 \\ -2 & 0 & 0 & 5 \\ 2 & 2 & 4 & 3 \end{bmatrix} \qquad P = \begin{bmatrix} 1 & 0 & 0 & 0 \\ 0 & 0 & 1 & 0 \\ 0 & 1 & 0 & 0 \\ 0 & 0 & 0 & 1 \end{bmatrix}$$

$$\downarrow$$

$$\begin{bmatrix} 2 & 3 & -1 & 2 \\ 1 & 1 & 5 & -3 \\ -2 & 0 & 0 & 5 \\ 2 & 2 & -6 & 9 \end{bmatrix}$$

$$\downarrow$$

$$\begin{bmatrix} 2 & 3 & -1 & 2 \\ 1 & 1 & 5 & -3 \\ 2 & 2 & -6 & 9 \\ -2 & 0 & 0 & 5 \end{bmatrix} \qquad P = \begin{bmatrix} 1 & 0 & 0 & 0 \\ 0 & 0 & 1 & 0 \\ 0 & 0 & 0 & 1 \\ 0 & 1 & 0 & 0 \end{bmatrix}$$

Thus $PA = \begin{bmatrix} 2 & 3 & -1 & 2 \\ 2 & 4 & 4 & -1 \\ 4 & 8 & 2 & 7 \\ -4 & -6 & 2 & 1 \end{bmatrix} = LU$

$$= \begin{bmatrix} 1 & 0 & 0 & 0 \\ 1 & 1 & 0 & 0 \\ 2 & 2 & 1 & 0 \\ -2 & 0 & 0 & 1 \end{bmatrix} \cdot \begin{bmatrix} 2 & 3 & -1 & 2 \\ 0 & 1 & 5 & -3 \\ 0 & 0 & -6 & 9 \\ 0 & 0 & 0 & 5 \end{bmatrix},$$

as the reader should verify by direct multiplication. ■

Example 6

Use the factorization in Example 5 to solve the system

$$AX = K = \begin{bmatrix} 4 \\ 6 \\ 9 \\ 0 \end{bmatrix}.$$

The system $LY = PK$ is

$$\begin{bmatrix} 1 & 0 & 0 & 0 \\ 1 & 1 & 0 & 0 \\ 2 & 2 & 1 & 0 \\ -2 & 0 & 0 & 1 \end{bmatrix} Y = \begin{bmatrix} 4 \\ 9 \\ 0 \\ 6 \end{bmatrix}, \quad \text{for which } Y = \begin{bmatrix} 4 \\ 5 \\ -18 \\ 14 \end{bmatrix},$$

and the system $UX = Y$ is

$$\begin{bmatrix} 2 & 3 & -1 & 2 \\ 0 & 1 & 5 & -3 \\ 0 & 0 & -6 & 9 \\ 0 & 0 & 0 & 5 \end{bmatrix} X = \begin{bmatrix} 4 \\ 5 \\ -18 \\ 14 \end{bmatrix}, \quad \text{for which } X = \begin{bmatrix} 36.7 \\ -22.6 \\ 7.2 \\ 2.8 \end{bmatrix}. \quad ■$$

In order to control round-off error in machine calculations, the partial-pivoting procedure, described at the beginning of this section, should be included in any general-purpose algorithm for computing the *LU* factorization or the inverse of a matrix. The row interchanges are handled as we have described above. We illustrate the calculations in our next example.

Example 7

Use partial pivoting to find a permutation matrix P so that $PA = LU$ for the matrix A of Example 1.

$$A = \begin{bmatrix} 2 & 3 & -1 & 2 \\ -4 & -6 & 2 & 1 \\ 2 & 4 & 4 & -1 \\ 4 & 8 & 2 & 7 \end{bmatrix} \xrightarrow[R_1 \leftrightarrow R_2]{} \begin{bmatrix} -4 & -6 & 2 & 1 \\ 2 & 3 & -1 & 2 \\ 2 & 4 & 4 & -1 \\ 4 & 8 & 2 & 7 \end{bmatrix}$$

$$P = \begin{bmatrix} 0 & 1 & 0 & 0 \\ 1 & 0 & 0 & 0 \\ 0 & 0 & 1 & 0 \\ 0 & 0 & 0 & 1 \end{bmatrix} \qquad \begin{matrix} R_2 \leftarrow 1/2R_1 + R_2 \\ R_3 \leftarrow 1/2R_1 + R_3 \\ R_4 \leftarrow R_1 + R_4 \end{matrix}$$

$$\begin{bmatrix} -4 & -6 & 2 & 1 \\ -.5 & 0 & 0 & 2.5 \\ -.5 & 1 & 5 & -.5 \\ -1 & 2 & 4 & 8 \end{bmatrix}$$

$$\downarrow \; R_2 \leftrightarrow R_4$$

$$P = \begin{bmatrix} 0 & 1 & 0 & 0 \\ 0 & 0 & 0 & 1 \\ 0 & 0 & 1 & 0 \\ 1 & 0 & 0 & 0 \end{bmatrix} \qquad \begin{bmatrix} -4 & -6 & 2 & 1 \\ -1 & 2 & 4 & 8 \\ -.5 & 1 & 5 & -.5 \\ -.5 & 0 & 0 & 2.5 \end{bmatrix}$$

$$\downarrow \; R_3 \leftarrow 1/2R_2 + R_3$$

$$\begin{bmatrix} -4 & -6 & 2 & 1 \\ -1 & 2 & 4 & 8 \\ -.5 & .5 & 3 & -4.5 \\ -.5 & 0 & 0 & 2.5 \end{bmatrix}$$

$$\text{Thus } PA = \begin{bmatrix} 0 & 1 & 0 & 0 \\ 0 & 0 & 0 & 1 \\ 0 & 0 & 1 & 0 \\ 1 & 0 & 0 & 0 \end{bmatrix} \begin{bmatrix} 2 & 3 & -1 & 2 \\ -4 & -6 & 2 & 1 \\ 2 & 4 & 4 & -1 \\ 4 & 8 & 2 & 7 \end{bmatrix}$$

$$= \begin{bmatrix} -4 & -6 & 2 & 1 \\ 4 & 8 & 2 & 7 \\ 2 & 4 & 4 & -1 \\ 2 & 3 & -1 & 2 \end{bmatrix} = LU$$

$$= \begin{bmatrix} 1 & 0 & 0 & 0 \\ -1 & 1 & 0 & 0 \\ -.5 & .5 & 1 & 0 \\ -.5 & 0 & 0 & 1 \end{bmatrix} \cdot \begin{bmatrix} -4 & -6 & 2 & 1 \\ 0 & 2 & 4 & 8 \\ 0 & 0 & 3 & -4.5 \\ 0 & 0 & 0 & 2.5 \end{bmatrix},$$

as you may verify by direct calculation. ■

We conclude this section by collecting, for easy reference, the operation counts for the matrix operations discussed in this chapter.

Operation	Number of Multiplications
$A \times B$, where A is $m \times n$ and B is $n \times r$	mnr
Reduction of $m \times n$ matrix to row echelon form ($m < n$)	$\dfrac{(m-1)}{6}(3mn - m^2 + 6n - 4m)$
Reduction of $m \times n$ matrix to reduced row echelon form	$\dfrac{(m-1)}{6}(6mn - 4m^2 + 6n - 4m)$
LU factorization of $n \times n$ nonsingular matrix	$\dfrac{n^3}{3} - \dfrac{n}{3}$
Solution of a system with unit lower triangular coefficient matrix	$\dfrac{n^2}{2} - \dfrac{n}{2}$
Solution of a system with upper triangular coefficient matrix	$\dfrac{n^2}{2} + \dfrac{n}{2}$
A^{-1} for $n \times n$ matrix A by Gauss-Jordan elimination	n^3

Exercise Set 1.8

1. Use Gaussian elimination with partial pivoting to solve for X.

(a) $\begin{bmatrix} 2 & 3 & -6 \\ 3 & -2 & 1 \\ 1 & -6 & 8 \end{bmatrix} X = \begin{bmatrix} -1 \\ 2 \\ 2 \end{bmatrix}$ (b) $\begin{bmatrix} 1 & 5 & 0 \\ 5 & 1 & 5 \\ 0 & 5 & 1 \end{bmatrix} X = \begin{bmatrix} 3 \\ 1 \\ 7 \end{bmatrix}$

(c) $\begin{bmatrix} 1 & 3 & 2 & 1 \\ 2 & 1 & 2 & 3 \\ -4 & -2 & 1 & 2 \\ 1 & 2 & 4 & 0 \end{bmatrix} X = \begin{bmatrix} 5 \\ 7 \\ 0 \\ 11 \end{bmatrix}$

(d) $\begin{bmatrix} .4095 & .0124 & .3678 & .2945 \\ .3654 & .1902 & .3718 & .0634 \\ .1748 & .4000 & .2708 & .3925 \\ .2251 & .3842 & .4015 & .1129 \end{bmatrix} X = \begin{bmatrix} 3.412 \\ 5.312 \\ 1.000 \\ -4.312 \end{bmatrix}$

2. Consider the system

$$\begin{bmatrix} -3 & 2 & -100{,}000 \\ 6 & 1 & 1 \\ 3 & -.5 & -9.2 \end{bmatrix} \begin{bmatrix} x_1 \\ x_2 \\ x_3 \end{bmatrix} = \begin{bmatrix} 100{,}000 \\ 14.0 \\ 13.7 \end{bmatrix}.$$

Assume that this system is solved on a digital computer which rounds each calculation to three significant digits.

(a) Find the solution if no pivoting is used.

(b) Find the solution using partial pivoting.

(c) Compare the solutions with the true solution $X = \begin{bmatrix} 2 \\ 3 \\ -1 \end{bmatrix}$.

(d) Multiply the first equation by 10^{-5} and then repeat steps (a) and (b). Are the results any better?

3. Find the inverses of the following matrices using Gauss-Jordan elimination with partial pivoting. Round C^{-1} to three significant figures and compute CC^{-1} and $C^{-1}C$.

$$A = \begin{bmatrix} 1 & 0 & 2 \\ -3 & 4 & 6 \\ -1 & -2 & 3 \end{bmatrix} \qquad B = \begin{bmatrix} -3 & -2 & 0 & 2 \\ 2 & 1 & 0 & -1 \\ 1 & 0 & 1 & 2 \\ 2 & 1 & -3 & 1 \end{bmatrix}$$

$$C = \begin{bmatrix} 2.35 & 4.12 & 3.17 & 4.31 \\ -4.27 & 8.36 & 7.15 & 9.87 \\ 5.87 & 0.00 & -5.89 & 18.0 \\ 11.3 & 14.7 & 11.2 & 5.75 \end{bmatrix}$$

Find the LU factorization of the following matrices (no row interchanges are required).

4. $A = \begin{bmatrix} 1 & 2 & 3 \\ 4 & 5 & 6 \\ 7 & 8 & 0 \end{bmatrix}$ 5. $B = \begin{bmatrix} 1 & 1 & 1 \\ 3 & -2 & -17 \\ 3 & 2 & -1 \end{bmatrix}$ 6. $C = \begin{bmatrix} 1 & 2 & 4 \\ 2 & 3 & 7 \\ 1 & 4 & 7 \end{bmatrix}$

7. $D = \begin{bmatrix} 1 & -1 & 2 & 3 \\ 2 & -1 & 0 & 2 \\ 4 & 1 & -11 & -1 \\ 1 & 2 & 33 & 83 \end{bmatrix}$

For the following matrices find a permutation matrix P and matrices L and U so that PA = LU.

8. $A = \begin{bmatrix} 1 & 2 & 3 \\ 2 & 4 & 5 \\ 3 & 5 & 6 \end{bmatrix}$ 9. $A = \begin{bmatrix} 2 & 4 & 3 & 2 \\ 3 & 6 & 5 & 2 \\ 2 & 5 & 2 & -3 \\ 4 & 5 & 14 & 14 \end{bmatrix}$

10. $A = \begin{bmatrix} 1 & 3 & 5 & 2 \\ 0 & -1 & 3 & 4 \\ 2 & 1 & 25 & 6 \\ 3 & 2 & 4 & 8 \end{bmatrix}$ 11. $A = \begin{bmatrix} 0 & 1 & 3 & 2 \\ 1 & -1 & 7 & -7 \\ -2 & 3 & -11 & 3 \\ 4 & 2 & 1 & 5 \end{bmatrix}$

For the following matrices use partial pivoting to find a permutation matrix P and matrices L and U so that PA = LU.

12. $A = \begin{bmatrix} 1 & 2 & 3 \\ 4 & 5 & 6 \\ 7 & 8 & 0 \end{bmatrix}$ **13.** $A = \begin{bmatrix} 1 & 2 & 3 \\ 2 & 4 & 5 \\ 3 & 5 & 6 \end{bmatrix}$

14. $A = \begin{bmatrix} 1 & -1 & 2 & 3 \\ 2 & -1 & 0 & 2 \\ 4 & 1 & -11 & -1 \\ 1 & 2 & 33 & 83 \end{bmatrix}$ **15.** $A = \begin{bmatrix} 2 & 4 & 3 & 2 \\ 3 & 6 & 5 & 2 \\ 2 & 5 & 2 & -3 \\ 4 & 5 & 14 & 14 \end{bmatrix}$

16. Solve the system $AX = \begin{bmatrix} 1 \\ 1 \\ 1 \end{bmatrix}$ by the following methods:

  (a) using the factorization from Exercise 4; and

  (b) using the factorization from Exercise 12.

17. Solve the system $DX = \begin{bmatrix} 1 \\ 2 \\ 3 \\ 4 \end{bmatrix}$ by the following methods:

  (a) using the factorization of Exercise 7; and

  (b) using the factorization of Exercise 14.

18. **(a)** Show that there is a one-to-one correspondence between $n \times n$ permutation matrices P and the row matrices $[1 \quad 2 \quad \cdots \quad n]P$.

  (b) Show how the observation in part (a) could be used to save storage space in the LU factorization algorithm with partial pivoting.

19 **(a)** Modify the computer program from Exercise 9 of Section 1.3 so that it will compute the LU factorization of A assuming that no pivoting is required.

  (b) Modify the program in part (a) to include partial pivoting.

Chapter 1   Review Exercises

1. Given $A = \begin{bmatrix} 2 & 1 & 1 \\ 2 & 3 & 2 \\ 1 & 1 & 2 \end{bmatrix}$ compute $A^3 - 7A^2 + 11A - 4I$.

2. List two major differences between matrix algebra and ordinary algebra.

3. If a matrix B satisfies $B^3 - 7B^2 + 7B + 3I = 0$, express B^{-1} as a polynomial in B.

4. Write a formula for the entry in the $(3, 5)$ position of $A(B + C)$, given that A, B, and C are $n \times n$ matrices.

5. Use Gauss-Jordan elimination to find the inverse of $M = \begin{bmatrix} 1 & 2 & 3 \\ 2 & 4 & 5 \\ 3 & 5 & 6 \end{bmatrix}$.

6. Find all solutions of the system
$$x_1 - x_2 + 2x_3 - x_4 = -1$$
$$2x_1 + x_2 - 2x_3 - 2x_4 = -2$$
$$-x_1 + 2x_2 - 4x_3 + x_4 = 1$$
$$2x_1 + 2x_2 - 4x_3 - 2x_4 = -2$$

7. Find all solutions of the homogeneous system
$$x + 2y + 3z = 0$$
$$4x + 5y + 6z = 0$$

8. Find the *LU* factorization of the matrix $A = \begin{bmatrix} 1 & 2 & 4 \\ 2 & 3 & 7 \\ 1 & 4 & 7 \end{bmatrix}$.

9. Find all solutions of the matrix equation $\begin{bmatrix} 1 & 2 \\ 3 & 5 \end{bmatrix} X = \begin{bmatrix} 7 & 3 & 4 \\ 2 & -1 & 0 \end{bmatrix}$.

10. Complete the following.

   (a) A^{-1} exists if and only if A is row equivalent to _____ .

   (b) A^{-1} exists if and only if A is a product of _____ .

   (c) A^{-1} exists if and only if $AX = 0$, implying _____ .

   (d) A^{-1} exists if and only if $AX = K$ has a _____ solution.

11. **(a)** Give an example of a 3×5 matrix A that is in row echelon form but not in reduced row echelon form.

   (b) Give an example of a 4×4 upper triangular matrix that is not a diagonal matrix.

   (c) Given $A = \begin{bmatrix} 2 & 1 \\ 0 & 2 \end{bmatrix}$, find all matrices $B = \begin{bmatrix} a & b \\ c & d \end{bmatrix}$ such that $AB = BA$.

12. **(a)** For $A = \begin{bmatrix} 1 & 2 \\ 3 & -4 \end{bmatrix}$, compute $A^2 - 5A + I$.

   (b) Write a formula for the element in the $(5, 7)$ position of $D = (2A + 3B)C$, if A, B, and C are 10 by 10 matrices.

   (c) Determine x so that
$$\begin{bmatrix} -3 & -2 & 0 & 2 \\ 2 & 1 & 0 & -1 \\ 1 & 0 & 1 & 2 \\ 2 & 1 & -3 & 1 \end{bmatrix}^{-1} = \frac{1}{8}\begin{bmatrix} 8 & 16 & 0 & 0 \\ -19 & -31 & 3 & 1 \\ x & -2 & 2 & -2 \\ -3 & -7 & 3 & 1 \end{bmatrix}.$$

13. Find the inverse of $B = \begin{bmatrix} 1 & 2 & 3 \\ 2 & 4 & 5 \\ 3 & 5 & 6 \end{bmatrix}$.

14. Given $A = \begin{bmatrix} 1 & -1 & 2 & -1 \\ 2 & 1 & -2 & -2 \\ -1 & 2 & -4 & 1 \\ 2 & 2 & -4 & -2 \end{bmatrix}$, $B = \begin{bmatrix} -1 \\ -2 \\ 1 \\ -2 \end{bmatrix}$, and $K = \begin{bmatrix} k_1 \\ k_2 \\ k_3 \\ k_4 \end{bmatrix}$,

   (a) reduce $[A \,|\, B \,|\, K]$ to row echelon form;

   (b) from your work in (a), find all X such that $AX = B$;

   (c) from your work in (a), find all K such that $AX = K$ is solvable; and

   (d) from your work in (a), find all Y such that $AY = 0$.

15. Complete the following statements.

   (a) A linear equation in n variables is an equation of the form

   _____ .

   (b) The matrices A and B are row equivalent if _____ .

   (c) An elementary matrix is a matrix which has been obtained from the identity matrix by _____ .

   (d) The matrix B is a multiplicative inverse of A if _____ .

16. Find the coefficient matrix and the augmented matrix for the system

$$\begin{aligned}
2x_1 + x_2 + 5x_3 + x_4 &= 5 \\
x_1 + x_2 - 3x_3 - 4x_4 &= -1 \\
3x_1 + 6x_2 - 2x_3 + x_4 &= 8 \\
2x_1 + 2x_2 + 2x_3 - 3x_4 &= 2 ,
\end{aligned}$$

and then find all solutions of the system.

17. Use Gaussian elimination to find all solutions of the system

$$\begin{aligned}
3x_2 + x_3 &= 2 \\
x_1 + x_2 + 2x_3 &= -1 \\
2x_1 - x_2 + 3x_3 &= -4 .
\end{aligned}$$

18. Let A be the augmented matrix for the system in Exercise 17. Reduce A to reduced row echelon form.

19. Find all values of λ such that the following system has a nontrivial solution:

$$\begin{aligned}
5x + 4y &= \lambda x \\
x + 8y &= \lambda y .
\end{aligned}$$

20. Compute $A(B + C)$ for the matrices $A = \begin{bmatrix} 1 & 5 & -2 \\ 0 & 3 & 4 \\ 1 & 5 & 3 \\ 7 & -2 & 2 \end{bmatrix}$,

$$B = \begin{bmatrix} 1 & 5 \\ 2 & -3 \\ 0 & 4 \end{bmatrix}, \quad \text{and} \quad C = \begin{bmatrix} -2 & 2 \\ 5 & 0 \\ 6 & -1 \end{bmatrix}.$$

21. For $J = \begin{bmatrix} 2 & 0 & 0 \\ 1 & 2 & 0 \\ 0 & 0 & 3 \end{bmatrix}$ it is true that $J^3 - 7J^2 + 16J - 12I = 0$.

Use this fact to find J^{-1}.

22. Find the inverse of the matrix $A = \begin{bmatrix} 3 & 1 & 3 \\ 4 & 1 & 3 \\ 3 & 1 & 4 \end{bmatrix}$.

2

Determinants

In several exercises of Chapter 1 we encountered the number ad − bc associated with the 2 × 2

matrix $A = \begin{bmatrix} a & b \\ c & d \end{bmatrix}$. In

particular, (see Exercise 7 in Section 1.4) we found that the system $AX = 0$ has a nonzero solution if and only if $ad - bc = 0$ and (see Exercise 9 in Section 1.5) that if $ad - bc \neq 0$, then

$$A^{-1} = \frac{1}{ad - bc} \begin{bmatrix} d & -b \\ -c & a \end{bmatrix}.$$

In Chapter 2 we will generalize these results by defining a scalar-valued function on square matrices, called the DETERMINANT, which is a rule that associates a number with a matrix. The definition we will use is recursive in nature; that is, the value of the function for an $n \times n$ matrix will depend on the value of the function for several $(n - 1) \times (n - 1)$ matrices. Recursive functions are particularly useful when one is evaluating a function by machine.

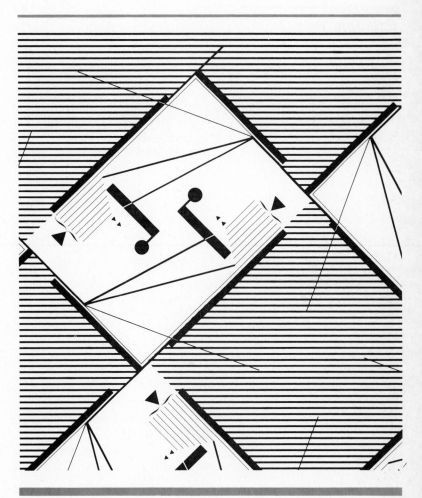

2.1   *Definitions and Examples*

Definition 2.1

Let A be an $n \times n$ matrix. The $(n - 1) \times (n - 1)$ submatrix of A obtained by deleting the ith row of A and the jth column of A is called the MINOR OF THE (i, j) POSITION OF A and will be denoted by either M_{ij} or $M_{ij}(A)$.

Example 1

$$\text{For } A = \begin{bmatrix} a_{11} & a_{12} & a_{13} & a_{14} \\ a_{21} & a_{22} & a_{23} & a_{24} \\ a_{31} & a_{32} & a_{33} & a_{34} \\ a_{41} & a_{42} & a_{43} & a_{44} \end{bmatrix},$$

$$M_{11} = \begin{bmatrix} a_{11} & a_{12} & a_{13} & a_{14} \\ a_{21} & a_{22} & a_{23} & a_{24} \\ a_{31} & a_{32} & a_{33} & a_{34} \\ a_{41} & a_{42} & a_{43} & a_{44} \end{bmatrix} = \begin{bmatrix} a_{22} & a_{23} & a_{24} \\ a_{32} & a_{33} & a_{34} \\ a_{42} & a_{43} & a_{44} \end{bmatrix}, \quad \text{while}$$

$$M_{34} = \begin{bmatrix} a_{11} & a_{12} & a_{13} & a_{14} \\ a_{21} & a_{22} & a_{23} & a_{24} \\ a_{31} & a_{32} & a_{33} & a_{34} \\ a_{41} & a_{42} & a_{43} & a_{44} \end{bmatrix} = \begin{bmatrix} a_{11} & a_{12} & a_{13} \\ a_{21} & a_{22} & a_{23} \\ a_{41} & a_{42} & a_{43} \end{bmatrix} \quad \text{and}$$

$$M_{22}(M_{11}(A)) = \begin{bmatrix} a_{22} & a_{24} \\ a_{42} & a_{44} \end{bmatrix}. \quad \blacksquare$$

Definition 2.2

For an $n \times n$ matrix A we define the DETERMINANT of A by

$$\det(A) = \sum_{j=1}^{n} a_{1j}(-1)^{1+j} \det(M_{1j}) \tag{2.1}$$

and

$$\det \begin{bmatrix} a_{11} & a_{12} \\ a_{21} & a_{22} \end{bmatrix} = a_{11}a_{22} - a_{12}a_{21}. \tag{2.2}$$

If we apply Definition 2.2 to a general 3×3 matrix A we obtain, using Equation 2.1,

$$\det(A) = \det \begin{bmatrix} a_{11} & a_{12} & a_{13} \\ a_{21} & a_{22} & a_{23} \\ a_{31} & a_{32} & a_{33} \end{bmatrix}$$

$$= a_{11}(-1)^{1+1} \det M_{11} + a_{12}(-1)^{1+2} \det M_{12} + a_{13}(-1)^{1+3} \det M_{13}$$

$$= a_{11} \det \begin{bmatrix} a_{22} & a_{23} \\ a_{32} & a_{33} \end{bmatrix} - a_{12} \det \begin{bmatrix} a_{21} & a_{23} \\ a_{31} & a_{33} \end{bmatrix} + a_{13} \det \begin{bmatrix} a_{21} & a_{22} \\ a_{31} & a_{32} \end{bmatrix},$$

and from Equation 2.2 we have the formula

$$\det(A) = a_{11}(a_{22}a_{33} - a_{23}a_{32}) - a_{12}(a_{21}a_{33} - a_{23}a_{31})$$
$$+ a_{13}(a_{21}a_{32} - a_{22}a_{31})$$

$$= a_{11}a_{22}a_{33} + a_{12}a_{23}a_{31} + a_{13}a_{21}a_{32} - a_{11}a_{23}a_{32} - a_{12}a_{21}a_{33}$$
$$- a_{13}a_{22}a_{31}, \qquad (2.3)$$

which consists of six terms, each the product of three numbers. Equation 2.3 is certainly not the sort of thing one should try to memorize. Fortunately there is a simple way of remembering Equation 2.3 that is illustrated in Figure 2.1. One proceeds by rewriting the first two columns of A to the right of A and then looking at products of elements on the same "slant." The algebraic sign of each product is as indicated in the diagram. Note that only complete slants are used; that is, each product contains three factors.

$$\det(A) = \quad$$

FIGURE 2.1

We will refer to this procedure as the *classical 3 × 3 trick*. It is like the trick suggested in Equation 2.2 for remembering the determinant of a 2 × 2 matrix. It is very important to resist the temptation to overgeneralize; despite the efforts of many generations of students, there is no 4 × 4 analog of the above trick!

Example 2

Use the classical 3 × 3 trick to evaluate

$$\det(A) = \det\begin{bmatrix} 1 & 0 & 2 \\ -3 & 4 & 6 \\ -1 & -2 & 3 \end{bmatrix}.$$

We rewrite the first two columns of A to obtain the array shown in Figure 2.2.

FIGURE 2.2

The positive products are $1 \times 4 \times 3 = 12$, $0 \times 6 \times (-1) = 0$, and $2 \times (-3) \times (-2) = 12$ while the negative products are $(-1) \times 4 \times 2 = -8$, $(-2) \times 6 \times 1 = -12$, and $3 \times (-3) \times 0 = 0$. Thus $\det(A) = 12 + 0 + 12 - (-8 - 12 + 0) = 44$. ■

For a 4×4 matrix A, Definition 2.2 would express $\det(A)$ as a linear combination of the determinants of four 3×3 matrices, each of which involves 6 terms. Altogether there would be 24 terms, each the product of 4 numbers. Thus the evaluation of a 4×4 determinant directly from the definition requires $72 = 3 \times (4!)$ multiplications.*

Example 3

Evaluate $\det(B) = \det \begin{bmatrix} -3 & -2 & 0 & 2 \\ 2 & 1 & 0 & -1 \\ 1 & 0 & 1 & 2 \\ 2 & 1 & -3 & 1 \end{bmatrix}$.

Directly from Equation 2.1 we have

$$\det(B) = -3 \det \begin{bmatrix} 1 & 0 & -1 \\ 0 & 1 & 2 \\ 1 & -3 & 1 \end{bmatrix} + 2 \det \begin{bmatrix} 2 & 0 & -1 \\ 1 & 1 & 2 \\ 2 & -3 & 1 \end{bmatrix}$$

$$+ 0 \det \begin{bmatrix} 2 & 1 & -1 \\ 1 & 0 & 2 \\ 2 & 1 & 1 \end{bmatrix} - 2 \det \begin{bmatrix} 2 & 1 & 0 \\ 1 & 0 & 1 \\ 2 & 1 & -3 \end{bmatrix},$$

and evaluating each of the 3×3 determinants directly from Equation 2.1 yields

$$\det(B) = -3 \left(1 \det \begin{bmatrix} 1 & 2 \\ -3 & 1 \end{bmatrix} - 0 \det \begin{bmatrix} 0 & 2 \\ 1 & 1 \end{bmatrix} - 1 \det \begin{bmatrix} 0 & 1 \\ 1 & -3 \end{bmatrix} \right)$$

$$+ 2 \left(2 \det \begin{bmatrix} 1 & 2 \\ -3 & 1 \end{bmatrix} - 0 \det \begin{bmatrix} 1 & 2 \\ 2 & 1 \end{bmatrix} - \det \begin{bmatrix} 1 & 1 \\ 2 & -3 \end{bmatrix} \right)$$

$$+ 0 - 2 \left(2 \det \begin{bmatrix} 0 & 1 \\ 1 & -3 \end{bmatrix} - \det \begin{bmatrix} 1 & 1 \\ 2 & -3 \end{bmatrix} + 0 \det \begin{bmatrix} 1 & 0 \\ 2 & 1 \end{bmatrix} \right)$$

$$= -3[(1 + 6) - 0 - (0 - 1)] + 2[2(1 + 6)$$

$$- 0 - (-3 - 2)] + 0 - 2[2(0 - 1) - (-3 - 2) + 0]$$

$$= -24 + 38 - 6 = 8. \quad \blacksquare$$

For an $n \times n$ matrix the evaluation procedure illustrated in Example 3 would involve the evaluation of a linear combination of

$$n, (n - 1) \times (n - 1) \text{ determinants},$$

which would lead to

$$n(n - 1), (n - 2) \times (n - 2) \text{ determinants},$$

---

*$n! = n(n - 1)(n - 2)(n - 3)\ldots(3)(2)(1)$ and is called n-FACTORIAL.

and then to

$$n(n - 1)(n - 2),\ (n - 3) \times (n - 3) \text{ determinants},$$

and eventually to

$$n(n - 1)(n - 2)\ldots 4 \times 3,\ 2 \times 2 \text{ determinants},$$

each of which involves two terms. Thus there would be $n!$ terms, each of which would involve the product of n numbers. Thus the evaluation of $\det(A)$, directly from Definition 2.2, would require $n! \times (n - 1)$ multiplications. This is a prohibitive amount of arithmetic for even moderately large n and very fast computers (see Exercise 13 in Section 2.1).

The recursive use of Definition 2.2, for a general $n \times n$ matrix A leads, as above, to an expression of the form

$$\det(A) = \sum \pm a_{1i_1}a_{2i_2}\ldots a_{ni_n} \tag{2.4}$$

where the sum is taken over all $n!$ possible arrangements (called *permutations*) of the second subscripts. This sum consists of all possible products obtained by taking one factor from each row and each column of A.

The classical way of deciding on the algebraic sign of each term in Equation 2.4 is to count the number of *transpositions* (interchange of two subscripts) required to restore the second subscripts to natural order; if the number is even, assign a plus sign and if the number is odd, assign a minus sign. For example, the term

$$a_{13}a_{24}a_{31}a_{42}$$

from the expansion of a 4×4 determinant would be positive because the sequence of second subscripts $(3, 4, 1, 2)$ can be restored to natural order by two transpositions:

$$(\overbrace{3, 4}, 1, 2) \longrightarrow (1, \overbrace{4, 3}, 2) \longrightarrow (1, 2, 3, 4).$$

Equation 2.4, with the above rule for choosing the algebraic sign of each term, constitutes the classical definition of the determinant function. We mention it here for historical interest and possibly to connect this discussion with your previous experience with determinants. We will have no further use for Equation 2.4.

If the determinant function is to be of any practical use, then it is clear that the first order of business is to find a more economical way of evaluating $\det(A)$. In the following sections we will develop such a method; it is like our methods for systems of equations in that it reduces the given problem to an equivalent problem for which the evaluation is *easy*.

$$A \xrightarrow{simplify} B \xrightarrow{easy} \det(B) \longrightarrow \det(A)$$

In order to implement such a procedure we need to answer two questions.

> (1) When is the evaluation of $\det(B)$ *easy?*
> (2) How can we simplify a matrix to take advantage of the answer to question 1 above?

Many of the exercises of this section are designed to prepare you to understand the answers to these questions, which will be presented in the next section.

We conclude this section by repeating the definition of the transpose of a matrix that was given in Exercise 20 of Section 1.6.

Definition 2.3

For an $n \times m$ matrix A, the TRANSPOSE of A, denoted as A^T, is the $m \times n$ matrix obtained from A by interchanging rows and columns. More precisely, if $B = A^T$, then

$$b_{ij} = a_{ji}; \qquad i = 1, 2, \ldots, m; \quad \text{and } j = 1, 2, \ldots, n.$$

The elementary properties of the transpose are given in the next theorem.

Theorem 2.1

The matrix transpose has the following properties for all matrices A and B for which the indicated operations are defined.

(1) $(A^T)^T = A$
(2) $(A + B)^T = A^T + B^T$
(3) $(kA)^T = kA^T$
(4) $(AB)^T = B^T A^T$
(5) $(A^{-1})^T = (A^T)^{-1}$

The only one of these properties that is at all obscure is the fourth one. A formal proof involves just a straightforward calculation using the definitions of matrix multiplication and the transpose. The details are given below where we assume A is $(m \times n)$, B is $(n \times s)$, and use $(C)_{ij}$ for c_{ij} when the matrix C has a name that is more complicated than a single letter.

$$(AB)^T_{ij} = (AB)_{ji} = \sum_{k=1}^{n} a_{jk}b_{ki} = \sum_{k=1}^{n} (A^T)_{kj}(B^T)_{ik}$$

$$= \sum_{k=1}^{n} (B^T)_{ik}(A^T)_{kj} = (B^TA^T)_{ij} \quad \square$$

The transpose has many uses, as we will see in the following sections and chapters. A theoretically unimportant use that nonetheless has practical value is the convention of writing a column matrix as the transpose of a row matrix; that is,

$$\begin{bmatrix} x_1 \\ x_2 \\ \vdots \\ x_n \end{bmatrix} = [x_1 \quad x_2 \quad \cdots \quad x_n]^T.$$

For our current purposes, the most important property of the transpose is its relationship to the determinant.

Theorem 2.2

For any $n \times n$ matrix A, $\det(A^T) = \det(A)$.

The proof, which involves calculations like those in Exercises 6 and 7, will be omitted.

Many of the properties of the determinant function will be given in terms of the rows of the matrix. Theorem 2.2 is important because it will allow us to give a column analog of each row result that we obtain.

Exercise Set 2.1

1. For $A = \begin{bmatrix} 1 & 2 \\ 3 & 4 \end{bmatrix}$ and $B = \begin{bmatrix} 3 & 4 \\ -2 & 3 \end{bmatrix}$ compute $\det(A)$, $\det(B)$, $\det(AB)$, $\det(BA)$, $\det(A)\det(B)$, $\det(A^{-1})$ and $\det(A^T)$.

2. For $A = \begin{bmatrix} 2 & 3 & 4 \\ 4 & 3 & 1 \\ 1 & 2 & 4 \end{bmatrix}$ and $B = \begin{bmatrix} 2 & -1 & 5 \\ 3 & 2 & 4 \\ 8 & 0 & -2 \end{bmatrix}$ compute $\det(A)$, $\det(B)$, $\det(AB)$, $\det(A)\det(B)$, and $\det(A^T)$.

3. Use the classical 3×3 trick to evaluate these determinants.

   (a) $\det \begin{bmatrix} 1 & 2 & 3 \\ 4 & 5 & 6 \\ 7 & 8 & 9 \end{bmatrix}$ (b) $\det \begin{bmatrix} x-2 & -3 & -4 \\ -4 & x-3 & -1 \\ -1 & -2 & x-4 \end{bmatrix}$

   (c) $\det \begin{bmatrix} x & -1 & 0 \\ 0 & x & -1 \\ -2 & -1 & x+2 \end{bmatrix}$

4. Evaluate the following, directly from Definition 2.2.

(a) $\det\begin{bmatrix} 1 & 2 & 3 & 4 \\ -1 & 1 & 2 & 3 \\ 1 & -1 & 1 & 2 \\ -1 & 1 & -1 & 1 \end{bmatrix}$ (b) $\det\begin{bmatrix} 1 & -1 & 2 & 3 \\ 2 & 2 & 0 & 2 \\ 4 & 1 & -1 & 1 \\ 1 & 2 & 3 & 0 \end{bmatrix}$

(c) $\det\begin{bmatrix} 1 & 2 & 3 & 4 \\ 1 & -1 & 1 & 2 \\ -1 & 1 & 2 & 3 \\ -1 & 1 & -1 & 11 \end{bmatrix}$

5. Show that one obtains a definition equivalent to Definition 2.2 if Equation 2.2 is replaced by $\det[a] = a$ for a 1×1 matrix $[a]$.

6. For a general 3×3 matrix A show that
   (a) $\det(A)$ can be computed from the elements in the first column of A and their minors via
   $$\det(A) = a_{11} \det M_{11} - a_{21} \det M_2 + a_{31} \det M_{31}, \text{ and}$$
   (b) $\det(A)$ can be computed from the elements in the second row of A and their minors via
   $$\det(A) = -a_{21} \det M_{21} + a_{22} \det M_{22} - a_{23} \det M_{23}.$$

†7. For a general 4×4 matrix A,
   (a) express $\det(A)$ as a linear combination of 2×2 determinants by using Definition (2.2) and then using the first formula in Exercise 6 for each of the remaining 3×3 determinants;
   (b) express $a_{11} \det M_{11} - a_{21} \det M_{21} + a_{31} \det M_{31} - a_{41} \det M_{41}$ as a linear combination of 2×2 determinants and compare with part (a) to conclude that $\det(A)$ can be computed from the elements in the first column and their minors;
   (c) show, by induction, that for any $n \times n$ matrix
   $$\det(A) = \sum_{i=1}^{n} a_{i1}(-1)^{1+i} \det(M_{i1}); \quad \text{and}$$
   (d) use part (c) to prove; by induction, that $\det(A^T) = \det(A)$.

†8. Compute the determinants of the following matrices.

(a) $A = \begin{bmatrix} a & b \\ c & d \end{bmatrix}$ (b) $\begin{bmatrix} c & d \\ a & b \end{bmatrix} = A_{(R_1 \leftrightarrow R_2)}$

(c) $\begin{bmatrix} a & b \\ kc & kd \end{bmatrix} = A_{(R_2 \leftarrow kR_2)}$

(d) $\begin{bmatrix} a & b \\ c + ka & d + kb \end{bmatrix} = A_{(R_2 \leftarrow kR_1 + R_2)}$

†9. Let A be the general 3×3 matrix whose determinant is given in Equation 2.3. Evaluate the determinants of
   (a) $A_{(R_1 \leftrightarrow R_3)}$,   (b) $A_{(R_2 \leftarrow kR_2)}$, and   (c) $A_{(R_3 \leftarrow kR_2 + R_3)}$
   and express the answers in terms of $\det(A)$.

10. Show that a general 3×3 matrix $A = \begin{bmatrix} a & b & c \\ d & e & f \\ g & h & i \end{bmatrix}$ is invertible if

$a \neq 0$, $ae - db \neq 0$, and $\det(A) \neq 0$.

(*Hint:* Try to row reduce A to I without using row interchanges.)
Which of the three given conditions can be eliminated if row interchanges are allowed?

11. Consider the TRACE function on square matrices that can be defined recursively by the following:

(1) $\text{tr}(A) = a_{11} + \text{tr}(M_{11})$

(2) $\text{tr}\begin{bmatrix} a & b \\ c & d \end{bmatrix} = a + d$.

(a) For the matrices A and B of Exercise 2 compute $\text{tr}(A)$, $\text{tr}(B)$, $\text{tr}(AB)$, and $\text{tr}(BA)$.

(b) Find a formula for $\text{tr}(A)$ in terms of the elements of A.

(c) Prove in general that $\text{tr}(AB) = \text{tr}(BA)$.

(d) Show that $\text{tr}(P^{-1}AP) = \text{tr}(A)$ for any nonsingular matrix P.

12. Find a formula for the determinant of a general 4×4 matrix.

13. If a large digital computer can do a multiplication in 2.9×10^{-6} seconds, how long would it take to do the $49(50!)$ multiplications required to evaluate the determinant of a 50×50 matrix?

14. Evaluate the determinants of the following triangular matrices.

(a) $\begin{bmatrix} a & 0 \\ c & d \end{bmatrix}$ (b) $\begin{bmatrix} a & 0 & 0 \\ b & d & 0 \\ c & e & f \end{bmatrix}$ (c) $\begin{bmatrix} a & b & c \\ 0 & d & e \\ 0 & 0 & f \end{bmatrix}$

15. Find the most general 3×3 matrix A such that $A^T = A$.

16. For $X = \begin{bmatrix} x_1 \\ x_2 \\ x_3 \end{bmatrix}$ and $A = \begin{bmatrix} 1 & -2 & 4 \\ -2 & 5 & 3 \\ 4 & 3 & 0 \end{bmatrix}$ compute X^TAX.

17. (a) For $X = [-2 \quad 3 \quad 4]^T$, compute X^TX and XX^T.

(b) For any $n \times 1$ matrix X show that $X^TX \geq 0$.

18. Show that if either A or B is singular, then so is AB.
(*Hint:* Use part 3 of Theorem 1.9 to show that if B is singular, then so is AB. Then use the transpose to do the other case.)

19. Show that for any nonsingular matrix M, $(M^T)^{-1} = (M^{-1})^T$.

2.2

Evaluation of det(A): A Better Way

In this section we will use the Gaussian elimination techniques of the last chapter to describe an efficient algorithm for the evaluation of the determinant function. In the process we will discover several interesting and useful properties of the determinant function.

We begin by answering the first of the two questions raised in Section 2.1: "For what kind of matrices is the evaluation of det(A) easy?" As suggested in Exercise 14 of Section 2.1 the answer to this question is that it is easy to evaluate the determinant of a triangular matrix. This ease occurs because the minor of the $(1, 1)$ position of a triangular matrix is another triangular matrix. For a lower triangular matrix there is only one nonzero element in the first row, so Definition 2.1 yields

$$\det(T) = t_{11} \det M_{11}(T) = t_{11}t_{22} \det M_{11}(M_{11}(T)) = \ldots$$

$$= t_{11}t_{22} \ldots t_{n-2\,n-2} \det \begin{bmatrix} t_{n-1\,n-1} & 0 \\ t_{n\,n-1} & t_{nn} \end{bmatrix}$$

$$= t_{11}t_{22} \ldots t_{n-1\,n-1}t_{nn} .$$

For an upper triangular matrix U, U^T is lower triangular, so from Theorem 2.1 and the above calculation we have

$$\det(U) = \det(U^T) = u_{11}u_{22} \ldots u_{nn} .$$

Property 2.1

The determinant of a triangular matrix is the product of its diagonal elements.

Because we have an orderly way of reducing A to an upper triangular matrix (Gaussian elimination), and since the evaluation of the determinant of a triangular matrix is easy, we will have found the desired "better way" if we can determine the effect of elementary row operations on the determinant.

Property 2.2

A Type II elementary row operation changes the sign of the determinant; that is, if B is obtained from A by interchanging two rows, then $\det(B) = -\det(A)$.

A careful proof of this property is quite involved and will be omitted. We will illustrate it here for a general 3×3 matrix and the elementary row operation $(R_1 \leftrightarrow R_3)$. We use Equation 2.3 to evaluate the determinants:

$$\det(A) = \det \begin{bmatrix} a_{11} & a_{12} & a_{13} \\ a_{21} & a_{22} & a_{23} \\ a_{31} & a_{32} & a_{33} \end{bmatrix}$$

$$= a_{11}a_{22}a_{33} + a_{12}a_{23}a_{31} + a_{13}a_{21}a_{32} - a_{11}a_{23}a_{32} - a_{12}a_{21}a_{33}$$
$$- a_{13}a_{22}a_{31}$$

$$\det(B) = \det \begin{bmatrix} a_{31} & a_{32} & a_{33} \\ a_{21} & a_{22} & a_{23} \\ a_{11} & a_{12} & a_{13} \end{bmatrix}$$

$$= a_{31}a_{22}a_{13} + a_{32}a_{23}a_{11} + a_{33}a_{21}a_{12} - a_{11}a_{22}a_{33} - a_{12}a_{23}a_{31}$$
$$- a_{13}a_{21}a_{32}$$

$$= -\det(A).$$

If the first row of A is zero, then it follows immediately from Definition 2.2 that $\det(A) = 0$. It also follows from Property 2.2 that if any row of A is zero, then $\det(A) = 0$; because the zero row can be moved to the first row by only changing the sign of the determinant. If a matrix A has two equal rows, then $\det(A) = -\det(A)$, since interchanging the two equal rows changes the sign of $\det(A)$ but does not change A. It follows that $\det(A) = 0$ if A has two equal rows. This observation establishes the following property.

Property 2.3

If the matrix A has a zero row or two equal rows, then $\det(A) = 0$.

Our next definition is aimed primarily at simplifying the notation in the definition of the determinant function.

Definition 2.4

The COFACTOR of the (i,j) position of A is

$$\text{cof}_{ij} = \text{cof}_{ij}(A) = (-1)^{i+j} \det(M_{ij}(A)).$$

Note that $\text{cof}_{ij}(A)$ does not depend on the elements from the ith row of A or the jth column of A.

The definition of the determinant can now be rephrased as

$$\det(A) = a_{11}\text{cof}_{11} + a_{12}\text{cof}_{12} + \ldots + a_{1n}\text{cof}_{1n}; \qquad (2.5)$$

that is, $\det(A)$ is the sum of the products of the elements of the first row of A times their cofactors.

The following checkerboard diagram is helpful in determining the signs associated with the cofactors. We show it for a 5×5 matrix; the generalization to any other size matrix is clear.

$$\begin{bmatrix} + & - & + & - & + \\ - & + & - & + & - \\ + & - & + & - & + \\ - & + & - & + & - \\ + & - & + & - & + \end{bmatrix}$$

Our next result shows that in Equation 2.5 the first row of A may be replaced by any other row of A.

Property 2.4

LAPLACE EXPANSION: For any $n \times n$ matrix A and any $i = 1, 2, \ldots, n$, $\det(A)$ can be computed by taking the sum of the products of the elements in the ith row by their cofactors; that is,

$$\det(A) = \sum_{k=1}^{n} a_{ik}\mathrm{cof}_{ik} = a_{i1}\mathrm{cof}_{i1} + a_{i2}\mathrm{cof}_{i2} + \ldots + a_{in}\mathrm{cof}_{in}.$$

To verify this property we move $\mathrm{Row}_i(A)$ to the first position, without destroying the order of the other rows, by repeatedly interchanging it with the row directly above it until the original ith row reaches the first row. This requires $(i - 1)$ Type II operations and leads to the matrix

$$B = \begin{bmatrix} \mathrm{Row}_i(A) \\ \hline A \text{ with its} \\ i\text{th row} \\ \text{deleted} \end{bmatrix}$$

for which, by Property 2.2, $\det(A) = (-1)^{i-1} \det(B)$. Now, by Definition 2.2,

$$\det(B) = \sum_{k=1}^{n} b_{1k}\mathrm{cof}_{1k}(B) = \sum_{k=1}^{n} a_{ik}(-1)^{1+k} \det(M_{ik}(A)),$$

since $M_{ik}(A) = M_{1k}(B)$ by the way we constructed B. Finally,

$$\det(A) = (-1)^{i-1} \det(B) = (-1)^{i-1} \sum_{k=1}^{n} a_{ik}(-1)^{1+k} \det(M_{ik})$$

$$= \sum_{k=1}^{n} a_{ik}(-1)^{i+k} \det(M_{ik}) = \sum_{k=1}^{n} a_{ik}\mathrm{cof}_{ik}. \quad \square$$

Our next example shows how Property 2.4 is useful when a row of the matrix has many zeros.

Example 1

Evaluate $\det(A)$ for $A = \begin{bmatrix} 3 & 2 & 4 \\ 0 & 5 & 0 \\ 6 & 0 & 7 \end{bmatrix}$.

Since the second row of A is the simplest (because it has the most zeros) we use the Laplace expansion by the second row to obtain

$$\det(A) = -0\mathrm{cof}_{21} + 5\mathrm{cof}_{22} - 0\mathrm{cof}_{23} = 5\det\begin{bmatrix} 3 & 4 \\ 6 & 7 \end{bmatrix}$$

$$= 5(-3) = -15. \quad \blacksquare$$

We can now show how the other two types of elementary row operations effect the determinant. If we use a Type I operation to multiply each element of the ith row by k, thus obtaining a matrix B, then an expansion by the ith row of B yields at once that $\det(B) = k \det(A)$. This result is most often used in the following form.

Property 2.5

If B is obtained from A by dividing each element of the ith row by k (factoring k from the ith row), then $\det(A) = k \det(B)$.

Example 2

$$\det \begin{bmatrix} 5 & 7 & 9 \\ 3 & 6 & -9 \\ 4 & 2 & 6 \end{bmatrix} = 3 \det \begin{bmatrix} 5 & 7 & 9 \\ 1 & 2 & -3 \\ 4 & 2 & 6 \end{bmatrix} = 6 \det \begin{bmatrix} 5 & 7 & 9 \\ 1 & 2 & -3 \\ 2 & 1 & 3 \end{bmatrix},$$

where we have first factored a 3 from the second row of the matrix and then a 2 from the third row. ■

Before we can investigate the effect of a Type III operation on the determinant we need to establish the following property.

Property 2.6

If $i \neq j$, then the sum of the product of the elements of row i by the cofactors of row j is zero; that is,

$$\sum_{k=1}^{n} a_{ik} \operatorname{cof}_{jk}(A) = 0 \quad \text{if } i \neq j.$$

To see that this is so we let B be the matrix obtained from A by replacing $\operatorname{Row}_j(A)$ by $\operatorname{Row}_i(A)$. Thus $b_{jk} = a_{ik}$ and

Note that $\det(B) = 0$, since B has two equal rows, and that $\operatorname{cof}_{jk}(B) = \operatorname{cof}_{jk}(A)$. Now, using the Laplace expansion by the jth row, we have

$$0 = \det(B) = \sum_{k=1}^{n} b_{jk} \operatorname{cof}_{jk}(B) = \sum_{k=1}^{n} a_{ik} \operatorname{cof}_{jk}(A). \quad \square$$

We can now establish a property of determinants that is extremely useful for purposes of evaluation.

Property 2.7

If B is obtained from A by the Type III elementary row operation $(\mathrm{R}_j \leftarrow c\mathrm{R}_i + \mathrm{R}_j)$, then $\det(B) = \det(A)$.

This is a very convenient property since most of the elementary row operations used in the reduction of a matrix to triangular form are of Type III. To verify this property we compute $\det(B)$ using the

Laplace expansion by the jth row. Since the element in the (j, k) position of B is $ca_{ik} + a_{jk}$ we have

$$\det(B) = \sum_k b_{jk} \operatorname{cof}_{jk}(B) = \sum_k (ca_{ik} + a_{jk}) \operatorname{cof}_{jk}(A)$$

$$= c \sum_k a_{ik} \operatorname{cof}_{jk}(A) + \sum_k a_{jk} \operatorname{cof}_{jk}(A) = 0 + \det(A),$$

using Property 2.6 for the first sum and Property 2.4 for the second. $\square$

We now have all the information necessary for an efficient algorithm for evaluating $\det(A)$: we reduce A to upper triangular form using elementary row operations and then evaluate the determinant of the resulting triangular matrix. Most of the operations used will be of Type III which do not change the determinant.

Example 3

Evaluate $\det(A)$ where $A = \begin{bmatrix} 1 & -1 & 2 & 3 \\ 2 & 2 & 0 & 2 \\ 4 & 1 & 3 & -1 \\ 1 & 2 & 3 & 0 \end{bmatrix}$.

The first column is reduced by the elementary row operations $(R_2 \leftarrow -2R_1 + R_2)$, $(R_3 \leftarrow -4R_1 + R_3)$, and $(R_4 \leftarrow -R_1 + R_4)$, none of which change the determinant, so that we have

$$\det(A) = \det \begin{bmatrix} 1 & -1 & 2 & 3 \\ 0 & 4 & -4 & -4 \\ 0 & 5 & -5 & -13 \\ 0 & 3 & 1 & -3 \end{bmatrix} = 4 \det \begin{bmatrix} 1 & -1 & 2 & 3 \\ 0 & 1 & -1 & -1 \\ 0 & 5 & -5 & -13 \\ 0 & 3 & 1 & -3 \end{bmatrix},$$

where the operation $(R_2 \leftarrow 1/4 R_2)$ was used to produce a 1 in the $(2, 2)$ position. The operations $(R_3 \leftarrow -5R_2 + R_3)$ and $(R_4 \leftarrow -3R_2 + R_4)$ will now simplify the second column without changing the determinant. We then have

$$\det(A) = 4 \det \begin{bmatrix} 1 & -1 & 2 & 3 \\ 0 & 1 & -1 & -1 \\ 0 & 0 & 0 & -8 \\ 0 & 0 & 4 & 0 \end{bmatrix} = -4 \det \begin{bmatrix} 1 & -1 & 2 & 3 \\ 0 & 1 & -1 & -1 \\ 0 & 0 & 4 & 0 \\ 0 & 0 & 0 & -8 \end{bmatrix}.$$

The operation $(R_3 \leftrightarrow R_4)$ was used at the last step to yield the final upper triangular matrix. It now follows from Property 2.1 that

$$\det(A) = -4(-32) = 128 . \quad \blacksquare$$

The elimination method for evaluating $\det(A)$, illustrated in Example 3, is clearly more efficient than the direct use of the definition. The reduction process takes about $n^3/3$ multiplications compared with the $(n - 1)n!$ multiplications required using the definition. On a machine that takes 2.9×10^{-6} seconds per multiplication, evaluating a 50×50 determinant requires less than 2 seconds, while performing $49 \times 50!$ multiplications requires 1.37×10^{53} years. This is a most dramatic saving.

Each row property stated above has a column analog that follows from the corresponding row property and Theorem 2.1. We leave it to you to formulate these column properties. You are strongly advised to use only row operations for evaluation purposes, at least until you have become proficient with the row operations. Column operations may be useful for hand calculations, but they are not needed for machine calculations. Remember, column operations should not be used in solving linear systems by Gaussian elimination.

Exercise Set 2.2

Use Gaussian reduction to evaluate the determinants of the following matrices.

1. $A = \begin{bmatrix} -1 & 1 & 3 \\ 2 & 1 & -1 \\ 4 & 2 & 2 \end{bmatrix}$

2. $B = \begin{bmatrix} 1 & 7 & 3 \\ 3 & -1 & 2 \\ 9 & -3 & 6 \end{bmatrix}$

3. $C = \begin{bmatrix} 1 & -1 & 2 \\ -3 & -3 & 3 \\ 4 & 0 & 5 \end{bmatrix}$

4. $D = \begin{bmatrix} 1 & 2 & 3 & 4 \\ 1 & -1 & 1 & 2 \\ -1 & 1 & -1 & 11 \\ -1 & 1 & 2 & 3 \end{bmatrix}$

5. $E = \begin{bmatrix} 1 & 0 & 2 & -1 \\ 3 & -2 & 6 & 4 \\ 2 & 2 & -5 & 6 \\ 5 & 4 & 3 & 0 \end{bmatrix}$

6. $F = \begin{bmatrix} 3 & -2 & 1 & 6 & 0 \\ 5 & -4 & 2 & 11 & 1 \\ 1 & -6 & 3 & 10 & 4 \\ -2 & 2 & -1 & -5 & 3 \\ 7 & -7 & 3 & 16 & -2 \end{bmatrix}$

7. $G = \begin{bmatrix} 1 & 1 & 2 & 1 \\ -1 & 1 & 0 & 1 \\ 2 & 1 & 1 & 0 \\ 1 & 3 & 1 & 0 \end{bmatrix}$

8. Show that if any row of a square matrix A is a linear combination of the other rows of A, then $\det(A) = 0$.

†9. For the matrices A and B above show that
   (a) $\det(AB) = \det(A) \det(B)$; and
   (b) $\det(A + B) \neq \det(A) + \det(B)$.

10. State and prove column properties analogous to Properties 2.2, 2.3, 2.4, 2.5, 2.6, and 2.7.

11. Prove Property 2.2 assuming the Laplace expansion.
    [*Hint:* Let the elementary row operation be $(R_i \leftrightarrow R_j)$. Expand by any row p for which $p \neq i$ and $p \neq j$, and then use mathematical induction.]

†**12.** Show that if A is $n \times n$, then $\det(kA) = k^n \det(A)$.

13. What is the net effect of the following three elementary row operations?
$$R_1 \longleftrightarrow R_2, \qquad R_2 \longleftrightarrow R_j, \qquad R_2 \longleftrightarrow R_1$$

14. What is the net effect of the following four elementary row operations?
$$R_j \longleftarrow R_i + R_j, \qquad R_i \longleftarrow -R_j + R_i,$$
$$R_j \longleftarrow R_i + R_j, \qquad R_i \longleftarrow -R_i$$

15. How can one find $\det(A)$ from the LU factorization of A discussed in Section 1.8?

16. Modify the computer program of Exercise 9 in Section 1.3 (on page 34) to compute $\det(A)$.

17. (a) Show that the equation $\det \begin{bmatrix} x & y & 1 \\ a & b & 1 \\ c & d & 1 \end{bmatrix} = 0$ represents the straight

line through (a, b) and (c, d).

(b) What does the equation $\det \begin{bmatrix} x & y & z & 1 \\ 1 & 2 & 3 & 1 \\ 4 & 5 & 6 & 1 \\ 7 & 8 & 0 & 1 \end{bmatrix} = 0$ represent?

18. Compare the area of the triangle with vertices $(2, 2)$, $(5, 2)$, and $(4, 5)$

with the determinant of the matrix $A = \begin{bmatrix} 1 & 2 & 2 \\ 1 & 4 & 5 \\ 1 & 5 & 2 \end{bmatrix}$.

19. A matrix S is said to be SKEW SYMMETRIC if $S^T = -S$. Show that the determinant of a 3×3 skew-symmetric matrix must be 0.

20. Determine a scalar x so that $\det(A) = 1$.

(a) $A = \begin{bmatrix} 3 & 5 \\ -8 & x \end{bmatrix}$ **(b)** $A = \begin{bmatrix} 1 & 2 & 3 \\ -1 & 4 & 5 \\ 2 & 3 & x \end{bmatrix}$ **(c)** $A = \begin{bmatrix} 1 & x & 2 \\ 1 & 0 & x \\ 2 & 5 & 1 \end{bmatrix}$

21. Use a pocket calculator or MATMAN to evaluate the following determinants.

(a) $\det \begin{bmatrix} 1.21 & 3.52 & 4.81 \\ 3.56 & .00 & 7.10 \\ 4.36 & -2.07 & 5.23 \end{bmatrix}$ **(b)** $\det \begin{bmatrix} 1 & -.897 & .805 & -.721 \\ 1 & 1.23 & 1.52 & 2.30 \\ 1 & 2.45 & 6.01 & 14.7 \\ 1 & 8.12 & 65.9 & 535 \end{bmatrix}$

2.3   *Additional Properties of Determinants*

In this section we will develop and illustrate some of the classical properties of the determinant function which are not directly related to the efficient evaluation of the function.

If the first row of the matrix can be written as

$$\text{Row}_1(A) = hR + kS = h[r_1 \quad r_2 \quad \dots \quad r_n] + k[s_1 \quad s_2 \quad \dots \quad s_n],$$

then we have, directly from the definition of the determinant, that

$$\det(A) = \det\begin{bmatrix} hr_1 + ks_1 & \dots & hr_n + ks_n \\ \hline & A \text{ with its first} & \\ & \text{row deleted} & \end{bmatrix}$$

$$= \sum_{j=1}^{n} (hr_j + ks_j)\text{cof}_{1j} = \sum_{j=1}^{n} hr_j\,\text{cof}_{1j} + \sum_{j=1}^{n} ks_j\,\text{cof}_{1j}$$

$$= h\det(A') + k\det(A''),$$

where A' is the matrix obtained from A by replacing the first row by R and A'' is the matrix obtained from A by replacing the first row by S. The above argument can be applied to any row of the matrix A to yield the following property.

Property 2.8

The determinant is a linear function of its rows; for example, if $\text{Row}_i(A) = hR + kS$, then $\det(A) = h\det(A') + k\det(A'')$ where A' and A'' are obtained from A by replacing the ith row of A by R and S respectively.

Example 1

Consider the matrix $A = \begin{bmatrix} 4 & 6 & 7 \\ 3 & -14 & -10 \\ 8 & 9 & 3 \end{bmatrix}$.

We can write the second row of A as

$$[3 \quad -14 \quad -10] = 3[1 \quad 2 \quad 0] - 5[0 \quad 4 \quad 2]$$

so, from Property 2.8, we have

$$\det(A) = 3\det\begin{bmatrix} 4 & 6 & 7 \\ 1 & 2 & 0 \\ 8 & 9 & 3 \end{bmatrix} - 5\det\begin{bmatrix} 4 & 6 & 7 \\ 0 & 4 & 2 \\ 8 & 9 & 3 \end{bmatrix}.$$

You should check the validity of this expansion by directly evaluating the three determinants above. ■

We next investigate the relationship between the determinant and the invertibility of a square matrix A. If A is invertible, then we know, from Theorem 1.9, that A is row equivalent to the identity matrix I. Since no elementary row operation can change the determinant from 0 to 1, it follows that $\det(A) \neq 0$. Conversely, if A is not row equivalent to I, then its row echelon form must have a zero row and hence $\det(A) = 0$. It follows that if $\det(A) \neq 0$, then A is row equivalent to I and hence invertible. We state this result, which has applications in many areas, as our next theorem.

Theorem 2.3

An $n \times n$ matrix A is invertible if and only if $\det(A) \neq 0$.

The result of Theorem 2.3 can now be added to the list of conditions equivalent to the existence of A^{-1} given in Theorem 1.9.

Another useful property of the determinant function is the following multiplicative property.

Property 2.9

For any $n \times n$ matrices A and B, $\det(AB) = \det(A) \det(B)$.

Proof. If A is an elementary matrix, this result follows directly from Theorem 1.7 and Properties 2.2, 2.5, and 2.7. If A is invertible, then by Theorem 1.9 it can be expressed as a product of elementary matrices $A = E_1 E_2 \ldots E_k$. Then, by repeatedly using the fact that the theorem holds for elementary matrices, we have

$$
\begin{aligned}
\det(AB) &= \det(E_1 E_2 \ldots E_k B) \\
&= \det(E_1) \det(E_2 \ldots E_k B) \\
&= \ldots \\
&= \det(E_1) \det(E_2) \ldots \det(E_k) \det(B) \\
&= \det(E_1 E_2 \ldots E_k) \det(B) \\
&= \det(A) \det(B) .
\end{aligned}
$$

If A is singular, then so is AB (refer to Exercise 18 of Section 2.1) and in this case we have $\det(AB) = 0 = 0 \det(B) = \det(A) \det(B)$. $\square$

The proof of the last property shows how elementary matrices can be used as the building blocks for all matrices. It is often possible to establish a general result by reducing it to the case of elementary matrices.

In the last section we established that

$$
\det(A) = \sum_{j=1}^{n} a_{ij} \text{cof}_{ij} \quad \text{for} \quad i = 1, 2, \ldots, n
$$

and that

$$
0 = \sum_{j=1}^{n} a_{ij} \text{cof}_{kj} \quad \text{for} \quad i \neq k .
$$

These equations look very much like the equations defining matrix multiplication. In fact, if we define a matrix B by

$$
b_{ji} = \text{cof}_{ij}(A) ,
$$

then the first equation above says that each of the diagonal elements of AB is equal to $\det(A)$ while the second equation says that each nondiagonal entry of AB is zero. Together these two equations say that $AB = \det(A)I$. It follows easily, from the column versions of our row results, that we also have $BA = \det(A)I$. If $\det(A) \neq 0$, then we can divide both sides of the last equation by $\det(A)$ to obtain

$$\frac{1}{\det(A)}AB = I.$$

It follows from Theorem 1.9 that

$$A^{-1} = \frac{1}{\det(A)}B.$$

The matrix B is usually called the adjoint of A.

Definition 2.5

For an $n \times n$ matrix A, the ADJOINT OF A is the $n \times n$ matrix whose (i,j) entry is $\mathrm{cof}_{ji}(A)$. We will denote the adjoint of A by $\mathrm{Adj}(A)$.

It is useful to think of the adjoint matrix as being constructed from A by first replacing each element of A by its cofactor and then transposing the resulting matrix. We summarize the discussion preceding Definition 2.5 with our next theorem.

Theorem 2.4

For an $n \times n$ matrix A we have

$$A\,\mathrm{Adj}(A) = \mathrm{Adj}(A)A = \det(A)I,$$

and if $\det(A) \neq 0$,

$$A^{-1} = \frac{1}{\det(A)}\,\mathrm{Adj}(A).$$

Example 2

Use the adjoint formula to invert $A = \begin{bmatrix} 3 & 1 & -4 \\ 6 & 9 & -2 \\ 1 & 2 & 1 \end{bmatrix}$.

The adjoint matrix of A is

$$\mathrm{Adj}(A) = \begin{bmatrix} +\det\begin{bmatrix} 9 & -2 \\ 2 & 1 \end{bmatrix} & -\det\begin{bmatrix} 6 & -2 \\ 1 & 1 \end{bmatrix} & +\det\begin{bmatrix} 6 & 9 \\ 1 & 2 \end{bmatrix} \\ -\det\begin{bmatrix} 1 & -4 \\ 2 & 1 \end{bmatrix} & +\det\begin{bmatrix} 3 & -4 \\ 1 & 1 \end{bmatrix} & -\det\begin{bmatrix} 3 & 1 \\ 1 & 2 \end{bmatrix} \\ +\det\begin{bmatrix} 1 & -4 \\ 9 & -2 \end{bmatrix} & -\det\begin{bmatrix} 3 & -4 \\ 6 & -2 \end{bmatrix} & +\det\begin{bmatrix} 3 & 1 \\ 6 & 9 \end{bmatrix} \end{bmatrix}^{T}$$

$$= \begin{bmatrix} 13 & -8 & 3 \\ -9 & 7 & -5 \\ 34 & -18 & 21 \end{bmatrix}^{T} = \begin{bmatrix} 13 & -9 & 34 \\ -8 & 7 & -18 \\ 3 & -5 & 21 \end{bmatrix}.$$

A direct multiplication now establishes that

$$A\,\text{Adj}(A) = 19I\,.$$

It follows that $\det(A) = 19$ and that the inverse of A is

$$A^{-1} = \frac{1}{19}\begin{bmatrix} 13 & -9 & 34 \\ -8 & 7 & -18 \\ 3 & -5 & 21 \end{bmatrix}. \quad \blacksquare$$

We conclude this section by stating a classical formula for the solution of a system of linear equations in terms of determinants.

Theorem 2.5

CRAMER'S RULE: If A is a nonsingular $n \times n$ matrix, then the unique solution of the linear system $AX = K$ is given by

$$x_i = \frac{\det(A_i)}{\det(A)}; \quad i = 1, 2, \ldots, n\,,$$

where A_i is A with its ith column replaced by K.

The proof of this theorem amounts to writing the solution of the system as $X = A^{-1}K$ and using the adjoint formula for the inverse; the details are left to the exercises.

Example 3

Use Cramer's rule to solve the system $AX = K$ if

$$A = \begin{bmatrix} 1 & -2 & 3 \\ 3 & 0 & 4 \\ 2 & -1 & -2 \end{bmatrix} \quad \text{and} \quad K = \begin{bmatrix} 1 \\ 0 \\ -2 \end{bmatrix}.$$

$$x_1 = \frac{\det(A_1)}{\det(A)} = \frac{\det\begin{bmatrix} 1 & -2 & 3 \\ 0 & 0 & 4 \\ -2 & -1 & -2 \end{bmatrix}}{\det\begin{bmatrix} 1 & -2 & 3 \\ 3 & 0 & 4 \\ 2 & -1 & -2 \end{bmatrix}} = \frac{20}{-33} = \frac{-20}{33}$$

$$x_2 = \frac{\det(A_2)}{\det(A)} = \frac{\det\begin{bmatrix} 1 & 1 & 3 \\ 3 & 0 & 4 \\ 2 & -2 & -2 \end{bmatrix}}{-33} = \frac{4}{-33} = \frac{-4}{33}$$

$$x_3 = \frac{\det(A_3)}{\det(A)} = \frac{\det\begin{bmatrix} 1 & -2 & 1 \\ 3 & 0 & 0 \\ 2 & -1 & -2 \end{bmatrix}}{-33} = \frac{-15}{-33} = \frac{15}{33}. \quad \blacksquare$$

It cannot be emphasized too strongly that the results of the last two theorems do not provide efficient computational tools. To solve an $n \times n$ system using Cramer's rule requires the evaluation of $n + 1$ $n \times n$ determinants. The evaluation of a single $n \times n$ determinant, by the most efficient method available, is roughly equivalent to the solution of the entire system by the elimination methods of Chapter 1. Thus Cramer's rule would require nearly $n + 1$ times as much work as Gaussian elimination. Similarly, the adjoint formula for the inverse is nearly n^2 times as much work as the Gauss-Jordan method of Chapter 1. Moreover, if A is singular, then these methods provide no information about multiple solutions while the elimination methods of Chapter 1 find all solutions and flag the inconsistent cases. *These results have important theoretical applications, but they are not to be used as the basis for general-purpose computational algorithms.*

There are many applications of determinants to problems in geometry. In particular note the results in Exercises 17 and 18 of Section 2.2 and Exercises 8, 9, 10, and 17 of Section 2.3. We conclude this section by showing how determinants can be used to compute areas. Additional applications to geometry will be considered in the next chapter.

Example 4

Show that the area of the triangle with vertices at the points (x_1, y_1), (x_2, y_2), and (x_3, y_3) is

$$\pm \frac{1}{2} \det \begin{bmatrix} 1 & 1 & 1 \\ x_1 & x_2 & x_3 \\ y_1 & y_2 & y_3 \end{bmatrix}.$$

From Figure 2.3 we see that

$$\text{Area}(P_1 P_2 P_3) = \text{Area}(Q_3 Q_2 P_2 P_3) - \text{Area}(Q_3 Q_1 P_1 P_3)$$
$$- \text{Area}(Q_1 Q_2 P_2 P_1),$$

where $\text{Area}(A, B, C)$ is the area enclosed by the lines connecting points A, B, and C. Using the standard formula for the area of a trapezoid we have

$$\text{Area}(P_1 P_2 P_3) = \frac{(x_2 - x_3)(y_2 + y_3)}{2} - \frac{(x_1 - x_3)(y_3 + y_1)}{2}$$
$$- \frac{(x_2 - x_1)(y_1 + y_2)}{2}$$

$$= \frac{1}{2}\{x_2 y_2 + x_2 y_3 - x_3 y_2 - x_3 y_3 - x_1 y_3 - x_1 y_1$$
$$+ x_3 y_3 + x_3 y_1 - x_2 y_1 - x_2 y_2 + x_1 y_1 + x_1 y_2\}$$

$$= \frac{1}{2}\{(x_2 y_3 - x_3 y_2) - (x_1 y_3 - x_3 y_1)$$
$$+ (x_1 y_2 - x_2 y_1)\}$$

$$= \frac{1}{2} \det \begin{bmatrix} 1 & 1 & 1 \\ x_1 & x_2 & x_3 \\ y_1 & y_2 & y_3 \end{bmatrix}.$$

Note that if P_1 is above the line $P_2 P_3$, then the area is the negative of the determinant. ■

Exercise Set 2.3

1. Use the adjoint formula to find the inverses of the following matrices.

$$A = \begin{bmatrix} -1 & 1 & 3 \\ 2 & 1 & -1 \\ 4 & 2 & 2 \end{bmatrix} \qquad C = \begin{bmatrix} 1 & -1 & 2 \\ -3 & -3 & 3 \\ 4 & 0 & 5 \end{bmatrix}$$

$$D = \begin{bmatrix} 1 & 2 & 3 & 4 \\ 1 & -1 & 1 & 2 \\ -1 & 1 & -1 & 11 \\ -1 & 1 & 2 & 3 \end{bmatrix} \qquad E = \begin{bmatrix} 1 & 0 & 2 & -1 \\ 3 & -2 & 6 & 4 \\ 2 & 2 & -5 & 6 \\ 5 & 4 & 3 & 0 \end{bmatrix}$$

$$G = \begin{bmatrix} 1 & 1 & 2 & 1 \\ -1 & 1 & 0 & 1 \\ 2 & 1 & 1 & 0 \\ 1 & 3 & 1 & 0 \end{bmatrix}$$

2. Solve the system $CX = [1 \quad 2 \quad 3]^T$ by Cramer's rule, where

$$C = \begin{bmatrix} 1 & -1 & 2 \\ -3 & -3 & 3 \\ 4 & 0 & 5 \end{bmatrix} \text{ is the matrix of Exercise 3 in Section 2.2.}$$

3. Use Cramer's rule to find x_3 (the third component of the matrix X) for the system

$$GX = \begin{bmatrix} 1 \\ 0 \\ 1 \\ 0 \end{bmatrix}, \text{ where } G = \begin{bmatrix} 1 & 1 & 2 & 1 \\ -1 & 1 & 0 & 1 \\ 2 & 1 & 1 & 0 \\ 1 & 3 & 1 & 0 \end{bmatrix}.$$

✓ 4. (a) Prove that if $\det(A) = \pm 1$ and all the entries of A are integers, then all the entries of A^{-1} are integers.

   (b) Let $AX = B$ be a system of n linear equations in n unknowns with integer coefficients and integer constants. Prove that if $\det(A) = \pm 1$, then the solution X has integer entries.

5. Use Cramer's rule to solve for z without solving for x, y, and w.

$$\begin{aligned} 4x + y + z + w &= 6 \\ 7x + 3y - 5z + 8w &= -3 \\ 3x + 7y - z + w &= 1 \\ x + y + z + 2w &= 3 \end{aligned}$$

6. (a) If A is partitioned as $A = \begin{bmatrix} A_{11} & 0 \\ 0 & A_{22} \end{bmatrix}$, where A_{11} and A_{22} are square,

   show that $\det(A) = \det(A_{11}) \det(A_{22})$.

   (*Hint:* Verify the factorization $A = \begin{bmatrix} A_{11} & 0 \\ 0 & I \end{bmatrix} \begin{bmatrix} I & 0 \\ 0 & A_{22} \end{bmatrix}$, and then use Property 2.9.)

   (b) Show that $\det \begin{bmatrix} A_{11} & A_{12} \\ 0 & A_{22} \end{bmatrix} = \det(A_{11}) \det(A_{22})$.

7. If A and D are square and $\det(A) \neq 0$, verify that

$$\begin{bmatrix} I & 0 \\ -CA^{-1} & I \end{bmatrix} \begin{bmatrix} A & B \\ C & D \end{bmatrix} = \begin{bmatrix} A & B \\ 0 & D - CA^{-1}B \end{bmatrix},$$

   and then show that

$$\det \begin{bmatrix} A & B \\ C & D \end{bmatrix} = \det(A) \det(D - CA^{-1}B).$$

8. (a) Prove that the equation of the line through the distinct points (a_1, b_1) and (a_2, b_2) can be written as

$$\det \begin{bmatrix} x & y & 1 \\ a_1 & b_1 & 1 \\ a_2 & b_2 & 1 \end{bmatrix} = 0.$$

   (b) Recall that any equation of the form $ax + by + cz = d$ represents a plane in space and then prove that the equation of the plane through the noncollinear points (a_1, b_1, c_1), (a_2, b_2, c_2), and (a_3, b_3, c_3) can be written as

$$\det \begin{bmatrix} x & y & z & 1 \\ a_1 & b_1 & c_1 & 1 \\ a_2 & b_2 & c_2 & 1 \\ a_3 & b_3 & c_3 & 1 \end{bmatrix} = 0.$$

9. (a) Prove that (x_1, y_1), (x_2, y_2), and (x_3, y_3) are collinear points if and only if $\det \begin{bmatrix} x_1 & y_1 & 1 \\ x_2 & y_2 & 1 \\ x_3 & y_3 & 1 \end{bmatrix} = 0.$

(b) Show that the points (x_1, y_1, z_1), (x_2, y_2, z_2), (x_3, y_3, z_3) and (x_4, y_4, z_4) are coplanar (that is, lie on the same plane) if and only if

$$\det \begin{bmatrix} x_1 & y_1 & z_1 & 1 \\ x_2 & y_2 & z_2 & 1 \\ x_3 & y_3 & z_3 & 1 \\ x_4 & y_4 & z_4 & 1 \end{bmatrix} = 0.$$

10. For the triangle below, use trigonometry to show that

$$b \cos \psi + c \cos \theta = a$$
$$c \cos \theta + a \cos \psi = b$$
$$a \cos \theta + b \cos \theta = c$$

and then apply Cramer's rule to deduce the law of cosines:

$$\cos \theta = \frac{b^2 + c^2 - a^2}{2bc}.$$

11. For the following matrices find the size of the largest submatrix with nonzero determinant.

$$A = \begin{bmatrix} 1 & -1 & 1 & 1 \\ 1 & 2 & -1 & -1 \\ 2 & -2 & 1 & -1 \\ 0 & -3 & -1 & -1 \end{bmatrix} \quad \text{and} \quad N = \begin{bmatrix} 0 & 0 & 0 & 0 \\ 1 & 0 & 0 & 0 \\ 0 & 1 & 0 & 0 \\ 0 & 0 & 1 & 0 \end{bmatrix}$$

12. (a) Show that the Gaussian elimination method for evaluating $\det(A)$ requires $(n^3 + 2n - 3)/3$ multiplications if A is $n \times n$.

(b) Find the number of multiplications required to solve an $n \times n$ system by Cramer's rule.

(c) Find the number of multiplications required to find the inverse of an $n \times n$ matrix using the adjoint formula.

13. Show that $\det(V) = (x_2 - x_1)(x_3 - x_1)(x_4 - x_1)(x_3 - x_2)(x_4 - x_3)$ if

$$V = \begin{bmatrix} 1 & 1 & 1 & 1 \\ x_1 & x_2 & x_3 & x_4 \\ x_1^2 & x_2^2 & x_3^2 & x_4^2 \\ x_1^3 & x_2^3 & x_3^3 & x_4^3 \end{bmatrix}.$$

This matrix is called the VANDERMONDE MATRIX.

[*Hint:* Use the elementary row operations $(R_4 \leftarrow -x_1 R_3 + R_4)$,

$(R_3 \leftarrow -x_1R_2 + R_3)$, $(R_2 \leftarrow -x_1R_1 + R_2)$ followed by an expansion by the first column.]

14. Evaluate $\det(\lambda I - C)$, where $C = \begin{bmatrix} 0 & 0 & 0 & -a_0 \\ 1 & 0 & 0 & -a_1 \\ 0 & 1 & 0 & -a_2 \\ 0 & 0 & 1 & -a_3 \end{bmatrix}$.

    [*Hint:* Use the elementary row operations $(R_1 \leftarrow \lambda R_2 + R_1)$, $(R_1 \leftarrow \lambda^2 R_3 + R_1)$, and $(R_1 \leftarrow \lambda^3 R_4 + R_1)$ followed by an expansion by the first row.]

15. Use Theorem 2.3 to find all values of λ for which $(\lambda I - A)$ is singular if A is the given matrix.

    (a) $A = \begin{bmatrix} 5 & 1 \\ 4 & 8 \end{bmatrix}$ **(b)** $A = \begin{bmatrix} 2 & 1 & 1 \\ 2 & 3 & 2 \\ 1 & 1 & 2 \end{bmatrix}$

16. Use the result of Example 4 on page 111 to find the area of the triangle with the following vertices.

    (a) $(1, 1)$, $(4, 5)$, and $(3, 2)$

    (b) $(1, 1)$, $(4, 5)$, and $(3, 4)$

17. What do the following formulas represent?

    (a) $\det \begin{bmatrix} 1 & x & y & x^2 + y^2 \\ 1 & x_1 & y_1 & x_1^2 + y_1^2 \\ 1 & x_2 & y_2 & x_2^2 + y_2^2 \\ 1 & x_3 & y_3 & x_3^2 + y_3^2 \end{bmatrix} = 0$

    (b) $\det \begin{bmatrix} 1 & x & y & x^2 & y^2 \\ 1 & x_1 & y_1 & x_1^2 & y_1^2 \\ 1 & x_2 & y_2 & x_2^2 & y_2^2 \\ 1 & x_3 & y_3 & x_3^2 & y_3^2 \\ 1 & x_4 & y_4 & x_4^2 & y_4^2 \end{bmatrix} = 0$

    (c) $\det \begin{bmatrix} 1 & x & y & x^2 & x^3 \\ 1 & x_1 & y_1 & x_1^2 & x_1^3 \\ 1 & x_2 & y_2 & x_2^2 & x_2^3 \\ 1 & x_3 & y_3 & x_3^2 & x_3^3 \\ 1 & x_4 & y_4 & x_4^2 & x_4^3 \end{bmatrix} = 0$

2.4

*Eigenvalues and Eigenvectors**

Many applications (such as those in Exercises 20 and 21 of this section) require that one find a nonzero matrix X such that $AX = \lambda X$, where A is a given $n \times n$ matrix and λ is a scalar. This problem is called the algebraic eigenvalue problem and is the second most fre-

---

*This material can be delayed until the beginning of Chapter 5 if desired.

quently encountered matrix problem; the first is the solution of a system of linear equations. Additional motivation for this problem will be provided in Chapter 5, where we will use the techniques of this section to find a nonsingular matrix P such that $P^{-1}AP$ is a diagonal matrix. The computational tools are all available now and provide an important application which uses both determinants and the techniques of Chapter 1. Some of you may find immediate use for this material in other areas. You will no doubt note that many of these ideas have already been introduced in earlier exercises.

Definition 2.5

Let A be any $n \times n$ matrix. A nonzero $n \times 1$ matrix X such that $AX = \lambda X$ is called an EIGENVECTOR of A and the scalar λ is called an EIGENVALUE of A, associated with X.

Example 1

For $A = \begin{bmatrix} 2 & 6 \\ 1 & 3 \end{bmatrix}$, $X = \begin{bmatrix} 4 \\ 2 \end{bmatrix}$, and $Z = \begin{bmatrix} 1 \\ 1 \end{bmatrix}$, we have

$$AX = \begin{bmatrix} 20 \\ 10 \end{bmatrix} = 5 \begin{bmatrix} 4 \\ 2 \end{bmatrix} = 5X \quad \text{and} \quad AZ = \begin{bmatrix} 8 \\ 4 \end{bmatrix} \neq \lambda \begin{bmatrix} 1 \\ 1 \end{bmatrix},$$

so X is an eigenvector of A associated with the eigenvalue $\lambda = 5$ and Z is not an eigenvector of A. We leave it to you to verify that

$Y = \begin{bmatrix} 3 \\ -1 \end{bmatrix}$ is an eigenvector of A associated with $\lambda = 0$. ■

The defining equation, $AX = \lambda X$, is clearly equivalent to $\lambda IX = AX$ or to $(\lambda I - A)X = 0$. This last equation sheds some light on the problem of how to find eigenvectors, since it is now clear that eigenvectors are solutions of certain homogeneous systems of linear equations. If we knew the eigenvalues, we could solve the homogeneous systems $(\lambda I - A)X = 0$ for the eigenvectors. If we knew the eigenvectors, then we could find the eigenvalues by comparing the entries of X and AX. If we know neither, and this is the usual case, then we can use Theorem 1.9 and Theorem 2.3 to observe the following.

The homogeneous system $(\lambda I - A)X = 0$ has a nonzero solution if and only if $\det(\lambda I - A) = 0$, that is, the matrix $\lambda I - A$ is singular. The scalar λ is an eigenvalue of A if and only if $\det(\lambda I - A) = 0$.

For the matrix A of Example 1 we see that

$$\det(5I - A) = \det\begin{bmatrix} 3 & -6 \\ -1 & 2 \end{bmatrix} = 0 \quad \text{and}$$

$$\det(0I - A) = \det\begin{bmatrix} -2 & -6 \\ -1 & -3 \end{bmatrix} = 0.$$

These calculations show that $\lambda = 5$ and $\lambda = 0$ are eigenvalues of A. Note that $\lambda = 3$ is not an eigenvalue of A since

$$\det(3I - A) = \det\begin{bmatrix} -1 & -6 \\ -1 & 0 \end{bmatrix} = -6 \neq 0$$

If we think of λ as a parameter, then $\det(\lambda I - A)$ is a polynomial in λ called the CHARACTERISTIC POLYNOMIAL of A (recall Exercises 14 and 15 of Section 2.3 and 3 of Section 2.1). For the matrix A of Example 1,

$$\det(\lambda I - A) = \det\begin{bmatrix} \lambda - 2 & -6 \\ -1 & \lambda - 3 \end{bmatrix}$$

$$= (\lambda - 2)(\lambda - 3) - 6$$

$$= \lambda^2 - 5\lambda.$$

The problem of finding the eigenvalues and eigenvectors of a matrix A can thus be reduced to the following three steps.

---

(1) Find the characteristic polynomial $c(\lambda) = \det(\lambda I - A)$.
(2) Find the eigenvalues of A by finding the roots λ_j of the characteristic equation $c(\lambda) = 0$.
(3) For each eigenvalue, find the associated eigenvector of A by solving the homogeneous systems $(\lambda_j I - A)X = 0$.

If the matrix A is large, then each of these three steps can involve a substantial amount of calculation. We now illustrate these calculations for a 3 × 3 matrix.

Example 2

Find the eigenvalues and eigenvectors of

$$A = \begin{bmatrix} 0 & 1 & 0 \\ 0 & 0 & 1 \\ 2 & 1 & -2 \end{bmatrix}.$$

Step 1: The characteristic polynomial of A is

$$\det(\lambda I - A) = \det\left(\lambda \begin{bmatrix} 1 & 0 & 0 \\ 0 & 1 & 0 \\ 0 & 0 & 1 \end{bmatrix} - \begin{bmatrix} 0 & 1 & 0 \\ 0 & 0 & 1 \\ 2 & 1 & -2 \end{bmatrix}\right)$$

$$= \det \begin{bmatrix} \lambda & -1 & 0 \\ 0 & \lambda & -1 \\ -2 & -1 & \lambda + 2 \end{bmatrix}$$

$$= \lambda \det \begin{bmatrix} \lambda & -1 \\ -1 & \lambda + 2 \end{bmatrix} + \det \begin{bmatrix} 0 & -1 \\ -2 & \lambda + 2 \end{bmatrix}$$

$$+ 0 \det \begin{bmatrix} 0 & \lambda \\ -2 & -1 \end{bmatrix}$$

$$= \lambda(\lambda^2 + 2\lambda - 1) - 2$$

$$= \lambda^3 + 2\lambda^2 - \lambda - 2 .$$

Step 2: From the easily verified factorization

$$\lambda^3 + 2\lambda^2 - \lambda - 2 = (\lambda - 1)(\lambda + 1)(\lambda + 2),$$

we see that the roots of the characteristic equation $c(\lambda) = 0$ are $\lambda_1 = 1$, $\lambda_2 = -1$, and $\lambda_3 = -2$. These roots are the eigenvalues of A.

Step 3: In order to find an eigenvector for $\lambda_1 = 1$ we solve the homogeneous system $(\lambda_1 I - A)X = (I - A)X = 0$, for which the augmented matrix and reduction to reduced row echelon form follow:

$$\begin{bmatrix} 1 & -1 & 0 & 0 \\ 0 & 1 & -1 & 0 \\ -2 & -1 & 3 & 0 \end{bmatrix} \longrightarrow \begin{bmatrix} 1 & -1 & 0 & 0 \\ 0 & 1 & -1 & 0 \\ 0 & -3 & 3 & 0 \end{bmatrix} \longrightarrow \begin{bmatrix} 1 & 0 & -1 & 0 \\ 0 & 1 & -1 & 0 \\ 0 & 0 & 0 & 0 \end{bmatrix}.$$

The system equivalent to this final matrix is

$$x_1 - x_3 = 0$$
$$x_2 - x_3 = 0 .$$

There are many solutions to this system; if we set $x_3 = 1$, we find $x_1 = x_2 = 1$ so that

$$X_1 = \begin{bmatrix} 1 & 1 & 1 \end{bmatrix}^T \text{ is an eigenvector of } A \text{ for } \lambda_1 = 1 .$$

Note that cX_1 is also an eigenvector, for $\lambda = 1$, for any $c \neq 0$. In order to find an eigenvector of A, for $\lambda_2 = -1$, we solve the system $(-I - A)X = 0$ via the row reduction

$$\begin{bmatrix} -1 & -1 & 0 & 0 \\ 0 & -1 & -1 & 0 \\ -2 & -1 & 1 & 0 \end{bmatrix} \longrightarrow \begin{bmatrix} 1 & 1 & 0 & 0 \\ 0 & -1 & -1 & 0 \\ 0 & 1 & 1 & 0 \end{bmatrix}$$

$$\longrightarrow \begin{bmatrix} 1 & 0 & -1 & 0 \\ 0 & 1 & 1 & 0 \\ 0 & 0 & 0 & 0 \end{bmatrix} \longrightarrow \begin{cases} x_1 - x_3 = 0 \\ x_2 + x_3 = 0 \end{cases}.$$

If we set $x_3 = 1$, we have $x_1 = 1$ and $x_2 = -1$ as one solution. Thus $X_2 = [1 \quad -1 \quad 1]^T$ is an eigenvector of A, for $\lambda_2 = -1$. Similar calculations, which you should carry out in detail, show that $X_3 = [1 \quad -2 \quad 4]^T$ is an eigenvector of A, for $\lambda_3 = -2$. ■

Several comments about the solution process illustrated in Example 2 are in order.

(1) For small matrices, it is best to evaluate the determinant, which defines the characteristic equation, by repeated use of the Laplace expansion (choose the row with the most zeros) rather than using elementary row operations, as recommended for matrices of constants. If A is $n \times n$, then the characteristic equation is a polynomial equation of degree n, of the form

$$\det(\lambda I - A) = \lambda^n - \text{tr}(A)\lambda^{n-1} + \ldots + (-1)^n \det(A) = 0.$$

The observation that the coefficient of λ^{n-1} is $-\text{tr}(A) = -(a_{11} + a_{22} + \ldots + a_{nn})$ is a useful check.* Note that the characteristic polynomial is always a MONIC polynomial; that is, the coefficient of λ^n is always 1.

(2) Finding all the roots of a polynomial equation is in general quite difficult since no general formulas exist for n greater than 4; for $n = 3$ and $n = 4$ these formulas are very complicated, difficult to use, and almost impossible to remember. If a monic polynomial equation with integer coefficients has any rational roots, then they must be integers that are factors of the constant term. This restriction provides a reasonably small set of candidates for roots; these candidates may then be checked individually. For hand calculations, the algebraic technique of synthetic division is an efficient way of doing the polynomial evaluations necessary to check the various candidates. If you are not familiar with synthetic division, please work Exercise 18 before you proceed further. If this "guess and check" method does not locate the roots, then one must resort to numerical approximation methods such as the bisection method or Newton's method (see Exercise 24 of this section).

---

*The trace function was defined in Exercise 11 of Section 2.1.

(3) Eigenvectors are solutions of homogeneous systems and as such are never unique. The set of eigenvectors of A, associated with a given eigenvalue, is closed under addition and scalar multiplication. For small problems with integer data, it is frequently possible to choose the eigenvectors to have integer entries.

(4) Polynomial equations with real coefficients may very well have some complex roots. If there are complex roots, then they occur in conjugate pairs of the form $a \pm bi$ where the complex number $i = \sqrt{-1}$ satisfies $i^2 = -1$. If an eigenvalue is complex, then the corresponding eigenvector will also have complex entries. We will, for computational ease, restrict our examples and exercises largely to problems with real eigenvalues, but you should be alerted to the fact that many important applications, including most oscillatory phenomena, involve complex eigenvalues.

Example 3

Find the eigenvalues and all eigenvectors of

$$B = \begin{bmatrix} 5 & 4 & 2 \\ 4 & 5 & 2 \\ 2 & 2 & 2 \end{bmatrix}.$$

The characteristic polynomial is $\det(\lambda I - B)$

$$= \det \begin{bmatrix} \lambda - 5 & -4 & -2 \\ -4 & \lambda - 5 & -2 \\ -2 & -2 & \lambda - 2 \end{bmatrix}$$

$$= (\lambda - 5)[(\lambda - 5)(\lambda - 2) - 4] + 4[-4(\lambda - 2) - 4]$$
$$- 2[8 + 2(\lambda - 5)]$$

$$= \lambda^3 - 12\lambda^2 + 21\lambda - 10 = c(\lambda).$$

The only rational possibilities for the eigenvalues are factors of the constant term: ± 1, ± 2, ± 5, and ± 10. From the synthetic divisions

$$
\begin{array}{r}
10)\ 1 \quad -12 \quad 21 \quad -10 \\
10 \quad -20 \quad 10 \\
\hline
1 \quad -2 \quad 1 \quad 0
\end{array}
\qquad \text{and} \qquad
\begin{array}{r}
2)\ 1 \quad -12 \quad 21 \quad -10 \\
2 \quad -20 \quad 2 \\
\hline
1 \quad -10 \quad 1 \quad -8
\end{array}
$$

we see that $c(10) = 0$ and $c(2) = -8$. Thus 10 is an eigenvalue of B but 2 is not. Moreover, from the first division we can see that the characteristic equation of B factors as

$$\lambda^3 - 12\lambda^2 + 21\lambda - 10 = (\lambda - 10)(\lambda^2 - 2\lambda + 1)$$
$$= (\lambda - 10)(\lambda - 1)(\lambda - 1) = 0$$

The eigenvalues of B are thus $\lambda_1 = 10$ and $\lambda_2 = \lambda_3 = 1$. The augmented matrix for the system $(10I - B)X = 0$ is

$$\begin{bmatrix} 5 & -4 & -2 & 0 \\ -4 & 5 & -2 & 0 \\ -2 & -2 & 8 & 0 \end{bmatrix} \longrightarrow \begin{bmatrix} 1 & 1 & -4 & 0 \\ 0 & 9 & -18 & 0 \\ 0 & 0 & 0 & 0 \end{bmatrix}$$

$$\longrightarrow \begin{bmatrix} 1 & 0 & -2 & 0 \\ 0 & 1 & -2 & 0 \\ 0 & 0 & 0 & 0 \end{bmatrix} \longrightarrow \begin{cases} x_1 - 2x_3 = 0 \\ x_2 - 2x_3 = 0 \end{cases}.$$

From this row reduction we see that the general solution of the system $(10I - B)X = 0$; that is, the most general eigenvector of B associated with the eigenvalue 10, is

$$X = \begin{bmatrix} 2c \\ 2c \\ c \end{bmatrix} = c \begin{bmatrix} 2 \\ 2 \\ 1 \end{bmatrix}.$$

In order to find the eigenvectors associated with the repeated eigenvalue $\lambda = 1$ we consider the system $(I - B)X = 0$ for which the augmented matrix and row reduction are as follows:

$$\begin{bmatrix} -4 & -4 & -2 & 0 \\ -4 & -4 & -2 & 0 \\ -2 & -2 & -1 & 0 \end{bmatrix} \longrightarrow \begin{bmatrix} 2 & 2 & 1 & 0 \\ 0 & 0 & 0 & 0 \\ 0 & 0 & 0 & 0 \end{bmatrix} \longrightarrow 2x_1 + 2x_2 + x_3 = 0.$$

The general solution of this 1×3 system, which involves two arbitrary constants, is

$$X = \begin{bmatrix} c_1 \\ c_2 \\ -2c_1 - 2c_2 \end{bmatrix} = c_1 \begin{bmatrix} 1 \\ 0 \\ -2 \end{bmatrix} + c_2 \begin{bmatrix} 0 \\ 1 \\ -2 \end{bmatrix}$$

and these are all of the eigenvectors of B for $\lambda = 1$. ■

Note that in Example 3 we found a set of eigenvectors with two degrees of freedom, associated with the double eigenvalue $\lambda = 1$. *This frequently happens, but is not always the case.* Note that the matrix

$$J = \begin{bmatrix} 3 & 1 & 0 \\ 0 & 3 & 1 \\ 0 & 0 & 3 \end{bmatrix}$$

has $\lambda = 3$ as a triple root, but the set of eigenvectors has only one degree of freedom (see Exercise 22). We will return to the study of eigenvalues and eigenvectors in Chapter 5.

Exercise Set 2.4

1. Determine if either X or Y is an eigenvector for A and if so, determine the eigenvalue.

$$\text{(a)}A = \begin{bmatrix} 11 & -9 \\ 4 & -2 \end{bmatrix}, \quad X = \begin{bmatrix} 1 \\ 1 \end{bmatrix}, \quad Y = \begin{bmatrix} 1 \\ 2 \end{bmatrix}$$

$$\text{(b)}A = \begin{bmatrix} 2 & 1 & 1 \\ 2 & 3 & 2 \\ 1 & 1 & 2 \end{bmatrix}, \quad X = \begin{bmatrix} -1 \\ 1 \\ 0 \end{bmatrix}, \quad Y = \begin{bmatrix} -1 \\ 0 \\ 1 \end{bmatrix}$$

For the matrices in Exercises 2–10 find the characteristic equation, the real eigenvalues, and the corresponding eigenvectors.

2. $\begin{bmatrix} 1 & 3 \\ -1 & 5 \end{bmatrix}$ 3. $\begin{bmatrix} 16 & -14 \\ -14 & 9 \end{bmatrix}$ 4. $\begin{bmatrix} 2 & 1 & 1 \\ 2 & 3 & 2 \\ 1 & 1 & 2 \end{bmatrix}$

5. $\begin{bmatrix} 1 & 0 & -4 \\ 0 & 5 & 4 \\ -4 & 4 & 3 \end{bmatrix}$ 6. $\begin{bmatrix} 5 & -1 & 1 \\ -1 & 5 & -1 \\ 1 & -1 & 5 \end{bmatrix}$ 7. $\begin{bmatrix} 2 & 1 & 0 \\ 0 & 2 & 1 \\ 2 & 1 & 0 \end{bmatrix}$

8. $\begin{bmatrix} -3 & 3 & -2 \\ 3 & 2 & -3 \\ -2 & -3 & -3 \end{bmatrix}$ 9. $\begin{bmatrix} -1 & -1 & -3 \\ -1 & 1 & 1 \\ -3 & 1 & -1 \end{bmatrix}$ 10. $\begin{bmatrix} 2 & 1 & 5 \\ 1 & -2 & -1 \\ 5 & -1 & 2 \end{bmatrix}$

11. Consider the matrices A of Exercises 2, 8, 9, and 10. Let P be a matrix whose columns are the eigenvectors of A. Compute $P^{-1}AP$.

12. Recall the definition of tr(A) from Exercise 11 of Section 2.1 and verify that $\det(\lambda I - A) = \lambda^n - \text{tr}(A)\lambda^{n-1} + \ldots + (-1)^n \det(A)$ for a general 3×3 matrix A.

13. What are the eigenvalues of
    (a) a diagonal matrix?
    (b) a triangular matrix?

14. Show that $\lambda = 0$ is an eigenvalue of a matrix A if and only if A is singular.

15. The characteristic polynomial of the matrix $B = \begin{bmatrix} 0 & 1 & 0 \\ 0 & 0 & 1 \\ 3 & -5 & 2 \end{bmatrix}$ is
    $\lambda^3 - 2\lambda^2 + 5\lambda - 3$. Show that $B^3 - 2B^2 + 5B - 3I = 0$.

16. Show that A and $B = P^{-1}AP$ have the same characteristic polynomial.

17. By multiplying the equation $AX = \lambda X$ on the left by A, show that if λ is an eigenvalue of A, then λ^2 is an eigenvalue of A^2. What is the corresponding eigenvector? What are the eigenvalues of A^3?

†18. Consider the polynomial function
    $$p(r) = a_k\lambda^k + \ldots + a_2\lambda^2 + a_1\lambda + a_0.$$

(a) Show that the value $p(r)$ may be evaluated by the algorithm

(1) $b_{k-1} = a_k$
(2) For $i = k - 2, k - 3, \ldots, 3, 2, 1$ $b_i = b_{i+1}r + a_{i+1}$
(3) $p(r) = b_0 r + a_0$.

(*Hint:* Rewrite $p(r)$ in the form

$p(r) = [(\ldots ([(a_k r + a_{k-1})r + a_{k-2}]r + a_{k-3})r + \ldots + a_2)r + a_1]r + a_0.)$

(b) Explain why this is an efficient way to evaluate $p(r)$ for a given number r, especially on a pocket calculator with at least one storage location.

(c) Show that $p(\lambda) = (\lambda - r)(b_{k-1}\lambda^{k-1} + \ldots + b_1\lambda + b_0) + p(r)$.

(d) Show that the algorithm can be arranged in the following way for hand calculation.

$$r)\ a_k \qquad a_{k-1} \quad a_{k-2} \quad \cdots \quad a_2 \quad a_1 \quad a_0$$
$$rb_{k-1} \quad rb_{k-2} \quad \cdots \quad rb_2 \quad rb_1 \quad rb_0$$
$$\overline{\qquad b_{k-1} \quad b_{k-2} \quad b_{k-3} \qquad b_1 \quad b_0 \quad rb_0 + a_0}$$

(e) This process is called SYNTHETIC DIVISION. For $p(\lambda) = \lambda^3 - 6\lambda^2 + 9\lambda - 4$, evaluate $p(3)$, $p(5)$, and $p(1)$. Find the quotient when $p(\lambda)$ is divided by $\lambda - 1$.

(f) Delete the subscripts on the b_i in the above algorithm in part (a). What does the resulting algorithm do?

19. Use Exercise 6 in Section 2.3 to find the characteristic polynomial of

$$A = \begin{bmatrix} 2 & 0 & 0 & 0 \\ 1 & 2 & 0 & 0 \\ 0 & 0 & 0 & 1 \\ 0 & 0 & -6 & 5 \end{bmatrix}.$$

†20. Let A be an $n \times n$ matrix and let $Y(t)$ be an $n \times 1$ matrix of unknown functions. Show that

$$Y(t) = Xe^{\lambda t} \text{ is a solution of } \frac{d}{dt}Y(t) = AY(t)$$

if and only if $AX = \lambda X$.

†21. As we saw in Example 5 of Section 1.5, population growth can be modeled by a matrix difference equation of the form

$$X^{k+1} = AX^k$$

where A is a fixed matrix, characteristic of the species and its environment, and X^k is the population distribution at the kth. time step. A population distribution X^k is said to be STABLE if $X^{k+1} = \lambda X^k$ for some scalar λ. Show that the population distribution X is stable if and only if X is an eigenvector of A.

22. Show that the characteristic polynomial of $J = \begin{bmatrix} 3 & 1 & 0 \\ 0 & 3 & 1 \\ 0 & 0 & 3 \end{bmatrix}$ is $(\lambda - 3)^3$

and that the set of eigenvectors of J has only one degree of freedom.

23. Show that if $P^{-1}AP = D$ is a diagonal matrix, then the columns of P are eigenvectors of A. (*Hint:* Look at the equation $AP = PD$ column by column.)

24. NEWTON'S METHOD for finding a solution of the equation $f(x) = 0$ is

(1) x_0 arbitrary;

(2) for $i = 0, 1, 2, 3, 4, \ldots$ $x_{i+1} = x_i - \dfrac{f(x_i)}{f'(x_i)}$.

The sequence $\{x_i\}$ will converge to a root of $f(x) = 0$ if x_0 is sufficiently close to a root. Normally the sequence will converge to the root which is nearest to x_0.

(a) Use Newton's method to find the roots of $x^2 + 5x - 11 = 0$.

(b) Use Newton's method to find the roots of $x^4 + 9x^3 + 11x^2 - 34x - 22 = 0$.

(c) Use Newton's method to find the eigenvalues of the matrices

$$A = \begin{bmatrix} 1 & 1 & 3 \\ 1 & -2 & 1 \\ 3 & 1 & 3 \end{bmatrix} \quad \text{and} \quad B = \begin{bmatrix} 2 & -1 & 1 \\ -1 & 2 & -1 \\ 1 & -1 & 1 \end{bmatrix}.$$

(d) Delete all the subscripts in the Newton algorithm. Show that the resulting algorithm is equivalent to the original. What does this implementation of the algorithm mean for storage in a computer program?

Chapter 2

Review Exercises

1. (a) Evaluate the determinants of

$$A = \begin{bmatrix} 1 & 1 & 2 & 1 \\ 1 & 1 & 0 & -1 \\ 0 & 1 & 1 & 2 \\ 0 & 3 & 1 & 1 \end{bmatrix} \quad \text{and} \quad B = \begin{bmatrix} 1 & 1 & 1 & 1 \\ 1 & 1 & 2 & 1 \\ 1 & 3 & 1 & 1 \\ 4 & 1 & 1 & 1 \end{bmatrix}.$$

(b) Find the entry in the $(3, 2)$ position of the adjoint of A.

2. Given $A = \begin{bmatrix} 7 & 4 \\ -1 & 2 \end{bmatrix}$, find the characteristic polynomial, the eigenvalues, and the eigenvectors.

3. Find the characteristic equation of $M = \begin{bmatrix} 1 & 1 & 1 \\ 0 & 2 & 1 \\ 1 & 0 & 1 \end{bmatrix}$.

4. Find the adjoint of $C = \begin{bmatrix} 3 & 2 & -1 \\ 1 & 6 & 3 \\ 2 & -4 & 0 \end{bmatrix}$ and use it to find C^{-1}.

5. Use Cramer's rule to solve the linear system

$$\begin{aligned} x - 2y + 3z &= 1 \\ 2x - y - 2z &= -2 \\ 3x \qquad + 4z &= 0. \end{aligned}$$

6. Show that the eigenvalues of $A = \begin{bmatrix} a & b \\ b & c \end{bmatrix}$ must be real numbers.

7. Are the following statements true or false?
   (a) $\det(AB) = \det(A) \det(B)$
   (b) $\det(A + B) = \det(A) + \det(B)$
   (c) $\det(kA) = k \det(A)$
   (d) $\det(kA) = k^n \det(A)$
   (e) $\det(A^{-1}) = \dfrac{1}{\det(A)}$
   (f) $\det(A^T) = \det(A)$
   (g) If A is row equivalent to I, then $\det(A) = 1$.
   (h) If $\det(A) = 0$, then A is nonsingular.
   (i) If $\det(A) = 0$, then $AX = 0$ has a nonzero solution.
   (j) $\det(\mathrm{Adj}(A)) = \det(A)^n$ $\quad n-1$

8. Let A be a matrix with the property that the sum of the elements in each row is 0. Show that $\det(A) = 0$.

9. Use the adjoint formula to show that the inverse of an invertible upper triangular matrix is also upper triangular.

10. Use Cramer's rule to solve for x_1 and y_1 in terms of x and y:
$$x = x_1 \cos \theta - y_1 \sin \theta$$
$$y = x_1 \sin \theta + y_1 \cos \theta.$$

11. Show that the area of the parallelogram with vertices at $(0,0)$, (a, b), (c, d), and $(a + c, b + d)$ is the absolute value of the determinant of the matrix $\begin{bmatrix} a & b \\ c & d \end{bmatrix}$.

12. (a) Evaluate $\det(B)$ given that $B = \begin{bmatrix} 1 & 2 & 3 & 1 \\ 2 & 5 & 3 & 2 \\ -3 & 9 & 2 & 1 \\ 1 & 6 & 5 & 2 \end{bmatrix}$.

   (b) Find the entry in the $(1, 4)$ position of the adjoint of B.

13. (a) Find the characteristic equation of $A = \begin{bmatrix} 1 & 0 & 1 \\ 0 & 1 & 2 \\ 1 & 2 & 5 \end{bmatrix}$.

   (b) Find an eigenvector of A associated with the eigenvalue $\lambda = 6$.

3

Vector Spaces

Many measurable quantities that occur in applications can be described by a single scalar; among these quantities are temperature, mass, height, and military time. Other quantities, like force, velocity, acceleration, wind, size of a matrix, and civilian time cannot be described by a single number; force, velocity, acceleration, and wind require the specification of both a magnitude and a direction, while the size of a matrix requires two numbers and civilian time requires a number and either an A.M. or P.M. specification. Quantities that require more than one scalar for their definition are called VECTOR QUANTITIES. As a mathematics student, you have most likely encountered this notion before, perhaps in an elementary physics course or in a beginning calculus course. In this chapter we treat vectors from an algebraic, rather than a geometric, point of view. The techniques and results of Chapters 1 and 2 will be

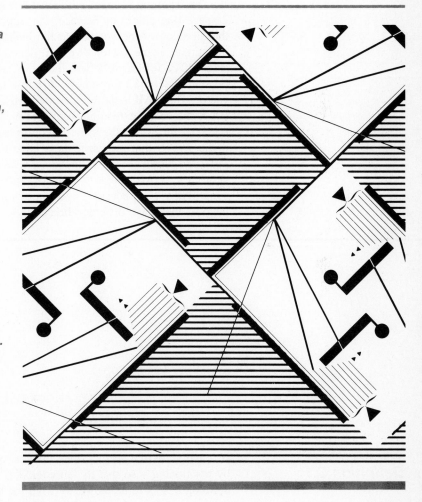

used extensively in Chapter 3, but the emphasis shifts from performing calculations to interpreting the results of calculations. We begin with a brief discussion of two- and three-dimensional vectors from the geometric point of view. This discussion will provide some geometric insight into, and a source of terminology for, the more general discussion that follows.

3.1 $\mathcal{R}^2$ and $\mathcal{R}^3$: Old Friends

We can represent vectors as directed line segments, or arrows, in either 2-space (the plane) or 3-space; the direction of the arrow represents the direction of the vector and the length of the arrow represents the magnitude of the vector. We will denote vectors in two ways: either as boldface lowercase letters like **u**, **v**, **w**, and **x**, or as two boldface uppercase letters representing the initial and terminal points of the vector. Thus, **AB** is the vector extending from point A to point B while **BA** is the vector extending from B to A. Two vectors are defined to be EQUAL if they have the same direction and the same magnitude, even if the arrows that represent them have different initial points (called *tails*) and terminal points (known as *heads*).* Hence, in Figure 3.1 the vectors **u** = **AB** and **v** = **CD** are equal, the vectors x_1 and x_2 are equal, but **v** $\neq$ x_1 because, even though the two vectors have the same lengths, they have different directions.

FIGURE 3.1

Definition 3.1

If **u** and **v** are two vectors, then the SUM of the two vectors is obtained by positioning **v** so that its tail coincides with the head of **u** and then defining **u** + **v** to be the vector from the initial point of **u** to the terminal point of **v**, as shown in Figure 3.2.

---

*See Exercise 20 of this section.

FIGURE 3.2

Definition 3.1 is frequently referred to as the PARALLELOGRAM LAW FOR VECTOR ADDITION because, as shown in Figure 3.3, **u + v** is the diagonal of the parallelogram with adjacent sides **u** and **v**. This diagram also shows that vector addition is commutative; that is,

$$\mathbf{u} + \mathbf{v} = \mathbf{v} + \mathbf{u}.$$

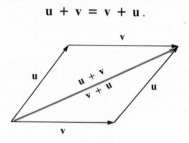

FIGURE 3.3

The geometric definition of vector addition given in Definition 3.1 can be motivated by the physical experiment pictured in Figure 3.4(a). In this experiment, force $\mathbf{F}_3$ balances the total effect of the forces $\mathbf{F}_1$ and $\mathbf{F}_2$. The magnitudes of the forces are read from the spring balances and the forces are plotted as in Figure 3.4(b), where the lengths of the arrows represent the magnitudes of the forces. If good equipment is used and the plotting is done carefully, you find that the force $\mathbf{F}_3$ is equal in magnitude and opposite in direction to the diagonal of the parallelogram whose sides are the vector representatives of the forces $\mathbf{F}_1$ and $\mathbf{F}_2$.

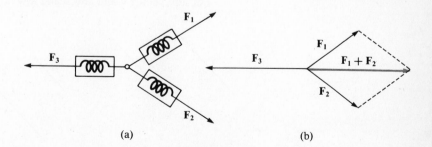

(a)   (b)

FIGURE 3.4

A second physical example of vector addition is described in Figure 3.5. If vector $\mathbf{w_T}$ represents the true wind and $\mathbf{w_B}$ represents the wind generated by the motion of a sailboat, then the *apparent* wind $\mathbf{w_A}$ (the wind actually encountered by the boat) can be represented by the sum of the vectors $\mathbf{w_T}$ and $\mathbf{w_B}$. Note that $\mathbf{w_A}$ has greater magnitude than $\mathbf{w_T}$ and that the length of $\mathbf{w_A}$ becomes much greater than the length of $\mathbf{w_T}$ as $\mathbf{w_B}$ increases. This observation helps to explain why iceboats are able to travel much faster than the true wind. Essentially the same phenomenon is observed when riding a bicycle or driving a car in a strong crosswind. In these cases, the wind (air resistance) felt by the cyclist or the car comes from much further forward than the true wind and is of greater strength.

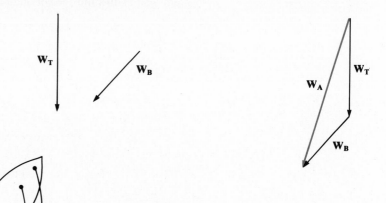

FIGURE 3.5

For any vector $\mathbf{v}$ we define $-\mathbf{v}$ to be the vector obtained from $\mathbf{v}$ by interchanging its head and tail. Clearly,

$$\mathbf{v} + (-\mathbf{v}) = \mathbf{0},$$

where $\mathbf{0}$ is the ZERO VECTOR, a vector with length 0.

For any vector $\mathbf{v}$ and any scalar k we define the vector $k\mathbf{v}$ to be the

FIGURE 3.6

vector whose length is $|k|$ times the length of **v** and whose direction is either the same as that of **v** if $0 < k$ or the same as that of $-\mathbf{v}$ if $k < 0$. The vector $k\mathbf{v}$ is called a SCALAR MULTIPLE of **v**. Figure 3.6 illustrates these ideas. Note that the vector $-\mathbf{v}$ is $(-1)\mathbf{v}$.

By the DIFFERENCE $\mathbf{v} - \mathbf{u}$ we mean $\mathbf{v} + (-\mathbf{u})$. Figure 3.7 shows two ways of obtaining $\mathbf{v} - \mathbf{u}$: the first method uses the parallelogram law to compute $\mathbf{v} - \mathbf{u} = \mathbf{v} + (-\mathbf{u})$ while the second method uses head-to-tail positioning to find $\mathbf{v} - \mathbf{u}$ as the vector that must be added to **u** to produce **v**.

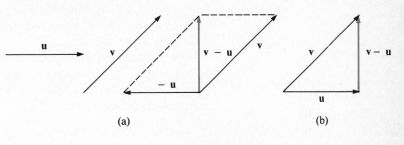

(a)　　　　　　　　　　　(b)

FIGURE 3.7

Example 1

Given the vectors **u**, **v**, and **w** of Figure 3.8(a), show how to determine constants a and b so that $\mathbf{w} = a\mathbf{u} + b\mathbf{v}$.

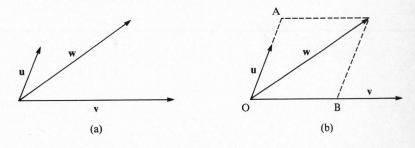

(a)　　　　　　　　　　　(b)

FIGURE 3.8

We first draw a line through the head of **w**, parallel to **v**, until it meets the extension of **u**. We next draw a line through the head of **w**, parallel to **u**, until it meets **v** extended, as shown in Figure 3.8(b). Since vectors $\mathbf{OA} + \mathbf{OB} = \mathbf{w} = a\mathbf{u} + b\mathbf{v}$, $\mathbf{OA}$ and $\mathbf{OB}$ are the vectors $a\mathbf{u}$ and $b\mathbf{v}$ respectively. The scalar a can now be determined by comparing the lengths of **u** and $a\mathbf{u}$; similarly, b can be determined by comparing the lengths of **v** and $b\mathbf{v}$. It is easy to see that similar arguments would work for other orientations of the three vectors. ■

The discussion in Example 1 is a geometric proof of the following assertion.

> If the vectors **u** and **v** are not parallel, then any vector in their plane can be written as a linear combination of **u** and **v**.

Later in this section we will provide algebraic techniques for handling this problem and will also give generalizations of the preceding assertion.

The LENGTH of the vector **v** will be denoted by $\|\mathbf{v}\|$. The length of the vector is also called the NORM of the vector. A vector of length one is called a UNIT VECTOR. Note that for any nonzero vector **u** the vector $(1/\|\mathbf{u}\|)\mathbf{u}$ is a unit vector.

We will now define a scalar-valued product of two vectors called the *dot product*. This product is useful in problems involving the angle between vectors, in projections of one vector onto another, and in computing lengths of vectors.

Definition 3.2

For any pair of vectors **u** and **v** we define the DOT PRODUCT by

$$\mathbf{u} \cdot \mathbf{v} = \|\mathbf{u}\|\,\|\mathbf{v}\|\cos\theta,$$

where θ is the smaller angle between **u** and **v**, as shown in Figure 3.9.

FIGURE 3.9

Since $\cos 90 = \cos \pi/2 = 0$, it follows that the vectors **u** and **v** are perpendicular if and only if $\mathbf{u} \cdot \mathbf{v} = 0$. Moreover, since $\cos 0 = 1$ it follows that $\mathbf{u} \cdot \mathbf{u} = \|\mathbf{u}\|^2$ or that

$$\|\mathbf{u}\| = \sqrt{(\mathbf{u} \cdot \mathbf{u})}.$$

The dot product is very useful in problems involving projections. Let **u** and **v** be positioned as shown in Figure 3.10 and let **w** be the orthogonal projection of **u** on **v**. Then $\cos\theta = \|\mathbf{w}\|/\|\mathbf{u}\|$, so

$$\|\mathbf{w}\| = \|\mathbf{u}\|\cos\theta = \frac{\mathbf{u} \cdot \mathbf{v}}{\|\mathbf{v}\|}$$

is the length of the orthogonal projection of the vector **u** on **v**. More-over, the vector **w** is $\|\mathbf{w}\|$ times the unit vector in the direction of **v**; that is,

$$\mathbf{w} = \|\mathbf{w}\| \left(\frac{\mathbf{v}}{\|\mathbf{v}\|} \right) = \left(\frac{\mathbf{u} \cdot \mathbf{v}}{\|\mathbf{v}\|} \right) \left(\frac{\mathbf{v}}{\|\mathbf{v}\|} \right) = \left(\frac{\mathbf{u} \cdot \mathbf{v}}{\mathbf{v} \cdot \mathbf{v}} \right) \mathbf{v}$$

is the VECTOR PROJECTION of **u** onto **v**.

FIGURE 3.10

All of the ideas discussed above for two-dimensions (the plane) are equally applicable to vectors in three-dimensions (space); the ideas are the same; only the pictures are harder to draw.

Every vector is equal to a vector whose tail is at the origin, as shown in Figure 3.11. Thus there is an obvious correspondence be-tween a vector and the coordinates of its head. This correspondence is both the key to algebraic treatment of vectors and the springboard to useful generalizations of the vector concept.

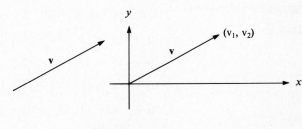

FIGURE 3.11

If **v** is the vector from the origin to the point (v_1, v_2) and **u** is the vector from the origin to the point (u_1, u_2) then $\mathbf{v} + \mathbf{u}$ is the vector from the origin to the point $(v_1 + u_1, v_2 + u_2)$, as is shown in Fig-ure 3.12 on the following page. Moreover, $k\mathbf{v}$ is the vector from the origin to the point (kv_1, kv_2).

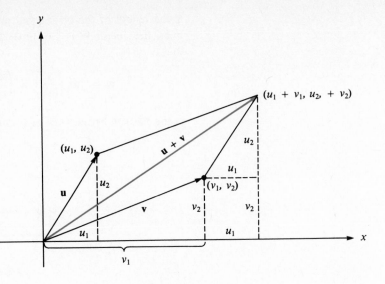

FIGURE 3.12

Note that we have associated vectors in the plane with 1×2 matrices and have observed that vector addition, by the parallelogram law, corresponds to matrix addition. Moreover, scalar multiplication of vectors corresponds to scalar multiplication of matrices. Again, these observations are as valid in three dimensions as they are in two. From now on we will write $\mathbf{v} = [a \quad b \quad c]$ to indicate that $\mathbf{v}$ is the vector from the origin to the point (a, b, c).

Example 2

For the vectors $\mathbf{u} = [1 \quad 3 \quad -2]$ and $\mathbf{v} = [2 \quad 0 \quad 4]$:

$$\mathbf{u} + \mathbf{v} = [1 \quad 3 \quad -2] + [2 \quad 0 \quad 4] = [3 \quad 3 \quad 2];$$

$$2\mathbf{u} + 3\mathbf{v} = 2[1 \quad 3 \quad -2] + 3[2 \quad 0 \quad 4] = [8 \quad 6 \quad 8];$$

$$\mathbf{u} - \mathbf{v} = [1 \quad 3 \quad -2] - [2 \quad 0 \quad 4] = [-1 \quad 3 \quad -6]; \quad \text{and}$$

$$\mathbf{u} - \mathbf{u} = [1 \quad 3 \quad -2] - [1 \quad 3 \quad -2] = [0 \quad 0 \quad 0] = \mathbf{0} \quad \blacksquare$$

Although every vector is equal to a vector with its tail at the origin, it is convenient to be able to handle vectors without first making the translation to the origin. Consider the vector $\mathbf{AB}$ where the coordinates of A are (a_1, a_2, a_3) and the coordinates of B are (b_1, b_2, b_3). As you can see from Figure 3.13, $\mathbf{OA} + \mathbf{AB} = \mathbf{OB}$, so

$$\mathbf{AB} = \mathbf{OB} - \mathbf{OA} = [b_1 \quad b_2 \quad b_3] - [a_1 \quad a_2 \quad a_3]$$
$$= [b_1 - a_1 \quad b_2 - a_2 \quad b_3 - a_3].$$

Thus the algebraic representative of the vector **AB** is obtained by subtracting the coordinates of A from the coordinates of B.

FIGURE 3.13

The distance formula of analytic geometry, or the Pythagorean theorem, allows us to compute the length of a vector algebraically. In two dimensions we have

$$\|\mathbf{u}\| = \sqrt{(u_1^2 + u_2^2)} \quad \text{if } \mathbf{u} = [u_1 \quad u_2],$$

and in three dimensions we have

$$\|\mathbf{v}\| = \sqrt{(v_1^2 + v_2^2 + v_3^2)} \quad \text{if } \mathbf{v} = [v_1 \quad v_2 \quad v_3]$$

as shown in Figure 3.14.

FIGURE 3.14

The dot product can be easily computed from the algebraic representation of vectors:

if $\mathbf{v} = [v_1 \quad v_2 \quad v_3]$ and $\mathbf{u} = [u_1 \quad u_2 \quad u_3]$, then

$$\mathbf{v} \cdot \mathbf{u} = v_1 u_1 + v_2 u_2 + v_3 u_3. \tag{3.1}$$

The result in Equation 3.1 follows from the law of cosines (recall Exercise 10 of Section 2.3) in a straightforward way. If **u** and **v** are positioned as shown in Figure 3.15, then the law of cosines yields

$$\|\mathbf{v} - \mathbf{u}\|^2 = \|\mathbf{v}\|^2 + \|\mathbf{u}\|^2 - 2\|\mathbf{v}\|\,\|\mathbf{u}\|\cos\theta,$$

from which we obtain

$$(v_1 - u_1)^2 + (v_2 - u_2)^2 + (v_3 - u_3)^2 =$$
$$(v_1^2 + v_2^2 + v_3^2) + (u_1^2 + u_2^2 + u_3^2) - 2(\mathbf{v}\cdot\mathbf{u}).$$

After some algebraic simplification this reduces to

$$-2v_1u_1 - 2v_2u_2 - 2v_3u_3 = -2(\mathbf{v}\cdot\mathbf{u}) \quad \text{or} \quad \mathbf{v}\cdot\mathbf{u} = v_1u_1 + v_2u_2 + v_3u_3.$$

FIGURE 3.15

Equation 3.1 allows us to compute dot products conveniently and thus to handle many vector problems involving angles and distances.

Example 3

Find the angle between the vectors $\mathbf{v} = [1 \quad 2]$ and $\mathbf{u} = [1 \quad 1]$.

We have $\|\mathbf{v}\| = \sqrt{5}$, $\|\mathbf{u}\| = \sqrt{2}$, and $\mathbf{v}\cdot\mathbf{u} = 1\times 1 + 2\times 1 = 3$, so it follows from Definition 3.2 that

$$\cos\theta = \frac{\mathbf{u}\cdot\mathbf{v}}{\|\mathbf{v}\|\,\|\mathbf{u}\|} = \frac{3}{\sqrt{10}}. \qquad \blacksquare$$

Example 4

Find the projection of the vector $\mathbf{u} = [5 \quad 2 \quad 1]$ on $\mathbf{v} = [3 \quad 4 \quad 12]$.

As shown in Figure 3.10, the projection of **u** on **v** is the vector

$$\mathbf{w} = \left(\frac{\mathbf{u}\cdot\mathbf{v}}{\mathbf{v}\cdot\mathbf{v}}\right)\mathbf{v}.$$

From Equation 3.1 we have $\mathbf{u}\cdot\mathbf{v} = 15 + 8 + 12 = 35$ and $\mathbf{v}\cdot\mathbf{v} = 9 + 16 + 144 = 169$. Thus, $\mathbf{w} = (35/169)\mathbf{v} = (35/169)[3 \quad 4 \quad 12]$.

The length of **w** is the scalar projection of **u** on **v**. It is

$$\|\mathbf{w}\| = \left(\frac{\mathbf{u}\cdot\mathbf{v}}{\mathbf{v}\cdot\mathbf{v}}\right)\|\mathbf{v}\| = \left(\frac{\mathbf{u}\cdot\mathbf{v}}{\mathbf{v}\cdot\mathbf{v}}\right)\sqrt{\mathbf{v}\cdot\mathbf{v}} = \frac{\mathbf{u}\cdot\mathbf{v}}{\sqrt{\mathbf{v}\cdot\mathbf{v}}} = \frac{35}{13}. \qquad \blacksquare$$

In general we shall denote the set of all $m \times n$ real matrices by $\mathcal{R}^{m \times n}$, so that we can summarize the above discussion by saying that the set of vectors in the plane behaves algebraically like $\mathcal{R}^{1 \times 2}$ and that the set of vectors in 3-space behaves algebraically like $\mathcal{R}^{1 \times 3}$.

We could just as well have used column matrices as row matrices to represent vectors; in fact, in most of what follows it will be more convenient to use column matrices than row matrices.

Definition 3.3

We shall call the set of all $n \times 1$ real matrices EUCLIDEAN n-SPACE and will denote it by $\mathcal{R}^n$.

We will study $\mathcal{R}^n$ in the next section; it will be important to keep clearly in mind that we are generalizing the familiar spaces $\mathcal{R}^2$ and $\mathcal{R}^3$ for which we have geometric interpretations.

Example 5

Find scalars a and b so that $\mathbf{w} = [-5 \quad 15] = a\mathbf{u} + b\mathbf{v}$, where $\mathbf{u} = [1 \quad 1]$ and $\mathbf{v} = [2 \quad -3]$.

This problem could be solved geometrically, as in Example 1. Here we give a more efficient algebraic solution. Since we want

$$a\mathbf{u} + b\mathbf{v} = [a + 2b \quad a - 3b] = [-5 \quad 15],$$

we must have

$$a + 2b = -5$$
$$a - 3b = 15,$$

from which it follows easily that $a = 3$ and $b = -4$. ■

Exercise Set 3.1

1. Sketch the following vectors with initial point at the origin and then sketch them in red with initial point at [3 4].
   (a) $\mathbf{u} = [2 \quad 5]$ (b) $\mathbf{v} = [-5 \quad -4]$ (c) $\mathbf{w} = [3 \quad -2]$

2. Find the vector $\mathbf{AB}$ if
   (a) A is $(3, 5)$ and B is $(2, 8)$;
   (b) A is $(3, 1, 4)$ and B is $(6, 4, 5)$; or
   (c) A is $(2, -1, 7)$ and B is $(-3, 4, 3)$.

3. Find a point Q so that the vector $\mathbf{PQ}$ is parallel to $\mathbf{w}$ if
   (a) $\mathbf{w} = [2 \quad 3]$ and P is $(4, 2)$;
   (b) $\mathbf{w} = [3 \quad -2 \quad 0]$ and P is $(1, 1, 2)$; or
   (c) $\mathbf{w} = [-2 \quad 3 \quad 5]$ and P is $(2, -2, -3)$.

4. Show that the line segments AB and PQ are parallel if and only if the vectors $\mathbf{AB}$ and $\mathbf{PQ}$ are scalar multiples of each other.

5. **(a)** Show that the midpoint M of a line segment PQ is determined by $\mathbf{OM} = \mathbf{OP} + (1/2)\mathbf{PQ}$.

   (b) If P is (x_1, x_2, x_3) and Q is (y_1, y_2, y_3), show that the midpoint is $((x_1 + y_1)/2, (x_2 + y_2)/2, (x_3 + y_3)/2)$.

   (c) Find the midpoint of PQ if P is $(3, 2, 7)$ and Q is $(5, -2, 3)$.

   (d) Find the midpoint of PQ if P is $(7, -5)$ and Q is $(4, -2)$.

6. Use the diagram below to show that vector addition is associative; that is, $\mathbf{u} + (\mathbf{v} + \mathbf{w}) = (\mathbf{u} + \mathbf{v}) + \mathbf{w}$.

$$(\mathbf{u} + \mathbf{v}) + \mathbf{w} = \mathbf{u} + (\mathbf{v} + \mathbf{w})$$

7. A cyclist with a long scarf rides due east at 15 mph. There is a cross wind from the north at 20 mph. In what direction does the scarf point? What is the magnitude of the wind felt by the cyclist?

8. For $\mathbf{u}$ and $\mathbf{v}$ as given below, find $\|\mathbf{u}\|$, $\mathbf{u} \cdot \mathbf{v}$, the unit vector in the direction of $\mathbf{u}$, and $\cos \theta$, where θ is the angle between $\mathbf{u}$ and $\mathbf{v}$.

   (a) $\mathbf{u} = [3 \quad 4]$, $\mathbf{v} = [4 \quad -3]$

   (b) $\mathbf{u} = [3 \quad -4 \quad 12]$, $\mathbf{v} = [5 \quad 2 \quad 3]$

9. For $\mathbf{u}$ and $\mathbf{v}$ in 3-space,

   (a) show geometrically that $\|\mathbf{u} + \mathbf{v}\| \neq \|\mathbf{u}\| + \|\mathbf{v}\|$; and

   (b) determine how $|\mathbf{u} \cdot \mathbf{v}|$ and $\|\mathbf{u}\| \|\mathbf{v}\|$ compare.

10. Find the vector projection of $\mathbf{u}$ on $\mathbf{v}$ and the length of the projection if

   (a) $\mathbf{u} = [5 \quad 2]$, $\mathbf{v} = [3 \quad 4]$;

   (b) $\mathbf{u} = [1 \quad 7 \quad 4]$, $\mathbf{v} = [3 \quad 4 \quad -12]$; or

   (c) $\mathbf{u} = [3 \quad 5 \quad 2]$, $\mathbf{v} = [1 \quad 1 \quad 1]$.

11. Express $\mathbf{w}$ as a linear combination of $\mathbf{u}$ and $\mathbf{v}$ if

   (a) $\mathbf{w} = [-4 \quad 1]$, $\mathbf{u} = [1 \quad 5]$, $\mathbf{v} = [2 \quad 3]$; or

   (b) $\mathbf{w} = [3 \quad 5]$, $\mathbf{u} = [1 \quad 0]$, $\mathbf{v} = [0 \quad 1]$.

12. Express $\mathbf{w}$ as a linear combination of $\mathbf{x}$, $\mathbf{y}$, and $\mathbf{z}$ if

   (a) $\mathbf{w} = [2 \quad 1 \quad 7]$, $\mathbf{x} = [1 \quad 0 \quad 1]$, $\mathbf{y} = [0 \quad 1 \quad 2]$, $\mathbf{z} = [-1 \quad 3 \quad 0]$;

   (b) $\mathbf{w} = [a \quad b \quad c]$, $\mathbf{x} = [1 \quad 0 \quad 0]$, $\mathbf{y} = [0 \quad 1 \quad 0]$, $\mathbf{z} = [0 \quad 0 \quad 1]$; or

   (c) $\mathbf{w} = [5 \quad 7 \quad 8]$, $\mathbf{x} = [1 \quad 2 \quad 3]$, $\mathbf{y} = [4 \quad 5 \quad 6]$, $\mathbf{z} = [7 \quad 8 \quad 9]$.

13. For the vectors $\mathbf{x}$, $\mathbf{y}$, and $\mathbf{z}$ of Exercise 12(c) show that $\mathbf{z}$ is in the plane determined by $\mathbf{x}$ and $\mathbf{y}$ by finding scalars a and b so that $\mathbf{z} = a\mathbf{x} + b\mathbf{y}$.

14. Let **u** and **v** be nonzero vectors in 2- or 3-space. If $k = \|\mathbf{u}\|$ and $h = \|\mathbf{v}\|$, show geometrically that the vector $\mathbf{w} = (k\mathbf{v} + h\mathbf{u})$ bisects the angle between **u** and **v**.

15. Show that any vector $\mathbf{u} = [u_1 \quad u_2 \quad u_3]$ in 3-space can be written as a linear combination of the unit vectors $\mathbf{i} = [1 \quad 0 \quad 0]$, $\mathbf{j} = [0 \quad 1 \quad 0]$, and $\mathbf{k} = [0 \quad 0 \quad 1]$.

16. For vectors $\mathbf{u} = [u_1 \quad u_2 \quad u_3]$ and $\mathbf{v} = [v_1 \quad v_2 \quad v_3]$ in 3-space, the VECTOR CROSS PRODUCT is defined by

$$\mathbf{u} \times \mathbf{v} = [u_2v_3 - u_3v_2, u_3v_1 - u_1v_3, u_1v_2 - u_2v_1].$$

(a) Show that

$$\mathbf{u} \times \mathbf{v} = \left[\det \begin{bmatrix} u_2 & u_3 \\ v_2 & v_3 \end{bmatrix} \quad -\det \begin{bmatrix} u_1 & u_3 \\ v_1 & v_3 \end{bmatrix} \quad \det \begin{bmatrix} u_1 & u_2 \\ v_1 & v_2 \end{bmatrix} \right]$$

$$= \det \begin{bmatrix} \mathbf{i} & \mathbf{j} & \mathbf{k} \\ u_1 & u_2 & u_3 \\ v_1 & v_2 & v_3 \end{bmatrix},$$

where $\mathbf{i} = [1 \quad 0 \quad 0]$, $\mathbf{j} = [0 \quad 1 \quad 0]$, and $\mathbf{k} = [0 \quad 0 \quad 1]$ are unit vectors in the coordinate (x, y, and z) directions. Note that the final form for the cross product is not a proper determinant since the elements of the first row are vectors, not scalars. This compact form for the cross product is a useful mnenomic device.

(b) For $\mathbf{u} = [1 \quad 2 \quad 3]$ and $\mathbf{v} = [3 \quad -1 \quad 4]$, compute $\mathbf{u} \times \mathbf{v}$ and $\mathbf{v} \times \mathbf{u}$.

(c) Show that $\mathbf{u} \cdot (\mathbf{u} \times \mathbf{v}) = \mathbf{v} \cdot (\mathbf{u} \times \mathbf{v}) = 0$ and interpret this result geometrically.

(d) Show that $\|\mathbf{u} \times \mathbf{v}\|^2 = \|\mathbf{u}\|^2\|\mathbf{v}\|^2 - (\mathbf{u} \cdot \mathbf{v})^2$.

17. Show, from the result of Exercise 16(d), that $\|\mathbf{u} \times \mathbf{v}\| = \|\mathbf{u}\|\|\mathbf{v}\| \sin \theta$, and hence that $\|\mathbf{u} \times \mathbf{v}\|$ is the area of the parallelogram with **u** and **v** as adjacent sides.

18. Show that $|\mathbf{u} \cdot (\mathbf{v} \times \mathbf{w})|$ is the volume of the parallelepiped with sides **u**, **v**, and **w**.

19. Given the vectors in the diagram below, find each of the following resultant vectors.

(a) $\mathbf{a} + \mathbf{c}$ **(d)** $\mathbf{d} + \mathbf{b}$ **(g)** $\mathbf{a} - \mathbf{b} + \mathbf{e} + \mathbf{f}$

(b) $\mathbf{a} - \mathbf{c}$ **(e)** $\mathbf{f} + \mathbf{b}$ **(h)** $\mathbf{d} - \mathbf{e} + \mathbf{a}$

(c) $\mathbf{d} + \mathbf{e} + \mathbf{a}$ **(f)** $2\mathbf{c} - \mathbf{b}$ **(i)** $\mathbf{a} + \mathbf{c} + \mathbf{e} + \mathbf{g}$

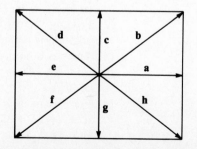

20. Define the following relation on arrows in the plane: $\mathbf{u} \sim \mathbf{v}$ if and only if $\mathbf{u}$ and $\mathbf{v}$ have the same length and direction.

    (a) Show that $\sim$ is an equivalence relation; that is, show that $\sim$ is reflexive, symmetric, and transitive.

    (b) Show that every equivalence class $\{\mathbf{v} \mid \mathbf{v} \sim \mathbf{u}\}$ contains precisely one arrow with tail at the origin.

    (c) Compare equality of these equivalence classes with equality of vectors as defined in the text.

21. Which of the following statements are correct? Explain your reasoning.

    (a) $\mathcal{R}^2$ is a subset of $\mathcal{R}^3$.

    (b) $\mathcal{R}^3$ has a subset that is "like" $\mathcal{R}^2$.

    (c) $\mathcal{R}^3$ has many subsets that are "like" $\mathcal{R}^2$.

22. Show that any plane in $\mathcal{R}^3$ can be described as
    $$\mathcal{W} = \{\mathbf{u} + \mathbf{v_p} \mid \mathbf{u} \text{ in } \mathcal{H}\},$$
    where $\mathcal{H}$ is a plane through the origin and $\mathbf{v_p} = \mathbf{OP}$, where P is a fixed point on the plane $\mathcal{W}$.

3.2   Euclidean n-Space and Its Subspaces

In the last section we defined $\mathcal{R}^n$ to be the set of all $n \times 1$ matrices and took a brief look at $\mathcal{R}^2$ and $\mathcal{R}^3$ from a geometric viewpoint. In this section we will study $\mathcal{R}^n$, the natural generalization of $\mathcal{R}^2$ and $\mathcal{R}^3$. Since it is not possible to deal with $\mathcal{R}^n$ geometrically, we have no choice but to proceed algebraically, although we will use geometric terminology and insight, from the cases $n = 2$ and $n = 3$. We will refer to the elements of $\mathcal{R}^n$ as either points or vectors.

Since the vector addition and scalar multiplication in $\mathcal{R}^n$ are matrix addition and matrix scalar multiplication, we already know quite a bit about the algebraic structure of $\mathcal{R}^n$. The pertinent results from Chapter 1 are collected in the following theorem.

Theorem 3.1

If $\mathbf{u}$, $\mathbf{v}$, and $\mathbf{w}$ are any elements of $\mathcal{R}^n$ and a and b are any scalars, then $\mathbf{u} + \mathbf{v}$ and $a\mathbf{u}$ are in $\mathcal{R}^n$ and the following properties hold.

(1) For any vectors $\mathbf{u}$, $\mathbf{v}$, and $\mathbf{w}$ in $\mathcal{R}^n$, $\mathbf{u} + (\mathbf{v} + \mathbf{w}) = (\mathbf{u} + \mathbf{v}) + \mathbf{w}$. (Vector addition is associative.)

(2) For any vectors $\mathbf{u}$ and $\mathbf{v}$ in $\mathcal{R}^n$, $\mathbf{u} + \mathbf{v} = \mathbf{v} + \mathbf{u}$. (Vector addition is commutative.)

(3) There is a unique vector $\mathbf{0}$ in $\mathcal{R}^n$ such that $\mathbf{u} + \mathbf{0} = \mathbf{u}$ for every $\mathbf{u}$ in $\mathcal{R}^n$. (The vector $\mathbf{0}$ is called the additive identity element of $\mathcal{R}^n$.)

(4) For each $\mathbf{u}$ in $\mathcal{R}^n$, there is a unique vector $-\mathbf{u}$ in $\mathcal{R}^n$ such that $\mathbf{u} + (-\mathbf{u}) = \mathbf{0}$. (The vector $-\mathbf{u}$ is called the additive inverse of $\mathbf{u}$.)

(5) For any vector **u** and any scalars h and k, $(h + k)\mathbf{u} = h\mathbf{u} + k\mathbf{u}$. (Scalar multiplication distributes over scalar addition.)

(6) For any scalar h and any vectors **u** and **v**, $h(\mathbf{u} + \mathbf{v}) = h\mathbf{u} + h\mathbf{v}$. (Scalar multiplication distributes over vector addition.)

(7) For any scalars h and k and any vector **u**, $h(k\mathbf{u}) = (hk)\mathbf{u}$. (Scalar multiplication is associative.)

(8) $1\mathbf{u} = \mathbf{u}$ for every **u** in $\mathcal{R}^n$. (The real number 1 is an identity element for scalar multiplication.)

The properties of $\mathcal{R}^n$ collected in Theorem 3.1 are all clear for $\mathcal{R}^n$ but should be carefully examined since they will be the basis for further generalizations of the vector concept.

The primary reason for our interest in $\mathcal{R}^n$ is that the solutions of the linear system $AX = K$ come from $\mathcal{R}^n$ if A is an $m \times n$ matrix. In that case A can be profitably viewed as defining a function (multiplication by A) from $\mathcal{R}^n$ to $\mathcal{R}^m$; we will pursue this point of view in Chapter 4.

Equation 3.1 provides a natural way to generalize the notion of the dot product from $\mathcal{R}^3$ to $\mathcal{R}^n$. The dot product was defined geometrically for $\mathcal{R}^2$ and $\mathcal{R}^3$, but for $\mathcal{R}^n$ we must give an algebraic definition.

Definition 3.4

The DOT PRODUCT, or EUCLIDEAN INNER PRODUCT, of

$$\mathbf{u} = [u_1 \quad u_2 \quad \cdots \quad u_n]^T \quad \text{and} \quad \mathbf{v} = [v_1 \quad v_2 \quad \cdots \quad v_n]^T$$

in $\mathcal{R}^n$ is defined by

$$\mathbf{u} \cdot \mathbf{v} = u_1 v_1 + u_2 v_2 + \ldots + u_n v_n.$$

Some texts use the notation $(\mathbf{u}, \mathbf{v})$ or $\langle \mathbf{u}, \mathbf{v} \rangle$ instead of $\mathbf{u} \cdot \mathbf{v}$ for the Euclidean inner product.

Example 1

For the vectors $\mathbf{u} = [1 \quad 2 \quad 3 \quad 4]^T$, $\mathbf{v} = [4 \quad -2 \quad 0 \quad 5]^T$, and $\mathbf{w} = [4 \quad 0 \quad 0 \quad -1]^T$ from $\mathcal{R}^4$ we have

$$\mathbf{u} \cdot \mathbf{v} = 1 \times 4 + 2 \times (-2) + 3 \times 0 + 4 \times 5 = 20;$$

$$\mathbf{u} \cdot \mathbf{u} = 1^2 + 2^2 + 3^2 + 4^2 = 30; \quad \text{and}$$

$$\mathbf{u} \cdot \mathbf{w} = 1 \times 4 + 2 \times 0 + 3 \times 0 + 4 \times (-1) = 0. \quad \blacksquare$$

Note that we can write the dot product $\mathbf{u} \cdot \mathbf{v}$ as the matrix product $\mathbf{u}^T\mathbf{v}$. Thus, we can use previously established properties of matrix multiplication and the transpose to establish some elementary properties of the dot product in $\mathcal{R}^n$.

Theorem 3.2

For **u**, **v**, and **w** in $\mathcal{R}^n$ and for any scalars a and b, the following statements are true:

(1) $\mathbf{u} \cdot \mathbf{v} = \mathbf{v} \cdot \mathbf{u}$;

(2) $\mathbf{u} \cdot (\mathbf{v} + \mathbf{w}) = \mathbf{u} \cdot \mathbf{v} + \mathbf{u} \cdot \mathbf{w}$;

(3) $\mathbf{u} \cdot (a\mathbf{v}) = (a\mathbf{u}) \cdot \mathbf{v} = a(\mathbf{u} \cdot \mathbf{v})$;

(4) $\mathbf{u} \cdot (a\mathbf{v} + b\mathbf{w}) = a(\mathbf{u} \cdot \mathbf{v}) + b(\mathbf{u} \cdot \mathbf{w})$;

(5) $\mathbf{u} \cdot \mathbf{0} = 0$; and

(6) $\mathbf{u} \cdot \mathbf{u} \geq 0$ with $\mathbf{u} \cdot \mathbf{u} = 0$ if and only if $\mathbf{u} = \mathbf{0}$.

For a vector $\mathbf{u}$ in $\mathcal{R}^3$ we know that the length of $\mathbf{u}$ is given by

$$\|\mathbf{u}\| = \sqrt{(\mathbf{u} \cdot \mathbf{u})} = \sqrt{(u_1^2 + u_2^2 + u_3^2)}.$$

For a vector $\mathbf{u}$ in $\mathcal{R}^n$ we can define the *length,* or *norm,* of $\mathbf{u}$ by generalizing the preceding formula.

Definition 3.5

For the vector $\mathbf{u}$ in $\mathcal{R}^n$ the LENGTH, or NORM, of $\mathbf{u}$ is defined by

$$\|\mathbf{u}\| = \sqrt{(\mathbf{u} \cdot \mathbf{u})} = \sqrt{(u_1^2 + u_2^2 + \ldots + u_n^2)} = \sqrt{\Sigma u_i^2}.$$

We will call the vector $\mathbf{u}$ a UNIT VECTOR if $\|\mathbf{u}\| = 1$. Since $\|a\mathbf{u}\| = \sqrt{(a\mathbf{u}) \cdot (a\mathbf{u})} = |a|\sqrt{(\mathbf{u} \cdot \mathbf{u})} = |a| \|\mathbf{u}\|$, it follows that for any vector $\mathbf{v} \neq \mathbf{0}$, the vector $(1/\|\mathbf{v}\|)\mathbf{v}$ is a unit vector.

We can also generalize directly from $\mathcal{R}^3$ to define what it means for two vectors in $\mathcal{R}^n$ to be perpendicular or orthogonal.

Definition 3.6

The vectors $\mathbf{u}$ and $\mathbf{v}$ from $\mathcal{R}^n$ are ORTHOGONAL if $\mathbf{u} \cdot \mathbf{v} = 0$.

We could carry the generalization further by defining the angle between two vectors in R^n by $\cos \theta = (\mathbf{u} \cdot \mathbf{v})/\|\mathbf{u}\| \|\mathbf{v}\|$ and by defining the distance between two points, P and Q, of $\mathcal{R}^n$ as $\|\mathbf{Q} - \mathbf{P}\|$.

Example 2

Consider the vectors $\mathbf{u} = [2 \quad -1 \quad 0 \quad 3]^T$ and $\mathbf{v} = [3 \quad 0 \quad 5 \quad -2]^T$ from $\mathcal{R}^4$.

We have

$$\mathbf{u} \cdot \mathbf{u} = 4 + 1 + 0 + 9 = 14,$$

so that

$$\|\mathbf{u}\| = \sqrt{14}$$

and

$$\left(\frac{1}{\sqrt{14}}\right)\mathbf{u} = \left(\frac{1}{\sqrt{14}}\right)[2 \quad -1 \quad 0 \quad 3]^T$$

is a unit vector. Moreover,

$$\mathbf{u} \cdot \mathbf{v} = 6 + 0 + 0 - 6 = 0,$$

so the vectors $\mathbf{u}$ and $\mathbf{v}$ are orthogonal. ■

We will have many occasions to consider certain subsets of $\mathcal{R}^n$. Of particular interest to us are subsets associated with a linear system $AX = K$. If A is an $m \times n$ matrix, then the following subsets will be of particular interest:

(1) the *solution space* $\{X \mid AX = K\} \subseteq \mathcal{R}^n$;

(2) the *null space* $\{X \mid AX = 0\} \subseteq \mathcal{R}^n$;

(3) the *column space* $\{K \mid AX = K$ is solvable$\} \subseteq \mathcal{R}^m$; and

(4) the *eigenspace* $\{Y \mid AY = \lambda Y$ for a fixed real $\lambda\} \subseteq \mathcal{R}^n$.

For example, let us consider the subset of $\mathcal{R}^3$ defined by

$$\mathcal{W} = \{[x \quad y \quad 0]^T\} = \{[x \quad y \quad z]^T \mid z = 0\}.$$

Although it is not correct to say that $\mathcal{W} = \mathcal{R}^2$, it should be clear geometrically that $\mathcal{W}$ and $\mathcal{R}^2$ are very much alike. In fact, $\mathcal{W}$ is a copy of $\mathcal{R}^2$ embedded in $\mathcal{R}^3$. In this case, $\mathcal{W}$ is said to be a subspace of $\mathcal{R}^3$.

Definition 3.7

A nonempty subset $\mathcal{W}$ of $\mathcal{R}^n$ is called a SUBSPACE of $\mathcal{R}^n$ if $\mathcal{W}$ is closed under vector addition and scalar multiplication; that is,

if **u** and **v** are in $\mathcal{W}$, then so are **u** + **v** and k**u** ,

where k is any scalar.

Example 3

Is $\mathcal{W} = \left\{ X \text{ in } \mathcal{R}^3 \,\middle|\, \begin{bmatrix} 1 & 0 & 2 \\ 0 & 1 & -3 \\ 0 & 0 & 0 \end{bmatrix} X = \begin{bmatrix} 3 \\ -1 \\ 0 \end{bmatrix} \right\}$ a subspace of $\mathcal{R}^3$?

Since the coefficient matrix of the system is already in reduced row echelon form we see at once that the general solution of the system, and hence a typical element of the set $\mathcal{W}$, is given by

$$X = \begin{bmatrix} 3 - 2c \\ -1 + 3c \\ c \end{bmatrix} = \begin{bmatrix} 3 \\ -1 \\ 0 \end{bmatrix} + c \begin{bmatrix} -2 \\ 3 \\ 1 \end{bmatrix}.$$

If we take $c = 1$ and then $c = 2$ we find that the two solutions

$$X_1 = \begin{bmatrix} 1 \\ 2 \\ 1 \end{bmatrix} \quad \text{and} \quad X_2 = \begin{bmatrix} -1 \\ 5 \\ 2 \end{bmatrix}$$

are elements of the set $\mathcal{W}$. In order to determine whether or not $X_1 + X_2$ is in $\mathcal{W}$, we compute

$$X_1 + X_2 = \begin{bmatrix} 0 \\ 7 \\ 3 \end{bmatrix} \quad \text{and} \quad \begin{bmatrix} 1 & 0 & 2 \\ 0 & 1 & -3 \\ 0 & 0 & 0 \end{bmatrix} \begin{bmatrix} 0 \\ 7 \\ 3 \end{bmatrix} = \begin{bmatrix} 6 \\ -2 \\ 0 \end{bmatrix} \neq \begin{bmatrix} 3 \\ -1 \\ 0 \end{bmatrix}.$$

Thus, $X_1 + X_2$ is not a solution of the given system and hence not an element of the set $\mathcal{W}$. This calculation shows that the set $\mathcal{W}$ is not closed under vector addition and therefore that $\mathcal{W}$ is not a subspace of $\mathcal{R}^3$. You should check that $2X_1$ is not in $\mathcal{W}$ and hence that $\mathcal{W}$ also fails to be closed under scalar multiplication. ▬

The two closure conditions of Definition 3.7 can be replaced by the condition that $\mathcal{W}$ be nonempty and closed under arbitrary linear combinations; that is, for any $\mathbf{u}$ and $\mathbf{v}$ in $\mathcal{W}$ and any a and b in $\mathcal{R}$, $a\mathbf{u} + b\mathbf{v}$ is in $\mathcal{W}$. Note that if $\mathbf{u}$ is in $\mathcal{W}$, then $\mathcal{W}$ must also contain the vectors $-\mathbf{u} = (-1)\mathbf{u}$ and $\mathbf{0} = \mathbf{u} - \mathbf{u} = 0\mathbf{u}$.

If A is any $m \times n$ matrix, then the set of all solutions of the homogeneous system $AX = 0$ is a very important example of a subspace. This subspace is called the NULL SPACE of A and will be denoted by $\mathcal{NS}(A)$:

$$\mathcal{NS}(A) = \{\mathbf{u} \text{ in } \mathcal{R}^n\} \mid A\mathbf{u} = \mathbf{0}\}.$$

Theorem 3.3

For any $m \times n$ matrix A, the null space of A is a subspace of $\mathcal{R}^n$.

Proof.   To prove this theorem we need to show that $\mathcal{NS}(A)$ is closed under addition and scalar multiplication. Thus, suppose that $\mathbf{u}$ and $\mathbf{v}$ are in $\mathcal{NS}(A)$, which means that $A\mathbf{u} = \mathbf{0}$ and $A\mathbf{v} = \mathbf{0}$. To decide whether $\mathbf{u} + \mathbf{v}$ is in $\mathcal{NS}(A)$ we compute

$$A(\mathbf{u} + \mathbf{v}) = A\mathbf{u} + A\mathbf{v} = \mathbf{0} + \mathbf{0} = \mathbf{0},$$

where the first equality is a consequence of the fact that matrix multiplication distributes over matrix addition. This calculation shows that $\mathbf{u} + \mathbf{v}$ is also in $\mathcal{NS}(A)$ and, hence, that $\mathcal{NS}(A)$ is closed under vector addition. Similarly, to show that $\mathcal{NS}(A)$ is closed under scalar multiplication we let k be any scalar and compute

$$A(k\mathbf{u}) = k(A\mathbf{u}) = k\mathbf{0} = \mathbf{0}.$$

Thus, $k\mathbf{u}$ is in $\mathcal{NS}(A)$ and we have shown that $\mathcal{NS}(A)$ is closed under scalar multiplication. This completes the demonstration that $\mathcal{NS}(A)$ is a subspace of $\mathcal{R}^n$. □

The key to success in deciding whether a subset $\mathcal{W}$ of $\mathcal{R}^n$ is also a subspace is to clearly understand the criteria for membership in the set $\mathcal{W}$; in Theorem 3.3 the criteria for membership in $\mathcal{NS}(A)$ was that $\mathbf{u}$ was in $\mathcal{NS}(A)$ if and only if $A\mathbf{u} = \mathbf{0}$.

Note that one example of vectors $\mathbf{u}$ and $\mathbf{v}$ in $\mathcal{W}$, for which $\mathbf{u} + \mathbf{v}$ is not in $\mathcal{W}$, was enough to show, in Example 3, that $\mathcal{W}$ fails to be a subspace. To show that a set $\mathcal{W}$ is a subspace requires that we show that $\mathbf{u} + \mathbf{v}$ is in $\mathcal{W}$ *for every choice* of $\mathbf{u}$ and $\mathbf{v}$ in $\mathcal{W}$.

Example 4

Determine whether $\mathcal{W} = \{\mathbf{u} \text{ in } R^3 \mid \|\mathbf{u}\| = 1\}$ is a subspace of $\mathcal{R}^3$.

The vectors $\mathbf{u} = [1 \quad 0 \quad 0]^T$ and $\mathbf{v} = [0 \quad 1 \quad 0]^T$ are each members of $\mathcal{W}$, but $\mathbf{u} + \mathbf{v} = [1 \quad 1 \quad 0]^T$ is not in $\mathcal{W}$ because

$$\|\mathbf{u} + \mathbf{v}\| = \sqrt{(1^2 + 1^2)} = \sqrt{2} \neq 1.$$

Thus, $\mathcal{W}$ is not closed under addition and hence is not a subspace of $\mathcal{R}^3$.

Alternately, one could observe that if $\mathbf{u}$ is in $\mathcal{W}$ then so is $-\mathbf{u}$, but $\mathbf{u} + (-\mathbf{u}) = \mathbf{0}$ is not in $\mathcal{W}$ since $\|\mathbf{0}\| = 0 \neq 1$.

A third alternative for this example is to show that if $\mathbf{u}$ is in $\mathcal{W}$, then $k\mathbf{u}$ is in $\mathcal{W}$ only if $k = \pm 1$ and, hence, that $\mathcal{W}$ is not closed under all scalar multiplications. ■

Example 5

Is $\mathcal{W} = \{\mathbf{u} = [0 \quad u_2 \quad u_3 \quad u_2 + u_3]^T\}$ a subspace of $\mathcal{R}^4$?

The membership criteria for the set $\mathcal{W}$ is not as clearly indicated here as it was in Example 4. For this set $\mathcal{W}$, $\mathbf{u} = [u_1 \quad u_2 \quad u_3 \quad u_4]^T$ is in $\mathcal{W}$ if and only if $u_1 = 0$ and $u_4 = u_2 + u_3$. Thus, $\mathbf{v}_1 = [0 \quad 2 \quad 3 \quad 5]^T$ and $\mathbf{v}_2 = [0 \quad 0 \quad 3 \quad 3]^T$ are in $\mathcal{W}$ but $\mathbf{v}_3 = [1 \quad 2 \quad 3 \quad 5]^T$ and $\mathbf{v}_4 = [0 \quad 4 \quad 2 \quad 7]^T$ are not in $\mathcal{W}$. Observe that $\mathbf{v}_1 + \mathbf{v}_2 = [0 \quad 2 \quad 6 \quad 8]^T$ is also in $\mathcal{W}$, which suggests, *but does not prove*, that $\mathcal{W}$ is closed under vector addition. To prove that $\mathcal{W}$ is a subspace we suppose that $\mathbf{u} = [u_1 \quad u_2 \quad u_3 \quad u_4]^T$ and $\mathbf{v} = [v_1 \quad v_2 \quad v_3 \quad v_4]^T$ are any two vectors in $\mathcal{W}$. Therefore, by the membership criteria for the set $\mathcal{W}$,

$$u_1 = v_1 = 0, \quad u_4 = u_2 + u_3, \quad \text{and} \quad v_4 = v_2 + v_3.$$

Now the vector $\mathbf{x} = \mathbf{u} + \mathbf{v} = [u_1 + v_1 \quad u_2 + v_2 \quad u_3 + v_3 \quad u_4 + v_4]^T$. To see if $\mathbf{x}$ is in $\mathcal{W}$ we check the first and fourth components:

$$x_1 = u_1 + v_1 = 0 \quad \text{since} \quad u_1 = 0 \quad \text{and} \quad v_1 = 0,$$

while

$$x_4 = u_4 + v_4 = (u_2 + u_3) + (v_2 + v_3) = (u_2 + v_2) + (u_3 + v_3)$$
$$= x_2 + x_3, \quad \text{since} \quad u_2 + u_3 = u_4 \quad \text{and} \quad v_2 + v_3 = v_4.$$

Therefore, $\mathbf{x}$ is in $\mathcal{W}$ and we have shown that $\mathcal{W}$ is closed under vector addition. Similarly, $\mathbf{y} = k\mathbf{u} = [ku_1 \quad ku_2 \quad ku_3 \quad ku_4]^T$ is in $\mathcal{W}$ because

$$y_1 = ku_1 = k0 = 0 \quad \text{and} \quad y_4 = k(u_2 + u_3) = ku_2 + ku_3 = y_2 + y_3.$$

Thus, $\mathcal{W}$ is also closed under scalar multiplication, and it follows from Definition 3.7 that $\mathcal{W}$ is a subspace of $\mathcal{R}^4$. ■

Exercise Set 3.2

1. For the vectors $\mathbf{u} = [3 \quad 0 \quad 1 \quad 2]^T$ and $\mathbf{v} = [2 \quad 0 \quad 1 \quad 1]^T$ in $\mathcal{R}^4$:
   (a) compute $\|\mathbf{u}\|$, $\mathbf{u} \cdot \mathbf{v}$, $\|\mathbf{v}\|$, and $\|\mathbf{u} + \mathbf{v}\|$; and
   (b) find a vector orthogonal to $\mathbf{u}$.

2. For the vectors $\mathbf{u}$ and $\mathbf{v}$ in $\mathcal{R}^n$ show that
   (a) $(\mathbf{u} + \mathbf{v}) \cdot (\mathbf{u} + \mathbf{v}) = \mathbf{u} \cdot \mathbf{u} + 2(\mathbf{u} \cdot \mathbf{v}) + \mathbf{v} \cdot \mathbf{v}$;
   (b) $(\mathbf{u} + \mathbf{v}) \cdot (\mathbf{u} - \mathbf{v}) = \|\mathbf{u}\|^2 - \|\mathbf{v}\|^2$; and
   (c) $\|\mathbf{u}\|^2 + \|\mathbf{v}\|^2 = \|\mathbf{u} - \mathbf{v}\|^2$ if and only if $\mathbf{u} \perp \mathbf{v}$.

†3. (a) Prove that $a\mathbf{0} = \mathbf{0}$ and $0\mathbf{u} = \mathbf{0}$ for every scalar a and every vector $\mathbf{u}$ in $\mathcal{R}^n$.
   (b) Prove that $a\mathbf{u} = \mathbf{0}$ implies that either the vector $\mathbf{u} = \mathbf{0}$ or the scalar $a = 0$.
   (c) Prove that $a\mathbf{u} = b\mathbf{u}$ if and only if either the vector $\mathbf{u} = \mathbf{0}$ or the scalar $a = b$.

4. Show that neither of the following subsets of $\mathcal{R}^4$ is a subspace.
   (a) $\{[x_1 \quad 1 + x_1 \quad x_3 \quad 5]^T \,|\, x_1 \text{ and } x_3 \text{ are real}\}$
   (b) $\{\mathbf{x} = [x_1 \quad x_2 \quad x_3 \quad x_4]^T \,|\, x_1 + x_3 \leq x_4\}$

5. Show that each of the following subsets of $\mathcal{R}^4$ is a subspace.
   (a) $\{\mathbf{u} \text{ in } \mathcal{R}^4 \,|\, A\mathbf{u} = 2\mathbf{u}\}$ (A is a fixed 4×4 matrix)
   (b) $\{\mathbf{x} = [x_1 \quad x_2 \quad x_3 \quad x_4]^T \,|\, x_1 + x_2 + x_4 = 0\}$

→ †6. Let $\mathbf{u}$, $\mathbf{v}$, and $\mathbf{w}$ be any three vectors from $\mathcal{R}^n$. Show that $\{a\mathbf{u} + b\mathbf{v} + c\mathbf{w} \,|\, a, b, \text{ and } c \text{ real numbers}\}$ is a subspace of $\mathcal{R}^n$.

7. Which of the following subsets of $\mathcal{R}^4$ are subspaces?
   (a) $\{[x_1 \quad x_2 \quad x_1 + 3x_2 \quad 5x_1 - 7x_2]^T \,|\, x_1, x_2 \text{ real}\}$
   (b) $\{\mathbf{u} \text{ in } \mathcal{R}^4 \,|\, \mathbf{u} \cdot \mathbf{v} = 0\}$ ($\mathbf{v}$ fixed in $\mathcal{R}^4$)
   (c) $\{\mathbf{x} = [x_1 \quad x_2 \quad x_3 \quad x_4]^T \,|\, x_1^2 + x_2^2 = 25\}$
   (d) $\{\mathbf{x} = [x_1 \quad x_2 \quad x_3 \quad x_4]^T \,|\, x_1x_2x_3x_4 = 0\}$
   (e) $\{\mathbf{x} = [x_1 \quad x_2 \quad x_3 \quad x_4]^T \,|\, x_3/x_4 = 5x_2/x_1\}$
   (f) $\{[a \quad a^{1/2} \quad \sin a \quad 2^a]^T \,|\, 0 < a\}$

8. Is the dot product in $\mathcal{R}^n$ associative?

†9. Let $\mathcal{W}$ be a subspace of $\mathcal{R}^n$. Show that
$$\mathcal{W}^\perp = \{\mathbf{u} \text{ in } \mathcal{R}^n \,|\, \mathbf{u} \cdot \mathbf{v} = 0 \text{ for every } \mathbf{v} \text{ in } \mathcal{W}\}$$
is a subspace of $\mathcal{R}^n$.

10. For vectors $\mathbf{u}$ and $\mathbf{v}$ in $\mathcal{R}^n$ we can generalize the results in $\mathcal{R}^2$ to define the ORTHOGONAL PROJECTION of $\mathbf{u}$ on $\mathbf{v}$ to be the vector
$$\mathbf{w} = \left(\frac{\mathbf{u} \cdot \mathbf{v}}{\mathbf{v} \cdot \mathbf{v}}\right)\mathbf{v}.$$
   (a) Find the projection of $\mathbf{u} = [1 \quad 2 \quad 4 \quad -2]^T$ on $\mathbf{v} = [5 \quad 2 \quad -1 \quad 0]^T$.
   (b) Find the length of the projection of $\mathbf{x} = [1 \quad 1 \quad 1 \quad 1 \quad 1]^T$ on $\mathbf{y} = [1 \quad 0 \quad 2 \quad 0 \quad 2]^T$.

3.3   *Subspaces of $\mathcal{R}^n$ — Continued*

An important and useful way of describing subspaces of $\mathcal{R}^n$ is found in the next definition.

Definition 3.8

If $\mathcal{S} = \{\mathbf{u}_1, \mathbf{u}_2, \ldots, \mathbf{u}_s\}$ is any set of vectors from $\mathcal{R}^n$, then the SUB-SPACE SPANNED by the set $\mathcal{S}$ is defined by

$$\text{SPAN}\{\mathbf{u}_1, \mathbf{u}_2, \ldots, \mathbf{u}_s\} = \left\{ \sum_{k=1}^{s} a_k \mathbf{u}_k \mid a_k \text{ real} \right\};$$

that is, the subspace consists of all possible linear combinations of the vectors of the set $\mathcal{S}$.

We noted, following Example 1 in Section 3.1, that for $\mathbf{u}$ and $\mathbf{v}$ in $\mathcal{R}^3$ the plane containing $\mathbf{u}$ and $\mathbf{v}$ consists of all linear combinations of $\mathbf{u}$ and $\mathbf{v}$; that is, it is precisely the subspace $\text{SPAN}\{\mathbf{u}, \mathbf{v}\}$. Thus, we can think of the "subspace spanned by a set of vectors in $\mathcal{R}^m$" as a generalization of the "plane generated by a pair of vectors in $\mathcal{R}^3$." Another useful point of view is to think of $\text{SPAN}\{\mathbf{u}_1, \mathbf{u}_2, \ldots, \mathbf{u}_n\}$ as the "smallest" subspace of $\mathcal{R}^n$ containing the vectors $\mathbf{u}_1, \mathbf{u}_2, \ldots, \mathbf{u}_n$.

To determine if a given vector $\mathbf{w}$ is an element of the subspace $\text{SPAN}\{\mathbf{u}_1, \mathbf{u}_2, \ldots, \mathbf{u}_n\}$, we need to decide whether the vector $\mathbf{w}$ can be written as a linear combination of the vectors of the spanning set. This determination normally leads to a question about the solvability of a system of linear equations; such questions can be answered efficiently using the results of Chapter 1.

Example 1

Is the vector $\mathbf{v}_1 = \begin{bmatrix} 4 \\ 7 \\ 6 \end{bmatrix}$ in $\mathcal{W} = \text{SPAN}\left\{ \mathbf{u}_1 = \begin{bmatrix} 1 \\ 2 \\ 1 \end{bmatrix}, \mathbf{u}_2 = \begin{bmatrix} 2 \\ 3 \\ 4 \end{bmatrix} \right\}$?

To determine whether $\mathbf{v}_1$ is in $\mathcal{W}$ we need to know if there are scalars x and y such that $x\mathbf{u}_1 + y\mathbf{u}_2 = \mathbf{v}_1$; that is, we want

$$x\begin{bmatrix} 1 \\ 2 \\ 1 \end{bmatrix} + y\begin{bmatrix} 2 \\ 3 \\ 4 \end{bmatrix} = \begin{bmatrix} 4 \\ 7 \\ 6 \end{bmatrix}, \text{ which is equivalent to } \begin{bmatrix} x + 2y \\ 2x + 3y \\ x + 4y \end{bmatrix} = \begin{bmatrix} 4 \\ 7 \\ 6 \end{bmatrix}.$$

Thus, to determine whether the vector $\mathbf{v}_1$ is in the subspace $\text{SPAN}\{\mathbf{u}_1, \mathbf{u}_2\}$, we need to decide whether or not the linear system

$$AX = \begin{bmatrix} 1 & 2 \\ 2 & 3 \\ 1 & 4 \end{bmatrix} \begin{bmatrix} x \\ y \end{bmatrix} = \begin{bmatrix} 4 \\ 7 \\ 6 \end{bmatrix} = \mathbf{v}_1$$

has a solution. Row reducing the augmented matrix of this system, we obtain

$$\begin{bmatrix} 1 & 2 & 4 \\ 2 & 3 & 7 \\ 1 & 4 & 6 \end{bmatrix} \longrightarrow \begin{bmatrix} 1 & 2 & 4 \\ 0 & -1 & -1 \\ 0 & 2 & 2 \end{bmatrix} \longrightarrow \begin{bmatrix} 1 & 1 & 4 \\ 0 & 1 & 1 \\ 0 & 0 & 0 \end{bmatrix} \longrightarrow \begin{bmatrix} 1 & 0 & 2 \\ 0 & 1 & 1 \\ 0 & 0 & 0 \end{bmatrix}$$

and hence determine that $x = 2$ and $y = 1$ is the unique solution of the system. Thus, $\mathbf{v}_1 = 2\mathbf{u}_1 + \mathbf{u}_2$, which shows that $\mathbf{v}_1$ is an element of the subspace $\mathcal{W} = \text{SPAN}\{\mathbf{u}_1, \mathbf{u}_2\}$. ∎

If we wish to test several vectors for membership in the subspace $\text{SPAN}\{\mathbf{u}_1, \mathbf{u}_2, \ldots, \mathbf{u}_n\}$, then we can efficiently use the methods of Section 1.4 for solving sets of linear systems with the same coefficient matrix.

Example 2

Using the information given in Example 1, which of $\mathbf{v}_1, \mathbf{v}_2 = \begin{bmatrix} 1 \\ 1 \\ 3 \end{bmatrix}$,

$\mathbf{v}_3 = \begin{bmatrix} 2 \\ 9 \\ 5 \end{bmatrix}$, and $\mathbf{v}_4 = \begin{bmatrix} 0 \\ 1 \\ -2 \end{bmatrix}$ are in $\text{SPAN}\{\mathbf{u}_1, \mathbf{u}_2\}$?

To decide if the $\mathbf{v}_i$ are in $\text{SPAN}\{\mathbf{u}_1, \mathbf{u}_2\}$ we need to investigate the solvability of the systems

$$AX = \begin{bmatrix} 1 & 2 \\ 2 & 3 \\ 1 & 4 \end{bmatrix} \begin{bmatrix} x \\ y \end{bmatrix} = \mathbf{v}_i; \quad i = 1, 2, 3, 4.$$

We have here a set of four systems of linear equations (the first system of which has already been solved), all with the same coefficient matrix. Using the techniques of Section 1.4, we row reduce the four-fold augmented matrix:

$$\begin{bmatrix} 1 & 2 & 4 & 1 & 2 & 0 \\ 2 & 3 & 7 & 1 & 9 & 1 \\ 1 & 4 & 6 & 3 & 5 & -2 \end{bmatrix} \longrightarrow \begin{bmatrix} 1 & 2 & 4 & 1 & 2 & 0 \\ 0 & -1 & -1 & -1 & 5 & 1 \\ 0 & 2 & 2 & 2 & 3 & -2 \end{bmatrix}$$

$$\longrightarrow \begin{bmatrix} 1 & 2 & 4 & 1 & 2 & 0 \\ 0 & 1 & 1 & 1 & -5 & -1 \\ 0 & 0 & 0 & 0 & 13 & 0 \end{bmatrix}.$$

The nonzero entry (13) in the (3, 5) position of the last matrix indicates that the system $AX = \mathbf{v}_3$ is not solvable and hence that $\mathbf{v}_3$ is not an element of $\text{SPAN}\{\mathbf{u}_1, \mathbf{u}_2\}$. Continuing the above row reduction we obtain the reduced row echelon matrix

$$\begin{bmatrix} 1 & 0 & 2 & -1 & 0 & 2 \\ 0 & 1 & 1 & 1 & 0 & -1 \\ 0 & 0 & 0 & 0 & 1 & 0 \end{bmatrix}.$$

From the third column of the last matrix we conclude, as above, that

$$\mathbf{v}_1 = 2\mathbf{u}_1 + \mathbf{u}_2;$$

from the fourth column we obtain

$$\mathbf{v}_2 = -\mathbf{u}_1 + \mathbf{u}_2;$$

and from the last column we conclude

$$\mathbf{v}_4 = 2\mathbf{u}_1 - \mathbf{u}_2. \quad \blacksquare$$

There are several subspaces that are useful in the study of a given $m \times n$ matrix A and linear systems with coefficient matrix A. We have already mentioned the nullspace of A, and we now define another subspace called the *column space* of A.

Definition 3.9

If A is an $m \times n$ matrix, the COLUMN SPACE of A is the subspace of $\mathcal{R}^m$ spanned by the columns of A. We will denote the column space of A by $\mathscr{CS}(A)$, so that

$$\mathscr{CS}(A) = \text{SPAN}\{\text{Col}_1(A), \text{Col}_2(A), \ldots, \text{Col}_n(A)\}.$$

For reasons that will become clear in Chapter 4, $\mathscr{CS}(A)$ is sometimes called the RANGE of A.

Recall from Chapter 1 that the columns of the matrix product AB are linear combinations of the columns of A. A typical element of the column space of A is $x_1\text{Col}_1(A) + x_2\text{Col}_2(A) + \ldots + x_n\text{Col}_n(A)$, which can be written in the form $A\mathbf{X}$, where $\mathbf{X}$ is the element of $\mathcal{R}^n$ made up of the coefficients in the linear combination. Thus, for an $m \times n$ matrix A,

$$\mathscr{CS}(A) = \{A\mathbf{X} \mid \mathbf{X} \text{ in } \mathcal{R}^n\}. \tag{3.2}$$

Observe that the linear system $A\mathbf{X} = \mathbf{K}$ has a solution if and only if $\mathbf{K}$ is an element of the column space of A. We state this important result as our next theorem.

Theorem 3.4

The column space of A consists of precisely those vectors $\mathbf{K}$ for which the linear system $A\mathbf{X} = \mathbf{K}$ has a solution; that is,

$$\mathscr{CS}(A) = \{\mathbf{K} \text{ in } \mathcal{R}^m \mid A\mathbf{X} = \mathbf{K} \text{ is solvable}\}.$$

Example 3

Find an alternate description of the column space of

$$A = \begin{bmatrix} 1 & -2 & 1 & -4 \\ 1 & 3 & 7 & 2 \\ 1 & -12 & -11 & -16 \\ 2 & -14 & -10 & -20 \end{bmatrix}.$$

Using Theorem 3.4 we can find the desired description for the column space of A by determining those vectors $\mathbf{K}$ in $\mathcal{R}^4$ for which the linear system $A\mathbf{X} = \mathbf{K}$ is solvable. The augmented matrix of this system is

$$[A \mid \mathbf{K}] = \begin{bmatrix} 1 & -2 & 1 & -4 & k_1 \\ 1 & 3 & 7 & 2 & k_2 \\ 1 & -12 & -11 & -16 & k_3 \\ 2 & -14 & -10 & -20 & k_4 \end{bmatrix}.$$

From the straightforward row reduction

$$\begin{bmatrix} 1 & -2 & 1 & -4 & k_1 \\ 1 & 3 & 7 & 2 & k_2 \\ 1 & -12 & -11 & -16 & k_3 \\ 2 & -14 & -10 & -20 & k_4 \end{bmatrix}$$

$$\longrightarrow \begin{bmatrix} 1 & -2 & 1 & -4 & k_1 \\ 0 & 5 & 6 & 6 & k_2 - k_1 \\ 0 & -10 & -12 & -12 & k_3 - k_1 \\ 0 & -10 & -12 & -12 & k_4 - 2k_1 \end{bmatrix}$$

$$\longrightarrow \begin{bmatrix} 1 & -2 & 1 & -4 & k_1 \\ 0 & 5 & 6 & 6 & k_2 - k_1 \\ 0 & 0 & 0 & 0 & k_3 - k_1 + 2(k_2 - k_1) \\ 0 & 0 & 0 & 0 & k_4 - 2k_1 + 2(k_2 - k_1) \end{bmatrix}$$

we see that the system $AX = K$ has a solution if and only if the elements in the $(3,5)$ and $(4,5)$ positions of the last matrix are zero, that is, if

$$k_3 - 3k_1 + 2k_2 = 0 \quad \text{and} \quad k_4 - 4k_1 + 2k_2 = 0 \,.$$

We have thus shown that

$$\mathscr{CS}(A) = \{\mathbf{K} \text{ in } \mathscr{R}^4 \,|\, -3k_1 + 2k_2 + k_3 = 0 \quad \text{and}$$
$$-4k_1 + 2k_2 + k_4 = 0\} \,.$$

It is worth noting that $\mathscr{CS}(A) = \mathscr{NS}(B)$, where

$$B = \begin{bmatrix} -3 & 2 & 1 & 0 \\ -4 & 2 & 0 & 1 \end{bmatrix}. \quad \blacksquare$$

It is a fact, although we will not prove it here, that every subspace of $\mathscr{R}^n$ is the null space of some matrix. This result would follow, as in Example 3, if we could show that any subspace of $\mathscr{R}^n$ had a spanning set; this fact will be demonstrated in Section 3.6.

Theorem 3.5

If $\mathscr{W}$ is any subspace of $\mathscr{R}^n$, then there exists an integer s and an $s \times n$ matrix A such that $\mathscr{W} = \mathscr{NS}(A)$.

Exercise Set 3.3

1. Show that $\mathbf{w} = [0 \quad 1 \quad -1]^T$ is in the subspace of $\mathscr{R}^3$ spanned by $\mathbf{u} = [1 \quad 2 \quad 0]^T$ and $\mathbf{v} = [3 \quad 5 \quad 1]^T$.

2. Show that every vector in $\mathscr{R}^2$ is a linear combination of the vectors $[1 \quad 2]^T$ and $[5 \quad 7]^T$.

3. Show that every vector in $\mathscr{R}^3$ is a linear combination of the vectors

$$\mathbf{u} = \begin{bmatrix} 1 \\ 0 \\ 1 \end{bmatrix}, \quad \mathbf{v} = \begin{bmatrix} 1 \\ 1 \\ 1 \end{bmatrix}, \quad \text{and} \quad \mathbf{w} = \begin{bmatrix} 0 \\ 1 \\ 1 \end{bmatrix}.$$

4. Find a spanning set for the null spaces of the following matrices.

(a) $\begin{bmatrix} 3 & 2 & 16 & 5 \\ 0 & 2 & 10 & 8 \\ 1 & 1 & 7 & 3 \end{bmatrix}$ (b) $\begin{bmatrix} 2 & 1 & 3 \\ 1 & 2 & 0 \\ 0 & 1 & 1 \end{bmatrix}$ (c) $\begin{bmatrix} 1 & 1 & -3 & 1 \\ 2 & 0 & 1 & -1 \\ 1 & 3 & -10 & 4 \end{bmatrix}$

5. Determine which of the following vectors belong to SPAN$\{\mathbf{u}_1, \mathbf{u}_2\}$:

$$\mathbf{v}_1 = \begin{bmatrix} 1 \\ 3 \\ 4 \end{bmatrix}, \quad \mathbf{v}_2 = \begin{bmatrix} 1 \\ 4 \\ 7 \end{bmatrix}, \quad \mathbf{v}_3 = \begin{bmatrix} 3 \\ 5 \\ 7 \end{bmatrix}, \text{ where } \mathbf{u}_1 = \begin{bmatrix} 1 \\ 2 \\ 3 \end{bmatrix}, \quad \text{and} \quad \mathbf{u}_2 = \begin{bmatrix} 0 \\ 1 \\ 2 \end{bmatrix}.$$

6. Determine which of the following vectors belong to SPAN$\{\mathbf{x}, \mathbf{y}, \mathbf{z}\}$:

$$\mathbf{v}_1 = \begin{bmatrix} 0 \\ 1 \\ 4 \\ 3 \end{bmatrix}, \quad \mathbf{v}_2 = \begin{bmatrix} 1 \\ 0 \\ 3 \\ 2 \end{bmatrix}, \quad \text{and} \quad \mathbf{v}_3 = \begin{bmatrix} 1 \\ -1 \\ 3 \\ 1 \end{bmatrix}, \text{ where }$$

$$\mathbf{x} = \begin{bmatrix} 0 \\ 1 \\ 2 \\ 3 \end{bmatrix}, \quad \mathbf{y} = \begin{bmatrix} 1 \\ -1 \\ 2 \\ -1 \end{bmatrix}, \quad \text{and } \mathbf{z} = \begin{bmatrix} -1 \\ 1 \\ 0 \\ 1 \end{bmatrix}.$$

7. Prove that the following two sets span the same subspace of $\mathcal{R}^3$:

   (a) $\mathbf{u}_1 = [1 \;\; -1 \;\; 2]^T$ and $\mathbf{u}_2 = [3 \;\; 0 \;\; 1]^T$

   (b) $\mathbf{v}_1 = [-1 \;\; -2 \;\; 3]^T$ and $\mathbf{v}_2 = [3 \;\; 3 \;\; -4]^T$

   (*Hint:* Show that $\mathbf{u}_1$ and $\mathbf{u}_2$ are in SPAN$\{\mathbf{v}_1, \mathbf{v}_2\}$ and that $\mathbf{v}_1$ and $\mathbf{v}_2$ are in SPAN$\{\mathbf{u}_1, \mathbf{u}_2\}$.)

8. Prove that the only nontrivial ($\neq \{\mathbf{0}\}$ or $\mathcal{R}^3$) subspaces of $\mathcal{R}^3$ are lines and planes through the origin.

†9. Find a matrix B such that $\mathcal{NS}(B) = \mathcal{CS}(A)$ if A is the matrix given.

   (a) $A = \begin{bmatrix} 1 & 2 & 3 \\ 4 & 5 & 6 \\ 7 & 8 & 9 \end{bmatrix}$ (b) $A = \begin{bmatrix} 1 & 1 & 4 \\ -1 & 0 & 2 \\ 3 & 1 & 3 \\ 0 & 0 & -6 \end{bmatrix}$

   (c) $A = \begin{bmatrix} 1 & 0 & 0 & 1 \\ 1 & 1 & -1 & 0 \\ 1 & 2 & -2 & -1 \\ 1 & 3 & -3 & -2 \end{bmatrix}$

†10. Show that if $\mathbf{u}_1 + 2\mathbf{u}_2 - 3\mathbf{u}_3 + 4\mathbf{u}_4 = 0$, then

$$\text{SPAN}\{\mathbf{u}_1, \mathbf{u}_2, \mathbf{u}_3, \mathbf{u}_4\} = \text{SPAN}\{\mathbf{u}_2, \mathbf{u}_3, \mathbf{u}_4\}.$$

11. (a) Show that elementary column operations on a matrix A do not change the column space of A.

   (*Hint:* Recalling Exercise 21 of Section 1.7, show that $\mathcal{CS}(AQ) \subseteq \mathcal{CS}(A)$ and, if Q^{-1} exists, that $\mathcal{CS}(A) \subseteq \mathcal{CS}(AQ)$.)

   (b) How do elementary row operations affect the column space of A?

   (*Hint:* Consider $A = \begin{bmatrix} 1 & 3 \\ 2 & 6 \end{bmatrix}$ and the matrix B obtained from A by the operation $R_2 \leftarrow -2R_1 + R_2$. What do $\mathcal{CS}(A)$ and $\mathcal{CS}(B)$ represent geometrically in R^2?)

3.4   *Vector Spaces—The General Concept*

In this section we will introduce and study a class of algebraic systems called vector spaces. The abstract vector concept, which we study here, is a fruitful generalization of the vectors in $\mathcal{R}^2$, $\mathcal{R}^3$, and $\mathcal{R}^n$ that we studied in the last three sections. The generalization of the vector concept that we undertake here is much like the multistep generalization of the number concept that you have been exposed to throughout your mathematical education. For most readers, studying this concept will be a new mathematical experience because we will define our abstract vectors by *how they behave,* rather than by what they are; this is consistent with the way most people define political terms such as "liberal" or "conservative." Generalizations of this type are most useful if there are many interesting and important examples, making it worthwhile to study all the systems at the same time. It is essential to have a familiar, motivating example to provide insight and a source of terminology.

The starting place for our generalized definition of a vector is the list of properties of $\mathcal{R}^n$ contained in Theorem 3.1. Thus, $\mathcal{R}^n$ will automatically satisfy our general definition and will have the distinction of being our "motivating example." The abstract systems we are going to study will be systems that behave "like $\mathcal{R}^n$," even if their elements are very different from those of $\mathcal{R}^n$.

In order for a system to behave "like $\mathcal{R}^n$," there must be defined, within the system, a way of combining two elements to get another element of the system, like vector addition in $\mathcal{R}^n$. There must also be a way of combining real numbers and elements of the system to get another element of the system, like scalar multiplication in $\mathcal{R}^n$. The operations in the system must also be like those in $\mathcal{R}^n$ in the sense that they satisfy the properties listed in Theorem 3.1. Thus, we are led to the following formal definition.

Definition 3.10

A REAL VECTOR SPACE $(\mathcal{V}, +, \cdot)$ is a set $\mathcal{V}$ of objects that we will call vectors and denote by boldface letters like **u**, **v**, **w**, and **x**, along with two ways of combining them called vector addition ($+$) and scalar multiplication ($\cdot$), which satisfy the following ten axioms.

(1) For any **u** and **v** in $\mathcal{V}$, **u** + **v** is also an element of $\mathcal{V}$. (The set $\mathcal{V}$ is closed under vector addition.)

(2) For any vectors **u**, **v**, and **w** in $\mathcal{V}$, **u** + (**v** + **w**) = (**u** + **v**) + **w**. (Vector addition is associative.)

(3) For any vectors **u** and **v** in $\mathcal{V}$, **u** + **v** = **v** + **u**. (Vector addition is commutative.)

(4) There is a unique vector **0** in $\mathcal{V}$ such that **u** + **0** = **u** for

every **u** in $\mathcal{V}$. (The vector **0** is called the additive identity element of $\mathcal{V}$.)

(5) For each **u** in $\mathcal{V}$, there is a unique vector $(-\mathbf{u})$ in $\mathcal{V}$ such that $\mathbf{u} + (-\mathbf{u}) = \mathbf{0}$. (The vector $-\mathbf{u}$ is called the additive inverse of **u**.)

(6) For any vector **u** and any real number k, $k \cdot \mathbf{u}$ is in $\mathcal{V}$. (The set $\mathcal{V}$ is closed under scalar multiplication.)

(7) For any vector **u** and any scalars h and k, $(h + k) \cdot \mathbf{u} = h \cdot \mathbf{u} + k \cdot \mathbf{u}$. (Scalar multiplication distributes over scalar addition.)

(8) For any scalar h and any vectors **u** and **v**, $h \cdot (\mathbf{u} + \mathbf{v}) = h \cdot \mathbf{u} + h \cdot \mathbf{v}$. (Scalar multiplication distributes over vector addition.)

(9) For any scalars h and k and any vector **u**, $h \cdot (k \cdot \mathbf{u}) = (hk) \cdot \mathbf{u}$. (Scalar multiplication is associative.)

(10) $1 \cdot \mathbf{u} = \mathbf{u}$ for every **u** in $\mathcal{V}$. (The real number 1 is an identity element for scalar multiplication.)

We call $(\mathcal{V}, +, \cdot)$ a *real* vector space because the scalars we use are real numbers; there are further generalizations to complex vector spaces and beyond, but we will restrict ourselves to real scalars in this text. We will adopt the common practice of referring to the vector space $(\mathcal{V}, +, \cdot)$ as the vector space $\mathcal{V}$, when there is no danger of confusion about the operations to be used. Also, we will normally use juxtaposition to indicate scalar multiplication; that is, we will write $k\mathbf{u}$ instead of $k \cdot \mathbf{u}$.

In order to test whether or not a given algebraic system is a real vector space, we must first identify a suitable vector addition and a suitable scalar multiplication. Then we must verify that each of the ten axioms of Definition 3.10 is satisfied. We will now give some examples of systems which have the desired structure and, hence, are real vector spaces. Additional examples are contained in the exercises.

Example 1

$\mathcal{R}^n = \mathcal{R}^{n \times 1}$ is our motivating example. Theorem 3.1 shows that $\mathcal{R}^n$ is a real vector space for any positive integer n. For $n = 1$, $\mathcal{R}^n = \mathcal{R}$, so $\mathcal{R}$ is itself a real vector space. ■

Example 2

Any subspace of $\mathcal{R}^n$ is a real vector space since the closure axioms (1 and 6) are satisfied as a consequence of Definition 3.7 and all of the remaining properties are inherited from $\mathcal{R}^n$. ■

Example 3

$\mathcal{R}^{m \times n}$, the set of all real $m \times n$ matrices, is a vector space by Theorem 1.4. An important special case is $\mathcal{R}^{1 \times n}$, a space that is "very much like" $\mathcal{R}^n = \mathcal{R}^{n \times 1}$. We will call the elements of $\mathcal{R}^{1 \times n}$ ROW VECTORS. ■

Example 4

The set $\mathcal{P} = \{a_0 + a_1x + \ldots + a_nx^n \mid n \text{ is finite}\}$ of all polynomials with addition and scalar multiplication defined as in elementary algebra. In other words,

$$\text{if } \mathbf{p} = a_0 + a_1x + a_2x^2 + \ldots + a_nx^n \text{ and}$$
$$\mathbf{q} = b_0 + b_1x + b_2x^2 + \ldots + b_nx^n,$$

then the sum of the polynomials $\mathbf{p}$ and $\mathbf{q}$ is defined to be

$$\mathbf{p} + \mathbf{q} = (a_0 + b_0) + (a_1 + b_1)x + (a_2 + b_2)x^2$$
$$+ \ldots + (a_n + b_n)x^n,$$

and scalar multiplication is defined by

$$k\mathbf{p} = ka_0 + ka_1x + ka_2x^2 + \ldots + ka_nx^n.$$

Note that the definition of addition of polynomials does not require that $\mathbf{p}$ and $\mathbf{q}$ have the same degree since we can always add terms with zero coefficients. The operations in this set are familiar to you and are known to satisfy all of the axioms of Definition 3.10. Thus $\mathcal{P}$ is an example of a real vector space.   ■

Our next example is also a familiar one because the elements of the set are functions, the objects studied in trigonometry, analytic geometry, and calculus.

Example 5

Let $\mathcal{F} = \mathcal{F}(a, b)$ be the set of all real-valued functions defined on the interval $[a, b] = \{x \mid a \leq x \leq b\}$.

In dealing with the vector space $\mathcal{F}$ it is important to distinguish carefully between a function

$$\mathbf{f} = \{(x, f(x)) \mid a \leq x \leq b\}$$

and the value, $f(x)$, of the function at x. Equality in $\mathcal{F}$ is defined by

$$\mathbf{f} = \mathbf{g} \text{ if and only if } f(x) = g(x) \text{ for all } x \text{ in } [a, b].$$

Addition in $\mathcal{F}$ is defined by

$$(\mathbf{f} + \mathbf{g})(x) = f(x) + g(x),$$

and scalar multiplication by

$$(a\mathbf{f})(x) = a(f(x)).$$

The zero vector in $\mathcal{F}$ is the function $\mathbf{0}$ whose value is 0 for every x and the additive inverse of $\mathbf{f}$ is the function $-\mathbf{f}$ which satisfies

$$(-\mathbf{f})(x) = -(f(x)).$$

It is routine to verify any of the remaining vector-space axioms for the set $\mathcal{F}$. For example, to show that addition in $\mathcal{F}$ is associative, we calculate

$$((\mathbf{f} + \mathbf{g}) + \mathbf{h})(x) = (\mathbf{f} + \mathbf{g})(x) + h(x) = [f(x) + g(x)] + h(x)$$

and

$$(\mathbf{f} + (\mathbf{g} + \mathbf{h}))(x) = f(x) + (\mathbf{g} + \mathbf{h})(x) = f(x) + [g(x) + h(x)],$$

and then observe that the last two expressions are equal by the associativity of addition in $\mathcal{R}$. Thus, $\mathbf{f} + (\mathbf{g} + \mathbf{h}) = (\mathbf{f} + \mathbf{g}) + \mathbf{h}$. ■

Note that the elements of the vector space $\mathcal{P}$ of Example 4 can be viewed as polynomial functions and, hence, as elements of the vector space $\mathcal{F}$ of Example 5. Thus, $\mathcal{F}$ contains $\mathcal{P}$ as a subset.

Example 6

Consider $\mathcal{V} = \{[a \quad b] \mid a \text{ and } b \text{ in } \mathcal{R}\}$, the set of all 1×2 real matrices, with addition defined by $[a \quad b] \oplus [c \quad d] = [a + c + 2 \quad b + d]$ and with scalar multiplication defined by $k : [a \quad b] = [ka \quad b]$. Is $(V, \oplus, :)$ a real vector space?

To answer this question we need to either establish that the ten axioms of Definition 3.10 are satisfied ($\mathcal{V}$ is a vector space) or find one of the ten axioms which is not satisfied ($\mathcal{V}$ is not a vector space). The closure properties (1 and 6) are clearly satisfied. To check the associativity of $\oplus$ we compute

$$[a \quad b] \oplus ([c \quad d] \oplus [e \quad f]) = [a \quad b] \oplus [c + e + 2 \quad d + f]$$
$$= [a + (c + e + 2) + 2 \quad b + (d + f)]$$

and

$$([a \quad b] \oplus [c \quad d]) \oplus [e \quad f] = [a + c + 2 \quad b + d] + [e \quad f]$$
$$= [(a + c + 2) + e + 2 \quad (b + d) + f].$$

These vectors are the same because addition in $\mathcal{R}$ is associative, so we have established that $\oplus$ is associative in $\mathcal{V}$ (Axiom 2). Since $[a \quad b] \oplus [c \quad d] = [a + c + 2 \quad b + d] = [c + a + 2 \quad d + b] = [c \quad d] \oplus [a \quad b]$, we see that $\oplus$ is also commutative in $\mathcal{V}$ (Axiom 3). In order to find an identity element for $\oplus$ we let $\mathbf{z} = [x \quad y]$ and try to determine x and y so that $[a \quad b] \oplus [x \quad y] = [a \quad b]$. This statement is equivalent to $[a + x + 2 \quad b + y] = [a \quad b]$, or to

$$a + x + 2 = a \quad \text{and} \quad b + y = b.$$

It follows that $x = -2$ and $y = 0$. Thus, $\mathbf{z} = [-2 \quad 0]$ is the identity element for $\oplus$ and Axiom 4 is satisfied. To verify Axiom 5 we need to find scalars x and y so that

$$[a \quad b] + [x \quad y] = \mathbf{z} = [-2 \quad 0].$$

This expression is the same as $[a + x + 2 \quad b + y] = [-2 \quad 0]$, so we must have $x = -4 - a$ and $y = -b$. Thus, $[-4 - a \quad -b] = -[a \quad b]$ is the additive inverse of $[a \quad b]$. So far we have shown that $(\mathcal{V}, \oplus, :)$ satisfies all the additive properties of a vector space. In order to check Axiom 7 we compute $(h + k):[a \quad b] = [(h + k)a \quad b]$ and $h:[a \quad b] \oplus k:[a \quad b] = [ha \quad b] \oplus [ka \quad b] = [ha + ka + 2, b + b]$. Since $(h + k):[a \quad b] \neq h:[a \quad b] \oplus k:[a \quad b]$, we see that Axiom 7 fails to hold and, hence, that $(\mathcal{V}, \oplus, :)$ is not a real vector space. You should check that Axiom 8 also fails to hold and that Axioms 9 and 10 are both satisfied for this system. Thus, $(\mathcal{V}, \oplus, :)$ fails to be a vector space because scalar multiplication does not distribute over scalar addition and scalar multiplication does not distribute over vector addition. ▪

In our next theorem we show that some familiar algebraic properties hold for arbitrary vector spaces. These results are certainly obvious for all of the examples already encountered; the question before us now is whether or not they can be deduced from the axioms of Definition 3.10. We illustrate the strength of the abstract approach by showing that the results hold in general. The proofs are not difficult, but they require careful attention if you have not dealt previously with arguments of this kind.

Theorem 3.6

If **u** is any element of the vector space $\mathcal{V}$ and k is any real number, then the following assertions are true.

(1) $0\mathbf{u} = \mathbf{0}$.
(2) $k\mathbf{0} = \mathbf{0}$.
(3) $(-1)\mathbf{u} = -\mathbf{u}$.
(4) If $k\mathbf{u} = \mathbf{0}$, then either $\mathbf{u} = \mathbf{0}$ or $k = 0$.

Proof.   We will prove the first assertion carefully and outline the proofs of the other parts. For parts 2, 3, and 4 you should supply a reason for each step in the argument.

(1) Since $0 + 0 = 0$, we can write $0\mathbf{u} = (0 + 0)\mathbf{u}$ and then use Axiom 7 to write

$$0\mathbf{u} = 0\mathbf{u} + 0\mathbf{u}.$$

Now, by Axiom 5, there is a vector $-0\mathbf{u}$ in V such that $0\mathbf{u} + (-0\mathbf{u}) = \mathbf{0}$. Adding this vector to both sides of the last equation yields

$$\mathbf{0} = 0\mathbf{u} + (-0\mathbf{u}) = (0\mathbf{u} + 0\mathbf{u}) + (-0\mathbf{u}),$$

which, by the associativity of addition in $\mathcal{V}$, can be rewritten as

$$\mathbf{0} = 0\mathbf{u} + (0\mathbf{u} + (-0\mathbf{u})) = 0\mathbf{u} + \mathbf{0} = 0\mathbf{u}.$$

Thus, $0\mathbf{u} = \mathbf{0}$ as desired.

(2) $k\mathbf{0} = k(\mathbf{0} + \mathbf{0}) = k\mathbf{0} + k\mathbf{0}$

$$\mathbf{0} = k\mathbf{0} + (-k\mathbf{0}) = (k\mathbf{0} + k\mathbf{0}) + (-k\mathbf{0})$$
$$= k\mathbf{0} + (k\mathbf{0} + (-k\mathbf{0})) = k\mathbf{0}$$

(3) $\mathbf{u} + (-1)\mathbf{u} = 1\mathbf{u} + (-1)\mathbf{u} = (1 - 1)\mathbf{u} = 0\mathbf{u} = \mathbf{0}$. Therefore, $(-1)\mathbf{u} = -\mathbf{u}$.

(4) If $k\mathbf{u} = \mathbf{0}$ and $k \neq 0$, then
$$\mathbf{0} = (1/k)\mathbf{0} = (1/k)(k\mathbf{u}) = ((1/k)k)\mathbf{u} = 1\mathbf{u} = \mathbf{u}. \quad \square$$

Example 7

Let $\mathbf{u}$, $\mathbf{v}$, and $\mathbf{w}$ be elements of the real vector space $\mathcal{V}$. Use the axioms of Definition 3.10 and the results of Theorem 3.6 to simplify

$$3\mathbf{u} + 2\mathbf{v} - 7\mathbf{w} = 5\mathbf{v} + 3\mathbf{u} - 8\mathbf{w}.$$

Adding $-(5\mathbf{v} + 3\mathbf{u} - 8\mathbf{w}) = -5\mathbf{v} - 3\mathbf{u} + 8\mathbf{w}$ to both sides of the given equation yields

$$3\mathbf{u} + 2\mathbf{v} - 7\mathbf{w} - 5\mathbf{v} - 3\mathbf{u} + 8\mathbf{w} = \mathbf{0}$$

or, using the commutativity and distributivity properties,

$$(3 - 3)\mathbf{u} + (2 - 5)\mathbf{v} + (-7 + 8)\mathbf{w} = \mathbf{0},$$

which is the same as

$$0\mathbf{u} + (-3)\mathbf{v} + (1)\mathbf{w} = \mathbf{0}.$$

Thus, by Theorem 3.6 and Axiom 10 we have

$$-3\mathbf{v} + \mathbf{w} = \mathbf{0} \quad \text{or} \quad \mathbf{w} = 3\mathbf{v}. \quad \blacksquare$$

Exercise Set 3.4

In Exercises 1–10 a set of objects $\mathcal{V}$, an addition, and a scalar multiplication are specified. Determine whether $\mathcal{V}$ is a real vector space with the operations given. Note that in some cases familiar symbols have nonstandard meanings.

1. $\mathcal{V}_1 = \{(a, b)\}$, with $(a, b) + (c, d) = (a + c, b + d)$ and $k(a, b) = (ka, 0)$

2. $\mathcal{V}_2 = \{(a, b)\}$, with $(a, b) + (c, d) = (a + c, b + d)$ and $k(a, b) = (0, 0)$

3. $\mathcal{V}_3 = \{(a, b)\}$, with $(a, b) + (c, d) = (a + c, b + d)$ and $k(a, b) = (2ka, 2kb)$

4. $\mathcal{V}_4 = \{n \times n \text{ diagonal matrices}\}$, with $A + B = AB$, and ordinary scalar multiplication

5. $\mathcal{V}_5 = \{A \text{ in } \mathcal{R}^{n \times n} | A^T = A\}$, with the usual matrix addition and scalar multiplication

6. $\mathcal{V}_6 = \{(a, b)\}$, with $(a, b) + (c, d) = (a + c + 1, b + d - 1)$ and $k(a, b) = (ka, kb)$

7. $\mathcal{V}_7 = \{a \mid a \geq 0\}$, with $a + b = ab$ and $ka = a^k$

8. $\mathcal{V}_8 = \{(a, b, c) \mid a \geq 0, b \geq 0\}$, with usual matrix operations

9. $\mathcal{V}_9 = \{p(x) \text{ in } \mathcal{P} \mid \text{degree of } p(x) \text{ is greater than } 5\}$

10. $\mathcal{V}_{10} = \{\text{subsets of } \mathcal{R}^3\}$ with $S_1 + S_2 = S_1 \cup S_2$ and $kS = S$

In Exercises 11–18 a subset $\mathcal{W}$ of a vector space $\mathcal{V}$ is given. Determine whether $\mathcal{W}$ is closed under addition and scalar multiplication.

11. $\mathcal{W}_1 = \{a_1\mathbf{u}_1 + a_2\mathbf{u}_2 \mid \mathbf{u}_1 \text{ and } \mathbf{u}_2 \text{ fixed in } \mathcal{V}; \quad a_1 \text{ and } a_2 \text{ real}\}$

12. $\mathcal{W}_2 = \{\mathbf{f} \text{ in } \mathcal{F}[a, b] \mid f(a) = 0\}$

13. $\mathcal{W}_3 = \{(x_1, x_2)^T \text{ in } \mathcal{R}^2 \mid x_1 \geq 0\}$

14. $\mathcal{W}_4 = \{p(x) \text{ in } \mathcal{P} \mid \text{degree of } p(x) \text{ is even}\}$

15. $\mathcal{W}_5 = \{[x_1 \quad x_2 \quad x_3]^T \text{ in } \mathcal{R}^3 \mid x_1^2 + x_2^2 + x_3^2 = 1\}$

16. $\mathcal{W}_6 = \{A \text{ in } \mathcal{R}^{n \times n} \mid AB = BA\}$ (B fixed in $\mathcal{R}^{n \times n}$)

17. $\mathcal{W}_7 = \{\mathbf{f} \text{ in } \mathcal{F} \mid f \text{ has a second derivative}\}$

18. $\mathcal{W}_8 = \{\text{eigenvectors of } A \text{ in } \mathcal{R}^n\}$ where A is a fixed $n \times n$ matrix

19. Verify that vector addition in $\mathcal{F} = \mathcal{F}[a, b]$ is commutative and that scalar multiplication distributes over vector addition.

20. Let $\mathcal{S}$ be any set. Show that the set of all real-valued functions with domain $\mathcal{S}$ is a real vector space.

21. Show how to define vector addition and scalar multiplication so that any set with a single element becomes a real vector space.

22. Let $\mathbf{u}$ and $\mathbf{v}$ be elements of an arbitrary real vector space. Prove that if $4\mathbf{u} + 3\mathbf{v} = 3\mathbf{v} + 2\mathbf{u}$, then $\mathbf{u} = \mathbf{0}$.

23. Let $\mathbf{u}$, $\mathbf{v}$, and $\mathbf{w}$ be elements of a real vector space $\mathcal{V}$. Show that if $\mathbf{u} + \mathbf{v} = \mathbf{w} + \mathbf{v}$, then $\mathbf{u} = \mathbf{w}$.

24. Prove that $-(-\mathbf{u}) = \mathbf{u}$ for any vector $\mathbf{u}$.

25. Show that Definition 3.10 may be weakened by deleting the word "unique" in Axioms 4 and 5.

†26. Let $\mathcal{V}$ be any real vector space and let $\mathcal{V}^{m \times n}$ be the set of all $m \times n$ matrices with entries from $\mathcal{V}$. Show that $\mathcal{V}^{m \times n}$ is a real vector space.

3.5   *Subspaces*

Many important examples of vector spaces are obtained by considering subspaces of the vector spaces encountered in the last section. We have considered the notion of a subspace of $\mathcal{R}^n$ in Sections 3.2 and 3.3; the general definition follows. Note that this definition is not just a rewrite of the one given in Section 3.2. The equivalence of the two definitions is established in Theorem 3.7.

Definition 3.11

A subset $\mathcal{W}$ of the vector space $(\mathcal{V}, +, \cdot)$ is called a SUBSPACE of $\mathcal{V}$ if $(\mathcal{W}, +, \cdot)$ is itself a vector space, with the operations $+$ and $\cdot$ inherited from $\mathcal{V}$.

To check whether a given subset $\mathcal{W}$ of the vector space $\mathcal{V}$ is a subspace, it is not necessary to check all ten axioms. Since the operations in $\mathcal{W}$ are inherited from $\mathcal{V}$, all the manipulative properties hold in the larger set $\mathcal{V}$ and hence also in the smaller set $\mathcal{W}$. It is therefore only necessary to check the closure properties (1 and 6) and the existence properties (4 and 5). Thus, $\mathcal{W}$ is a subspace of $\mathcal{V}$ if

(1) for every $\mathbf{u}$ and $\mathbf{v}$ in $\mathcal{W}$, $\mathbf{u} + \mathbf{v}$ is in $\mathcal{W}$;
(2) for every $\mathbf{u}$ in $\mathcal{W}$ and every a in $\mathcal{R}$, $a\mathbf{u}$ is in $\mathcal{W}$;
(3) for any $\mathbf{u}$ in $\mathcal{W}$, $-\mathbf{u}$ is in $\mathcal{W}$; and
(4) the zero vector $\mathbf{0}$ is in $\mathcal{W}$.

If any one of these four properties fails to hold, then $\mathcal{W}$ is not a subspace of $\mathcal{V}$. Note that property 4 implies that a subspace $\mathcal{W}$ is nonempty. Many interesting subsets of the vector space $\mathcal{V}$ fail to be subspaces because property 4 fails; it is always worth checking this property first. For example, this argument shows that a plane in $\mathcal{R}^3$ is not a subspace unless it contains the origin, and that the set of all solutions of the nonhomogeneous system $AX = K$, $K \neq 0$ is not a subspace.

Since $-\mathbf{u} = (-1)\mathbf{u}$ and $\mathbf{0} = 0\mathbf{u}$, we see that the existence properties (3 and 4) follow from the closure properties, provided that $\mathcal{W}$ is nonempty. If we combine the two closure properties into a single condition, then we have the following theorem which provides an efficient criterion for determining if a given subset of a vector space is a subspace. Note that this theorem also establishes the equivalence of the definition of a subspace given here and the one given in Section 3.2 for the special case $\mathcal{V} = \mathcal{R}^n$.

Theorem 3.7

Let $\mathcal{V}$ be a real vector space. The subspaces of $\mathcal{V}$ are precisely those nonempty subsets of $\mathcal{V}$ that are closed under arbitrary linear combinations; that is, a nonempty subset $\mathcal{W}$ is a subspace of $\mathcal{V}$ if and only if $a\mathbf{u} + b\mathbf{v}$ is in $\mathcal{W}$ for all vectors $\mathbf{u}$ and $\mathbf{v}$ in $\mathcal{W}$ and all scalars a and b in $\mathcal{R}$.

Example 1

Is $\mathcal{W} = \{A \text{ in } \mathcal{R}^{3 \times 3} \mid \det(A) = 0\}$ a subspace of $\mathcal{R}^{3 \times 3}$?

In general, $\det(A + B) \neq \det(A) + \det(B)$, so we do not expect this set to be closed under addition. In fact, the matrices

$$A = \begin{bmatrix} 1 & 0 & 0 \\ 0 & 1 & 0 \\ 0 & 0 & 0 \end{bmatrix} \quad \text{and} \quad B = \begin{bmatrix} 0 & 0 & 0 \\ 0 & 1 & 0 \\ 0 & 0 & 1 \end{bmatrix} \text{ are in } \mathcal{W}$$

since $\det(A) = \det(B) = 0$, but $\det(A + B) = \det\begin{bmatrix} 1 & 0 & 0 \\ 0 & 2 & 0 \\ 0 & 0 & 1 \end{bmatrix} = 2 \neq 0$.

This shows that $A + B$ is not in $\mathcal{W}$ and hence that $\mathcal{W}$ is not a subspace of the vector space $\mathcal{R}^{3\times3}$. This example generalizes easily to show that the set of all $n \times n$ singular matrices is not a subspace of the vector space $\mathcal{R}^{n\times n}$. ▪

The next example requires a little knowledge of calculus. If you have never studied calculus, please go on to Example 3.

Example 2

Is $\mathcal{W} = \{\mathbf{f} \text{ in } \mathcal{F}(a, b) \mid \mathbf{f} \text{ has a continuous second derivative and } f''(x) + 25f(x) = 0\}$ a subspace of the function space $\mathcal{F}(a, b)$?

If $\mathbf{f}$ and $\mathbf{g}$ are elements of $\mathcal{W}$, then $f''(x) + 25f(x) = 0$ and $g''(x) + 25g(x) = 0$. We now let h and k be any scalars and try to determine if $h\mathbf{f} + k\mathbf{g}$ is in $\mathcal{W}$. To do this we compute

$$(hf(x) + kg(x))'' + 25(hf(x) + kg(x))$$

$$= (hf''(x) + kg''(x)) + 25hf(x) + 25kg(x)$$

$$= h(f''(x) + 25f(x)) + k(g''(x) + 25g(x))$$

$$= h0 + k0 = 0.$$

Thus $h\mathbf{f} + k\mathbf{g}$ is in $\mathcal{W}$ and, from Theorem 3.7, we conclude that $\mathcal{W}$ is a subspace of the function space $\mathcal{F}$. It is easy to verify that $\cos 5x$ and $\sin 5x$ are elements of $\mathcal{W}$; in fact, $\mathcal{W}$ can be shown to consist of all functions $\mathbf{h}$ such that $h(x) = a \cos 5x + b \sin 5x$. ▪

This last example, along with Theorem 3.3, points out one reason for the interest in subspaces: frequently the set of all solutions of an equation or system of equations can be shown to be a subspace of some vector space. It is precisely the linear homogeneous equations that have this property. This fact motivates the general definition of linearity to be given in the next chapter.

Example 3

Is $\mathcal{P}_5 = \{p(x) = a_0 + a_1x + \ldots + a_nx^n \mid n \leq 5\}$ a subspace of $\mathcal{P}$?

The elements of $\mathcal{P}_5$ are the polynomials of degree 5 or less. Since multiplication by a scalar cannot increase the degree of a polynomial and since the sum of two polynomials of degree 5 or less cannot have degree greater than 5, it follows from Theorem 3.7 that $\mathcal{P}_5$ is a subspace of $\mathcal{P}$. ▪

The notion of the subspace spanned by a given set of vectors, which we have already considered for the special case $\mathscr{R}^n$ (see Definition 3.8), generalizes naturally to an arbitrary vector space $\mathscr{V}$.

Definition 3.12

Let $\mathscr{V}$ be an arbitrary real vector space and let $\mathscr{S} = \{\mathbf{u}_1, \mathbf{u}_2, \ldots, \mathbf{u}_s\}$ be an arbitrary set of vectors from $\mathscr{V}$. The subspace spanned by the set $\mathscr{S}$ is defined by

$$\text{SPAN}\{\mathbf{u}_1, \mathbf{u}_2, \ldots, \mathbf{u}_s\} = \left\{ \sum_{k=1}^{s} a_k \mathbf{u}_k \,\middle|\, a_k \text{ real} \right\};$$

that is, the subspace consists of all possible linear combinations of the vectors of the set $\mathscr{S}$. A set $\mathscr{S}$ of vectors is said to span the subspace $\mathscr{W}$ if every element of $\mathscr{W}$ can be written as a linear combination of the vectors of the set $\mathscr{S}$.

The calculation

$$r\left(\sum a_k \mathbf{u}_k \right) + s\left(\sum b_k \mathbf{u}_k \right) = \sum (ra_k + sb_k)\mathbf{u}_k$$

shows that a linear combination, of two linear combinations of the $\mathbf{u}_i$, is another linear combination of the $\mathbf{u}_i$. It follows, from Theorem 3.7, that $\text{SPAN}\{\mathbf{u}_1, \ldots, \mathbf{u}_s\}$ is indeed a subspace of $\mathscr{V}$.

It is sometimes useful to think of $\text{SPAN}\{\mathbf{u}_1, \ldots, \mathbf{u}_s\}$ as the "smallest" subspace of $\mathscr{V}$ containing the spanning set; that is, if $\mathscr{W}$ is any other subspace of $\mathscr{V}$ containing all the $\mathbf{u}_i$, then $\text{SPAN}\{\mathbf{u}_1, \ldots, \mathbf{u}_s\}$ is a subspace of $\mathscr{W}$ (see Exercise 11 of this section). Recall that for $\mathscr{V} = \mathscr{R}^3$, $\text{SPAN}\{\mathbf{u}, \mathbf{v}\}$ can be thought of as the plane through the origin determined by the vectors $\mathbf{u}$ and $\mathbf{v}$.

Example 4

For the subspace $\mathscr{P}_5$ of $\mathscr{P}$, discussed in Example 3, it should be apparent that the set of polynomials (vectors in $\mathscr{P}$, if you please)

$$\mathscr{S} = \{1, x, x^2, x^3, x^4, x^5\}$$

spans $\mathscr{P}_5$. To see this we observe that a typical element of $\mathscr{P}_5$ is

$$a_0 + a_1 x + a_2 x^2 + a_3 x^3 + a_4 x^4 + a_5 x^5,$$

which is clearly a linear combination of the elements of the set $\mathscr{S}$. ∎

It is generally helpful to be able to find a spanning set for a given subspace; this is particularly true if the subspace consists of all solutions of a given problem. The subspace of Example 2 can be shown to be $\text{SPAN}\{\cos 5x, \sin 5x\}$ and this description of that subspace describes it clearly, even for those of you who have not studied calculus.

Finding a spanning set, for a subspace consisting of all solutions of a given problem, amounts to solving the problem. Be careful not to overgeneralize; not all problems have solution sets that are subspaces.

Example 5

Find a spanning set for $\mathscr{NS}(A) = \{X \mid AX = 0\}$ if

$$A = \begin{bmatrix} 1 & 0 & 2 & -4 \\ 0 & 1 & 5 & 3 \\ 0 & 0 & 0 & 0 \\ 0 & 0 & 0 & 0 \end{bmatrix}.$$

Finding the elements of the null space of A amounts to finding all solutions of the homogeneous linear system $AX = 0$. Since A is already in echelon form the solution of the system $AX = 0$ is easy. The nonzero equations of the system are

$$x_1 + 2x_3 - 4x_4 = 0$$

$$x_2 + 5x_3 + 3x_4 = 0$$

and, if we set $x_3 = c_1$ and $x_4 = c_2$, we obtain as the general solution,

$$X = \begin{bmatrix} -2c_1 + 4c_2 \\ -5c_1 - 3c_2 \\ c_1 \\ c_2 \end{bmatrix} = c_1 \begin{bmatrix} -2 \\ -5 \\ 1 \\ 0 \end{bmatrix} + c_2 \begin{bmatrix} 4 \\ -3 \\ 0 \\ 1 \end{bmatrix}.$$

This expression for the solution shows that $\mathscr{NS}(A) = \text{SPAN}\{X_1, X_2\}$

where $X_1 = \begin{bmatrix} -2 \\ -5 \\ 1 \\ 0 \end{bmatrix}$ and $X_2 = \begin{bmatrix} 4 \\ -3 \\ 0 \\ 1 \end{bmatrix}.$ ∎

Example 6

Which of the functions $\sin^2 x$, $\sin x$, and $\cos^2 x$ are in the subspace of $\mathscr{F}$ spanned by $\{1, \sin 2x, \cos 2x\}$?

From the familiar trigonometric identities

$$\sin^2 x = \frac{(1 - \cos 2x)}{2} \quad \text{and} \quad \cos^2 x = \frac{(1 + \cos 2x)}{2},$$

it follows at once that both $\sin^2 x$ and $\cos^2 x$ are in the subspace $\text{SPAN}\{1, \cos 2x, \sin 2x\}$. In order to determine whether $\sin x$ is in this subspace we will try to find scalars a, b, and c so that

$$\sin x = a + b \cos 2x + c \sin 2x. \tag{3.3}$$

Equation 3.3 is a functional identity and must hold for all values of x. For each value of x the functional equation reduces to a linear equation; in particular for

$$x = 0, \qquad 0 = a + b;$$

$$x = \frac{\pi}{2}, \qquad 1 = a - b; \quad \text{and}$$

$$x = \frac{\pi}{6}, \qquad \frac{1}{2} = a + \frac{b}{2} + \frac{\sqrt{3}c}{2}.$$

These three equations are easily solved to yield $a = 1/2$, $b = -1/2$, and $c = 1/(2\sqrt{3})$. Thus, the only possible solution of Equation 3.3 is

$$\sin x = \frac{1}{2} - \frac{\cos 2x}{2} + \frac{\sin 2x}{2\sqrt{3}}.$$

Now we need to ask if this last equation is valid for all values of x as required. For $x = \pi/4$ this equation leads to

$$.707\ldots = \frac{\sqrt{2}}{2} \overset{?}{=} \frac{1}{2} - 0 + \frac{1}{2\sqrt{3}} = .788\ldots,$$

which is clearly not valid. Thus it is not possible to find scalars a, b, and c so that Equation 3.3 is satisfied for all x. Therefore, $\sin x$ is *not* an element of the subspace SPAN$\{1, \cos 2x, \sin 2x\}$. ■

Example 7

Show that the polynomials

$$h_1(x) = x^2 - 3x + 2, \quad h_2(x) = x^2 - 2x, \quad \text{and} \quad h_3(x) = x^2 - x$$

span the subspace

$$\mathcal{P}_2 = \{a_0 + a_1x + a_2x^2 \,|\, a_i \text{ real}\}.$$

We need to demonstrate that any element of $\mathcal{P}_2$ can be expressed as a linear combination of $h_1(x)$, $h_2(x)$, and $h_3(x)$. Thus for $a_0 + a_1x + a_2x^2$, an arbitrary element of $\mathcal{P}_2$, we seek scalars c_1, c_2, and c_3 such that

$$c_1h_1(x) + c_2h_2(x) + c_3h_3(x) = a_0 + a_1x + a_2x^2$$

This equation is equivalent to

$$c_1(x^2 - 3x + 2) + c_2(x^2 - 2x) + c_3(x^2 - x) = a_0 + a_1x + a_2x^2$$

or

$$(c_1 + c_2 + c_3)x^2 + (-3c_1 - 2c_2 - c_3)x + 2c_1 = a_0 + a_1x + a_2x^2.$$

From the last equation we see that c_1, c_2, and c_3 must satisfy

$$
\begin{aligned}
2c_1 &= a_0 \\
-3c_1 - 2c_2 - c_3 &= a_1 \\
c_1 + c_2 + c_3 &= a_2
\end{aligned}
\quad \text{or} \quad
\begin{bmatrix} 2 & 0 & 0 \\ -3 & -2 & -1 \\ 1 & 1 & 1 \end{bmatrix}
\begin{bmatrix} c_1 \\ c_2 \\ c_3 \end{bmatrix}
=
\begin{bmatrix} a_0 \\ a_1 \\ a_2 \end{bmatrix}.
$$

Since the coefficient matrix of this system is nonsingular (its determinant is $-2 \neq 0$) it follows that this system has a unique solution for every choice of a_0, a_1, and a_2. Thus

$$
\mathcal{P}_2 = \mathrm{SPAN}\{h_1(x), h_2(x), h_3(x)\}. \quad \blacksquare
$$

In Sections 3.2 and 3.3 we introduced two subspaces of $\mathcal{R}^n$: $\mathcal{NS}(A)$ and $\mathcal{CS}(A)$, which are related to a given matrix A. A third subspace associated with the matrix A is the row space of A.

Definition 3.13

If A is an $m \times n$ matrix, then the ROW SPACE of A is the subspace of $R^{1 \times n}$ spanned by the rows of A; that is,

$$
\mathcal{RS}(A) = \mathrm{SPAN}\{\mathrm{Row}_1(A), \mathrm{Row}_2(A), \ldots, \mathrm{Row}_m(A)\}.
$$

Example 8

Which of the row vectors $\mathbf{r}_1 = [1 \ \ 1 \ \ 1 \ \ 1]$, $\mathbf{r}_2 = [1 \ \ 2 \ \ 1 \ \ 1]$, and $\mathbf{r}_3 = [2 \ \ 4 \ \ 2 \ \ -2]$ are in the row space of

$$
A = \begin{bmatrix} 1 & 2 & 3 & -3 \\ 2 & 4 & 3 & 0 \\ 4 & 8 & 2 & 8 \end{bmatrix} ?
$$

To determine if $\mathbf{r}_1$ is in $\mathcal{RS}(A)$ we try to find scalars x, y, and z such that

$$
\mathbf{r}_1 = x\mathrm{Row}_1(A) + y\mathrm{Row}_2(A) + z\mathrm{Row}_3(A).
$$

This statement is equivalent to

$$
\begin{aligned}
\mathbf{r}_1 = [1 \ \ 1 \ \ 1 \ \ 1] &= x[1 \ \ 2 \ \ 3 \ \ -3] + y[2 \ \ 4 \ \ 3 \ \ 0] \\
&\quad + z[4 \ \ 8 \ \ 2 \ \ 8] \\
&= [x + 2y + 4z \quad 2x + 4y + 8z \quad 3x + 3y + 2z \quad -3x + 8z].
\end{aligned}
$$

Thus x, y, and z must satisfy the linear system

$$
\begin{aligned}
x + 2y + 4z &= 1 \\
2x + 4y + 8z &= 1 \\
3x + 3y + 2z &= 1 \\
-3x \qquad + 8z &= 1
\end{aligned}
\quad \text{or} \quad
\begin{bmatrix} 1 & 2 & 4 \\ 2 & 4 & 8 \\ 3 & 3 & 2 \\ -3 & 0 & 8 \end{bmatrix}
\begin{bmatrix} x \\ y \\ z \end{bmatrix}
=
\begin{bmatrix} 1 \\ 1 \\ 1 \\ 1 \end{bmatrix}
$$

which, we observe, can be written as $A^T X = \mathbf{r}_1^T$. Similarly, to determine if $\mathbf{r}_2$ and $\mathbf{r}_3$ are in $\mathcal{RS}(A)$, we must try to solve the linear systems

$$A^T X = \mathbf{r}_i^T; \; i = 2, 3 \, .$$

We can solve these three systems with the single row reduction

$$\begin{bmatrix} 1 & 2 & 4 & 1 & 1 & 2 \\ 2 & 4 & 8 & 1 & 2 & 4 \\ 3 & 3 & 2 & 1 & 1 & 2 \\ -3 & 0 & 8 & 1 & 1 & -2 \end{bmatrix} \longrightarrow$$

$$\begin{bmatrix} 1 & 2 & 4 & 1 & 1 & 2 \\ 0 & 0 & 0 & -1 & 0 & 0 \\ 0 & -3 & -10 & -2 & -2 & -4 \\ 0 & 6 & 20 & 4 & 4 & 4 \end{bmatrix} \longrightarrow$$

$$\begin{bmatrix} 1 & 2 & 4 & 1 & 1 & -2 \\ 0 & 3 & 10 & 2 & 2 & 4 \\ 0 & 0 & 0 & -1 & 0 & 0 \\ 0 & 0 & 0 & 0 & 0 & -4 \end{bmatrix} .$$

From this reduction we see that only the second system is consistent. Thus $\mathbf{r}_2$ is an element of $\mathcal{RS}(A)$ while $\mathbf{r}_1$ and $\mathbf{r}_3$ are not in $\mathcal{RS}(A)$. There are many solutions of the second system; among them are $z = 2$, $y = -6$, and $x = 5$. Thus,

$$\mathbf{r}_2 = [1 \quad 2 \quad 1 \quad 1] = 5\text{Row}_1(A) - 6\text{Row}_2(A) + 2\text{Row}_3(A) \, ,$$

as you should verify directly. ■

The preceding examples have illustrated the following observations.

(1) Finding spanning sets usually requires solving a linear system.
(2) Determining if a given vector is in a subspace, which is described by a spanning set, usually requires investigating whether a certain linear system has a solution.

We conclude this section with a theorem about row equivalence that will be useful in later sections.

Theorem 3.8

If A and B are row-equivalent matrices, then $\mathscr{RS}(A) = \mathscr{RS}(B)$.

Proof. If $A \underset{R}{\approx} B$, then from Theorem 1.8 we know that there exists a nonsingular matrix P such that $PA = B$. Since the rows of B are linear combinations of the rows of A (see Equation 1.10), it follows at once that each row of B is in $\mathscr{RS}(A)$, and hence that $\mathscr{RS}(B) \subseteq \mathscr{RS}(A)$. Since P is nonsingular we can write $A = P^{-1}B$ and observe that the rows of A are linear combinations of the rows of B. Thus we also have $\mathscr{RS}(A) \subseteq \mathscr{RS}(B)$; it follows that $\mathscr{RS}(A) = \mathscr{RS}(B)$. $\square$

If a subspace $\mathscr{W}$ of $\mathscr{R}^{1 \times n}$ can be described as the rowspace of some matrix A, then Theorem 3.8 and the row-reduction techniques of Chapter 1 can be used to find the simplest matrix R such that $\mathscr{W} = \mathscr{RS}(A) = \mathscr{RS}(R)$.

Example 9

Use Theorem 3.8 to obtain an alternate description of the row space of the matrix A of Example 8.

From the row reduction

$$A = \begin{bmatrix} 1 & 2 & 3 & -3 \\ 2 & 4 & 3 & 0 \\ 4 & 8 & 2 & 8 \end{bmatrix} \longrightarrow \begin{bmatrix} 1 & 2 & 3 & -3 \\ 0 & 0 & -3 & 6 \\ 0 & 0 & -10 & 20 \end{bmatrix} \longrightarrow$$

$$\begin{bmatrix} 1 & 2 & 0 & 3 \\ 0 & 0 & 1 & -2 \\ 0 & 0 & 0 & 0 \end{bmatrix} = R$$

it follows, using Theorem 3.8, that

$$\begin{aligned}
\mathscr{RS}(A) = \mathscr{RS}(R) &= \text{SPAN}\{[1 \quad 2 \quad 0 \quad 3], [0 \quad 0 \quad 1 \quad -2]\} \\
&= \{[a \quad 2a \quad b \quad 3a - 2b] \,|\, a \quad b \text{ in } R\} \\
&= \{[x_1 \quad x_2 \quad x_3 \quad x_4] \,|\, x_2 = 2x_1 \text{ and } x_4 = 3x_1 - 2x_2\}.
\end{aligned}$$

From the last characterization of $\mathscr{RS}(A)$ it is easy to see that $\mathbf{r}_1$ and $\mathbf{r}_3$ are not in $\mathscr{RS}(A)$, while $\mathbf{r}_2$ is. ■

Exercise Set 3.5

1. Which of the following subsets of $\mathscr{R}^{1 \times 3}$ are subspaces? Interpret each of these subsets geometrically.

   (a) $\{[x_1 \quad x_2 \quad x_3] \,|\, 3x_1 - 5x_2 + 6x_3 = 0\}$
   (b) $\{[x_1 \quad x_2 \quad x_3] \,|\, x_1^2 + x_2^2 + x_3^2 = 1\}$
   (c) $\{[x_1 \quad x_2 \quad 3x_1 + 5x_2] \,|\, x_1, x_2 \text{ real}\}$
   (d) $\{[x_1 \quad x_2 \quad x_3] \,|\, x_1 + x_2 + x_3 = 5\}$
   (e) $\{[x_1 \quad x_2 \quad x_3] \,|\, x_1 + x_2 > x_3\}$
   (f) $\{[x_1 \quad x_2 \quad x_3] \,|\, x_1^2 + x_2^2 + x_3^2 = 0\}$

2. Which of the following subsets of $\mathcal{R}^{3\times3}$ are subspaces?

(a) $\{A \mid A^T = A\}$

(b) $\{A \mid \text{tr}(A) = 0\}$ (Recall Exercise 11 in Section 2.1.)

(c) $\{A \mid \det(A) = 1\}$

(d) $\{A \mid AB = BA\}$ (B is a fixed 3×3 matrix.)

(e) $\{A \mid a_{31} + a_{22} + a_{13} = 0\}$

(f) $\{A \mid a_{21} + a_{22} + a_{23} = 5\}$

3. Which of the following subsets of $\mathcal{P}$ are subspaces?

(a) $\{p(x) \mid \text{constant term is } 0\}$

(b) $\{p(x) \mid \text{all coefficients are positive}\}$

(c) $\{p(x) \mid \text{the sum of the coefficients is zero}\}$

(d) $\{p(x) \mid p(x) \text{ is a monic polynomial}\}$ (A polynomial is MONIC if the coefficient of the highest power of x is $+1$.)

4. Which of the following subsets of $\mathcal{F}$ are subspaces?

(a) $\{\mathbf{f} \mid f(2) = 0\}$

(b) $\{\mathbf{f} \mid f(x) \text{ is a polynomial in } x\}$

(c) $\{\mathbf{f} \mid \mathbf{f} \text{ is a nondecreasing function}\}$

(d) $\{\mathbf{f} \mid \mathbf{f} \text{ is continuous}\} = \mathscr{C}^0[a, b]$

(e) $\left\{\mathbf{f} \text{ in } \mathscr{C}^0[a, b] \,\middle|\, \int_a^b f(x)\, dx = 0\right\}$

(f) $\{\mathbf{f} \mid f''(x) - 7f'(x) = \sin 3x\}$

5. Determine which of the following vectors from $\mathcal{R}^{1\times4}$ are elements of the subspace $\mathcal{W}$ spanned by the set

$$\{\mathbf{u_1} = [1 \quad 1 \quad 1 \quad 1], \mathbf{u_2} = [3 \quad -2 \quad 0 \quad 0], \mathbf{u_3} = [2 \quad -5 \quad -4 \quad 11]\}.$$

You should be able to do all six parts with a single row reduction.

(a) $\mathbf{a} = [2 \quad 3 \quad 0 \quad 13]$ (b) $\mathbf{b} = [2 \quad 0 \quad 0 \quad 3]$

(c) $\mathbf{c} = [0 \quad 0 \quad 0 \quad 4]$ (d) $\mathbf{d} = [4 \quad -1 \quad 1 \quad 1]$

(e) $\mathbf{e} = [1 \quad 1 \quad 1 \quad -1]$ (f) $\mathbf{f} = [0 \quad 5 \quad 4 \quad -1]$

6. (a) Show that the function $\mathbf{f}$, for which $f(x) = x^2 + 2$, is not in SPAN$\{\cos x, \sin x\}$.

(b) Show that $\sin(x + 5)$ is in SPAN$\{\cos x, \sin x\}$.

(c) Is $\sin 2x$ in SPAN$\{\cos x, \sin x\}$?

7. Determine which of the following polynomials are elements of the subspace of $\mathcal{P}$ spanned by the set

$$\{p_1(x) = x^3 + 2x^2 + 1,\ p_2(x) = x^2 - 2,\ p_3(x) = x^3 + x\}.$$

(a) $x^2 - x + 3$ (b) $x^2 - 2x + 1$ (c) $4x^3 - 3x + 5$

(d) $x^4 + 1$ (e) $-x^3 + 5x^2 - 2x - 2$ (f) $x - 5$

8. Let A be any $n \times n$ matrix, let P be any $m \times n$ matrix, and let Q be any $n \times m$ matrix.

(a) Show that $\mathcal{NS}(A) \subseteq \mathcal{NS}(PA)$ with equality if P is invertible.

(b) Use Equation 1.10 to show that $\mathcal{RS}(PA) \subseteq \mathcal{RS}(A)$ with equality if P is invertible.

(c) Use Equation 1.9 to show that $\mathcal{CS}(AQ) \subseteq \mathcal{CS}(A)$ with equality if Q is invertible.

9. Let $\mathcal{W}_1$ and $\mathcal{W}_2$ be subspaces of the vector space $\mathcal{V}$. Which of the following subsets of $\mathcal{V}$ are subspaces of $\mathcal{V}$?

(a) $\mathcal{W}_1 \cap \mathcal{W}_2 = \{\mathbf{u} \mid \mathbf{u} \text{ in } \mathcal{W}_1 \text{ and } \mathbf{u} \text{ in } \mathcal{W}_2\}$

(b) $\mathcal{W}_1 \cup \mathcal{W}_2 = \{\mathbf{u} \mid \mathbf{u} \text{ in } \mathcal{W}_1 \text{ or } \mathbf{u} \text{ in } \mathcal{W}_2\}$

(c) $\mathcal{W}_1 + \mathcal{W}_2 = \{\mathbf{u}_1 + \mathbf{u}_2 \mid \mathbf{u}_1 \text{ in } \mathcal{W}_1 \text{ and } \mathbf{u}_2 \text{ in } \mathcal{W}_2\}$

10. Find a matrix in reduced row echelon form whose rowspace is the same as $\mathcal{RS}(A)$ if A is one of the following matrices. In each case provide several alternate descriptions of $\mathcal{RS}(A)$.

(a) $A = \begin{bmatrix} 1 & 2 & 3 \\ 4 & 5 & 6 \\ 7 & 8 & 9 \end{bmatrix}$ (b) $A = \begin{bmatrix} 1 & 1 & 2 & 6 \\ 1 & -1 & -4 & -8 \\ 3 & -2 & 5 & 11 \\ 2 & 5 & -2 & 3 \end{bmatrix}$

(c) $A = \begin{bmatrix} 3 & -1 & 2 & 1 \\ 2 & 1 & 1 & 1 \\ 1 & -3 & 0 & 1 \end{bmatrix}$

†11. Let $\mathcal{W}$ be any subspace of $\mathcal{V}$ containing the vectors $\{\mathbf{u}_1, \ldots, \mathbf{u}_k\}$. Show that SPAN$\{\mathbf{u}_1, \ldots, \mathbf{u}_k\}$ is a subspace of $\mathcal{W}$ and hence that it is the smallest subspace of $\mathcal{V}$ containing $\{\mathbf{u}_1, \ldots, \mathbf{u}_k\}$.

12. Find spanning sets for the null spaces of the following matrices.

$A = \begin{bmatrix} 1 & -2 & 0 & 4 \\ 0 & 0 & 1 & -3 \\ 0 & 0 & 0 & 0 \end{bmatrix}$ $B = [2 \quad 5 \quad 6 \quad 7]$ $C = \begin{bmatrix} 1 & 1 & 1 & 1 \\ 2 & 2 & 2 & 3 \\ 3 & 3 & 3 & 4 \end{bmatrix}$

3.6   *Linear Independence and Dependence*

In Example 8 of the last section we showed that the row vector $\mathbf{r}_2 = [1 \quad 2 \quad 1 \quad 1]$ is an element of the subspace of $\mathcal{R}^{1 \times 4}$ spanned by the vectors $\mathbf{u}_1 = [1 \quad 2 \quad 3 \quad -3]$, $\mathbf{u}_2 = [2 \quad 4 \quad 3 \quad 0]$, and $\mathbf{u}_3 = [4 \quad 8 \quad 2 \quad 8]$. In fact, we showed that

$$\mathbf{r}_2 = 5\mathbf{u}_1 - 6\mathbf{u}_2 + 2\mathbf{u}_3.$$

You can check directly that we also have

$$\mathbf{r}_2 = \frac{7}{3}\mathbf{u}_1 - \frac{8}{3}\mathbf{u}_2 + \mathbf{u}_3.$$

From these last two equations it follows that

$$5\mathbf{u}_1 - 6\mathbf{u}_2 + 2\mathbf{u}_3 = \frac{7}{3}\mathbf{u}_1 - \frac{8}{3}\mathbf{u}_2 + \mathbf{u}_3$$

or, equivalently, that

$$\mathbf{u}_3 = \frac{-8}{3}\mathbf{u}_1 + \frac{10}{3}\mathbf{u}_2 .$$

It follows, as in Exercise 10 of Section 3.3, that the spanning set $\{\mathbf{u}_1, \mathbf{u}_2, \mathbf{u}_3\}$ is larger than necessary and that SPAN$\{\mathbf{u}_1, \mathbf{u}_2, \mathbf{u}_3\}$ = SPAN$\{\mathbf{u}_1, \mathbf{u}_2\}$. In general, if the set of vectors $\{\mathbf{u}_1, \mathbf{u}_2, \ldots, \mathbf{u}_k\}$ spans the vector space $\mathcal{V}$, then we know, from Definition 3.12, that every vector $\mathbf{v}$ in $\mathcal{V}$ can be expressed as a linear combination of the $\mathbf{u}_i$:

$$\mathbf{v} = a_1\mathbf{u}_1 + a_2\mathbf{u}_2 + \ldots + a_k\mathbf{u_k} . \tag{3.4}$$

We begin this section by investigating the uniqueness of the co-efficients a_i in Equation 3.4. Suppose that in addition to Equation 3.4 we also had

$$\mathbf{v} = b_1\mathbf{u}_1 + b_2\mathbf{u}_2 + \ldots + b_k\mathbf{u_k} ,$$

where $a_i \neq b_i$ for at least one i. Simple algebraic manipulation, using the properties of vector algebra in Definition 3.10 now leads to

$$\mathbf{0} = \mathbf{v} - \mathbf{v} = (a_1 - b_1)\mathbf{u}_1 + (a_2 - b_2)\mathbf{u}_2 + \ldots + (a_k - b_k)\mathbf{u_k} .$$

Since some of the coefficients in the last expression are different from zero, the $\mathbf{u}_i$ are dependent in the sense that at least one of them can be expressed as a linear combination of the others. We will use the above discussion as motivation for the following important definition.

Definition 3.14

Let $\mathcal{S} = \{\mathbf{u}_1, \mathbf{u}_2, \ldots, \mathbf{u}_k\}$ be a set of vectors from the vector space $\mathcal{V}$. The set $\mathcal{S}$ is said to be LINEARLY DEPENDENT if there exist scalars a_i, not all zero, such that

$$a_1\mathbf{u}_1 + a_2\mathbf{u}_2 + \ldots + a_k\mathbf{u_k} = \mathbf{0} . \tag{3.5}$$

The set $\mathcal{S}$ is LINEARLY INDEPENDENT if it is not linearly dependent.

Since $a\mathbf{u} + b\mathbf{v} = \mathbf{0}$ $(a \neq 0)$ implies that $\mathbf{u} = -(b/a)\mathbf{v}$, it follows that two vectors are linearly dependent if and only if one of them is a scalar multiple of the other; this case is always easy to determine.

Note that if $\mathbf{x}$, $\mathbf{y}$, and $\mathbf{z}$ are three vectors in $\mathcal{R}^3$, then a nontrivial relation of the form $a\mathbf{x} + b\mathbf{y} + c\mathbf{z} = \mathbf{0}$ means that one of the vectors is in the plane determined by the other two. In particular, if $a \neq 0$, then $\mathbf{x} = (-b/a)\mathbf{y} + (-c/a)\mathbf{z}$ is in the plane determined by $\mathbf{y}$ and $\mathbf{z}$. Thus we can view linear dependence as being a generalization of the notion of three vectors being coplanar in 3-space, or two vectors being collinear in $\mathcal{R}^2$ or $\mathcal{R}^3$.

Normally we think of linear independence as being a favorable condition and linear dependence as being unfavorable. In order to investigate the linear independence or dependence of a given set of vectors we try to deduce from Equation 3.5 whether the coefficients must all be zero. This process normally leads to a homogeneous system of linear equations, for which we need to determine whether a nonzero solution exists. Recall from our earlier work that the homogeneous system $AX = 0$ has a nonzero solution if any one of the following conditions hold:

(1) A has fewer rows than columns;
(2) A is row equivalent to a matrix with fewer nonzero rows than columns:
(3) A is square and singular; or
(4) A is square and $\det(A) = 0$.

Example 1

Determine if the following vectors from $\mathcal{R}^4$ are linearly independent or linearly dependent.

$$\mathbf{u_1} = \begin{bmatrix} 1 \\ 2 \\ -1 \\ 2 \end{bmatrix} \quad \mathbf{u_2} = \begin{bmatrix} -2 \\ -5 \\ 3 \\ 0 \end{bmatrix} \quad \mathbf{u_3} = \begin{bmatrix} 1 \\ 0 \\ 1 \\ 10 \end{bmatrix}$$

The vector equation $a_1\mathbf{u_1} + a_2\mathbf{u_2} + a_3\mathbf{u_3} = \mathbf{0}$ is equivalent to

$$a_1 \begin{bmatrix} 1 \\ 2 \\ -1 \\ 2 \end{bmatrix} + a_2 \begin{bmatrix} -2 \\ -5 \\ 3 \\ 0 \end{bmatrix} + a_3 \begin{bmatrix} 1 \\ 0 \\ 1 \\ 10 \end{bmatrix} = \begin{bmatrix} 0 \\ 0 \\ 0 \\ 0 \end{bmatrix}.$$

By comparing the elements of the vectors above we obtain the linear system

$$\begin{aligned} a_1 - 2a_2 + a_3 &= 0 \\ 2a_1 - 5a_2 &= 0 \\ -a_1 + 3a_2 + a_3 &= 0 \\ 2a_1 + 10a_3 &= 0 \end{aligned} \quad \text{or} \quad \begin{bmatrix} 1 & -2 & 1 \\ 2 & -5 & 0 \\ -1 & 3 & 1 \\ 2 & 0 & 10 \end{bmatrix} \begin{bmatrix} a_1 \\ a_2 \\ a_3 \end{bmatrix} = \begin{bmatrix} 0 \\ 0 \\ 0 \\ 0 \end{bmatrix}.$$

The question of the independence or dependence of the $\mathbf{u_i}$ has now been reduced to the question of whether this system has a nonzero solution; if it does, the vectors are dependent, and if it does not, the vectors are independent. To decide if this system has any nonzero solutions we row reduce the augmented matrix as follows:

$$\begin{bmatrix} 1 & -2 & 1 & 0 \\ 2 & -5 & 0 & 0 \\ -1 & 3 & 1 & 0 \\ 2 & 0 & 10 & 0 \end{bmatrix} \longrightarrow \begin{bmatrix} 1 & -2 & 1 & 0 \\ 0 & -1 & -2 & 0 \\ 0 & 1 & 2 & 0 \\ 0 & 4 & 8 & 0 \end{bmatrix} \longrightarrow \begin{bmatrix} 1 & -2 & 1 & 0 \\ 0 & 1 & 2 & 0 \\ 0 & 0 & 0 & 0 \\ 0 & 0 & 0 & 0 \end{bmatrix}.$$

The homogeneous system associated with the final matrix has only two equations in three unknowns and is therefore certain to have a nonzero solution; $a_1 = 5$, $a_2 = 2$, and $a_3 = -1$ is one such solution. Thus the vectors $\mathbf{u}_1$, $\mathbf{u}_2$, and $\mathbf{u}_3$ are linearly dependent. You should verify directly that $5\mathbf{u}_1 + 2\mathbf{u}_2 - \mathbf{u}_3 = \mathbf{0}$. ▬

Example 2

Determine if the following polynomials (vectors from $\mathscr{P}$) are linearly independent or linearly dependent:

$$p(x) = x^2 - 1, \quad q(x) = x^2 + x - 2, \quad \text{and} \quad r(x) = x^2 + 3x + 2.$$

The vector equation $ap(x) + bq(x) + cr(x) = 0$ is equivalent to

$$a(x^2 - 1) + b(x^2 + x - 2) + c(x^2 + 3x + 2) = 0,$$

which, after some rearranging, becomes

$$(-a - 2b + 2c) + (b + 3c)x + (a + b + c)x^2 = 0 = 0 + 0x + 0x^2.$$

By comparing coefficients on the two sides of the above equation we see that this polynomial equation is equivalent to the homogeneous system

$$
\begin{aligned}
-a - 2b + 2c &= 0 \\
b + 3c &= 0 \quad \text{or} \\
a + b + c &= 0
\end{aligned}
\qquad
\begin{bmatrix} -1 & -2 & 2 \\ 0 & 1 & 3 \\ 1 & 1 & 1 \end{bmatrix}
\begin{bmatrix} a \\ b \\ c \end{bmatrix}
=
\begin{bmatrix} 0 \\ 0 \\ 0 \end{bmatrix}.
$$

We have again reduced the vector question to an equivalent question about a system of linear equations. To decide if this system has nonzero solutions we row reduce the augmented matrix:

$$
\begin{bmatrix} -1 & -2 & 2 & 0 \\ 0 & 1 & 3 & 0 \\ 1 & 1 & 1 & 0 \end{bmatrix}
\rightarrow
\begin{bmatrix} 1 & 2 & -2 & 0 \\ 0 & 1 & 3 & 0 \\ 0 & -1 & 3 & 0 \end{bmatrix}
\rightarrow
\begin{bmatrix} 1 & 2 & -2 & 0 \\ 0 & 1 & 3 & 0 \\ 0 & 0 & 6 & 0 \end{bmatrix}.
$$

From the last matrix we observe that the system has only the zero solution. Hence the polynomials $p(x)$, $q(x)$, and $r(x)$ are linearly independent polynomials (vectors in $\mathscr{P}$). ▬

Example 3

It is easy to see that the nonzero rows of either a row echelon matrix or a reduced row echelon matrix are always linearly independent. For example, for the row echelon matrix

$$
B = \begin{bmatrix} 1 & 3 & 5 & 0 & 9 \\ 0 & 0 & 1 & 6 & 8 \\ 0 & 0 & 0 & 1 & 6 \\ 0 & 0 & 0 & 0 & 0 \end{bmatrix},
$$

the equation

$$x_1\text{Row}_1(B) + x_2\text{Row}_2(B) + x_3\text{Row}_3(B) = [0 \quad 0 \quad 0 \quad 0 \quad 0]$$

or

$$x_1[1 \quad 3 \quad 5 \quad 0 \quad 9] + x_2[0 \quad 0 \quad 1 \quad 6 \quad 8]$$
$$+ x_3[0 \quad 0 \quad 0 \quad 1 \quad 6] = [0 \quad 0 \quad 0 \quad 0 \quad 0]$$

yields at once that $x_1 = x_2 = x_3 = 0$. Therefore the nonzero rows of B are linearly independent. ∎

Example 4

Determine if the following matrices (vectors from $\mathcal{R}^{2\times 2}$) are linearly independent or dependent.

$$M_1 = \begin{bmatrix} 1 & 2 \\ 3 & 4 \end{bmatrix} \qquad M_2 = \begin{bmatrix} 1 & -1 \\ -1 & 1 \end{bmatrix} \qquad M_3 = \begin{bmatrix} 1 & 3 \\ 0 & 4 \end{bmatrix}$$

$$M_4 = \begin{bmatrix} 0 & 2 \\ 2 & 0 \end{bmatrix} \qquad M_5 = \begin{bmatrix} 1 & 1 \\ 2 & 3 \end{bmatrix}$$

The vector equation

$$x_1 M_1 + x_2 M_2 + x_3 M_3 + x_4 M_4 + x_5 M_5 = 0$$

is equivalent to

$$x_1 \begin{bmatrix} 1 & 2 \\ 3 & 4 \end{bmatrix} + x_2 \begin{bmatrix} 1 & -1 \\ -1 & 1 \end{bmatrix} + x_3 \begin{bmatrix} 1 & 3 \\ 0 & 4 \end{bmatrix}$$
$$+ x_4 \begin{bmatrix} 0 & 2 \\ 2 & 0 \end{bmatrix} + x_5 \begin{bmatrix} 1 & 1 \\ 2 & 3 \end{bmatrix} = \begin{bmatrix} 0 & 0 \\ 0 & 0 \end{bmatrix}$$

or
$$\begin{bmatrix} x_1 + x_2 + x_3 + x_5 & 2x_1 - x_2 + 3x_3 + 2x_4 + x_5 \\ 3x_1 - x_2 + 2x_4 + 2x_5 & 4x_1 + x_2 + 4x_3 + 3x_5 \end{bmatrix} = \begin{bmatrix} 0 & 0 \\ 0 & 0 \end{bmatrix}.$$

This last matrix equation leads to a system of four homogeneous equation in five unknowns:

$$x_1 + x_2 + x_3 + \qquad\qquad x_5 = 0$$
$$2x_1 - x_2 + 3x_3 + 2x_4 + x_5 = 0$$
$$3x_1 - x_2 + \qquad 2x_4 + 2x_5 = 0$$
$$4x_1 + x_2 + 4x_3 + \qquad 3x_5 = 0.$$

Since this system has fewer equations than unknowns it is certain to have a nonzero solution. Thus the five given matrices are linearly dependent vectors in $\mathcal{R}^{2\times 2}$. ∎

Example 5

Determine if the following functions (vectors from $\mathcal{F}(1, \infty)$) are linearly independent or dependent.

$$f(x) = \sqrt{4x} \qquad g(x) = \frac{12}{x} \qquad h(x) = 2^x$$

The vector equation

$$a\mathbf{f} + b\mathbf{g} + c\mathbf{h} = \mathbf{0}$$

amounts to

$$af(x) + bg(x) + ch(x)$$
$$= a\sqrt{4x} + b\frac{12}{x} + c2^x$$
$$= 0 \quad \text{for all } x \text{ in } [1, \infty).$$

We obtain linear equations for a, b, and c by choosing particular values for x in the last equation:

$$\begin{array}{ll} x = 1: & 2a + 12b + 2c = 0 \\ x = 2: \sqrt{8}a + 6b + 4c = 0 & \text{or} \\ x = 4: & 4a + 3b + 16c = 0 \end{array} \qquad \begin{bmatrix} 2 & 12 & 2 \\ \sqrt{8} & 6 & 4 \\ 4 & 3 & 16 \end{bmatrix} \begin{bmatrix} a \\ b \\ c \end{bmatrix} = \begin{bmatrix} 0 \\ 0 \\ 0 \end{bmatrix}.$$

This homogeneous system has no nontrivial solution since

$$\det \begin{bmatrix} 2 & 12 & 2 \\ \sqrt{8} & 6 & 4 \\ 4 & 3 & 16 \end{bmatrix} = 312 - 186\sqrt{8} \neq 0.$$

Therefore, the functions $\mathbf{f}$, $\mathbf{g}$, and $\mathbf{h}$ are linearly independent vectors from $\mathcal{F}$. ■

As illustrated in the above examples, the procedure for testing a set of vectors $\{\mathbf{u}_1, \ldots, \mathbf{u}_k\}$ for linear independence or dependence requires three main steps.

---

(1) Suppose that $x_1\mathbf{u}_1 + x_2\mathbf{u}_2 + \ldots + x_k\mathbf{u}_k = \mathbf{0}$.

(2) Interpret the vector equation in step 1 as a homogeneous system of linear equations $AX = 0$. (The technique depends on the nature of the vector space.)

(3) Using the techniques of Chapter 1, determine if the system in step 2 has a nonzero solution. If the system has only the trivial solution, then the $\mathbf{u}_i$ are linearly independent; otherwise they are linearly dependent.

The method just described cannot be used to establish the linear dependence of a set of functions (vectors from $\mathscr{F}$), since in this case it must be shown that the equation in step 1 holds for an infinite number of values of x. As illustrated in Example 4, the method works well for establishing the linear independence of a set of functions. You may consult a differential-equations text for a more complete discussion of linear dependence of functions.

We conclude this section with a theorem that lists some useful facts about linear independence and linear dependence.

Theorem 3.9

The following properties hold in any vector space $\mathscr{V}$.

(1) Any subset of an independent set is independent.
(2) Any set with a dependent subset is dependent.
(3) If the set $\{\mathbf{u}_1, \ldots, \mathbf{u}_k\}$ is independent and $\mathbf{u}_{k+1}$ is not in SPAN$\{\mathbf{u}_1, \ldots, \mathbf{u}_k\}$, then the set $\{\mathbf{u}_1, \ldots, \mathbf{u}_k, \mathbf{u}_{k+1}\}$ is linearly independent.
(4) If $\{\mathbf{u}_1, \mathbf{u}_2, \ldots, \mathbf{u}_n\}$ is a set of vectors such that $\mathbf{u}_1 \neq \mathbf{0}$ and $\mathbf{u}_{j+1}$ is not in SPAN$\{\mathbf{u}_1, \ldots, \mathbf{u}_j\}$ for $j = 1, 2, 3, \ldots, n - 1$, then $\{\mathbf{u}_1, \mathbf{u}_2, \ldots, \mathbf{u}_n\}$ is an independent set of vectors.

We will prove the third property; proofs of the other properties are left to the exercises.

Proof.   The set $\{\mathbf{u}_1, \ldots, \mathbf{u}_k, \mathbf{u}_{k+1}\}$ is either linearly independent or linearly dependent. If it is dependent, then there exist scalars, not all zero, such that

$$a_1\mathbf{u}_1 + \ldots + a_k\mathbf{u}_k + a_{k+1}\mathbf{u}_{k+1} = \mathbf{0}.$$

If $a_{k+1}\mathbf{u}_{k+1} = \mathbf{0}$, then this equation says that $\{\mathbf{u}_1, \ldots, \mathbf{u}_k\}$ is a dependent set; this is a contradiction. Thus, $a_{k+1} \neq 0$ and it follows that

$$\mathbf{u}_{k+1} = (-1/a_{k+1})(a_1\mathbf{u}_1 + \ldots + a_k\mathbf{u}_k) \text{ is in SPAN}\{\mathbf{u}_1, \ldots, \mathbf{u}_k\}.$$

This is again a contradiction, so the assumption that the set $\{\mathbf{u}_1, \ldots, \mathbf{u}_k, \mathbf{u}_{k+1}\}$ is linearly dependent must be wrong. The only remaining possibility is that the set is linearly independent. This completes the proof of the third assertion of the theorem. □

From the second conclusion of Theorem 3.9 it follows at once that any set of vectors containing the zero vector must be linearly dependent.

Example 6

As an application of the fourth conclusion of Theorem 3.9, show how to identify an independent subset $\mathscr{S}_1$ of a set $\mathscr{S} = \{\mathbf{u}_1, \ldots, \mathbf{u}_m\}$ such that SPAN$\{\mathscr{S}\} = $ SPAN$\{\mathscr{S}_1\}$.

We will construct $\mathcal{S}_1$ from $\mathcal{S}$ by working through the list of elements of $\mathcal{S}$, omitting any vector which can be expressed as a linear combination of those that precede it in the list. Thus we can construct $\mathcal{S}_1$ using the following algorithm.

If $\mathbf{u}_1 \neq \mathbf{0}$, include $\mathbf{u}_1$ in $\mathcal{S}_1$.

For $j = 1, 2, \ldots, m - 1$

If $\mathbf{u}_{j+1}$ is not in $\mathrm{SPAN}\{\mathcal{S}_1\} = \mathrm{SPAN}\{\mathbf{u}_1, \ldots, \mathbf{u}_j\}$, then add $\mathbf{u}_{j+1}$ to $\mathcal{S}_1$.

By Theorem 3.9, the set $\mathcal{S}_1$ will be linearly independent. Since deleting dependent vectors from the spanning set will not make the subspace any smaller, we also have $\mathrm{SPAN}\{\mathcal{S}_1\} = \mathrm{SPAN}\{\mathcal{S}\}$. ▄

Exercise Set 3.6

Determine if the sets of vectors in Exercises 1–14 are linearly independent or linearly dependent. In each case first rephrase the vector question as an equivalent question about a system of linear equations.

1. $\mathbf{x}_1 = [1 \quad -1 \quad 2 \quad 0]$, $\mathbf{x}_2 = [1 \quad 1 \quad 2 \quad 0]$, and $\mathbf{x}_3 = [4 \quad 1 \quad 2 \quad 1]$ in $\mathcal{R}^{1 \times 4}$

2. $\mathbf{y}_1 = [0 \quad 1 \quad 2 \quad 1]^T$, $\mathbf{y}_2 = [1 \quad 2 \quad 3 \quad 1]^T$, and $\mathbf{y}_3 = [1 \quad 6 \quad 11 \quad 5]^T$, $\mathbf{y}_4 = [-1 \quad 0 \quad 1 \quad 1]^T$ in $\mathcal{R}^4$

3. $p_1(x) = x^3 + 2x + 5$, $p_2(x) = 3x^2 + 2$, $p_3(x) = 6x$, and $p_4(x) = 2$ in $\mathcal{P}$.

4. The rows of the matrix $A = \begin{bmatrix} 1 & 1 & 3 \\ 5 & -1 & 2 \\ -3 & 0 & 4 \end{bmatrix}$

5. The columns of the matrix $B = \begin{bmatrix} 1 & 2 & 1 & 3 \\ 4 & -1 & 5 & 8 \\ 0 & 2 & 3 & 7 \\ 3 & 6 & -1 & 3 \end{bmatrix}$

6. $\sin x$, $\cos x$, and $e^{(x/\pi)}$ in $\mathcal{F}$

7. $\sqrt{x}$, x^2, and $|x|$ in $\mathcal{F}(0, \infty)$

8. $\sin(3x + 7)$, $\cos 3x$, and $\sin 3x$ in $\mathcal{F}$

9. $q_1(x) = 3 + x + x^2$, $q_2(x) = 2 - x + 5x^2$, and $q_3(x) = 4 - 3x^2$ in $\mathcal{P}$

10. x^2, $x^2 - 4x$, $x^2 + 3x$, and $x^2 - x - 12$ in $\mathcal{P}$

11. $\mathbf{z}_1 = [1 \quad 2 \quad 1 \quad -2]$, $\mathbf{z}_2 = [0 \quad 1 \quad 1 \quad 0]$, $\mathbf{z}_3 = [0 \quad 2 \quad 3 \quad 1]$, and $\mathbf{z}_4 = [1 \quad 0 \quad -1 \quad 2]$ in $\mathcal{R}^{1 \times 4}$

12. $\mathbf{u}_1 = [1 \quad 3 \quad 3]^T$, $\mathbf{u}_2 = [0 \quad 1 \quad 4]^T$, $\mathbf{u}_3 = [5 \quad 6 \quad 3]^T$, and $\mathbf{u}_4 = [7 \quad 2 \quad 3]^T$ in $\mathcal{R}^3$

13. $\begin{bmatrix} 1 & 0 \\ 1 & 0 \end{bmatrix}$, $\begin{bmatrix} 1 & 2 \\ 0 & 0 \end{bmatrix}$, $\begin{bmatrix} 5 & -1 \\ 4 & 0 \end{bmatrix}$, and $\begin{bmatrix} -1 & 7 \\ 6 & 0 \end{bmatrix}$ in $\mathcal{R}^{2\times2}$

14. $\begin{bmatrix} 1 & 0 & 0 \\ 0 & 0 & 0 \end{bmatrix}$, $\begin{bmatrix} 1 & 1 & 0 \\ 0 & 0 & 0 \end{bmatrix}$, $\begin{bmatrix} 1 & 1 & 1 \\ 0 & 0 & 0 \end{bmatrix}$, $\begin{bmatrix} 0 & 0 & 0 \\ 1 & 0 & 0 \end{bmatrix}$, $\begin{bmatrix} 0 & 0 & 0 \\ 1 & 1 & 0 \end{bmatrix}$,

and $\begin{bmatrix} 1 & 1 & 1 \\ 1 & 1 & 1 \end{bmatrix}$ in $\mathcal{R}^{2\times3}$

15. Show that if **u**, **v**, and **w** are linearly independent, then so are the vectors $x_1 = u + v$, $x_2 = v + w$, $x_3 = w + u$

16. Which of the eight defining properties of a vector space are used in deducing from $v = a_1u_1 + a_2u_2 + \ldots + a_nu_n = b_1u_1 + b_2u_2 + \ldots + b_nu_n$ that $0 = v - v = (a_1 - b_1)u_1 + (a_2 - b_2)u_2 + \ldots + (a_n - b_n)u_n$?

17. Consider the three functions **f**, **g**, and **h** in $\mathcal{F}$:

$$f(t) = \cos^2 t, \quad g(t) = \sin^2 t, \quad \text{and } h(t) = \sin 2t.$$

Assume that $x_1f + x_2g + x_3h = 0$; that is,

$$x_1 \cos^2 t + x_2 \sin^2 t + x_3 \sin 2t = 0 \text{ for all } t.$$

Differentiate this relation twice to obtain two more linear relations. Write these three linear equations in matrix form and show that the determinant of the coefficient matrix is not zero for any t. What can we now conclude about the independence or dependence of the functions **f**, **g**, and **h**?

18. Complete the proof of Theorem 3.9.

19. Show that any four polynomials from $\mathcal{P}_2$ are linearly dependent.

†20. Find a set of linearly independent 3×3 matrices that is as large as possible.

21. Find an independent subset $\mathcal{S}_1$ of $\mathcal{S} = \{[1 \quad 2 \quad 4 \quad 1], [2 \quad 3 \quad 7 \quad 1], [1 \quad 4 \quad 6 \quad 3]\}$ such that SPAN$\{\mathcal{S}_1\}$ = SPAN$\{\mathcal{S}\}$.

22. Find an independent subset $\mathcal{S}_1$ of $\mathcal{S} = \{\sin^2 x, \cos^2 x, \cos 2x, e^x\}$ such that SPAN$\{\mathcal{S}_1\}$ = SPAN$\{\mathcal{S}\}$.

†23. Show that the columns of an $n \times n$ matrix A are linearly independent if and only if A is nonsingular.

†24. Show that the rows of a matrix A are linearly independent if and only if the columns of A^T are linearly independent. Then show that the rows of A are linearly independent if and only if A is nonsingular.

3.7

Basis and Dimension

At the beginning of Section 3.6 we showed that if the vectors of a spanning set for the vector space $\mathcal{V}$ are linearly independent, then every vector in the space is uniquely expressible as a linear combination of the spanning vectors. Many problems dealing with vector spaces are greatly simplified if we can find a linearly independent spanning set for the space. We shall show that this has the effect of introducing a coordinate system in the vector space $\mathcal{V}$.

Definition 3.15

A BASIS for the vector space $\mathcal{V}$ is a linearly independent set of vectors which span $\mathcal{V}$. If $\mathcal{V}$ has a basis containing a finite number of vectors, then $\mathcal{V}$ is said to be FINITE DIMENSIONAL. If $\{v_1, v_2, \ldots, v_n\}$ is a basis for $\mathcal{V}$, we will write

$$\mathcal{V} = \text{BASIS}\{v_1, v_2, \ldots, v_n\}.$$

We summarize the discussion at the beginning of Section 3.6 with the following theorem, the converse of which is left to the exercises.

Theorem 3.10

If the vectors $\{v_1, v_2, \ldots, v_n\}$ form a basis for $\mathcal{V}$, then every vector in $\mathcal{V}$ can be expressed uniquely as a linear combination of the v_i. Conversely, if every vector in $\mathcal{V}$ can be expressed uniquely as a linear combination of the v_i, then the vectors $\{v_1, v_2, \ldots, v_n\}$ form a basis for $\mathcal{V}$.

The way in which a basis for $\mathcal{V}$ leads to a coordinate system in the vector space $\mathcal{V}$ is described in the next definition.

Definition 3.16

If the set $\mathcal{B} = \{v_1, v_2, \ldots, v_n\}$ is a basis for $\mathcal{V}$ and

$$u = a_1v_1 + a_2v_2 + \ldots + a_nv_n$$

is any vector from $\mathcal{V}$, then the uniquely determined scalars a_i are called the COORDINATES OF u RELATIVE TO THE BASIS $\mathcal{B}$. The column matrix $[a_1 \quad a_2 \quad \ldots \quad a_n]^T$ is called the COORDINATE MATRIX OF u RELATIVE TO THE BASIS $\mathcal{B}$, or the $\mathcal{B}$-COORDINATE MATRIX OF u, and will be denoted by $[u]_{\mathcal{B}}$.

Note that the $\mathcal{B}$-coordinate matrix of a vector would be changed if we rearranged the elements of the basis $\mathcal{B}$, (see Exercise 18 following this section). Whenever we refer to a basis $\mathcal{B} = \{v_1, v_2, \ldots, v_n\}$ for $\mathcal{V}$ we will mean that the vectors are to be taken in the indicated order; thus every basis which we consider is assumed to be an ordered basis.

In general, finding the coordinates of a vector with respect to a given basis involves solving a system of linear equations. By Theorem 3.10 this system must have a unique solution. These systems must therefore have square, nonsingular coefficient matrices.

In order to establish that a given set of vectors is a basis for the vector space $\mathcal{V}$ we will normally use the converse part of Theorem 3.10. This use involves showing that a certain linear system has a unique solution for all possible right-hand sides. Recall that the system $AX = K$, where A is square, has a unique solution if and only if the coefficient matrix A is nonsingular. Among the conditions equivalent to the nonsingularity of the matrix A are $A \underset{R}{\sim} I$ and $\det(A) \neq 0$.

Example 1

The vectors $\mathbf{e}_1 = \begin{bmatrix} 1 \\ 0 \\ 0 \\ 0 \end{bmatrix}$, $\mathbf{e}_2 = \begin{bmatrix} 0 \\ 1 \\ 0 \\ 0 \end{bmatrix}$, $\mathbf{e}_3 = \begin{bmatrix} 0 \\ 0 \\ 1 \\ 0 \end{bmatrix}$, and $\mathbf{e}_4 = \begin{bmatrix} 0 \\ 0 \\ 0 \\ 1 \end{bmatrix}$

form a basis for $\mathcal{R}^4$ since any element $\mathbf{x}$ of $\mathcal{R}^4$ can be written uniquely as

$$\mathbf{x} = \begin{bmatrix} x_1 \\ x_2 \\ x_3 \\ x_4 \end{bmatrix} = x_1\mathbf{e}_1 + x_2\mathbf{e}_2 + x_3\mathbf{e}_3 + x_4\mathbf{e}_4 . \quad \blacksquare$$

The vectors $\mathbf{e}_i$ of Example 1 will be referred to as the *standard basis* for $\mathcal{R}^4$. Note that each $\mathbf{e}_i$ is a unit vector. The obvious generalization to $\mathcal{R}^n$ will be called the STANDARD BASIS for $\mathcal{R}^n$. We will reserve the notation $\mathbf{e}_i$ for these vectors. Note that if $\mathcal{S}$ is the standard basis for $\mathcal{R}^n$, then $\mathbf{x}$ and its standard coordinate matrix are equal; that is,

$$[\mathbf{x}]_{\mathcal{S}} = \mathbf{x} \text{ for every } \mathbf{x} \text{ in } \mathcal{R}^n .$$

Example 2

Show that $\mathcal{B} = \left\{ \mathbf{u}_1 = \begin{bmatrix} 1 \\ 3 \\ -2 \end{bmatrix}, \mathbf{u}_2 = \begin{bmatrix} -3 \\ -12 \\ 10 \end{bmatrix}, \mathbf{u}_3 = \begin{bmatrix} -2 \\ -6 \\ 5 \end{bmatrix} \right\}$ is a basis for

$\mathcal{R}^3$ and find $[\mathbf{e}_1]_{\mathcal{B}}$, the $\mathcal{B}$-coordinate matrix of $\mathbf{e}_1$.

In order to show that $\mathcal{B}$ is a basis for $\mathcal{R}^3$ we will show that any vector in $\mathcal{R}^3$ can be written in a unique way as a linear combination of

the $\mathbf{u}_i$. Thus we let $\mathbf{w} = \begin{bmatrix} a \\ b \\ c \end{bmatrix}$ be arbitrary in $\mathcal{R}^3$ and try to find scalars

x_i such that $x_1\mathbf{u}_1 + x_2\mathbf{u}_2 + x_3\mathbf{u}_3 = \mathbf{w}$, or equivalently

$$x_1\begin{bmatrix}1\\3\\-2\end{bmatrix}+x_2\begin{bmatrix}-3\\-12\\10\end{bmatrix}+x_3\begin{bmatrix}-2\\-6\\5\end{bmatrix}=\begin{bmatrix}a\\b\\c\end{bmatrix}.$$

This vector equation is equivalent to the system of linear equations

$$x_1 - 3x_2 - 2x_3 = a$$
$$3x_1 - 12x_2 - 6x_3 = b$$
$$-2x_1 + 10x_2 + 5x_3 = c.$$

In matrix form this system of equations is

$$AX=\begin{bmatrix}1&-3&-2\\3&-12&-6\\-2&10&5\end{bmatrix}\begin{bmatrix}x_1\\x_2\\x_3\end{bmatrix}=\begin{bmatrix}a\\b\\c\end{bmatrix}=\mathbf{w}.$$

Note that the vectors $\mathbf{u}_i$ are the columns of the matrix A. The row reduction

$$\begin{bmatrix}1&-3&-2&1\\3&-12&-6&0\\-2&10&5&0\end{bmatrix}\longrightarrow\begin{bmatrix}1&-3&-2&1\\0&-3&0&-3\\0&4&1&2\end{bmatrix}\longrightarrow$$

$$\begin{bmatrix}1&-3&-2&1\\0&1&1&-1\\0&4&1&2\end{bmatrix}\longrightarrow\begin{bmatrix}1&0&1&-2\\0&1&1&-1\\0&0&-3&6\end{bmatrix}\longrightarrow$$

$$\begin{bmatrix}1&0&0&0\\0&1&0&1\\0&0&1&-2\end{bmatrix}$$

shows that the coefficient matrix A is nonsingular and also solves the system in the special case $\mathbf{w}=\mathbf{e}_1$. Since A^{-1} exists, the system $AX=\mathbf{w}$ has a unique solution for any $\mathbf{w}$ in $\mathcal{R}^3$ and, by Theorem 3.10, it follows that $\mathcal{B}$ is a basis for $\mathcal{R}^3$.

The above row reduction also shows that $\mathbf{e}_1=\mathbf{u}_2-2\mathbf{u}_3$ and it follows that the $\mathcal{B}$-coordinate matrix of $\mathbf{e}_1$ is

$$[\mathbf{e}_1]_{\mathcal{B}}=\begin{bmatrix}0\\1\\-2\end{bmatrix}.\quad\blacksquare$$

Example 3

Show that $\{x^2-3x+2, x^2-2x, x^2-x\}$ is a basis for $\mathcal{P}_2$ and find the coordinate matrix of $x+2$ with respect to this basis.

In Example 7 of Section 3.5 we showed that every element of $\mathcal{P}_2$ is uniquely expressible as a linear combination of these three polynomials. It follows from Theorem 3.10 that these vectors form a basis for $\mathcal{P}_2$. In order to find the coordinates of $x + 2$ with respect to this basis, we need to find scalars a, b, and c so that

$$a(x^2 - 3x + 2) + b(x^2 - 2x) + c(x^2 - x) = x + 2.$$

This last equation is equivalent to

$$(a + b + c)x^2 + (-3a - 2b - c)x + 2a = x + 2.$$

Comparing coefficients, we see that the scalars a, b, and c must satisfy

$$\begin{aligned} 2a &= 2 \\ -3a - 2b - c &= 1 \\ a + b + c &= 0 \end{aligned} \quad \text{or} \quad \begin{bmatrix} 2 & 0 & 0 \\ -3 & -2 & -1 \\ 1 & 1 & 1 \end{bmatrix} \begin{bmatrix} a \\ b \\ c \end{bmatrix} = \begin{bmatrix} 2 \\ 1 \\ 0 \end{bmatrix}.$$

From the row reduction

$$\begin{bmatrix} 2 & 0 & 0 & 2 \\ -3 & -2 & -1 & 1 \\ 1 & 1 & 1 & 0 \end{bmatrix} \longrightarrow \begin{bmatrix} 1 & 0 & 0 & 1 \\ 2 & 1 & 0 & -1 \\ 1 & 1 & 1 & 0 \end{bmatrix} \longrightarrow$$

$$\begin{bmatrix} 1 & 0 & 0 & 1 \\ 2 & 1 & 0 & -1 \\ -1 & 0 & 1 & 1 \end{bmatrix} \longrightarrow \begin{bmatrix} 1 & 0 & 0 & 1 \\ 0 & 1 & 0 & -3 \\ 0 & 0 & 1 & 2 \end{bmatrix}$$

we see again that the coefficient matrix is invertible and that the solution of the above system is $a = 1$, $b = -3$, and $c = 2$. These are the desired coordinates. If we call this the $\mathcal{H}$ basis for $\mathcal{P}_2$, then $[x + 2]_{\mathcal{H}} = [1 \quad -3 \quad 2]^T$ is the $\mathcal{H}$-coordinate matrix of $x + 2$. ∎

Another basis for the vector space $\mathcal{P}_2$ of Example 3 is $\{1, x, x^2\}$. (This might well be called the standard basis for $\mathcal{P}_2$.) Note that the number of vectors in the two bases is the same. That this is true in general is an immediate consequence of our next theorem.

Theorem 3.11

If $\{v_1, v_2, \ldots, v_k\}$ is a basis for the vector space $\mathcal{V}$, then any set of vectors from $\mathcal{V}$ having more than k elements must be linearly dependent.

Proof.   Let $\{w_1, w_2, \ldots, w_r\}$ be any set of $r > k$ vectors from $\mathcal{V}$ and suppose that

$$x_1 w_1 + x_2 w_2 + \ldots + x_r w_r = 0. \tag{3.6}$$

In order to show that a nonzero solution exists for Equation 3.6 we express each of the w_j as a linear combination of the v_i:

$$w_j = a_{1j} v_1 + a_{2j} v_2 + \ldots + a_{kj} v_k; \quad j = 1, 2, \ldots, r.$$

If we substitute these expressions for the w_j into Equation 3.6 we obtain

$$x_1(a_{11} v_1 + \ldots + a_{k1} v_k) + x_2(a_{12} v_1 + \ldots + a_{k2} v_k) + \ldots \\ + x_r(a_{1r} v_1 + \ldots + a_{kr} v_k) = 0,$$

which, after some rearrangement, reduces to

$$(a_{11}x_1 + \ldots + a_{1r}x_r)\mathbf{v_1} + (a_{21}x_1 + \ldots + a_{2r}x_r)\mathbf{v_2} + \ldots$$
$$+ (a_{k1}x_1 + \ldots + a_{kr}x_r)\mathbf{v_k} = 0.$$

Since the $\mathbf{v_i}$ are assumed to form a basis for $\mathcal{V}$, they must be linearly independent. Thus, each of the coefficients in the last equation must be zero. This observation leads to the homogeneous system $AX = 0$, where the coefficient matrix A is $k \times r$. Since $k < r$, this homogeneous system has fewer equations than unknowns and hence is certain to have a nonzero solution. It follows that the $\mathbf{w_i}$ are linearly dependent as claimed and the proof is complete. $\square$

Theorem 3.11 shows that any four vectors from $\mathcal{R}^3$ must be linearly dependent, that any $n + 1$ vectors from $\mathcal{R}^n$ must be linearly dependent, and that any ten matrices from $\mathcal{R}^{3 \times 3}$ must be linearly dependent.

Suppose now that we have two different bases for $\mathcal{V}$:

$$\mathcal{B} = \{\mathbf{u_1}, \mathbf{u_2}, \ldots, \mathbf{u_n}\} \quad \text{and} \quad \mathcal{B}' = \{\mathbf{v_1}, \mathbf{v_2}, \ldots, \mathbf{v_k}\}.$$

If $n < k$, then Theorem 3.11 says that $\mathcal{B}'$ is dependent; this is a contradiction, so we must have $n \geq k$. Similarly, assuming that $k < n$ leads to the conclusion that $\mathcal{B}$ is dependent; from this contradiction we conclude that $k \leq n$. It follows that $k = n$; that is, the two bases must contain the same number of vectors. We state this important result as our next theorem.

Theorem 3.12

Any two bases for a vector space must contain the same number of vectors; that is, if $\{\mathbf{v_1}, \mathbf{v_2}, \ldots, \mathbf{v_n}\}$ is a basis for the vector space $\mathcal{V}$, then any other basis of $\mathcal{V}$ must contain precisely n vectors.

The number of vectors in a basis of $\mathcal{V}$ is an important property of the vector space $\mathcal{V}$. Theorem 3.12 shows that this number does not depend on how the basis is chosen.

Definition 3.17

If the vector space $\mathcal{V}$ has a finite basis, then the number of vectors in a basis is called the DIMENSION of $\mathcal{V}$ and is denoted by $\dim(\mathcal{V})$. If $\mathcal{V}$ does not have a finite basis, then we will say that $\mathcal{V}$ is INFINITE DIMENSIONAL and write $\dim(\mathcal{V}) = \infty$.

Example 1 shows that $\dim(\mathcal{R}^4) = 4$, and the obvious generalization to $\mathcal{R}^n$ shows that $\dim(\mathcal{R}^n) = n$. Example 3 shows that $\dim(\mathcal{P}_2) = 3$, and a similar argument shows that $\dim(\mathcal{P}_k) = k + 1$. In the exercises you are asked to show that $\dim(\mathcal{R}^{m \times n}) = mn$ and that $\dim(\mathcal{F}) = \infty$.

The simplest example of an infinite dimensional vector space is the vector space $\mathcal{P}$ of all polynomials of finite degree. To see that this space does not have a finite basis we suppose that

$$\{p_1(x), p_2(x), \ldots, p_k(x)\}$$

is a finite basis for $\mathcal{P}$. Note that each $p_i(x) \neq 0$, let d_i be the degree of $p_i(x)$ and let m be the maximum of the d_i, $i = 1, 2, \ldots, k$. Since no element of SPAN$\{p_1(x), \ldots, p_k(x)\}$ can have degree greater than m, it is clear that x^{m+1} is in $\mathcal{P}$ but not in SPAN$\{p_1(x), \ldots, p_k(x)\}$. Thus SPAN$\{p_1(x), \ldots, p_k(x)\} \neq \mathcal{P}$. It follows that $\mathcal{P}$ does not have a finite spanning set and hence is not finite dimensional.

In order to find the dimension of a subspace of a vector space $\mathcal{V}$ one normally has to find a basis for that subspace.

Example 4

Find a basis for the subspace $\mathcal{W}$ of $R^{1 \times 4}$ spanned by

$$\mathbf{u}_1 = [1 \quad 2 \quad 3 \quad 1], \quad \mathbf{u}_2 = [3 \quad 2 \quad 1 \quad 1],$$
$$\mathbf{u}_3 = [0 \quad 2 \quad 4 \quad 1], \quad \text{and} \quad \mathbf{u}_4 = [1 \quad 1 \quad 1 \quad 1].$$

We give two solutions; the first uses part 3 of Theorem 3.9.

First solution: The spanning set can fail to be a basis only if the spanning vectors are linearly dependent. Since $\mathbf{u}_1 \neq k\mathbf{u}_2$, $\{\mathbf{u}_1, \mathbf{u}_2\}$ is an independent set. We ask next if $\mathbf{u}_3$ is in SPAN$\{\mathbf{u}_1, \mathbf{u}_2\}$; that is, can we find scalars a and b so that

$$\mathbf{u}_3 = [0 \quad 2 \quad 4 \quad 1] = a[1 \quad 2 \quad 3 \quad 1] + b[3 \quad 2 \quad 1 \quad 1]$$
$$= a\mathbf{u}_1 + b\mathbf{u}_2 ?$$

This is equivalent to the system

$$
\begin{aligned}
a + 3b &= 0 \\
2a + 2b &= 2 \\
3a + b &= 4 \\
a + b &= 1
\end{aligned}
\quad \text{or} \quad
\begin{bmatrix} 1 & 3 \\ 2 & 2 \\ 3 & 1 \\ 1 & 1 \end{bmatrix}
\begin{bmatrix} a \\ b \end{bmatrix}
=
\begin{bmatrix} 0 \\ 2 \\ 4 \\ 1 \end{bmatrix}.
$$

From the row reduction

$$
\begin{bmatrix} 1 & 3 & 0 \\ 2 & 2 & 2 \\ 3 & 1 & 4 \\ 1 & 1 & 1 \end{bmatrix}
\longrightarrow
\begin{bmatrix} 1 & 3 & 0 \\ 0 & -4 & 2 \\ 0 & -8 & 4 \\ 0 & -2 & 1 \end{bmatrix}
\longrightarrow
\begin{bmatrix} 1 & 0 & 3/2 \\ 0 & 2 & -1 \\ 0 & 0 & 0 \\ 0 & 0 & 0 \end{bmatrix},
$$

we see that $a = 3/2$ and $b = -1/2$ is the solution of the system and hence $\mathbf{u}_3 = 3/2\mathbf{u}_1 - 1/2\mathbf{u}_2$ is in SPAN$\{\mathbf{u}_1, \mathbf{u}_2\}$. Thus SPAN$\{\mathbf{u}_1, \mathbf{u}_2, \mathbf{u}_3\}$ = SPAN$\{\mathbf{u}_1, \mathbf{u}_2\}$ so we can discard $\mathbf{u}_3$ from the spanning set for $\mathcal{W}$. Finally, we ask if $\mathbf{u}_4$ is in SPAN$\{\mathbf{u}_1, \mathbf{u}_2\}$. If $\mathbf{u}_4 = [1 \quad 1 \quad 1 \quad 1]$ = $c[1 \quad 2 \quad 3 \quad 1] + d[3 \quad 2 \quad 1 \quad 1] = c\mathbf{u}_1 + d\mathbf{u}_2$, then

$$
\begin{aligned}
c + 3d &= 1 \\
2c + 2d &= 1 \\
3c + d &= 1 \\
c + d &= 1
\end{aligned}
\quad \text{or} \quad
\begin{bmatrix} 1 & 3 \\ 2 & 2 \\ 3 & 1 \\ 1 & 1 \end{bmatrix}
\begin{bmatrix} c \\ d \end{bmatrix}
=
\begin{bmatrix} 1 \\ 1 \\ 1 \\ 1 \end{bmatrix}.
$$

The row reduction

$$\begin{bmatrix} 1 & 3 & 1 \\ 2 & 2 & 1 \\ 3 & 1 & 1 \\ 1 & 1 & 1 \end{bmatrix} \longrightarrow \begin{bmatrix} 1 & 3 & 1 \\ 0 & -4 & -1 \\ 0 & -8 & -2 \\ 0 & -2 & 0 \end{bmatrix} \longrightarrow \begin{bmatrix} 1 & 3 & 1 \\ 0 & 2 & 0 \\ 0 & 0 & -1 \\ 0 & 0 & -2 \end{bmatrix}$$

shows that this system is inconsistent and hence that u_4 is not in SPAN$\{u_1, u_2\}$. It follows from Theorem 3.9 that the set $\{u_1, u_2, u_4\}$ is linearly independent and hence forms a basis for $\mathcal{W}$.

Second solution: This solution uses Theorem 3.8 and the observation that the nonzero rows of an echelon matrix are independent. Since $\mathcal{W} = \mathcal{RS}(A)$ where

$$A = \begin{bmatrix} 1 & 2 & 3 & 1 \\ 3 & 2 & 1 & 1 \\ 0 & 2 & 4 & 1 \\ 1 & 1 & 1 & 1 \end{bmatrix},$$

and since elementary row operations do not change the row space, we see from the row reduction

$$A = \begin{bmatrix} 1 & 2 & 3 & 1 \\ 3 & 2 & 1 & 1 \\ 0 & 2 & 4 & 1 \\ 1 & 1 & 1 & 1 \end{bmatrix} \longrightarrow \begin{bmatrix} 1 & 2 & 3 & 1 \\ 0 & -4 & -8 & -2 \\ 0 & -1 & -2 & 0 \\ 0 & -1 & -2 & 0 \end{bmatrix} \longrightarrow$$

$$\begin{bmatrix} 1 & 2 & 3 & 1 \\ 0 & 1 & 2 & 0 \\ 0 & 0 & 0 & 1 \\ 0 & 0 & 0 & 0 \end{bmatrix} \longrightarrow \begin{bmatrix} 1 & 0 & -1 & 0 \\ 0 & 1 & 2 & 0 \\ 0 & 0 & 0 & 1 \\ 0 & 0 & 0 & 0 \end{bmatrix} = B$$

that a basis for the row space of A can be obtained from the nonzero rows of B. Thus $v_1 = [1 \quad 0 \quad -1 \quad 0]$, $v_2 = [0 \quad 1 \quad 2 \quad 0]$, and $v_3 = [0 \quad 0 \quad 0 \quad 1]$ form a basis for $\mathcal{W}$. Note that this basis is different from the one obtained in the first solution. The technique used in the second solution produces a basis which is "simplest" in the sense that its vectors have many zero components. It is particularly easy to find coordinates with respect to such an "echelon" basis, since the system to be solved will be triangular. ▄▄

Example 5

Find a basis for the null space of $A = \begin{bmatrix} 3 & -1 & -1 & -3 \\ -2 & 2 & -2 & 2 \\ -1 & -1 & 3 & 1 \end{bmatrix}$.

Since $\mathcal{NS}(A)$ consists of all solutions of the homogeneous system $A\mathbf{X} = 0$ we need to find the general solution of that system. From the row reduction

$$[A\,|\,0] = \begin{bmatrix} 3 & -1 & -1 & -3 & 0 \\ -2 & 2 & -2 & 2 & 0 \\ -1 & -1 & 3 & 1 & 0 \end{bmatrix} \longrightarrow \begin{bmatrix} 0 & -4 & 8 & 0 & 0 \\ 0 & 4 & -8 & 0 & 0 \\ -1 & -1 & 3 & 1 & 0 \end{bmatrix}$$

$$\longrightarrow \begin{bmatrix} 1 & 0 & -1 & -1 & 0 \\ 0 & 1 & -2 & 0 & 0 \\ 0 & 0 & 0 & 0 & 0 \end{bmatrix} = [B\,|\,0]$$

we see that $\mathbf{X}$ is in $\mathcal{NS}(A)$ if and only $B\mathbf{X} = 0$; that is,

$$x_1 - x_3 - x_4 = 0$$

$$x_2 - 2x_3 = 0.$$

The most general solution of this system is

$$\mathbf{X} = \begin{bmatrix} c_1 + c_2 \\ 2c_1 \\ c_1 \\ c_2 \end{bmatrix} = c_1 \begin{bmatrix} 1 \\ 2 \\ 1 \\ 0 \end{bmatrix} + c_2 \begin{bmatrix} 1 \\ 0 \\ 0 \\ 1 \end{bmatrix} = c_1\mathbf{X_1} + c_2\mathbf{X_2},$$

where c_1 and c_2 are arbitrary. From this last equation we see that

$$\mathcal{NS}(A) = \mathrm{SPAN}\{\mathbf{X_1}, \mathbf{X_2}\}$$

and, since $\mathbf{X_1}$ and $\mathbf{X_2}$ are clearly independent, they form a basis for $\mathcal{NS}(A)$. ∎

In Chapter 1 we observed that the number of degrees of freedom in the general solution of the linear system $AX = K$ was equal to the number of unknowns less the number of nonzero rows in the echelon matrix. If we interpret this result in terms of the concepts introduced in this chapter, observing that the number of degrees of freedom in the solution is $\dim(\mathcal{NS}(A))$ and the number of nonzero rows in the echelon form is $\dim(\mathcal{RS}(A))$, we have the following theorem.

Theorem 3.13

If A is an $m \times n$ matrix, then $\dim(\mathcal{NS}(A)) = n - \dim(\mathcal{RS}(A))$.

The dimension of the row space of A is frequently called the RANK of A and denoted $\mathrm{rank}(A)$. Theorem 3.13 can then be rephrased as

$$\dim(\mathcal{NS}(A)) + \mathrm{rank}(A) = \text{number of columns of } A.$$

The normal way of computing the rank of a matrix is to reduce the matrix to echelon form and then count the number of nonzero rows. For the matrix A of Example 5 we see that $\mathrm{rank}(A) = 2$, since the echelon matrix B has two nonzero rows. Since the number of columns of A is four, it follows that $\dim(\mathcal{NS}(A)) = 4 - 2 = 2$, which is the number of arbitrary constants in the general solution of the homogeneous system $AX = 0$.

For an $n \times n$ matrix A we know that A^{-1} exists if and only if $\mathcal{NS}(A) = \{0\}$. It then follows from the last equation that A^{-1} exists if and only if rank$(A) = n$. We conclude this section by expanding Theorem 1.9 to include all the conditions so far discussed that are equivalent to the nonsingularity of the matrix A.

Theorem 3.14

For an $n \times n$ matrix A the following statements are equivalent:

(1) A is nonsingular;
(2) A has either a right or a left inverse;
(3) $\mathcal{NS}(A) = \{0\}$;
(4) A is row equivalent to I;
(5) A is a product of elementary matrices;
(6) $\det(A) \neq 0$;
(7) rank$(A) = n$;
(8) The rows of A are linearly independent;
(9) The columns of A are linearly independent; and
(10) $AX = K$ has a unique solution for every K.

Exercise Set 3.7

1. Which of the following sets is a basis for $\mathcal{R}^n$?
   (a) $[2 \quad 2 \quad 2]^T$, $[1 \quad 1 \quad 0]^T$, $[3 \quad 0 \quad 0]^T$; $n = 3$
   (b) $[1 \quad 2 \quad 3]^T$, $[4 \quad 5 \quad 6]^T$, $[7 \quad 8 \quad 9]^T$; $n = 3$
   (c) $[1 \quad 6 \quad 4]^T$, $[1 \quad -2 \quad -5]^T$, $[2 \quad 4 \quad -1]^T$; $n = 3$
   (d) $[3 \quad -3 \quad -1]^T$, $[-3 \quad -3 \quad 3]^T$, $[-1 \quad 3 \quad 3]^T$; $n = 3$
   (e) $[1 \quad 1 \quad 1 \quad -3]^T$, $[3 \quad -2 \quad -17 \quad 16]^T$, $[3 \quad 2 \quad -1 \quad -4]^T$, $[1 \quad 1 \quad 1 \quad 1]^T$; $n = 4$
   (f) $[1 \quad 2 \quad 4 \quad 1 \quad 0]^T$, $[-1 \quad -1 \quad 1 \quad 2 \quad 0]^T$, $[2 \quad 0 \quad -11 \quad 3 \quad 0]^T$, $[3 \quad 2 \quad -1 \quad 8 \quad 0]^T$, $[0 \quad 0 \quad 0 \quad 1 \quad 1]^T$; $n = 5$

2. Which of the following sets is a basis for $\mathcal{P}_2$?
   (a) x^2, $x^2 + x$, $x^2 + x + 1$
   (b) $2x^2 - 3x - 1$, $4x^2 + x + 1$
   (c) $x^2 + 6x + 4$, $2x^2 + 4x - 1$, $x^2 - 2x - 5$, $3x - 5$
   (d) $x^2 - 3x + 2$, $x^2 - 4x + 3$, $x^2 - 5x + 6$

3. Show that the set
   $$\mathcal{B} = \{u_1 = [1 \quad -1 \quad 1 \quad -1], u_2 = [1 \quad -1 \quad 1 \quad 2],$$
   $$u_3 = [0 \quad 2 \quad -1 \quad 3], u_4 = [0 \quad 0 \quad 1 \quad 3]\}$$
   forms a basis for $\mathcal{R}^{1 \times 4}$ and find $[v_1]_{\mathcal{B}}$ and $[v_2]_{\mathcal{B}}$ where $v_1 = [2 \quad 0 \quad 2 \quad 7]$ and $v_2 = [2 \quad -2 \quad 4 \quad 7]$.

4. Show that the set $\mathcal{B} = \{x^2 - 5x + 6, x^2 - 6x + 8, x^2 - 7x + 12\}$ is a basis for $\mathcal{P}_2$ and find $[x^2 + x + 1]_{\mathcal{B}}$, $[x]_{\mathcal{B}}$ and $[1]_{\mathcal{B}}$.

5. Show that the following matrices form a basis for $\mathcal{R}^{2\times 2}$.

$$\mathcal{B} = \left\{ A_1 = \begin{bmatrix} 1 & 2 \\ 1 & 0 \end{bmatrix} \quad A_2 = \begin{bmatrix} 0 & 1 \\ 1 & 0 \end{bmatrix} \quad A_3 = \begin{bmatrix} 0 & 2 \\ 3 & 1 \end{bmatrix} \quad A_4 = \begin{bmatrix} -1 & 0 \\ 1 & -2 \end{bmatrix} \right\}$$

and find the $\mathcal{B}$-coordinates of I and $T = \begin{bmatrix} 0 & 5 \\ 7 & -3 \end{bmatrix}$.

6. Find the echelon basis $\mathcal{E}$ for the row space of $A = \begin{bmatrix} 1 & 2 & 4 & 1 \\ 2 & 3 & 7 & 1 \\ 1 & 4 & 6 & 3 \end{bmatrix}$ and

   for $\mathbf{u}_1 = [3 \quad -2 \quad 4 \quad -5]$ and $\mathbf{u}_2 = [1 \quad -1 \quad 1 \quad -2]$ find $[\mathbf{u}_1]_\mathcal{E}$ and $[\mathbf{u}_2]_\mathcal{E}$.

7. Find $\dim(\mathcal{RS}(A))$, $\dim(\mathcal{CS}(A))$ and $\dim(\mathcal{NS}(A))$ if

   $$A = \begin{bmatrix} 1 & 2 & -1 & 4 \\ -1 & -2 & 6 & -7 \\ 2 & 4 & 3 & 5 \end{bmatrix}.$$

8. Consider $\mathbf{w}_1 = [1 \quad 1 \quad 2 \quad 4]$, $\mathbf{w}_2 = [1 \quad -1 \quad -4 \quad 0]$, $\mathbf{w}_3 = [2 \quad 1 \quad 1 \quad 6]$, and $\mathbf{w}_4 = [2 \quad -1 \quad -5 \quad 2]$ from $\mathcal{R}^{1\times 4}$. Let $\mathcal{V}$ denote the subspace spanned by $\{\mathbf{w}_1, \mathbf{w}_2, \mathbf{w}_3, \mathbf{w}_4\}$. Find a matrix A such that $\mathcal{V} = \mathcal{RS}(A)$ and then, by reducing A to echelon form, find a basis for $\mathcal{V}$. Also, find a basis for $\mathcal{V}$ which is a subset of $\{\mathbf{w}_1, \mathbf{w}_2, \mathbf{w}_3, \mathbf{w}_4\}$.

9. Let $\mathcal{W}$ be the subspace of $\mathcal{R}^4$ spanned by $\mathbf{c}_1 = [0 \quad 1 \quad 2 \quad 1]^T$, $\mathbf{c}_2 = [1 \quad 2 \quad 3 \quad 1]^T$, $\mathbf{c}_3 = [3 \quad 6 \quad 9 \quad 3]^T$, and $\mathbf{c}_4 = [-2 \quad 0 \quad 2 \quad 2]^T$. Construct a matrix A such that $\mathcal{W} = \mathcal{CS}(A)$. By column reducing A, find a basis for $\mathcal{W}$. Explain why one cannot find a basis for $\mathcal{W}$ by row reducing the matrix A. Find a basis for $\mathcal{W}$ that is a subset of $\{\mathbf{c}_1, \mathbf{c}_2, \mathbf{c}_3, \mathbf{c}_4\}$.

10. (a) Show that $\dim(\mathcal{R}^{m\times n}) = mn$.

    (b) Show that $\dim(\mathcal{F}) = \infty$.

11. (a) Show that any independent subset of a vector space $\mathcal{V}$ can be extended to a basis for $\mathcal{V}$.

    (*Hint:* Let $(\mathbf{u}_1, \ldots, \mathbf{u}_k\}$ be independent and $\{\mathbf{v}_1, \ldots, \mathbf{v}_n\}$ be a basis. Show that $\{\mathbf{u}_1, \ldots, \mathbf{u}_k, \mathbf{v}_1, \ldots, \mathbf{v}_n\}$ spans $\mathcal{V}$. Then use the third property of Theorem 3.9.

    (b) Find a basis for $\mathcal{R}^3$ containing $\mathbf{u}_1 = [1 \quad 2 \quad 3]^T$.

12. (a) Show that a subspace $\mathcal{W}$ of a finite dimensional space $\mathcal{V}$ is finite dimensional with $\dim(W) < \dim(V)$.

    (b) Show that if a subspace $\mathcal{W}$ of a finite dimensional space $\mathcal{V}$ satisfies $\dim \mathcal{W} = \dim \mathcal{V}$, then $\mathcal{W} = \mathcal{V}$.

    (c) Show that if a subspace $\mathcal{W}$ of $\mathcal{V}$ is infinite dimensional, then so is $\mathcal{V}$.

13. Find the rank of the following matrices.

$$A = \begin{bmatrix} 1 & 3 \\ 4 & -1 \\ 3 & 7 \end{bmatrix} \quad B = \begin{bmatrix} 1 & 1 & 1 & 3 \\ 3 & -2 & -17 & 16 \\ 3 & 2 & -1 & -4 \end{bmatrix} \quad C = \begin{bmatrix} 1 & -1 & 2 & 3 \\ 2 & 2 & 0 & 2 \\ 4 & 1 & 3 & -1 \\ 1 & 2 & 3 & 0 \end{bmatrix}$$

14. (a) Show that every subspace of $\mathcal{R}^n$ has a finite basis.

    (b) Prove Theorem 3.5 on page 150.

15. Find a basis for the nullspace of the following matrices. In each case find dim($\mathcal{NS}(A)$) and dim($\mathcal{RS}(A)$).

    (a) $A = \begin{bmatrix} 1 & -3 & 4 & 3 & -2 \\ 7 & 4 & -2 & 6 & 1 \end{bmatrix}$ (b) $A = \begin{bmatrix} 1 & 1 & 1 & 3 \\ 3 & -2 & -17 & 16 \\ 3 & 2 & -1 & -4 \end{bmatrix}$

16. (a) Show that if dim($\mathcal{V}$) = n, and $\mathcal{S} = \{\mathbf{u}_1, \ldots, \mathbf{u}_n\}$ spans $\mathcal{V}$, then $\mathcal{S}$ is a basis for $\mathcal{V}$.

    (b) Show that if dim($\mathcal{V}$) = n, and $\mathcal{S} = \{\mathbf{v}_1, \ldots, \mathbf{v}_n\}$ is independent, then $\mathcal{S}$ is a basis for V.

17. Complete the proof of Theorem 3.10.

†18. Let $\mathcal{B} = \{\mathbf{u}_1, \mathbf{u}_2, \mathbf{u}_3, \mathbf{u}_4\}$ be a basis for a four dimensional subspace $\mathcal{W}$ and suppose that the $\mathcal{B}$-coordinate matrix of $\mathbf{v}$ is $[1 \quad 2 \quad -3 \quad 6]^T$. Find the $\mathcal{B}'$-coordinate of $\mathbf{v}$ with respect to the basis $\mathcal{B}' = \{\mathbf{u}_3, \mathbf{u}_1, \mathbf{u}_4, \mathbf{u}_2\}$.

3.8 *Orthogonality in $\mathcal{R}^n$*

In this section we will present some special results about the inner product in $\mathcal{R}^n$. Many of the results of this section can be generalized to more general vector spaces, but we leave that undertaking to more advanced texts. The exercises of this section contain some introduction to the more general concepts.

Recall, from Section 3.2, that we have defined the length, or norm, of a vector $\mathbf{u}$ in $\mathcal{R}^n$ by

$$\|\mathbf{u}\| = \sqrt{\mathbf{u} \cdot \mathbf{u}} = \sqrt{\sum u_i^2}$$

and have defined two vectors, $\mathbf{u}$ and $\mathbf{v}$, to be orthogonal if

$$\mathbf{u} \cdot \mathbf{v} = \sum_{i=1}^{n} u_i v_i = 0.$$

As the first theorem of this section we list some elementary, but important, properties of the length, or norm, of a vector. The proofs will be left to the exercises.

Theorem 3.15

For any vector $\mathbf{u}$ in $\mathcal{R}^n$ and any scalar k,

(1) $\|\mathbf{u}\| \geq 0$;

(2) $\|\mathbf{u}\| = 0$ if and only if $\mathbf{u} = \mathbf{0}$; and

(3) $\|k\mathbf{u}\| = |k| \|\mathbf{u}\|$.

For vectors $\mathbf{u}$ and $\mathbf{v}$ in $\mathcal{R}^3$ we have the alternate formula for the inner product

$$\mathbf{u} \cdot \mathbf{v} = \|\mathbf{u}\| \|\mathbf{v}\| \cos \theta .$$

Since $|\cos \theta| \leq 1$, it follows at once that

$$|\mathbf{u} \cdot \mathbf{v}| \leq \|\mathbf{u}\| \|\mathbf{v}\| .$$

This result is true for vectors in $\mathcal{R}^n$, although the above argument is not valid in the general case since we have no geometric way of defining angles and trigonometric functions in $\mathcal{R}^n$. These ideas can be defined algebraically in $\mathcal{R}^n$ using the above formulas; that is, we could define $\theta = \cos^{-1}(\mathbf{u} \cdot \mathbf{v}/\|\mathbf{u}\| \|\mathbf{v}\|)$, provided that we could show that $\mathbf{u} \cdot \mathbf{v}/\|\mathbf{u}\| \|\mathbf{v}\| \leq 1$.

Theorem 3.16

CAUCHY-SCHWARZ INEQUALITY: For any $\mathbf{u}$ and $\mathbf{v}$ in $\mathcal{R}^n$ we have $|\mathbf{u} \cdot \mathbf{v}| \leq \|\mathbf{u}\| \|\mathbf{v}\|$.

Proof. We first consider the case where $\mathbf{u} = [u_1, \ldots, u_n]^T$ and $\mathbf{v} = [v_1 \ldots v_n]^T$ are unit vectors so that

$$\|\mathbf{u}\|^2 = \sum u_i^2 = 1 \text{ and } \|\mathbf{v}\|^2 = \sum v_i^2 = 1 .$$

In this case we have

$$0 \leq \|\mathbf{u} - \mathbf{v}\|^2 = \sum (u_i - v_i)^2 = \sum u_i^2 - 2 \sum u_i v_i + \sum v_i^2$$
$$= 2 - 2\mathbf{u} \cdot \mathbf{v}$$

and

$$0 \leq \|\mathbf{u} + \mathbf{v}\|^2 = \sum (u_i + v_i)^2 = \sum u_i^2 + 2 \sum u_i v_i + \sum v_i^2$$
$$= 2 + 2\mathbf{u} \cdot \mathbf{v} .$$

It follows from the first equation that $\mathbf{u} \cdot \mathbf{v} \leq 1$ and from the second that $-1 \leq \mathbf{u} \cdot \mathbf{v}$; together these two inequalities yield $|\mathbf{u} \cdot \mathbf{v}| \leq 1$.

If $\mathbf{u}$ and $\mathbf{v}$ are both nonzero, then $\mathbf{u}/\|\mathbf{u}\|$ and $\mathbf{v}/\|\mathbf{v}\|$ are unit vectors, for which we have just shown that

$$\left| \left(\frac{\mathbf{u}}{\|\mathbf{u}\|} \right) \cdot \left(\frac{\mathbf{v}}{\|\mathbf{v}\|} \right) \right| = \left| \frac{\mathbf{u} \cdot \mathbf{v}}{\|\mathbf{u}\| \|\mathbf{v}\|} \right| \leq 1 .$$

It follows at once that

$$|\mathbf{u} \cdot \mathbf{v}| \leq \|\mathbf{u}\| \|\mathbf{v}\| .$$

Since the theorem is trivially true for either $\mathbf{u} = \mathbf{0}$ or $\mathbf{v} = \mathbf{0}$, the proof is complete. $\square$

An immediate consequence of the last theorem is a result that is usually referred to as the triangle inequality; in the case $n = 3$ it says that one side of a triangle is less than the sum of the other two sides; see Figure 3.16.

FIGURE 3.16

Theorem 3.17

TRIANGLE INEQUALITY: For any vectors $\mathbf{u}$ and $\mathbf{v}$ in $\mathscr{R}^n$ we have

$$\|\mathbf{u} + \mathbf{v}\| \le \|\mathbf{u}\| + \|\mathbf{v}\|.$$

Proof. Since $\|\mathbf{u} + \mathbf{v}\|^2 = (\mathbf{u} + \mathbf{v}) \cdot (\mathbf{u} + \mathbf{v}) = \mathbf{u} \cdot \mathbf{u} + \mathbf{u} \cdot \mathbf{v}$
$$+ \mathbf{v} \cdot \mathbf{u} + \mathbf{v} \cdot \mathbf{v}$$
$$= \|\mathbf{u}\|^2 + 2(\mathbf{u} \cdot \mathbf{v}) + \|\mathbf{v}\|^2$$

it follows from Theorem 3.16 that

$$\|\mathbf{u} + \mathbf{v}\|^2 \le \|\mathbf{u}\|^2 + 2\|\mathbf{u}\|\,\|\mathbf{v}\| + \|\mathbf{v}\|^2 = (\|\mathbf{u}\| + \|\mathbf{v}\|)^2.$$

Thus $\|\mathbf{u} + \mathbf{v}\| \le \|\mathbf{u}\| + \|\mathbf{v}\|$ as claimed. ☐

Definition 3.18

Two vectors, $\mathbf{u}$ and $\mathbf{v}$, from $\mathscr{R}^n$ are ORTHOGONAL if $\mathbf{u} \cdot \mathbf{v} = 0$. A set of vectors, $\mathscr{S} = \{\mathbf{u}_1, \mathbf{u}_2, \ldots, \mathbf{u}_k\}$ from $\mathscr{R}^n$ is said to be MUTUALLY ORTHOGONAL if the vectors are pairwise orthogonal, that is, $\mathbf{u}_i \cdot \mathbf{u}_j = 0$ for $i \ne j$. If the vectors of the mutually orthogonal set $\mathscr{S}$ are all unit vectors, then we say that the set is ORTHONORMAL; that is, $\mathbf{u}_i \cdot \mathbf{u}_j = 0$ for $i \ne j$ and $\mathbf{u}_i \cdot \mathbf{u}_i = 1$ for $i = 1, 2, \ldots, k$. An orthonormal set of vectors that also forms a basis for $\mathscr{R}^n$ is called an ORTHONORMAL BASIS for $\mathscr{R}^n$.

Example 1

The standard basis for $\mathscr{R}^n$, which we denote by

$$\mathbf{e}_1 = \begin{bmatrix} 1 \\ 0 \\ \vdots \\ 0 \end{bmatrix}, \mathbf{e}_2 = \begin{bmatrix} 0 \\ 1 \\ \vdots \\ 0 \end{bmatrix}, \ldots, \mathbf{e}_n = \begin{bmatrix} 0 \\ 0 \\ \vdots \\ 1 \end{bmatrix},$$

is easily seen to be an orthonormal basis. For $n = 3$ these vectors are frequently denoted $\mathbf{i}$, $\mathbf{j}$, and $\mathbf{k}$. ■

Example 2

In $\mathcal{R}^3$ the vectors $\mathbf{u}_1 = \begin{bmatrix} 1 \\ 1 \\ 1 \end{bmatrix}$, $\mathbf{u}_2 = \begin{bmatrix} 0 \\ 1 \\ -1 \end{bmatrix}$, and $\mathbf{u}_3 = \begin{bmatrix} -2 \\ 1 \\ 1 \end{bmatrix}$ are mutu-

ally orthogonal since

$$\mathbf{u}_1 \cdot \mathbf{u}_2 = (0 + 1 - 1) = 0, \quad \mathbf{u}_1 \cdot \mathbf{u}_3 = (-2 + 1 + 1) = 0, \quad \text{and}$$
$$\mathbf{u}_2 \cdot \mathbf{u}_3 = (0 + 1 - 1) = 0.$$

These vectors are not orthonormal since they are not unit vectors:

$$\mathbf{u}_1 \cdot \mathbf{u}_1 = 1 + 1 + 1 = 3 \neq 1, \quad \mathbf{u}_2 \cdot \mathbf{u}_2 = 2, \quad \text{and} \quad \mathbf{u}_3 \cdot \mathbf{u}_3 = 6.$$

We can easily modify the $\mathbf{u}_i$ to obtain an orthonormal set by dividing each of the vectors by its length. Thus, the vectors

$$\mathbf{v}_1 = \frac{\mathbf{u}_1}{\|\mathbf{u}_1\|} = \frac{1}{\sqrt{3}} \begin{bmatrix} 1 \\ 1 \\ 1 \end{bmatrix}, \quad \mathbf{v}_2 = \frac{1}{\sqrt{2}} \begin{bmatrix} 0 \\ 1 \\ -1 \end{bmatrix}, \quad \text{and} \quad \mathbf{v}_3 = \frac{1}{\sqrt{6}} \begin{bmatrix} -2 \\ 1 \\ 1 \end{bmatrix}$$

form an orthonormal set in $\mathcal{R}^3$. We leave it to you to verify that these vectors are linearly independent and hence form an orthonormal basis for $\mathcal{R}^3$. ∎

Three mutually orthogonal vectors in $\mathcal{R}^3$ are clearly not coplanar; thus they are linearly independent. In general, an orthonormal set of vectors will always be linearly independent. This is an immediate corollary of the following theorem.

Theorem 3.18

Any set of mutually orthogonal nonzero vectors from $\mathcal{R}^n$ is linearly independent.

Proof. Let $\mathcal{S} = \{\mathbf{u}_1, \mathbf{u}_2, \ldots, \mathbf{u}_n\}$ be a mutually orthogonal set of nonzero vectors, thus $\mathbf{u}_i \cdot \mathbf{u}_j = 0$ for $i \neq j$. In order to test these vectors for linear independence we suppose that

$$a_1\mathbf{u}_1 + a_2\mathbf{u}_2 + \ldots + a_n\mathbf{u}_n = \mathbf{0}.$$

Taking the inner product of both sides of the last equation with $\mathbf{u}_i$ yields, using the properties of the dot product in Theorem 3.2,

$$a_1(\mathbf{u}_i \cdot \mathbf{u}_1) + a_2(\mathbf{u}_i \cdot \mathbf{u}_2) + \ldots + a_n(\mathbf{u}_i \cdot \mathbf{u}_n) = \mathbf{u}_i \cdot \mathbf{0} = 0.$$

This equation, because of the orthogonality of the $\mathbf{u}_i$, reduces to

$$a_i(\mathbf{u}_i \cdot \mathbf{u}_i) = a_i\|\mathbf{u}_i\|^2 = 0.$$

Since $\mathbf{u}_i \neq \mathbf{0}$ implies $\|\mathbf{u}_i\| \neq 0$, it follows that $a_i = 0$. This calculation can be repeated for each $i = 1, 2, \ldots, n$ to show that all the $a_i = 0$. Therefore the vectors $\mathbf{u}_i$ are linearly independent. □

In many applications it is convenient to work with an orthonormal basis for the subspace of $\mathcal{R}^n$ under investigation. One of the advantages of an orthonormal basis is that it is very easy to compute coordinates. Thus let $\mathcal{B} = \{\mathbf{w}_1, \mathbf{w}_2, \ldots, \mathbf{w}_k\}$ be an orthonormal basis for a subspace $\mathcal{W}$ of $\mathcal{R}^n$ and let $\mathbf{z}$ be any vector in $\mathcal{W}$. We know that there exist unique scalars a_i such that

$$\mathbf{z} = a_1\mathbf{w}_1 + a_2\mathbf{w}_2 + \ldots + a_k\mathbf{w}_k.$$

We could find these coefficients by solving a system of linear equations, as illustrated in Example 2 of Section 3.6, or by using the technique used in proving Theorem 3.18. We adopt the second point of view and compute the inner product of both sides of the preceding equation with $\mathbf{w}_i$ to obtain

$$\mathbf{w}_i \cdot \mathbf{z} = \sum_{j=1}^{k} a_j(\mathbf{w}_i \cdot \mathbf{w}_j) = a_i(\mathbf{w}_i \cdot \mathbf{w}_i) = a_i; \quad i = 1, \ldots, k.$$

Thus we can find the coordinates of the vector $\mathbf{z}$ with respect to the orthonormal basis $\mathcal{B}$ by simply taking the inner product of $\mathbf{z}$ and the basis vectors. Thus we have

$$\mathbf{z} = \sum_{j=1}^{k} (\mathbf{z} \cdot \mathbf{w}_j)\mathbf{w}_j = (\mathbf{z} \cdot \mathbf{w}_1)\mathbf{w}_1 + \ldots + (\mathbf{z} \cdot \mathbf{w}_k)\mathbf{w}_k;$$

the $\mathcal{B}$-coordinate matrix of $\mathbf{z}$ is

$$[\mathbf{z}]_{\mathcal{B}} = [(\mathbf{z} \cdot \mathbf{w}_1) \quad (\mathbf{z} \cdot \mathbf{w}_2) \quad \ldots \quad (\mathbf{z} \cdot \mathbf{w}_k)]^T. \tag{3.7}$$

Example 3

Find the coordinates of the vector $\mathbf{z} = [1 \quad 2 \quad 3]^T$ with respect to the orthonormal basis of $\mathcal{R}^3$ obtained in Example 2.

Using the result in Equation 3.7 we see that the coordinates of $\mathbf{z}$ are

$$\mathbf{z} \cdot \mathbf{v}_1 = \left(\frac{1}{\sqrt{3}}\right)(1 + 2 + 3) = \frac{6}{\sqrt{3}};$$

$$\mathbf{z} \cdot \mathbf{v}_2 = \left(\frac{1}{\sqrt{2}}\right)(0 + 2 - 3) = \frac{-1}{\sqrt{2}}; \quad \text{and}$$

$$\mathbf{z} \cdot \mathbf{v}_3 = \left(\frac{1}{\sqrt{6}}\right)(-2 + 2 + 3) = \frac{3}{\sqrt{6}}.$$

Thus $\mathbf{z} = (6/\sqrt{3})\mathbf{v}_1 + (-1/\sqrt{2})\mathbf{v}_2 + (3/\sqrt{6})\mathbf{v}_3$. ∎

We will now show how to construct an orthonormal basis for any subspace of $\mathcal{R}^n$. The procedure is essentially a replacement process in which the vectors of one basis are replaced, one at a time, by the vectors of the orthonormal basis. The basic step is described below. If $\{\mathbf{u}_1, \ldots, \mathbf{u}_k\}$ is a mutually orthogonal set and $\mathbf{v}_{k+1}$ is not in $\mathrm{SPAN}\{\mathbf{u}_1, \ldots, \mathbf{u}_k\}$, then we try to find $\mathbf{u}_{k+1}$ in a vector $\mathrm{SPAN}\{\mathbf{u}_1, \ldots, \mathbf{u}_k, \mathbf{v}_{k+1}\}$, which is orthogonal to each of the $\mathbf{u}_i$. Thus we need to choose scalars a_i so that

$$\mathbf{u}_{k+1} = a_1\mathbf{u}_1 + a_2\mathbf{u}_2 + \ldots + a_k\mathbf{u}_k + \mathbf{v}_{k+1} \tag{3.8}$$

is orthogonal to each of the $\mathbf{u}_i$, $i = 1, 2, \ldots, k$. We determine the coefficients in Equation 3.8 by taking the inner product of Equation 3.8 with each of the orthogonal vectors $\mathbf{u}_i$, $i = 1, 2, \ldots, k$. This step leads to

$$0 = (\mathbf{u}_i \cdot \mathbf{u}_{k+1}) = \sum a_j(\mathbf{u}_i \cdot \mathbf{u}_j) + (\mathbf{u}_i \cdot \mathbf{v}_{k+1})$$

$$= a_i(\mathbf{u}_i \cdot \mathbf{u}_i) + (\mathbf{u}_i \cdot \mathbf{v}_{k+1}),$$

from which we obtain

$$a_i = -(\mathbf{u}_i \cdot \mathbf{v}_{k+1})/(\mathbf{u}_i \cdot \mathbf{u}_i); \quad i = 1, 2, \ldots, k. \tag{3.9}$$

This step determines the coefficients in Equation 3.8 and hence the vector $\mathbf{u}_{k+1}$.

The next theorem summarizes these calculations.

Theorem 3.19

If the vectors $\{\mathbf{u}_1, \ldots, \mathbf{u}_k\}$ are mutually orthogonal in $\mathcal{R}^n$ and $\mathbf{v}_{k+1}$ is not in $\mathrm{SPAN}\{\mathbf{u}_1, \ldots, \mathbf{u}_k\}$, then the vector

$$\mathbf{u}_{k+1} = \mathbf{v}_{k+1} - \sum_{i=1}^{k} \left(\frac{\mathbf{u}_i \cdot \mathbf{v}_{k+1}}{\mathbf{u}_i \cdot \mathbf{u}_i} \right) \mathbf{u}_i$$

is orthogonal to each of the $\mathbf{u}_i$.

It is worth noting the geometric interpretation of Equations 3.8 and 3.9 in the case $n = 3$ and $k = 2$. If $\mathbf{u}_1$ and $\mathbf{u}_2$ are orthogonal and $\mathbf{v}_3$ is not in the plane $\mathcal{P} = \mathrm{SPAN}\{\mathbf{u}_1, \mathbf{u}_2\}$, then a vector $\mathbf{u}_3$ orthogonal to both $\mathbf{u}_1$ and $\mathbf{u}_2$ is obtained by taking $\mathbf{u}_3$ to be the difference between $\mathbf{v}_3$ and the projection of $\mathbf{v}_3$ on the plane $\mathcal{P}$. Thus $\mathbf{u}_3 = \mathbf{v}_3 - (b_1\mathbf{u}_1 + b_2\mathbf{u}_2)$ where

$$b_1 = -a_1 = \frac{(\mathbf{u}_1 \cdot \mathbf{v}_3)}{(\mathbf{u}_1 \cdot \mathbf{u}_1)} \quad \text{and} \quad b_2 = -a_2 = \frac{(\mathbf{u}_2 \cdot \mathbf{v}_3)}{(\mathbf{u}_2 \cdot \mathbf{u}_2)}$$

are the scalar projections of $\mathbf{v}_3$ on $\mathbf{u}_1$ and $\mathbf{u}_2$ respectively. Equivalently, $\mathbf{u}_3$ is the projection of $\mathbf{v}_3$ on a line perpendicular to the plane determined by $\mathbf{u}_1$ and $\mathbf{u}_2$. See Figure 3.17.

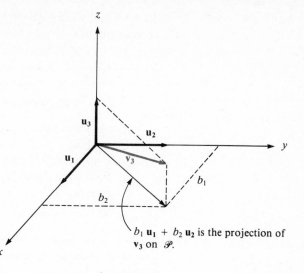

$b_1 \mathbf{u}_1 + b_2 \mathbf{u}_2$ is the projection of $\mathbf{v}_3$ on $\mathcal{P}$.

FIGURE 3.17

Example 4

Given the orthogonal vectors $\mathbf{u}_1 = [1 \quad 1 \quad 2]^T$ and $\mathbf{u}_2 = [-2 \quad 0 \quad 1]^T$ and the vector $\mathbf{v}_3 = [-1 \quad 1 \quad 4]^T$, which is not in SPAN$\{\mathbf{u}_1, \mathbf{u}_2\}$, find a vector $\mathbf{u}_3$ that is orthogonal to both $\mathbf{u}_1$ and $\mathbf{u}_2$.

Following the procedure described in Theorem 3.19, we write

$$\mathbf{u}_3 = a_1\mathbf{u}_1 + a_2\mathbf{u}_2 + \mathbf{v}_3$$

and take the dot product with $\mathbf{u}_1$ to obtain

$$0 = (\mathbf{u}_1 \cdot \mathbf{u}_3) = a_1(\mathbf{u}_1 \cdot \mathbf{u}_1) + a_2(\mathbf{u}_1 \cdot \mathbf{u}_2) + (\mathbf{u}_1 \cdot \mathbf{v}_3) = 6a_1 + 0 + 8,$$

from which it follows that $a_1 = -8/6$. Similarly,

$$0 = (\mathbf{u}_2 \cdot \mathbf{u}_3) = a_1(\mathbf{u}_2 \cdot \mathbf{u}_1) + a_2(\mathbf{u}_2 \cdot \mathbf{u}_2) + (\mathbf{u}_2 \cdot \mathbf{v}_3) = 0 + 5a_2 + 6,$$

from which we obtain $a_2 = -6/5$. Thus we conclude that

$$\mathbf{u}_3 = -\frac{4}{3}\mathbf{u}_1 - \frac{6}{5}\mathbf{u}_2 + \mathbf{v}_3$$

$$= -\left(\frac{4}{3}\right)[1 \quad 1 \quad 2]^T - \left(\frac{6}{5}\right)[-2 \quad 0 \quad 1]^T$$
$$+ [-1 \quad 1 \quad 4]^T$$

$$= \left(\frac{1}{15}\right)[1 \quad -5 \quad 2]^T.$$

Note that $15\mathbf{u}_3 = [1 \quad -5 \quad 2]^T$ is also orthogonal to $\mathbf{u}_1$ and $\mathbf{u}_2$. ∎

To find an orthonormal basis for a subspace $\mathcal{W}$ of $\mathcal{R}^n$, one may proceed as follows.

THE GRAM-SCHMIDT
ORTHOGONALIZATION PROCESS

(1) Find any basis $\{\mathbf{v}_1, \mathbf{v}_2, \ldots, \mathbf{v}_k\}$ for $\mathcal{W}$.
(2) Take $\mathbf{u}_1 = \mathbf{v}_1$.
(3) Find $\mathbf{u}_2$ in SPAN$\{\mathbf{u}_1, \mathbf{v}_2\}$ orthogonal to $\mathbf{u}_1$.
(4) Find $\mathbf{u}_3$ in SPAN$\{\mathbf{u}_1, \mathbf{u}_2, \mathbf{v}_3\}$ orthogonal to $\mathbf{u}_1$ and $\mathbf{u}_2$.
(5) Find $\mathbf{u}_4$ in SPAN$\{\mathbf{u}_1, \mathbf{u}_2, \mathbf{u}_3, \mathbf{v}_4\}$ orthogonal to $\mathbf{u}_1, \mathbf{u}_2$, and
   $\mathbf{u}_3$.
   $\vdots$

Stop when all $\mathbf{v}_k$ have been replaced by $\mathbf{u}_k$ and then divide each $\mathbf{u}_i$ by its length to obtain the desired orthonormal basis.

Example 5

Find an orthonormal basis for the column space of $A = \begin{bmatrix} 1 & 1 & 1 \\ 1 & 0 & 1 \\ 1 & 0 & 0 \\ 1 & 0 & 0 \end{bmatrix}$.

Let the columns of A be denoted by $\mathbf{v}_1, \mathbf{v}_2$, and $\mathbf{v}_3$. We set $\mathbf{u}_1 = \mathbf{v}_1$ and then try to find a_1 so that $\mathbf{u}_2 = a_1\mathbf{u}_1 + \mathbf{v}_2$ will be orthogonal to $\mathbf{u}_1$. Taking the dot product of this last equation with $\mathbf{u}_1$ yields

$$0 = \mathbf{u}_1 \cdot \mathbf{u}_2 = a_1(\mathbf{u}_1 \cdot \mathbf{u}_1) + (\mathbf{u}_1 \cdot \mathbf{v}_2) = 4a_1 + 1.$$

Therefore, $a_1 = -1/4$ and

$$\mathbf{u}_2 = -\frac{1}{4}\mathbf{u}_1 + \mathbf{v}_2 = \frac{-1}{4}[1 \quad 1 \quad 1 \quad 1]^T + [1 \quad 0 \quad 0 \quad 0]^T$$

$$= \frac{1}{4}[3 \quad -1 \quad -1 \quad -1]^T$$

is orthogonal to $\mathbf{u}_1$. We also have SPAN$\{\mathbf{u}_1, \mathbf{u}_2\}$ = SPAN$\{\mathbf{v}_1, \mathbf{v}_2\}$.
   Next we try to choose b_1 and b_2 so that

$$\mathbf{u}_3 = b_1\mathbf{u}_1 + b_2\mathbf{u}_2 + \mathbf{v}_3$$

is orthogonal to both $\mathbf{u}_1$ and $\mathbf{u}_2$:

$$0 = (\mathbf{u}_1 \cdot \mathbf{u}_3) = b_1(\mathbf{u}_1 \cdot \mathbf{u}_1) + b_2(\mathbf{u}_1 \cdot \mathbf{u}_2) + (\mathbf{u}_1 \cdot \mathbf{v}_3) = 4b_1 + 2$$

implies that $b_1 = -1/2$, and

$$0 = (\mathbf{u_2} \cdot \mathbf{u_3}) = b_1(\mathbf{u_2} \cdot \mathbf{u_1}) + b_2(\mathbf{u_2} \cdot \mathbf{u_2}) + (\mathbf{u_2} \cdot \mathbf{v_3}) = \frac{3b_2}{4} + \frac{1}{2}$$

implies that $b_2 = -2/3$. Thus, $\mathbf{u_3} = (-1/2)\mathbf{u_1} + (-2/3)\mathbf{u_2} + \mathbf{v_3} = -1/6[0 \quad -4 \quad 2 \quad 2]^T$ is orthogonal to both $\mathbf{u_1}$ and $\mathbf{u_2}$. Note that $-3\mathbf{u_3} = [0 \quad -2 \quad 1 \quad 1]^T$ is also orthogonal to both $\mathbf{u_1}$ and $\mathbf{u_2}$. The desired orthonormal basis can now be obtained by dividing each of the $\mathbf{u_i}$ by its length to obtain a unit vector. This step leads to the vectors

$$\mathbf{w_1} = \frac{\mathbf{u_1}}{\|\mathbf{u_1}\|} = \frac{1}{\sqrt{2}}[1 \quad 1 \quad 1 \quad 1]^T$$

$$\mathbf{w_2} = \frac{\mathbf{u_2}}{\|\mathbf{u_2}\|} = \frac{1}{\sqrt{12}}[3 \quad -1 \quad -1 \quad -1]^T$$

$$\mathbf{w_3} = \frac{\mathbf{u_3}}{\|\mathbf{u_3}\|} = \frac{-3\mathbf{u_3}}{\|-3\mathbf{u_3}\|} = \frac{1}{\sqrt{6}}[0 \quad -2 \quad 1 \quad 1]^T$$

which form an orthonormal basis for $\mathscr{CS}(A)$. ■

There are many possible variations on the basic Gram-Schmidt process. The way that we have described the process works well for hand calculations. Since scaling (scalar multiplication) does not affect the orthogonality, one is free to scale the vectors in any way that is convenient; for example, in Example 5 it would have been helpful to replace $\mathbf{u_2}$ by $4\mathbf{u_2}$. One can change the vectors to unit vectors at any time during the process. If one is working by machine, or writing a program, then it is best to normalize the vectors at each step. In this case Equation 3.9 reduces to $a_i = -(\mathbf{u_i} \cdot \mathbf{v_{k+1}})$.

A frequently occurring problem is to find an orthonormal basis for $\mathcal{R}^n$ that has a prescribed first vector. An attractive alternative to the Gram-Schmidt process in this case is described in the next theorem, the proof of which is left to the exercises.

Theorem 3.20

If $x_1 \neq 1$ and $[x_1 \quad x_2 \quad \cdots \quad x_n]^T = \begin{bmatrix} x_1 \\ Y \end{bmatrix}$ is a unit vector in $\mathcal{R}^n$, then

the columns of the matrix

$$H = \left[\begin{array}{c|c} x_1 & Y^T \\ \hline Y & I - \dfrac{1}{1 - x_1} YY^T \end{array} \right]$$

form an orthonormal basis for $\mathcal{R}^n$.

A matrix whose columns form an orthonormal basis for $\mathcal{R}^n$ is called an ORTHOGONAL MATRIX. Since the entry in the (i, j) position of the product A^TA is the inner product $(\text{Col}_i(A) \cdot \text{Col}_j(A))$, it follows that a matrix A is orthogonal if and only if $A^TA = I$.

Example 6

Use Theorem 3.20 to find an orthonormal basis for $\mathcal{R}^4$ containing the vector $\mathbf{u}_1 = (1/2)[1 \quad 1 \quad 1 \quad 1]^T$.

The given vector $\mathbf{u}_1$ is of the form $\begin{bmatrix} x_1 \\ Y \end{bmatrix}$ with $Y = \begin{bmatrix} 1/2 \\ 1/2 \\ 1/2 \end{bmatrix}$ and $x_1 = $

$1/2$. Thus, $YY^T = \begin{bmatrix} 1/4 & 1/4 & 1/4 \\ 1/4 & 1/4 & 1/4 \\ 1/4 & 1/4 & 1/4 \end{bmatrix}$, $\dfrac{1}{1-x_1} = 2$, and

$$I - \frac{1}{1-x_1}YY^T = \begin{bmatrix} 1 & 0 & 0 \\ 0 & 1 & 0 \\ 0 & 0 & 1 \end{bmatrix} - \frac{2}{4}\begin{bmatrix} 1 & 1 & 1 \\ 1 & 1 & 1 \\ 1 & 1 & 1 \end{bmatrix} = \frac{1}{2}\begin{bmatrix} 1 & -1 & -1 \\ -1 & 1 & -1 \\ -1 & -1 & 1 \end{bmatrix}.$$

Thus, the matrix of Theorem 3.20 is

$$H = \frac{1}{2}\begin{bmatrix} 1 & 1 & 1 & 1 \\ 1 & 1 & -1 & -1 \\ 1 & -1 & 1 & -1 \\ 1 & -1 & -1 & 1 \end{bmatrix}.$$

You should verify directly that H is an orthogonal matrix; that is, the columns of this matrix are an orthonormal set of vectors. ◼

Exercise Set 3.8

1. For $\mathbf{u} = [1 \quad 2 \quad 3 \quad -1 \quad 4]^T$ and $\mathbf{v} = [3 \quad 0 \quad -4 \quad 2 \quad 5]^T$, find $\mathbf{u} \cdot \mathbf{v}$, $\|\mathbf{u}\|$, $\|\mathbf{v}\|$, and $\|\mathbf{u} + \mathbf{v}\|$. Which is larger, $\|\mathbf{u}\| + \|\mathbf{v}\|$ or $\|\mathbf{u} + \mathbf{v}\|$? Find a scalar a in R such that $\|a\mathbf{u}\| = 1$. Determine h so that $\mathbf{y} = [1 \quad 1 \quad 1 \quad 1 \quad h]^T$ is orthogonal to $\mathbf{u}$. Determine a and b so that $\mathbf{w} = [1 \quad 1 \quad 1 \quad a \quad b]^T$ is orthogonal to both $\mathbf{u}$ and $\mathbf{v}$. Determine c so that $\mathbf{z} = c\mathbf{u} + \mathbf{v}$ is orthogonal to $\mathbf{v}$.

2. Which of the following sets are orthonormal?
   (a) $[1 \quad 0 \quad 1]^T/\sqrt{2}$, $[1 \quad 1 \quad -1]^T/\sqrt{3}$, $[-1 \quad 2 \quad 1]^T/\sqrt{6}$
   (b) $[1 \quad 0 \quad 0]^T$, $[0 \quad 1 \quad 1]^T$, $[0 \quad 0 \quad 1]^T$
   (c) $[1 \quad 1 \quad -2]^T/\sqrt{6}$, $[1 \quad -1 \quad 0]^T/\sqrt{6}$, $[1 \quad 1 \quad 1]^T/\sqrt{3}$
   (d) $[2 \quad -2 \quad 1]^T/3$, $[2 \quad 1 \quad -2]^T/3$, $[1 \quad 2 \quad 2]^T/3$
   (e) $[1 \quad 1 \quad 1 \quad 1]^T/2$, $[3 \quad -5 \quad 1 \quad 1]^T/6$, $[3 \quad 1 \quad -5 \quad 1]^T/6$, $[3 \quad 1 \quad 1 \quad -5]^T/6$

3. Find the eigenvectors of $A = \begin{bmatrix} 1 & 0 & 1 \\ 0 & 1 & 2 \\ 1 & 2 & 5 \end{bmatrix}$ and show that they are orthogonal.

4. Prove Theorem 3.15.

5. Let $\mathcal{B}$ be the orthonormal basis for R^4 given in Exercise 2(e). Find the coordinates, with respect to the $\mathcal{B}$ basis, of each of the following vectors.
   (a) $[1 \quad 0 \quad 0 \quad 0]^T$ (b) $[1 \quad 2 \quad 3 \quad 4]^T$ (c) $[-1 \quad 0 \quad 4 \quad 11]^T$

6. Show that if the matrix A is orthogonal, then $\det(A) = 1$ or -1.

7. Show that the vectors $\mathbf{u}$ and $\mathbf{v}$ in R^n are orthogonal if and only if $\|\mathbf{u} - \mathbf{v}\| = \|\mathbf{u} + \mathbf{v}\|$. Interpret geometrically for $n = 3$.

†**8.** Show that if P_1 and P_2 are orthogonal matrices, then so is $\begin{bmatrix} P_1 & 0 \\ 0 & P_2 \end{bmatrix}$.

†**9.** Show that if A is an $m \times n$ matrix, then A and $A^T A$ have the same rank. (*Hint:* Show that $\mathcal{NS}(A) = \mathcal{NS}(A^T A)$.)

10. Let $\{\mathbf{u}_1, \mathbf{u}_2, \ldots, \mathbf{u}_n\}$ be an orthonormal basis for $\mathcal{W}$. Show that for any $\mathbf{w}$ in $\mathcal{W}$
$$\|\mathbf{w}\|^2 = (\mathbf{w} \cdot \mathbf{u}_1)^2 + (\mathbf{w} \cdot \mathbf{u}_2)^2 + \ldots + (\mathbf{w} \cdot \mathbf{u}_n)^2.$$

11. Find an orthonormal basis for the subspace of $\mathcal{R}^3$ spanned by
$$\mathbf{u}_1 = \begin{bmatrix} 1 & 2 & 2 \end{bmatrix}^T \quad \text{and} \quad \mathbf{u}_2 = \begin{bmatrix} 7 & 1 & 0 \end{bmatrix}^T.$$

12. Find an orthonormal basis for the subspace of $\mathcal{R}^4$ spanned by
$$\mathbf{u}_1 = \begin{bmatrix} 1 & 2 & 1 & 1 \end{bmatrix}^T, \mathbf{u}_2 = \begin{bmatrix} 1 & -1 & 0 & 2 \end{bmatrix}^T \quad \text{and} \quad \mathbf{u}_3 = \begin{bmatrix} 2 & 0 & 1 & 1 \end{bmatrix}^T$$

13. Use the Gram-Schmidt process to find an orthonormal basis for $\mathcal{CS}(A)$ if A is the given matrix.

(a) $A = \begin{bmatrix} 1 & 0 & 0 \\ 1 & 1 & 0 \\ 1 & 1 & 1 \end{bmatrix}$ (b) $A = \begin{bmatrix} 3 & 0 & 5 \\ 0 & 1 & -2 \\ 4 & 1 & 1 \end{bmatrix}$

14. Find a unit vector in $\mathcal{R}^3$ orthogonal to both $\mathbf{u}_1 = \begin{bmatrix} 1 & -1 & 3 \end{bmatrix}^T$ and $\mathbf{u}_2 = \begin{bmatrix} 3 & 0 & -1 \end{bmatrix}^T$.

15. Find an orthonormal basis for $\mathcal{R}^4$ containing scalar multiples of $\mathbf{u}_1 = \begin{bmatrix} 1 & 2 & 1 & 1 \end{bmatrix}^T$ and $\mathbf{u}_2 = \begin{bmatrix} -1 & 0 & 2 & -1 \end{bmatrix}^T$.

16. Find an orthonormal basis for the following subspaces and extend it to an orthonormal basis for the whole space.

(a) SPAN$\{\begin{bmatrix} 1 & 1 & 0 \end{bmatrix}^T, \begin{bmatrix} 0 & 1 & 1 \end{bmatrix}^T\}$

(b) SPAN$\{\begin{bmatrix} 1 & 0 & 1 & 0 \end{bmatrix}^T, \begin{bmatrix} 1 & -1 & -1 & 1 \end{bmatrix}^T, \begin{bmatrix} 1 & 0 & 0 & 0 \end{bmatrix}^T\}$

(c) SPAN$\{\begin{bmatrix} 1 & 1 & 0 \end{bmatrix}^T, \begin{bmatrix} 2 & -1 & 3 \end{bmatrix}^T, \begin{bmatrix} 1 & -1 & 1 \end{bmatrix}^T\}$

(d) SPAN$\{\begin{bmatrix} 3 & 0 & 4 & 12 \end{bmatrix}^T, \begin{bmatrix} 1 & 0 & 1 & 0 \end{bmatrix}^T, \begin{bmatrix} 1 & 1 & 1 & 1 \end{bmatrix}^T\}$

17. Find an orthonormal basis for $\mathcal{NS}(A)$ if A is the matrix shown.

(a) $A = \begin{bmatrix} 1 & 0 & 5 & -2 & -1 \\ 0 & 1 & -2 & 4 & 2 \end{bmatrix}$ (b) $A = \begin{bmatrix} 1 & 0 & 0 & 3 & -4 \\ 0 & 1 & 0 & 4 & 2 \\ 0 & 0 & 1 & 5 & 0 \end{bmatrix}$

18. Show that every 2×2 orthogonal matrix is of the form
$$\begin{bmatrix} \cos\theta & \sin\theta \\ \sin\theta & -\cos\theta \end{bmatrix} \quad \text{or} \quad \begin{bmatrix} \cos\theta & -\sin\theta \\ \sin\theta & \cos\theta \end{bmatrix}.$$

19. Definition: The ORTHOGONAL COMPLEMENT of $\mathcal{W}$ is the set
$$\mathcal{W}^\perp = \{\mathbf{u} \text{ in } \mathcal{V} \mid \mathbf{u} \cdot \mathbf{w} = 0 \text{ for every } \mathbf{w} \text{ in } \mathcal{W}\}.$$

(a) Show that $\mathcal{W}^\perp$ is a subspace of $\mathcal{V}$.

(b) Show that for any matrix A, $\mathcal{NS}(A^T) = \mathcal{CS}(A)^\perp$.

(c) Find an orthonormal basis for the orthogonal complement of each of the subspaces in Exercise 16.

20. Use Theorem 3.20 to find an orthonormal basis for $\mathcal{R}^4$ containing each of the following.
    (a) $\mathbf{u_1} = [3 \quad 0 \quad 4 \quad 0]^T/5$
    (b) $\mathbf{u_1} = [2 \quad -3 \quad 1 \quad 4]^T/\sqrt{30}$ (use a calculator or MATALG)
    (c) $\mathbf{u_1} = [1 \quad -1 \quad 1 \quad -1]^T/2$

21. Define an inner product in $\mathcal{R}^{1 \times n}$ in the obvious way and then find an orthonormal basis for the row space of

$$A = \begin{bmatrix} 1 & 0 & 1 & 0 \\ 1 & -1 & -1 & 1 \\ 1 & 0 & 0 & 0 \end{bmatrix}.$$

22. Verify Theorem 3.20, by direct calculation; that is, show that $H^T H = I$. Note that $x_1{}^2 + Y^T Y = 1$.

23. Let $\mathbf{w}$ be a unit vector in $\mathcal{R}^n$. Show that $H = I - 2\mathbf{w}\mathbf{w}^T$ is an orthogonal matrix. What is H if $\mathbf{w} = [3/13 \quad 4/13 \quad 12/13]^T$?

24. Let $[x_1 \quad Y]^T$ be a unit vector in $\mathcal{R}^n$. Show that
$$\mathbf{w} = (1/\sqrt{(2(1 - x_1))})[x_1 - 1 \quad Y]^T$$
is also a unit vector in $\mathcal{R}^n$. Find $H = I - 2\mathbf{w}\mathbf{w}^T$ and compare with Theorem 3.20.

25. Explain why there is no loss of generality in choosing the coefficient of $\mathbf{v_{k+1}}$ in Equation 3.8 to be 1.

†26. Show that the Gram-Schmidt process, when applied to the column space of a nonsingular matrix A, yields a factorization of the form $A = QR$, where Q is orthogonal and R is upper triangular.

27. Find the QR factorization, described in Exercise 26, for each of the matrices of Exercise 13.

28. Write a computer program that will produce the QR factorization of a given matrix A.

29. For functions $\mathbf{f}$ and $\mathbf{g}$ in $\mathcal{V} = \mathcal{C}^0[a \quad b]$ (the vector space of all continuous functions on the interval $a \le x \le b$) define an inner product by

$$\mathbf{f} \cdot \mathbf{g} = \int_a^b f(x)g(x)\,dx\,.$$

Show that this inner product satisfies the conclusions of Theorem 3.2.

30. For $\mathbf{u} = [a_1 \quad a_2 \quad \ldots \quad a_n]$ in $\mathcal{R}^n$ define

$$\|\mathbf{u}\|_1 = \sum_{i=1}^{n} |a_i| \quad \text{and} \quad \|\mathbf{u}\|_\infty = \max_i\{|a_i|\}\,.$$

Show that for each of these norms, the conclusions of Theorems 3.15 and 3.17 are satisfied.

†31. Prove the following generalization of the Pythagorean theorem: for X and Y in $\mathcal{R}^n$, $\|X\|^2 + \|Y\|^2 = \|X - Y\|^2$ if and only if X and Y are orthogonal.

3.9   *Change of Basis**

Let $\mathcal{V}$ be an arbitrary n-dimensional vector space and let

$$\mathcal{B} = \{\mathbf{u_1}, \mathbf{u_2}, \ldots, \mathbf{u_n}\}$$

be a fixed basis for $\mathcal{V}$. For any vector $\mathbf{w}$ in $\mathcal{V}$ we can write

$$\mathbf{w} = a_1\mathbf{u_1} + a_2\mathbf{u_2} + \ldots + a_n\mathbf{u_n}$$

where the uniquely determined scalars a_i are called the $\mathcal{B}$-coordinates of $\mathbf{w}$ and the column matrix $[\mathbf{w}]_{\mathcal{B}} = [a_1 \quad a_2 \quad \ldots \quad a_n]^T$ is called the $\mathcal{B}$-coordinate matrix of $\mathbf{w}$. Finding coordinates usually involves solving linear systems; the ease of solution depends on the choice of the basis. Finding coordinates is very easy for the standard basis for $\mathcal{R}^n$, easy for any orthonormal basis for $\mathcal{R}^n$, and easy for an echelon basis for the row space of a matrix A.

We list some important elementary properties of the coordinate matrix in our next theorem. These properties will be referred to in later sections.

Theorem 3.21

If $\mathcal{B}$ is any basis for the finite dimensional vector space $\mathcal{V}$, then

(1) $[\mathbf{w}]_{\mathcal{B}} = [\mathbf{v}]_{\mathcal{B}}$ if and only if $\mathbf{w} = \mathbf{v}$;
(2) $[\mathbf{w} + \mathbf{v}]_{\mathcal{B}} = [\mathbf{w}]_{\mathcal{B}} + [\mathbf{v}]_{\mathcal{B}}$;
(3) $[k\mathbf{w}]_{\mathcal{B}} = k[\mathbf{w}]_{\mathcal{B}}$; and
(4) for each X in $\mathcal{R}^n$, there exists a unique vector $\mathbf{u}$ in $\mathcal{V}$ such that $[\mathbf{u}]_{\mathcal{B}} = X$.

Suppose now that we have two bases for the vector space $\mathcal{V}$:

$$\mathcal{B} = \{\mathbf{u_1}, \mathbf{u_2}, \ldots, \mathbf{u_n}\} \quad \text{and} \quad \mathcal{B}' = \{\mathbf{v_1}, \mathbf{v_2}, \ldots, \mathbf{v_n}\}.$$

We will refer to $\mathcal{B}$ as the *old basis* and $\mathcal{B}'$ as the *new basis*. For a typical vector $\mathbf{w}$ in $\mathcal{V}$ we have the two representations

$$\mathbf{w} = a_1\mathbf{u_1} + a_2\mathbf{u_2} + \ldots + a_n\mathbf{u_n} \quad \text{and}$$
$$\mathbf{w} = b_1\mathbf{v_1} + b_2\mathbf{v_2} + \ldots + b_n\mathbf{v_n},$$

---

*This section may be delayed until the beginning of Chapter 5.

as well as the two coordinate matrices

$$[\mathbf{w}]_{\mathscr{B}} = \begin{bmatrix} a_1 \\ a_2 \\ \cdot \\ \cdot \\ \cdot \\ a_n \end{bmatrix} \quad \text{and} \quad [\mathbf{w}]_{\mathscr{B}'} = \begin{bmatrix} b_1 \\ b_2 \\ \cdot \\ \cdot \\ \cdot \\ b_n \end{bmatrix}.$$

It would be extremely naive to expect that these two coordinate matrices are equal. For example, consider the two bases for $\mathscr{R}^2$:

$$\mathscr{B} = \left\{ \mathbf{u}_1 = \begin{bmatrix} 1 \\ 1 \end{bmatrix}, \mathbf{u}_2 = \begin{bmatrix} 1 \\ 0 \end{bmatrix} \right\} \quad \text{and} \quad \mathscr{B}' = \left\{ \mathbf{v}_1 = \begin{bmatrix} -1 \\ 1 \end{bmatrix}, \mathbf{v}_2 = \begin{bmatrix} 4 \\ -5 \end{bmatrix} \right\}.$$

For the vector $\mathbf{w} = [3 \quad 5]^T$, it is easy to check that

$$\mathbf{w} = 5\mathbf{u}_1 - 2\mathbf{u}_2 = -35\mathbf{v}_1 - 8\mathbf{v}_2.$$

Thus we have $[\mathbf{w}]_{\mathscr{B}} = \begin{bmatrix} 5 \\ -2 \end{bmatrix}$ and $[\mathbf{w}]_{\mathscr{B}'} = \begin{bmatrix} -35 \\ -8 \end{bmatrix} \neq [\mathbf{w}]_{\mathscr{B}}.$

We also have $[\mathbf{u}_1]_{\mathscr{B}} = \begin{bmatrix} 1 \\ 0 \end{bmatrix}$ and $[\mathbf{u}_1]_{\mathscr{B}'} = \begin{bmatrix} -9 \\ -2 \end{bmatrix} \neq [\mathbf{u}_1]_{\mathscr{B}}.$

We will now investigate how the two coordinate matrices above are related. Since we can change one column matrix into another column matrix by multiplying by an $n \times n$ matrix, let us try to find a matrix P such that

$$P[\mathbf{w}]_{\mathscr{B}} = [\mathbf{w}]_{\mathscr{B}'} \quad \text{for all vectors } \mathbf{w} \text{ in } \mathscr{V}. \qquad (3.10)$$

If Equation 3.10 is to hold for all vectors $\mathbf{w}$ in $\mathscr{V}$, it must hold for the vectors $\mathbf{u}_i$ which make up the $\mathscr{B}$ basis (the old basis). For these vectors we have $[\mathbf{u}_i]_{\mathscr{B}} = \mathbf{e}_i$ so that, if Equation 3.10 is to hold, we must have

$$P[\mathbf{u}_i]_{\mathscr{B}} = P\mathbf{e}_i = \text{Col}_i(P) = [\mathbf{u}_i]_{\mathscr{B}'}; \quad i = 1, 2, \ldots, n.$$

Thus, we see that the only matrix that satisfies Equation 3.10 *for all vectors* $\mathbf{w}$ in $\mathscr{V}$ is the matrix

$$P = \left[[\mathbf{u}_1]_{\mathscr{B}'} \quad [\mathbf{u}_2]_{\mathscr{B}'} \quad \cdots \quad [\mathbf{u}_n]_{\mathscr{B}'} \right],$$

whose columns are the coordinate matrices of the old basis vectors $\mathbf{u}_i$ with respect to the new basis.

If $\mathbf{w} = a_1\mathbf{u}_1 + a_2\mathbf{u}_2 + \ldots + a_n\mathbf{u}_n$ is any element of $\mathscr{V}$, then we have

$$P[\mathbf{w}]_{\mathscr{B}} = P\begin{bmatrix} a_1 \\ a_2 \\ \cdot \\ \cdot \\ \cdot \\ a_n \end{bmatrix} = \left[[\mathbf{u}_1]_{\mathscr{B}'} \quad [\mathbf{u}_2]_{\mathscr{B}'} \quad \cdots \quad [\mathbf{u}_n]_{\mathscr{B}'} \right] \begin{bmatrix} a_1 \\ a_2 \\ \cdot \\ \cdot \\ \cdot \\ a_n \end{bmatrix}$$

$$= a_1[\mathbf{u}_1]_{\mathscr{B}'} + a_2[\mathbf{u}_2]_{\mathscr{B}'} + \ldots + a_n[\mathbf{u}_n]_{\mathscr{B}'}.$$

From parts 2 and 3 of Theorem 3.21 it follows that this last expression is

$$[a_1\mathbf{u}_1 + a_2\mathbf{u}_2 + \ldots + a_n\mathbf{u}_n]_{\mathcal{B}'} = [\mathbf{w}]_{\mathcal{B}'} .$$

This calculation establishes that Equation 3.10 holds for every $\mathbf{w}$ in $\mathcal{V}$.

If $P[\mathbf{w}]_{\mathcal{B}} = [\mathbf{w}]_{\mathcal{B}'} = 0$, then it follows that $\mathbf{w} = \mathbf{0}$ and $[\mathbf{w}]_{\mathcal{B}} = 0$. Thus, from Theorem 1.9, it follows that the change-of-basis matrix P must be nonsingular. This observation completes the proof of the following theorem.

Theorem 3.22

If $\mathcal{B} = \{\mathbf{u}_1, \mathbf{u}_2, \ldots, \mathbf{u}_n\}$ and $\mathcal{B}' = \{\mathbf{v}_1, \mathbf{v}_2, \ldots, \mathbf{v}_n\}$ are two bases for a finite dimensional vector space $\mathcal{V}$, then the matrix

$$P = \begin{bmatrix} [\mathbf{u}_1]_{\mathcal{B}'} & [\mathbf{u}_2]_{\mathcal{B}'} & \cdots & [\mathbf{u}_n]_{\mathcal{B}'} \end{bmatrix},$$

whose columns are the new coordinates of the old basis vectors, is the unique nonsingular matrix P such that

$$P[\mathbf{w}]_{\mathcal{B}} = [\mathbf{w}]_{\mathcal{B}'} \text{ for all } \mathbf{w} \text{ in } \mathcal{V} .$$

We will refer to this matrix P as the $\mathcal{B}$ to $\mathcal{B}'$ CHANGE-OF-BASIS MATRIX. Note that P^{-1} must satisfy

$$[\mathbf{w}]_{\mathcal{B}} = P^{-1}[\mathbf{w}]_{\mathcal{B}'} \text{ for all } \mathbf{w} \text{ in } \mathcal{V}$$

so that P^{-1} is the $\mathcal{B}'$ to $\mathcal{B}$ change-of-basis matrix.

Example 1

Let $\mathcal{V} = \mathcal{R}^{1 \times 3}$. Find the $\mathcal{B}$ to $\mathcal{B}'$ change-of-basis matrix if

$$\mathcal{B} = \{\mathbf{u}_1 = [2 \quad 3 \quad 2], \mathbf{u}_2 = [7 \quad 10 \quad 6], \mathbf{u}_3 = [6 \quad 10 \quad 7]\}$$

and

$$\mathcal{B}' = \{\mathbf{v}_1 = [1 \quad 1 \quad 1], \mathbf{v}_2 = [0 \quad 1 \quad 1], \mathbf{v}_3 = [1 \quad 1 \quad 0]\} .$$

From Theorem 3.22 we know that the change-of-basis matrix we seek has as its columns the new coordinates of the old basis vectors; that is,

$$P = \begin{bmatrix} [\mathbf{u}_1]_{\mathcal{B}'} & [\mathbf{u}_2]_{\mathcal{B}'} & [\mathbf{u}_3]_{\mathcal{B}'} \end{bmatrix} .$$

In order to find P we need to express the $\mathbf{u}_i$ in terms of the $\mathbf{v}_i$. To find $[\mathbf{u}_1]_{\mathcal{B}'}$ we must solve the vector equation

$$x_1\mathbf{v}_1 + x_2\mathbf{v}_2 + x_3\mathbf{v}_3 = \mathbf{u}_1$$

or

$$x_1[1 \quad 1 \quad 1] + x_2[0 \quad 1 \quad 1] + x_3[1 \quad 1 \quad 0] = [2 \quad 3 \quad 2].$$

This equation is equivalent to the system

$$AX = \begin{bmatrix} 1 & 0 & 1 \\ 1 & 1 & 1 \\ 1 & 1 & 0 \end{bmatrix} \begin{bmatrix} x_1 \\ x_2 \\ x_3 \end{bmatrix} = \begin{bmatrix} 2 \\ 3 \\ 2 \end{bmatrix} = \mathbf{u}_1^T.$$

In order to find $[\mathbf{u}_2]_{\mathcal{B}'}$ and $[\mathbf{u}_3]_{\mathcal{B}'}$ we need to solve the systems

$$AX = \mathbf{u}_2^T \quad \text{and} \quad AX = \mathbf{u}_3^T.$$

Since these three systems have the same coefficient matrix, we can solve all three systems with the single row reduction

$$\begin{bmatrix} 1 & 0 & 1 & 2 & 7 & 6 \\ 1 & 1 & 1 & 3 & 10 & 10 \\ 1 & 1 & 0 & 2 & 6 & 7 \end{bmatrix} \longrightarrow \begin{bmatrix} 1 & 0 & 1 & 2 & 7 & 6 \\ 0 & 1 & 0 & 1 & 3 & 4 \\ 0 & 1 & -1 & 0 & -1 & 1 \end{bmatrix} \longrightarrow$$

$$\begin{bmatrix} 1 & 0 & 0 & 1 & 3 & 3 \\ 0 & 1 & 0 & 1 & 3 & 4 \\ 0 & 0 & 1 & 1 & 4 & 3 \end{bmatrix}.$$

From this row reduction it follows that $\mathbf{u}_1 = \mathbf{v}_1 + \mathbf{v}_2 + \mathbf{v}_3$, $\mathbf{u}_2 = 3\mathbf{v}_1 + 3\mathbf{v}_2 + 4\mathbf{v}_3$, and $\mathbf{u}_3 = 3\mathbf{v}_1 + 4\mathbf{v}_2 + 3\mathbf{v}_3$.

Thus, $[\mathbf{u}_1]_{\mathcal{B}'} = \begin{bmatrix} 1 \\ 1 \\ 1 \end{bmatrix}$, $[\mathbf{u}_2]_{\mathcal{B}'} = \begin{bmatrix} 3 \\ 3 \\ 4 \end{bmatrix}$, and $[\mathbf{u}_3]_{\mathcal{B}'} = \begin{bmatrix} 3 \\ 4 \\ 3 \end{bmatrix}$,

so that $P = \begin{bmatrix} 1 & 3 & 3 \\ 1 & 3 & 4 \\ 1 & 4 & 3 \end{bmatrix}$ is the $\mathcal{B}$ to $\mathcal{B}'$ change-of-basis matrix.

For the vector $\mathbf{w} = 3\mathbf{u}_1 + 5\mathbf{u}_2 - 6\mathbf{u}_3 = [5 \quad -1 \quad -6]^T$, we have $[\mathbf{w}]_{\mathcal{B}} = [3 \quad 5 \quad -6]^T$, so from Theorem 3.22 we must have

$$[\mathbf{w}]_{\mathcal{B}'} = P[\mathbf{w}]_{\mathcal{B}} = \begin{bmatrix} 1 & 3 & 3 \\ 1 & 3 & 4 \\ 1 & 4 & 3 \end{bmatrix} \begin{bmatrix} 3 \\ 5 \\ -6 \end{bmatrix} = \begin{bmatrix} 0 \\ -6 \\ 5 \end{bmatrix}.$$

Thus, we have $\mathbf{w} = -6\mathbf{v}_2 + 5\mathbf{v}_3 = [5 \quad -1 \quad -6]^T$, which checks with the above expression for $\mathbf{w}$. ■

Example 2

Consider the subspace $\mathcal{P}_2$ of the vector space $\mathcal{P}$ of all polynomials and the two bases

$\mathcal{S} = \{1, x, x^2\}$ and
$\mathcal{H} = \{h_1(x) = x^2 - 3x + 2, h_2(x) = x^2 - 2x, h_3(x) = x^2 - x\}$.

Find the $\mathcal{S}$ to $\mathcal{H}$ change-of-basis matrix Q and use it to express $p(x) = 12x^2 - 11x + 5$ as a linear combination of the $h_i(x)$.

The $\mathcal{S}$ to $\mathcal{H}$ change-of-basis matrix Q must satisfy $Q[p(x)]_{\mathcal{S}} = [p(x)]_{\mathcal{H}}$ and, from Theorem 3.22, is the matrix

$$Q = [[1]_{\mathcal{H}} \quad [x]_{\mathcal{H}} \quad [x^2]_{\mathcal{H}}].$$

Thus we need to express 1, x, and x^2 as linear combinations of the $h_i(x)$. If

$$a_1 h_1(x) + a_2 h_2(x) + a_3 h_3(x) = (1, x, \text{ or } x^2)$$

then

$$a_1(x^2 - 3x + 2) + a_2(x^2 - 2x) + a_3(x^2 - x)$$
$$= (2a_1) + (-3a_1 - 2a_2 - a_3)x + (a_1 + a_2 + a_3)x^2$$
$$= (1, x, \text{ or } x^2),$$

so that the a_i must satisfy the systems

$$\begin{bmatrix} 2 & 0 & 0 \\ -3 & -2 & -1 \\ 1 & 1 & 1 \end{bmatrix} \begin{bmatrix} a_1 \\ a_2 \\ a_3 \end{bmatrix} = \begin{bmatrix} 1 \\ 0 \\ 0 \end{bmatrix} \left(\begin{bmatrix} 0 \\ 1 \\ 0 \end{bmatrix} \text{ or } \begin{bmatrix} 0 \\ 0 \\ 1 \end{bmatrix} \right).$$

From the row reduction

$$\begin{bmatrix} 2 & 0 & 0 & 1 & 0 & 0 \\ -3 & -2 & -1 & 0 & 1 & 0 \\ 1 & 1 & 1 & 0 & 0 & 1 \end{bmatrix} \longrightarrow \begin{bmatrix} 1 & 0 & 0 & 1/2 & 0 & 0 \\ 0 & -2 & -1 & 3/2 & 1 & 0 \\ 0 & 1 & 1 & -1/2 & 0 & 1 \end{bmatrix}$$

$$\longrightarrow \begin{bmatrix} 1 & 0 & 0 & 1/2 & 0 & 0 \\ 0 & 1 & 0 & -1 & -1 & -1 \\ 0 & 0 & 1 & 1/2 & 1 & 2 \end{bmatrix}$$

we see that $1 = h_1(x) - h_2(x) + h_3(x)/3$, $x = -h_2(x) + h_3(x)$, and $x^2 = -h_2(x) + 2h_3(x)$. Thus,

$$[1]_H = \begin{bmatrix} 1/2 \\ -1 \\ 1/2 \end{bmatrix}, \quad [x]_H = \begin{bmatrix} 0 \\ -1 \\ 1 \end{bmatrix}, \quad \text{and} \quad [x^2]_H = \begin{bmatrix} 0 \\ -1 \\ 2 \end{bmatrix},$$

$$\text{so } Q = \begin{bmatrix} 1/2 & 0 & 0 \\ -1 & -1 & -1 \\ 1/2 & 1 & 2 \end{bmatrix}.$$

To find $[12x^2 - 11x + 5]_{\mathcal{H}}$ we first observe that

$$[12x^2 - 11x + 5]_{\mathcal{S}} = \begin{bmatrix} 5 \\ -11 \\ 12 \end{bmatrix} \text{ so } [12x^2 - 11x + 5]_{\mathcal{H}}$$

$$= Q \begin{bmatrix} 5 \\ -11 \\ 12 \end{bmatrix} = \begin{bmatrix} 5/2 \\ -6 \\ 31/2 \end{bmatrix}.$$

Thus, $12x^2 - 11x + 5 = (5/2)h_1(x) - 6h_2(x) + (31/2)h_3(x)$, as you should check by direct multiplication. ■

We conclude this section with a theorem about the change of basis matrix between two orthonormal bases of $\mathscr{R}^n$. The proof will be left to the exercises.

Theorem 3.23

The change-of-basis matrix P, between two orthonormal bases of $\mathscr{R}^n$, is an orthogonal matrix; that is, $P^T P = I$.

Exercise Set 3.9

1. Find the $\mathscr{B}$ to $\mathscr{B}'$ change-of-basis matrix P in the following cases.

   (a) $\mathscr{B} = \left\{ \mathbf{u_1} = \begin{bmatrix} 1 \\ 1 \end{bmatrix}, \mathbf{u_2} = \begin{bmatrix} 2 \\ 1 \end{bmatrix} \right\}$ and $\mathscr{B}' = \left\{ \mathbf{v_1} = \begin{bmatrix} 1 \\ 3 \end{bmatrix}, \mathbf{v_2} = \begin{bmatrix} 1 \\ 2 \end{bmatrix} \right\}$

   (b) $\mathscr{B} = \{\mathbf{u_1} = [-1 \ \ 0 \ \ 1], \mathbf{u_2} = [3 \ \ -2 \ \ 1], \mathbf{u_3} = [1 \ \ 6 \ \ -1]\}$
   $\mathscr{B}' = \{\mathbf{v_1} = [3 \ \ 1 \ \ -5], \mathbf{v_2} = [1 \ \ 1 \ \ -3], \mathbf{v_3} = [1 \ \ 0 \ \ -2]\}$

   (c) $\mathscr{B} = \{p_1(x) = 2x^2 + x + 1, p_2(x) = 2x^2 - x + 1, p_3(x) = x^2 + 2x + 1\}$
   $\mathscr{B}' = \{q_1(x) = 2x^2 + 3x - 7, q_2(x) = x^2 + 3x - 2, q_3(x) = x^2 + x\}$

   (d) $\mathscr{B} = \left\{ A_1 = \begin{bmatrix} 1 & 0 \\ 1 & 0 \end{bmatrix}, A_2 = \begin{bmatrix} 1 & 2 \\ 0 & 0 \end{bmatrix}, A_3 = \begin{bmatrix} 5 & -1 \\ 4 & 0 \end{bmatrix}, \right.$

   $\left. A_4 = \begin{bmatrix} -1 & 7 \\ 0 & 6 \end{bmatrix} \right\}$

   $\mathscr{B}' = \left\{ C_1 = \begin{bmatrix} 1 & 0 \\ 0 & 0 \end{bmatrix}, C_2 = \begin{bmatrix} 0 & 1 \\ 0 & 0 \end{bmatrix}, C_3 = \begin{bmatrix} 0 & 0 \\ 1 & 0 \end{bmatrix}, C_4 = \begin{bmatrix} 0 & 0 \\ 0 & 1 \end{bmatrix} \right\}$

   (e) $\mathscr{B} = \{\sin x, \cos x\}$, $\mathscr{B}' = \{\sin(x + 2), \cos(x + 3)\}$

2. Use the matrix from 1(a) to compute $[\mathbf{w}]_{\mathscr{B}'}$, where $\mathbf{w} = 3\mathbf{u_1} + 5\mathbf{u_2}$.

3. Use the matrix from 1(b) to compute $[\mathbf{w}]_{\mathscr{B}'}$, where $\mathbf{w} = [1 \ \ 2 \ \ 3]$.

4. Use the matrix from 1(c) to compute $[h(x)]_{\mathscr{B}'}$, where
   $h(x) = p_1(x) + 5p_2(x) - 7p_3(x)$.

5. Let $\mathscr{S}$ be the standard basis for R^2 and let $\mathscr{N}$ be the basis obtained by rotating the vectors of $\mathscr{S}$ through an angle θ. Find the $\mathscr{S}$ to $\mathscr{N}$ change-of-basis matrix and the $\mathscr{N}$ to $\mathscr{S}$ change-of-basis matrix.

6. Let $\mathscr{B}$, $\mathscr{B}'$, and $\mathscr{B}''$ be three bases for $\mathcal{V}$ and let P be the $\mathscr{B}$ to $\mathscr{B}'$ change-of-basis matrix and let Q be the $\mathscr{B}'$ to $\mathscr{B}''$ change-of-basis matrix. Show that QP is the $\mathscr{B}$ to $\mathscr{B}''$ change-of-basis matrix.

7. Let $\mathscr{B} = \{\mathbf{u_1}, \ldots, \mathbf{u_n}\}$ be an orthonormal basis for $\mathscr{R}^n$ and let $\mathscr{S}$ be the standard basis for $\mathscr{R}^n$. Find the $\mathscr{B}$ to $\mathscr{S}$ change-of-basis matrix and show that it is orthogonal. What is the $\mathscr{S}$ to $\mathscr{B}$ change-of-basis matrix?

$AA^T = I$

8. Let $\mathcal{B}$ be an orthonormal basis for $\mathcal{R}^n$.

    (a) Show that $\mathbf{u} \cdot \mathbf{v} = [\mathbf{u}]_{\mathcal{B}} \cdot [\mathbf{v}]_{\mathcal{B}}$ and, hence, that $\mathbf{u}$ and $\mathbf{v}$ are orthogonal if and only if $[\mathbf{u}]_{\mathcal{B}}$ and $[\mathbf{v}]_{\mathcal{B}}$ are orthogonal.

    (b) Use the result of part (a) to prove Theorem 3.23.

9. Let $\mathcal{B} = \{\mathbf{u_1}, \mathbf{u_2}, \mathbf{u_3}, \mathbf{u_4}\}$ and $\mathcal{B}' = \{\mathbf{u_3}, \mathbf{u_1}, \mathbf{u_4}, \mathbf{u_2}\}$ be two bases for a vector space V. Find the $\mathcal{B}$ to $\mathcal{B}'$ change-of-basis matrix and the $\mathcal{B}'$ to $\mathcal{B}$ change-of-basis matrix.

Chapter 3   *Review Exercises*

1. Find a basis for the null space of $A = \begin{bmatrix} 1 & 2 & -3 & 10 \\ 2 & 3 & 8 & 17 \\ 1 & 3 & -5 & 13 \end{bmatrix}$.

2. Let A be a 4×4 matrix whose characteristic polynomial is $(\lambda - 2)^2(\lambda + 7)$.

    (a) Show that $\mathcal{E} = \{X \text{ in } \mathcal{R}^4 \,|\, AX = 2X\}$ is a subspace of $\mathcal{R}^4$.

    (b) What are the possible values of $\dim(\mathcal{E})$?

3. Let A be an $m \times n$ matrix and let $K \neq 0$ be $m \times 3$. Let $W = \{X \,|\, AX = K\}$. Explain why W is not a subspace of $\mathcal{R}^{n \times 3}$.

4. For $\mathbf{u} = \begin{bmatrix} 3 \\ 4 \\ 12 \end{bmatrix}$ and $\mathbf{v} = \begin{bmatrix} 1 \\ 1 \\ 1 \end{bmatrix}$ in $\mathcal{R}^3$, find the following:

    (a) $\mathbf{u} \cdot \mathbf{v}$;   **(b)** $\|\mathbf{u}\|$;   **(c)** a unit vector parallel to $\mathbf{u}$; and

    (d) the length of the projection of $\mathbf{u}$ on $\mathbf{v}$.

5. Given: $\mathbf{u_1} = \begin{bmatrix} 1 \\ 2 \\ 1 \\ 1 \end{bmatrix}$, $\mathbf{u_2} = \begin{bmatrix} 1 \\ 0 \\ -2 \\ 1 \end{bmatrix}$ and $\mathbf{u_3} = \begin{bmatrix} 0 \\ 0 \\ 1 \\ 0 \end{bmatrix}$.

    (a) Which of these vectors are orthogonal?

    (b) Find an orthonormal basis for $\text{SPAN}\{\mathbf{u_1}, \mathbf{u_2}, \mathbf{u_3}\}$.

6. For questions (a)–(f) write an equivalent question of the form,

       "Does the system $\{\ \ \}$ have a $\{$nonzero, unique$\}$ solution?"

    Give the system explicitly and show where it came from. Do not try to answer the questions!

    (a) Is $x^2 + 5x - 11$ in $\text{SPAN}\{x^2 + 5x, 3x^2 - 11, 5x + 7\}$?

    (b) Is the set $\{[1 \quad 5 \quad 3], [2 \quad -1 \quad 4], [3 \quad 1 \quad 7]\}$ linearly independent?

    (c) Is the set $\left\{ \begin{bmatrix} 1 & 2 \\ 3 & 4 \end{bmatrix}, \begin{bmatrix} 5 & 6 \\ 7 & 8 \end{bmatrix}, \begin{bmatrix} 9 & 0 \\ 1 & 2 \end{bmatrix}, \begin{bmatrix} 3 & 4 \\ 5 & 6 \end{bmatrix} \right\}$ a basis for $\mathcal{R}^{2 \times 2}$?

    (d) Is $[1 \quad 3 \quad 7 \quad 9]$ in the row space of $A = \begin{bmatrix} 1 & 3 & 5 & 9 \\ 2 & 8 & 6 & 4 \\ 3 & 0 & 5 & 1 \end{bmatrix}$?

(e) Is the set $\{x^2 + 5x, 3x^2 - 11, 5x + 7\}$ a basis for $\mathcal{P}_2$?

(f) Is $[7 \quad 8 \quad 10]$ in SPAN$\{[1 \quad 5 \quad 3], [2 \quad -1 \quad 4], [3 \quad 1 \quad 7]\}$?

7. (a) Find a basis for the null space of $A = \begin{bmatrix} 2 & 4 & 0 & -2 \\ 1 & 2 & -1 & -2 \\ 1 & 2 & 3 & 2 \end{bmatrix}$.

(b) What is the rank of A?

8. Are statements (a) through (j) true or false?

(a) $A \underset{\mathcal{R}}{\sim} B \Rightarrow \mathcal{NS}(A) = \mathcal{NS}(B)$

(b) $A \underset{\mathcal{R}}{\sim} B \Rightarrow \mathcal{CS}(A) = \mathcal{CS}(B)$

(c) $A \underset{\mathcal{R}}{\sim} B \Rightarrow A$ and B have the same eigenvalues

(d) $A \underset{\mathcal{R}}{\sim} B \Rightarrow$ If $B = I$, then $\det(A) = 1$

(e) $A \underset{\mathcal{R}}{\sim} B \Rightarrow \mathcal{RS}(A) = \mathcal{RS}(B)$

(f) If A is an orthogonal matrix, then $\det(A) = \pm 1$.

(g) Any independent subset of $\mathcal{V}$ is a basis for $\mathcal{V}$.

(h) Any independent subset of $\mathcal{V}$ is a basis for a subspace of $\mathcal{V}$.

(i) Any spanning set contains a basis.

(j) If $\mathcal{W}_1$ and $\mathcal{W}_2$ are subspaces of $\mathcal{V}$, then so is $\mathcal{W}_1 \cap \mathcal{W}_2$.

9. For $\mathbf{u} = \begin{bmatrix} 1 \\ 3 \\ 5 \end{bmatrix}$ and $\mathbf{v} = \begin{bmatrix} 2 \\ -3 \\ 4 \end{bmatrix}$ in $\mathcal{R}^3$ find

(a) $\mathbf{u} \cdot \mathbf{v}$;

(b) $\|\mathbf{u}\|$;

(c) a unit vector in the direction of $\mathbf{u} + \mathbf{v}$; and

(d) the length of the projection of $\mathbf{u}$ on $\mathbf{v}$.

10. Let B be a fixed 3×3 matrix. Show that $\mathcal{W} = \{A \text{ in } R^{3\times3} \,|\, AB = 3A\}$ is a subspace of $\mathcal{R}^{3\times3}$.

11. Given $\mathbf{u}_1 = \begin{bmatrix} 1 \\ 1 \\ 1 \\ 1 \end{bmatrix}, \mathbf{u}_2 = \begin{bmatrix} 3 \\ 1 \\ 1 \\ -5 \end{bmatrix}$ and $\mathbf{u}_3 = \begin{bmatrix} 12 \\ 0 \\ 0 \\ 0 \end{bmatrix}$ in $\mathcal{R}^4$,

(a) show that $\mathbf{u}_1$ and $\mathbf{u}_2$ are orthogonal.

(b) use the Gram-Schmidt process to find an orthonormal basis for SPAN$\{\mathbf{u}_1, \mathbf{u}_2, \mathbf{u}_3\}$.

12. Find an orthonormal basis for the column space of $A = \begin{bmatrix} 2 & 4 \\ 0 & 1 \\ 1 & 0 \end{bmatrix}$.

4

Linear Transformations

*L*et $\mathcal{V}$ and $\mathcal{W}$ be arbitrary, real vector spaces and suppose that for every vector v in $\mathcal{V}$ there is defined a unique vector T(v) in $\mathcal{W}$. We describe this situation by saying that T is a vector-valued function of the vector variable v. We will denote such vector-valued functions by bold-face capital letters. For example, if $\mathcal{V} = \mathcal{W} = \mathcal{R}^3$, then

$$T\begin{bmatrix} x \\ y \\ z \end{bmatrix} = \begin{bmatrix} 2x + 3y - z \\ 2 + \sin(x + y) \\ 3|x - 3z| \end{bmatrix}$$

is a vector-valued function; we will give other examples later. In this chapter we will study the most important kind of vector-valued functions, those that are linear.

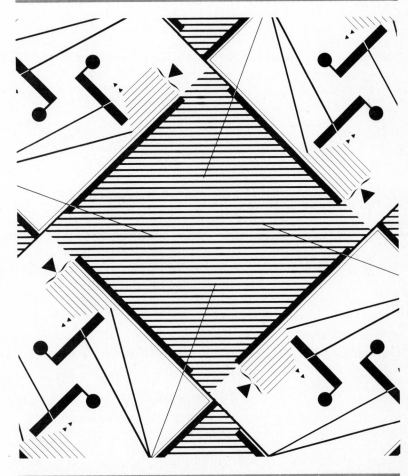

4.1

Definitions and Examples

We will use the following standard terminology, from algebra and calculus, in discussing functions from $\mathcal{V}$ to $\mathcal{W}$.

(1) $\mathcal{V}$ is called the DOMAIN of **T**.

(2) **T(v)** is the VALUE of **T** at **v** or the IMAGE of **v** under **T**.

(3) The RANGE of **T** is the set $\mathcal{R}(\mathbf{T}) = \mathbf{T}(\mathcal{V}) = \{\mathbf{T}(\mathbf{v}) \,|\, \mathbf{v} \text{ in } \mathcal{V}\} \subseteq \mathcal{W}$.

(4) **T** is ONE-TO-ONE if **T(u)** = **T(v)** implies **u** = **v**, or, equivalently, if **u** ≠ **v** implies **T(u)** ≠ **T(v)** (see Figure 4.1).

(5) **T** is ONTO $\mathcal{W}$ if $\mathcal{R}(\mathbf{T}) = \mathcal{W}$.

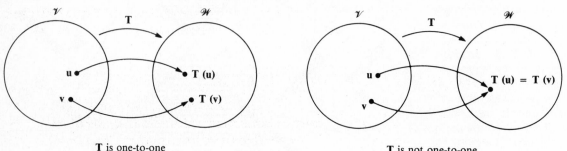

T is one-to-one   **T** is not one-to-one

FIGURE 4.1

We will not attempt to study all these functions; that would be an overly ambitious and difficult undertaking. We will restrict our attention to the linear functions, an important and useful subclass of all such functions.

Definition 4.1

Let $\mathcal{V}$ and $\mathcal{W}$ be vector spaces. A LINEAR TRANSFORMATION from $\mathcal{V}$ to $\mathcal{W}$ is a function **T**, from $\mathcal{V}$ to $\mathcal{W}$, that satisfies

$$\mathbf{T}(k\mathbf{u} + h\mathbf{v}) = k\mathbf{T}(\mathbf{u}) + h\mathbf{T}(\mathbf{v}) \tag{4.1}$$

for all **u** and **v** in $\mathcal{V}$ and all real scalars h and k. Note that Equation 4.1 is equivalent to the two conditions,

$$\mathbf{T}(\mathbf{u} + \mathbf{v}) = \mathbf{T}(\mathbf{u}) + \mathbf{T}(\mathbf{v}) \quad \text{and} \quad \mathbf{T}(k\mathbf{u}) = k\mathbf{T}(\mathbf{u}),$$

for all **u** and **v** in $\mathcal{V}$ and all k in $\mathcal{R}$.

Linear transformations are also called LINEAR FUNCTIONS or LINEAR MAPPINGS. In the special case in which $\mathcal{V} = \mathcal{W}$, a linear transformation is frequently called a LINEAR OPERATOR; in the special case where $\mathscr{W} = \mathscr{R}$, a linear transformation is often called a LINEAR FUNCTIONAL. In all of these situations *linear* means that Equation 4.1 is satisfied; that is, the functional value of a linear combination of vectors is the same linear combination of the functional values of those vectors.

You have already had a good bit of experience with linear transformations, as the following familiar examples will point out.

Example 1

Let $\mathscr{V} = \mathscr{W} = \mathscr{R}^n$ and let the function $\mathbf{T}$ be defined as multiplication by the matrix A; that is, $\mathbf{T}(\mathbf{u}) = A\mathbf{u}$. In this case the previously established fact that matrix multiplication distributes over matrix addition (see Theorem 1.5) yields

$$\mathbf{T}(k\mathbf{u} + h\mathbf{v}) = A(k\mathbf{u} + h\mathbf{v}) = k(A\mathbf{u}) + h(A\mathbf{v}) = k\mathbf{T}(\mathbf{u}) + h\mathbf{T}(\mathbf{v}),$$

and it follows at once that matrix multiplication is a linear function. To check whether or not $\mathbf{T}$ is one-to-one we let $\mathbf{u}$ and $\mathbf{v}$ be vectors in $\mathscr{V}$ such that $\mathbf{T}(\mathbf{u}) = \mathbf{T}(\mathbf{v})$. This is equivalent to $A\mathbf{u} = A\mathbf{v}$ or $A(\mathbf{u} - \mathbf{v}) = 0$. If A is invertible, it follows, from Theorem 3.14, that $\mathbf{u} - \mathbf{v} = \mathbf{0}$. Thus, $\mathbf{u} = \mathbf{v}$ and $\mathbf{T}$ is one-to-one. If A is singular, then the homogeneous system $A\mathbf{w} = 0$ has nonzero solutions and $\mathbf{T}$ is not one-to-one. Thus, $\mathbf{T}$ is one-to-one if and only if the matrix A is nonsingular. Since the $n \times n$ linear system $A\mathbf{u} = \mathbf{w}$ has a solution for every $\mathbf{w}$ in $\mathscr{W}$ if and only if A is nonsingular, it follows that $\mathbf{T}$ is onto $\mathscr{W}$ if and only if A is nonsingular. ∎

Example 2

Let $\mathscr{V} = \mathscr{W} = \mathscr{R}^{1 \times 3}$ and let $\mathbf{T}$ be the function defined by

$$\mathbf{T}[x \quad y \quad z] = [2x + 3y - z \quad 2 + x + y \quad 3|x - 3z|].$$

To get some feel for this function we will do some experimentation. If we let $\mathbf{u} = [1 \quad 2 \quad 3]$ and $\mathbf{v} = [-2 \quad 1 \quad 0]$ be a specific pair of vectors from $\mathscr{V}$, then by direct computation we have

$$\mathbf{T}(\mathbf{u}) = \mathbf{T}[1 \quad 2 \quad 3] = [2 + 6 - 3 \quad 2 + 1 + 2 \quad 3|1 - 9|]$$
$$= [5 \quad 5 \quad 24];$$

$$\mathbf{T}(\mathbf{v}) = \mathbf{T}[-2 \quad 1 \quad 0] = [-1 \quad 1 \quad 6];$$

$$\mathbf{T}(2\mathbf{v}) = \mathbf{T}[-4 \quad 2 \quad 0] = [-2 \quad 0 \quad 12];$$

$$\mathbf{T}(-2\mathbf{u}) = \mathbf{T}[-2 \quad -4 \quad -6] = [-10 \quad -4 \quad 48]; \text{ and}$$

$$\mathbf{T}(\mathbf{u} + \mathbf{v}) = \mathbf{T}[-1 \quad 3 \quad 3] = [4 \quad 4 \quad 30].$$

From these calculations we can see that

$$T(u + v) = [4 \quad 4 \quad 30] \neq T(u) + T(v)$$
$$= [5 \quad 5 \quad 24] + [-1 \quad 1 \quad 6] = [4 \quad 6 \quad 30];$$

$$T(2v) = [-2 \quad 0 \quad 12] \neq 2T(v) = 2[-1 \quad 1 \quad 6]$$
$$= [-2 \quad 2 \quad 12]; \text{ and}$$

$$T(-2u) = [-10 \quad -4 \quad 48] \neq -2T(u) = -2[5 \quad 5 \quad 24]$$
$$= [-10 \quad -10 \quad -48].$$

Any one of the last three equations is sufficient proof that T is not a linear transformation. Note that T is not onto $\mathcal{W}$ since no vector with a negative third component is in the range of T. ■

Example 2 illustrates an important point: to show that a function is not linear, it is sufficient to find only one special case in which Equation 4.1 fails to hold.

Example 3

Is the function T, from $\mathcal{R}^2$ to $\mathcal{R}^{1 \times 2}$, defined by

$$T\begin{bmatrix} x \\ y \end{bmatrix} = [x - 2y \quad 4x + 3y], \text{ a linear transformation?}$$

Proceeding as in the last example we let $u = \begin{bmatrix} 1 \\ 3 \end{bmatrix}$ and $v = \begin{bmatrix} 2 \\ -1 \end{bmatrix}$

be a particular pair of vectors from $\mathcal{R}^2$ and compute

$$T(u) = T\begin{bmatrix} 1 \\ 3 \end{bmatrix} = [-5 \quad 13], \ T(v) = T\begin{bmatrix} 2 \\ -1 \end{bmatrix} = [4 \quad 5];$$

$$T(2u) = T\begin{bmatrix} 2 \\ 6 \end{bmatrix} = [-10 \quad 26] = 2T(u); \text{ and}$$

$$T(u + v) = T\begin{bmatrix} 3 \\ 2 \end{bmatrix} = [-1 \quad 18] = T(u) + T(v).$$

These calculations suggest, *but do not prove,* that T is a linear trans-

formation. To establish this in general we let $u = \begin{bmatrix} u_1 \\ u_2 \end{bmatrix}$ and $v = \begin{bmatrix} v_1 \\ v_2 \end{bmatrix}$

be arbitrary vectors in $\mathcal{V}$ and compute

$$T(ku + hv) = T\begin{bmatrix} ku_1 + hv_1 \\ ku_2 + hv_2 \end{bmatrix}$$
$$= [(ku_1 + hv_1) - 2(ku_2 + hv_2) \quad 4(ku_1 + hv_1) + 3(ku_2 + hv_2)]$$

and $\qquad kT(u) + hT(v) = k[u_1 - 2u_2 \quad 4u_1 + 3u_2]$
$$+ h[v_1 - 2v_2 \quad 4v_1 + 3v_2].$$

Since these two vectors are equal, we have shown that Equation 4.1 holds for all vectors $\mathbf{u}$ and $\mathbf{v}$. Thus, $\mathbf{T}$ is a linear transformation. ■

Example 4

Let $\mathcal{V}$ be any n-dimensional real vector space and let

$$\mathcal{B} = \{\mathbf{u}_1, \mathbf{u}_2, \ldots, \mathbf{u_n}\}$$

be a basis for $\mathcal{V}$. If $\mathbf{w}$ is any vector in $\mathcal{V}$, then we can write

$$\mathbf{w} = a_1\mathbf{u}_1 + \ldots + a_n\mathbf{u_n}.$$

Let $[\mathbf{w}]_{\mathcal{B}} = [a_1 \;\; \ldots \;\; a_n]^T$ be the $\mathcal{B}$-coordinate matrix of the vector $\mathbf{w}$ and define a function $\mathbf{C}$ from $\mathcal{V}$ to $\mathcal{R}^n$ by $\mathbf{C}(\mathbf{w}) = [\mathbf{w}]_{\mathcal{B}}$.

It follows from Theorem 3.21 that the coordinate mapping $\mathbf{C} = [\;\;]_{\mathcal{B}}$ is a one-to-one linear transformation from $\mathcal{V}$ onto $\mathcal{R}^n$. This example will be important to us in later sections. ■

Example 5

Consider the transformation $\mathbf{P}$, from $\mathcal{R}^3$ to $\mathcal{R}^3$, defined by

$$\mathbf{P}\begin{bmatrix} x \\ y \\ z \end{bmatrix} = \begin{bmatrix} x \\ y \\ 0 \end{bmatrix} = \begin{bmatrix} 1 & 0 & 0 \\ 0 & 1 & 0 \\ 0 & 0 & 0 \end{bmatrix}\begin{bmatrix} x \\ y \\ z \end{bmatrix}.$$

Because it is defined by matrix multiplication, $\mathbf{P}$ is linear by Example 1. The range of this transformation is the xy-plane of $\mathcal{R}^3$ and hence the transformation is not onto $\mathcal{R}^3$; $\mathbf{e}_3 = [0 \;\; 0 \;\; 1]$ is not in the range of $\mathbf{P}$. Since $\mathbf{P}(\mathbf{e}_3) = \mathbf{P}(\mathbf{0}) = \mathbf{0}$, $\mathbf{P}$ is not one-to-one. This transformation can be described geometrically as a PROJECTION of $\mathcal{R}^3$ onto the xy-plane. ■

If you have studied calculus, you have already spent a good deal of time studying linear operators on vector spaces of functions. If you have not studied calculus, then you should disregard the next example.

Example 6

Let $\mathcal{V} = \mathcal{C}^1[a, b]$ be the subspace of $\mathcal{F}[a, b]$ which consists of all those functions from $\mathcal{F}[a, b]$ that have continuous derivatives on the interval $[a, b]$. Let $\mathbf{D}$ be the function on $\mathcal{V}$ defined by

$$\mathbf{D}(f(x)) = f'(x) = \frac{d}{dx}(f(x)).$$

It is shown in every beginning calculus course that $\mathbf{D}$ is linear; that is,

$$\mathbf{D}(af(x) + bg(x)) = af'(x) + bg'(x) = a\,\mathbf{D}(f(x)) + b\,\mathbf{D}(g(x)). \qquad \blacksquare$$

Many additional examples of linear transformations are contained in the exercises. We now list some elementary, but important, properties of any linear transformation.

Theorem 4.1

If $\mathbf{T}$ is a linear transformation from $\mathcal{V}$ to $\mathcal{W}$, then for all vectors $\mathbf{v}$ and $\mathbf{w}$ in $\mathcal{V}$ and all scalars a_i we have

(1) $\mathbf{T}(\mathbf{0}) = \mathbf{0}$;
(2) $\mathbf{T}(-\mathbf{v}) = -\mathbf{T}(\mathbf{v})$;
(3) $\mathbf{T}(\mathbf{v} - \mathbf{w}) = \mathbf{T}(\mathbf{v}) - \mathbf{T}(\mathbf{w})$;
(4) $\mathbf{T}(a_1\mathbf{v_1} + a_2\mathbf{v_2} + \ldots + a_k\mathbf{v_k}) = a_1\mathbf{T}(\mathbf{v_1}) + a_2\mathbf{T}(\mathbf{v_2}) + \ldots + a_k\mathbf{T}(\mathbf{v_k})$; and
(5) $\mathbf{T}$ is one-to-one if and only if $\mathbf{T}(\mathbf{u}) = \mathbf{0}$ implies $\mathbf{u} = \mathbf{0}$.

We will prove part 5; the other proofs are left to the exercises.

Proof. If $\mathbf{T}$ is one-to-one and $\mathbf{T}(\mathbf{u}) = \mathbf{0}$, then, since we also have $\mathbf{T}(\mathbf{0}) = \mathbf{0}$, it follows that $\mathbf{u} = \mathbf{0}$. Conversely, suppose that $\mathbf{x}$ and $\mathbf{y}$ are vectors such that $\mathbf{T}(\mathbf{x}) = \mathbf{T}(\mathbf{y})$. It follows that $\mathbf{T}(\mathbf{x}) - \mathbf{T}(\mathbf{y}) = \mathbf{0}$ and, by the linearity of $\mathbf{T}$, that $\mathbf{T}(\mathbf{x} - \mathbf{y}) = \mathbf{0}$. It follows from the hypothesis that $\mathbf{x} - \mathbf{y} = \mathbf{0}$ so that $\mathbf{x} = \mathbf{y}$ and $\mathbf{T}$ is one-to-one. $\square$

If $\mathcal{B} = \{\mathbf{u_1}, \mathbf{u_2}, \ldots, \mathbf{u_n}\}$ is a basis for the vector space $\mathcal{V}$ and $\mathbf{w}$ is any vector in $\mathcal{V}$, we know that we can express $\mathbf{w}$ as a linear combination of the basis vectors by solving a linear system:

$$\mathbf{w} = \sum_{i=1}^{n} a_i\mathbf{u_i} = a_1\mathbf{u_1} + \ldots + a_n\mathbf{u_n}.$$

If $\mathbf{T}$ is a linear transformation on $\mathcal{V}$, we can compute $\mathbf{T}(\mathbf{w})$ using

$$\mathbf{T}(\mathbf{w}) = \sum_{i=1}^{n} a_i\mathbf{T}(\mathbf{u_i}) = a_1\mathbf{T}(\mathbf{u_1}) + \ldots + a_n\mathbf{T}(\mathbf{u_n}).$$

This last equation shows that once we know the $\mathbf{T}(\mathbf{u_i})$, which are the images of the basis vectors, then we can compute the image, under $\mathbf{T}$, of any other vector in the space. Since $\mathbf{T}(\mathbf{w})$ is a typical vector in the range of $\mathbf{T}$, this equation also shows that the vectors $\mathbf{T}(\mathbf{u_i})$ span the range of $\mathbf{T}$.

We summarize this argument in the following theorem.

Theorem 4.2

A linear transformation on a finite dimensional vector space $\mathcal{V}$ is completely determined by its values for the vectors of any basis for $\mathcal{V}$. Moreover, if $\mathcal{B} = \{\mathbf{u_1}, \mathbf{u_2}, \ldots, \mathbf{u_n}\}$ is a basis for $\mathcal{V}$, then $\{\mathbf{T}(\mathbf{u_1}), \mathbf{T}(\mathbf{u_2}), \ldots, \mathbf{T}(\mathbf{u_n})\}$ is a spanning set for the range of $\mathbf{T}$.

Example 7

Let **T** be a linear transformation from $\mathcal{R}^3$ to $\mathcal{R}^3$ satisfying

$$\mathbf{T}\begin{bmatrix}1\\1\\1\end{bmatrix} = \begin{bmatrix}2\\3\\5\end{bmatrix}, \ \mathbf{T}\begin{bmatrix}1\\1\\0\end{bmatrix} = \begin{bmatrix}-2\\4\\3\end{bmatrix}, \text{ and } \mathbf{T}\begin{bmatrix}0\\1\\1\end{bmatrix} = \begin{bmatrix}0\\7\\5\end{bmatrix}.$$

Use Theorem 4.2 to find $\mathbf{T}\begin{bmatrix}-1\\3\\6\end{bmatrix}$.

It is easy to see that the vectors

$$\left\{ \mathbf{u_1} = \begin{bmatrix}1\\1\\1\end{bmatrix}, \mathbf{u_2} = \begin{bmatrix}1\\1\\0\end{bmatrix}, \mathbf{u_3} = \begin{bmatrix}0\\1\\1\end{bmatrix} \right\}$$

form a basis for $\mathcal{R}^3$. Thus, there must exist unique scalars a, b, and c such that

$$\mathbf{w} = \begin{bmatrix}-1\\3\\6\end{bmatrix} = a\begin{bmatrix}1\\1\\1\end{bmatrix} + b\begin{bmatrix}1\\1\\0\end{bmatrix} + c\begin{bmatrix}0\\1\\1\end{bmatrix}$$

$$= a\mathbf{u_1} + b\mathbf{u_2} + c\mathbf{u_3}.$$

The scalars a, b, and c must satisfy the linear system

$$\begin{bmatrix}1 & 1 & 0\\1 & 1 & 1\\1 & 0 & 1\end{bmatrix}\begin{bmatrix}a\\b\\c\end{bmatrix} = \begin{bmatrix}-1\\3\\6\end{bmatrix}.$$

The row reduction

$$\begin{bmatrix}1 & 1 & 0 & -1\\1 & 1 & 1 & 3\\1 & 0 & 1 & 6\end{bmatrix} \longrightarrow \begin{bmatrix}1 & 1 & 0 & -1\\0 & 0 & 1 & 4\\0 & -1 & 1 & 7\end{bmatrix} \longrightarrow$$

$$\begin{bmatrix}1 & 1 & 0 & -1\\0 & 1 & -1 & -7\\0 & 0 & 1 & 4\end{bmatrix} \longrightarrow \begin{bmatrix}1 & 0 & 0 & 2\\0 & 1 & 0 & -3\\0 & 0 & 1 & 4\end{bmatrix}$$

shows that $a = 2$, $b = -3$, and $c = 4$, so that $\mathbf{w} = 2\mathbf{u_1} - 3\mathbf{u_2} + 4\mathbf{u_3}$. We can now compute $\mathbf{T(w)}$ using the linearity of **T**:

$$\mathbf{T(w)} = \mathbf{T}(2\mathbf{u_1} - 3\mathbf{u_2} + 4\mathbf{u_3})$$

$$= 2\mathbf{T(u_1)} - 3\mathbf{T(u_2)} + 4\mathbf{T(u_3)}$$

$$= 2\begin{bmatrix}2\\3\\5\end{bmatrix} - 3\begin{bmatrix}-2\\4\\3\end{bmatrix} + 4\begin{bmatrix}0\\7\\5\end{bmatrix} = \begin{bmatrix}10\\22\\21\end{bmatrix}. \ \blacksquare$$

Exercise Set 4.1

1. Which of the following transformations are linear? In each case identify the domain of the function and the space W which contains the range.

   (a) $T\begin{bmatrix} x_1 \\ x_2 \\ x_3 \end{bmatrix} = \begin{bmatrix} x_1 + x_2 \\ 3x_1 - 7x_2 + x_3 \\ x_1 - 3x_2 + 2x_3 \end{bmatrix}$

   (b) $T\begin{bmatrix} x_1 \\ x_2 \\ x_3 \end{bmatrix} = \begin{bmatrix} 0 \\ 1 \\ x_1 + x_2 \end{bmatrix}$

   (c) $T\begin{bmatrix} x_1 \\ x_2 \\ x_3 \end{bmatrix} = \begin{bmatrix} x_1^2 \\ |x_2| \\ \sin x_3 \end{bmatrix}$

   (d) $T(a + bx + cx^2 + dx^3) = b + cx + dx^2$.

   (e) $T(y) = d^2y/dx^2 + 3y^2 - 6dy/dx$.

   (f) $T\begin{bmatrix} a & b \\ c & d \end{bmatrix} = \begin{bmatrix} a + b & 2b \\ c + d & 2d \end{bmatrix}$

2. Which of the following functions are linear? In each case identify the domain of the function and the space W which contains the range.

   (a) $T[x \quad y] = \begin{bmatrix} 2x \\ y \end{bmatrix}$

   (b) $T[x \quad y \quad z] = x^2 + 2y + z$

   (c) $T[x \quad y] = [x + y \quad 2x - y + 2]^T$

   (d) $T[x \quad y \quad z] = \begin{bmatrix} 2x + y & 3y \\ 4x - z & 2z + x \end{bmatrix}$

   (e) $T(a_0 + a_1x + a_2x^2) = a_0 + x + a_2x^2$

   (f) $T(A) = \det(A)$

   (g) $T(A) = \text{tr}(A)$ (Recall Exercise 11 of Section 2.1.)

   (h) $T\begin{bmatrix} a & b \\ c & d \end{bmatrix} = [a \quad b \quad c \quad d]$

   (i) $T(A) = A^TA$

3. Let T be a linear transformation from $\mathcal{R}^2$ to $\mathcal{R}^3$ such that $T(e_1) = [3 \quad 2 \quad 1]^T$ and $T(e_2) = [-2 \quad 4 \quad 3]^T$. Find $T(7e_1 + 5e_2)$, $T[3 \quad 2]^T$, and $T[6 \quad 5]^T$.

4. A linear transformation T, from $\mathcal{R}^3$ to $\mathcal{R}^{1\times3}$, satisfies

   $$T\begin{bmatrix} 1 \\ 0 \\ 0 \end{bmatrix} = [1 \quad 2 \quad 3], \quad T\begin{bmatrix} 1 \\ 1 \\ 0 \end{bmatrix} = [-1 \quad 4 \quad 2], \quad \text{and } T\begin{bmatrix} 1 \\ 1 \\ 1 \end{bmatrix} = [0 \quad 6 \quad 4].$$

   Find $T\begin{bmatrix} 3 \\ 2 \\ 2 \end{bmatrix}$, $T\begin{bmatrix} 0 \\ 0 \\ 1 \end{bmatrix}$, and $T\begin{bmatrix} 3 \\ 5 \\ -2 \end{bmatrix}$.

5. Let T be the linear operator on $\mathcal{R}^3$ defined by $T(u) = Au$, where

   $$A = \begin{bmatrix} 1 & -1 & 0 \\ -1 & 2 & 3 \\ 0 & 3 & 0 \end{bmatrix}. \text{ For } u = \begin{bmatrix} 1 \\ 2 \\ 3 \end{bmatrix}, \text{ compute } T(u), T(T(u)),$$

   $T(T(u) + 3u)$ and $(A^2 + 3A)u$.

6. Consider the transformation **U** from $\mathcal{R}^4$ to $\mathcal{R}^3$ defined by $\mathbf{U}(\mathbf{v}) = M\mathbf{v}$,

   where $M = \begin{bmatrix} 1 & 2 & 3 & 1 \\ 4 & 5 & 6 & 1 \\ 7 & 8 & 9 & 1 \end{bmatrix}$. Is **U** linear? Is it one-to-one? Is it onto $\mathcal{R}^3$?

7. Consider the transformation **S** from $\mathcal{R}^{m\times m}$ to $\mathcal{R}^{m\times m}$ defined by
   $\mathbf{S}(A) = U^{-1}AU$, where U is a fixed nonsingular matrix from $\mathcal{R}^{m\times m}$.
   Is **S** linear? Is it one-to-one? Is it onto $\mathcal{R}^{m\times m}$?

†8. (a) Let $\mathcal{V}$ be any real vector space and define the operator **I** on $\mathcal{V}$
   by $\mathbf{I}(\mathbf{v}) = \mathbf{v}$. The transformation **I** is called the IDENTITY MAPPING
   on $\mathcal{V}$. Show that **I** is linear. Describe **I** as a matrix multiplication
   in the case $\mathcal{V} = \mathcal{R}^n$.

   (b) Let $\mathcal{V}$ be any real vector space, let k be any scalar and define the
   scalar multiplication operator **k** on $\mathcal{V}$ by $\mathbf{k}(\mathbf{v}) = k\mathbf{v}$. Show that **k**
   is linear.

9. Show that a function **F**, from $\mathcal{R}$ to $\mathcal{R}$, is linear if and only if $\mathbf{F}(x) = kx$
   for $k = \mathbf{F}(1)$.

10. Can you find a one-to-one linear transformation from $\mathcal{R}^3$ onto $\mathcal{R}^4$? Can
    you find a one-to-one linear transformation from $\mathcal{R}^3$ onto $\mathcal{R}^2$?

11. Let **T** be a linear transformation on $\mathcal{R}^{2\times 2}$ such that

$$\mathbf{T}\begin{bmatrix} 1 & 0 \\ 0 & 0 \end{bmatrix} = \begin{bmatrix} 2 \\ 3 \end{bmatrix},\ \mathbf{T}\begin{bmatrix} 0 & 1 \\ 0 & 0 \end{bmatrix} = \begin{bmatrix} 5 \\ -7 \end{bmatrix},$$

$$\mathbf{T}\begin{bmatrix} 0 & 0 \\ 1 & 0 \end{bmatrix} = \begin{bmatrix} 1 \\ 1 \end{bmatrix},\ \mathbf{T}\begin{bmatrix} 0 & 0 \\ 0 & 1 \end{bmatrix} = \begin{bmatrix} 0 \\ 0 \end{bmatrix}.$$

Find $\mathbf{T}\begin{bmatrix} 3 & -11 \\ 4 & 2 \end{bmatrix}$ and $\mathbf{T}\begin{bmatrix} 5 & 0 \\ 0 & 7 \end{bmatrix}$.

12. Show that if $\{\mathbf{v}_1, \ldots, \mathbf{v}_k\}$ is a linearly independent set in $\mathcal{V}$ and $\mathcal{B}$ is
    any basis for $\mathcal{V}$, then the coordinate matrices

$$\{[\mathbf{v}_1]_{\mathcal{B}}, [\mathbf{v}_2]_{\mathcal{B}}, \ldots, [\mathbf{v}_k]_{\mathcal{B}}\}$$

form a linearly independent set in $\mathcal{R}^n$.

13. Prove parts 1–4 of Theorem 4.1.

14. Show geometrically that the function **R** on $\mathcal{R}^2$, defined by rotation
    through an angle θ, is a linear transformation on $\mathcal{R}^2$.

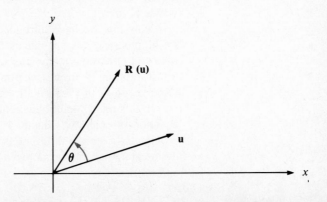

15. Let **T** be the linear transformation of Exercise 5. Find **x** in $\mathcal{R}^3$ such that (a) $\mathbf{T}(\mathbf{x}) = [1 \quad 0 \quad 0]^T$ and (b) $\mathbf{T}(\mathbf{x}) = [2 \quad 3 \quad -7]^T$.

16. Let **D** be the derivative operator on $\mathcal{P}$, the vector space of all polynomials of finite degree. Show that **D** is onto $\mathcal{P}$ but not one-to-one. Find an example of a linear operator on $\mathcal{P}$ that is one-to-one but not onto $\mathcal{P}$.

†17. Let $\mathcal{V} = \mathcal{W} = \mathcal{R}^n$ and let **T** be the linear transformation $\mathbf{T}(\mathbf{u}) = A\mathbf{u}$ where A is a fixed $n \times n$ matrix. Show that the following statements are equivalent:

(1) **T** is one-to-one;
(2) A is nonsingular; and
(3) **T** is onto $\mathcal{W}$.

18. Let **T** be a linear transformation from $\mathcal{V}$ to $\mathcal{W}$ and let
$$\mathcal{B} = \{\mathbf{u}_1, \mathbf{u}_2, \ldots, \mathbf{u}_n\}$$
be a basis for $\mathcal{V}$. Show that if
$$\mathbf{T}(\mathcal{B}) = \{\mathbf{T}(\mathbf{u}_1), \mathbf{T}(\mathbf{u}_2), \ldots, \mathbf{T}(\mathbf{u}_n)\}$$
is a basis for $\mathcal{W}$, then **T** is one-to-one and onto $\mathcal{W}$.

4.2   *The Range and Nullspace of a Linear Transformation*

Let us consider the linear transformation **T**, from $\mathcal{R}^n$ to $\mathcal{R}^m$, that is defined by multiplication by the $m \times n$ matrix A, that is, $\mathbf{T}(\mathbf{u}) = A\mathbf{u}$. Transformations of this type are called MATRIX TRANSFORMATIONS; they will serve as the motivating example for the study of linear transformations. Solving the linear system

$$AX = K \quad \text{or} \quad \mathbf{T}(X) = \mathbf{K},$$

with A as its coefficient matrix, can be viewed as the problem of finding all vectors **X** in $\mathcal{R}^n$, whose image under the transformation **T** is the vector **K** in $\mathcal{R}^m$.

Now that we have introduced the notion of linearity, we can describe precisely what it means for a given problem to be linear. The GENERAL LINEAR PROBLEM can be phrased as follows: given a linear transformation **T** from a vector space $\mathcal{V}$ to a vector space $\mathcal{W}$ and a fixed vector **w** in $\mathcal{W}$, find all vectors **u** in $\mathcal{V}$ such that $\mathbf{T}(\mathbf{u}) = \mathbf{w}$.

If **T** is matrix multiplication, then the linear problem is the linear-equations problem of Chapter 1; if **T** is differentiation, then the linear problem is a simple differential-equations problem that can be solved by integration. Other examples occur frequently in higher mathematics and in many areas of application.

The linear problem $T(u) = w$ is said to be HOMOGENEOUS if $w = 0$ and NONHOMOGENEOUS otherwise. In Figure 4.2 the vectors u_1, u_2, and u_3 are all solutions of the nonhomogeneous linear problem $T(x) = w$, while v_1, v_2, and 0 are solutions of the associated homogeneous problem $T(u) = 0$.

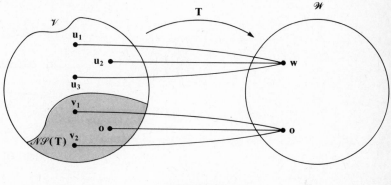

FIGURE 4.2

In the last chapter we introduced several subspaces associated with the matrix A. We can just as easily associate these subspaces with the linear transformation T defined by $T(u) = Au$. In particular, recall that the nullspace of the $m \times n$ matrix A and the column space of A were defined by

$$\mathscr{NS}(A) = \{X \text{ in } \mathscr{R}^n \,|\, AX = 0\} \quad \text{and} \quad \mathscr{CS}(A) = \{AX \,|\, X \text{ in } \mathscr{R}^n\}.$$

In terms of the matrix transformation T these definitions become

$$\mathscr{NS}(T) = \{u \text{ in } \mathscr{R}^n \,|\, T(u) = 0\} \quad \text{and} \quad \mathscr{CS}(A) = \{T(u) \,|\, u \text{ in } \mathscr{R}^n\}.$$

Note that in this case $\mathscr{CS}(A) = \mathscr{R}(T)$, which is the range of the transformation T. When dealing with linear transformations the nullspace is sometimes called the KERNEL of T; we will not use that terminology here.

These notions extend easily to the general case.

Definition 4.2

If T is a linear transformation from $\mathscr{V}$ to $\mathscr{W}$, then the NULLSPACE of T is defined by

$$\mathscr{NS}(T) = \{u \text{ in } \mathscr{V} \,|\, T(u) = 0\}.$$

The RANGE of T is defined by

$$\mathscr{R}(T) = T(\mathscr{V}) = \{T(u) \,|\, u \text{ in } \mathscr{R}^n\}.$$

The diagram in Figure 4.3 should be helpful in visualizing these subsets of $\mathcal{V}$ and $\mathcal{W}$. Recall, from part 5 of Theorem 4.1, that **T** is one-to-one if and only if $\mathcal{NS}(\mathbf{T})$ contains only the vector **0**.

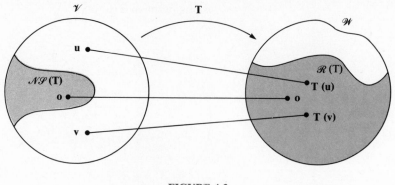

FIGURE 4.3

Theorem 4.3

If **T** is a linear transformation from $\mathcal{V}$ to $\mathcal{W}$, then the nullspace of **T** is a subspace of the domain $\mathcal{V}$ and the range of **T** is a subspace of $\mathcal{W}$.

Proof. If **u** and **v** are in $\mathcal{NS}(\mathbf{T})$, then $\mathbf{T}(\mathbf{u}) = \mathbf{T}(\mathbf{v}) = \mathbf{0}$. It follows that $\mathbf{T}(a\mathbf{u} + b\mathbf{v}) = a\mathbf{T}(\mathbf{u}) + b\mathbf{T}(\mathbf{v}) = \mathbf{0}$, so $a\mathbf{u} + b\mathbf{v}$ is also in $\mathcal{NS}(\mathbf{T})$. Therefore, by Theorem 3.7, $\mathcal{NS}(\mathbf{T})$ is a subspace of $\mathcal{V}$. To see that $\mathcal{R}(\mathbf{T})$ is a subspace of $\mathcal{W}$ we let $\mathbf{w}_1$ and $\mathbf{w}_2$ be any two vectors in $\mathcal{R}(\mathbf{T})$ and let a and b be any real numbers. We need to show that $a\mathbf{w}_1 + b\mathbf{w}_2$ is also in $\mathcal{R}(\mathbf{T})$. Since $\mathbf{w}_1$ and $\mathbf{w}_2$ are in $\mathcal{R}(\mathbf{T})$, there exist vectors $\mathbf{v}_1$ and $\mathbf{v}_2$ in $\mathcal{V}$ such that $\mathbf{T}(\mathbf{v}_1) = \mathbf{w}_1$ and $\mathbf{T}(\mathbf{v}_2) = \mathbf{w}_2$. In order to show that $a\mathbf{w}_1 + b\mathbf{w}_2$ is in $\mathcal{R}(\mathbf{T})$ we need to find a vector **x** in $\mathcal{V}$ such that $\mathbf{T}(\mathbf{x}) = a\mathbf{w}_1 + b\mathbf{w}_2$. The calculation

$$\mathbf{T}(a\mathbf{v}_1 + b\mathbf{v}_2) = a\mathbf{T}(\mathbf{v}_1) + b\mathbf{T}(\mathbf{v}_2) = a\mathbf{w}_1 + b\mathbf{w}_2$$

shows that we can choose $\mathbf{x} = a\mathbf{v}_1 + b\mathbf{v}_2$. Since **x** is in $\mathcal{V}$ the proof is completed. $\square$

Definition 4.3

For a linear transformation **T** we define the RANK OF **T** to be the dimension of the range of **T** and the NULLITY of **T** to be the dimension of the nullspace of **T**:

$$\text{rank}(\mathbf{T}) = \dim(\mathcal{R}(\mathbf{T})) \quad \text{and} \quad \text{nullity}(\mathbf{T}) = \dim(\mathcal{NS}(\mathbf{T})).$$

Example 1

Find bases for $\mathcal{NS}(\mathbf{T})$ and $\mathcal{R}(\mathbf{T})$ if $\mathbf{T}$ is the linear transformation on $\mathcal{R}^4$ defined by $\mathbf{T}(\mathbf{u}) = A\mathbf{u}$, with

$$A = \begin{bmatrix} 1 & -2 & 1 & -4 \\ 1 & 3 & 7 & 2 \\ 1 & -12 & -11 & -16 \\ 2 & -14 & -10 & -20 \end{bmatrix}.$$

In Example 2 of Section 3.3 we did the following row reduction:

$$[A\,|\,K] = \begin{bmatrix} 1 & -2 & 1 & -4 & k_1 \\ 1 & 3 & 7 & 2 & k_2 \\ 1 & -12 & -11 & -16 & k_3 \\ 2 & -14 & -10 & -20 & k_4 \end{bmatrix} \longrightarrow$$

$$\begin{bmatrix} 1 & -2 & 1 & -4 & k_1 \\ 0 & 5 & 6 & 6 & k_2 - k_1 \\ 0 & 0 & 0 & 0 & k_3 - 3k_1 + 2k_2 \\ 0 & 0 & 0 & 0 & k_4 - 4k_1 + 2k_2 \end{bmatrix}.$$

From this row reduction, with $K = 0$, we see that the homogeneous system $AX = 0$ is equivalent to the system

$$x_1 - 2x_2 + x_3 - 4x_4 = 0$$

$$5x_2 + 6x_3 + 6x_4 = 0.$$

If we set $x_3 = 5c_1$ and $x_4 = 5c_2$, it follows that the general solution of $AX = 0$ is

$$X = c_1 \begin{bmatrix} -17 \\ -6 \\ 5 \\ 0 \end{bmatrix} + c_2 \begin{bmatrix} 8 \\ -6 \\ 0 \\ 5 \end{bmatrix} = c_1 X_1 + c_2 X_2.$$

It follows that $\dim(\mathcal{NS}(\mathbf{T})) = 2$ and that a basis for $\mathcal{NS}(\mathbf{T})$ is

$$\mathcal{B}_1 = \{X_1 = [-17 \quad -6 \quad 5 \quad 0]^T, X_2 = [8 \quad -6 \quad 0 \quad 5]^T\}.$$

Since $\mathcal{R}(\mathbf{T}) = \mathcal{CS}(A)$ is the set of vectors K for which the system $AX = K$ has a solution, it follows from the above row reduction that K is in $\mathcal{R}(\mathbf{T})$ if and only if

$$k_3 - 3k_1 + 2k_2 = 0 \quad \text{and} \quad k_4 - 4k_1 + 2k_2 = 0.$$

The general solution of this system is easily seen to be

$$K = k_1 \begin{bmatrix} 1 \\ 0 \\ 3 \\ 4 \end{bmatrix} + k_2 \begin{bmatrix} 0 \\ 1 \\ -2 \\ -2 \end{bmatrix}$$

so rank($\mathbf{T}$) = 2 and a basis for $\mathscr{R}(\mathbf{T})$ is

$$\mathscr{B}_2 = \{[1 \quad 0 \quad 3 \quad 4]^T, [0 \quad 1 \quad -2 \quad -2]^T\}.$$

A second way of finding a basis for $\mathscr{R}(\mathbf{T})$ is to observe, as in the proof of Theorem 4.2, that if $\{\mathbf{u}_1, \mathbf{u}_2, \ldots, \mathbf{u}_n\}$ is a basis for $\mathscr{V}$, then the set $\{\mathbf{T}(\mathbf{u}_1), \mathbf{T}(\mathbf{u}_2), \ldots, \mathbf{T}(\mathbf{u}_n)\}$ spans the range of $\mathbf{T}$. Thus

$$\left\{ \mathbf{T}(\mathbf{e}_1) = \begin{bmatrix} 1 \\ 1 \\ 1 \\ 2 \end{bmatrix}, \mathbf{T}(\mathbf{e}_2) = \begin{bmatrix} -2 \\ 3 \\ -12 \\ -14 \end{bmatrix}, \mathbf{T}(\mathbf{e}_3) = \begin{bmatrix} 1 \\ 7 \\ -11 \\ -10 \end{bmatrix}, \mathbf{T}(\mathbf{e}_4) = \begin{bmatrix} -4 \\ 2 \\ -16 \\ -20 \end{bmatrix} \right\}$$

spans $\mathscr{R}(\mathbf{T})$. A basis could be obtained from this spanning set as illustrated in Example 4 of Section 3.7. In this case, any two independent vectors from the spanning set will form a basis for $\mathscr{R}(\mathbf{T})$. ■

For the case of a matrix transformation $\mathbf{T}$, with $\mathbf{T}(\mathbf{u}) = A\mathbf{u}$, we saw in Theorem 3.13 that

$$\dim(\mathscr{N\!S}(A)) + \text{rank}(A) = \text{number of columns of } A. \quad (4.2)$$

Note that we have defined the rank of a matrix A to be $\dim(\mathscr{R\!S}(A))$ while the rank of matrix transformation $\mathbf{T}$ is defined to be $\dim(\mathscr{R}(\mathbf{T})) = \dim(\mathscr{C\!S}(A))$. Our next theorem generalizes Equation 4.2 to any linear transformation on a finite dimensional vector space. A corollary (see Exercise 12 of this section) of this theorem is that $\dim(\mathscr{R\!S}(A)) = \dim(\mathscr{C\!S}(A))$ for any matrix A. The proof of this theorem is somewhat technical, but it is included because it is typical of many arguments about finite dimensional vector spaces.

Theorem 4.4

If $\mathscr{V}$ is any finite dimensional vector space and $\mathbf{T}$ is a linear transformation from $\mathscr{V}$ to the vector space $\mathscr{W}$, then

$$\dim(\mathscr{N\!S}(\mathbf{T})) + \dim(\mathbf{T}(\mathscr{V})) = \dim(\mathscr{V})$$

or

$$\text{nullity}(\mathbf{T}) + \text{rank}(\mathbf{T}) = \dim(\mathscr{V}).$$

Proof. Let $\mathscr{B}_1 = \{\mathbf{u}_1, \ldots, \mathbf{u}_k\}$ be a basis for $\mathscr{N\!S}(\mathbf{T})$. Since $\mathscr{N\!S}(\mathbf{T})$ is a subspace of $\mathscr{V}$, this basis for $\mathscr{N\!S}(\mathbf{T})$ can be extended to a basis for all of $\mathscr{V}$. Suppose that $\dim(\mathscr{V}) = n$ and let

$$\mathscr{B} = \{\mathbf{u}_1, \ldots, \mathbf{u}_k, \mathbf{u}_{k+1}, \ldots, \mathbf{u}_n\}$$

be such a basis for $\mathscr{V}$. Now a typical vector in the range of $\mathbf{T}$ is $\mathbf{w} = \mathbf{T}(\mathbf{u})$, where $\mathbf{u}$ is arbitrary in $\mathscr{V}$. Since $\mathscr{B}$ is a basis for $\mathscr{V}$, $\mathbf{u}$ can be written as a linear combination of the vectors of $\mathscr{B}$; that is,

$$\mathbf{u} = a_1\mathbf{u}_1 + \ldots + a_k\mathbf{u}_k + a_{k+1}\mathbf{u}_{k+1} + \ldots + a_n\mathbf{u}_n.$$

We then have, from the linearity of **T**,

$$\mathbf{w} = \mathbf{T(u)} = (a_1\mathbf{T(u_1)} + \ldots + a_k\mathbf{T(u_k)})$$
$$+ (a_{k+1}\mathbf{T(u_{k+1})} + \ldots + a_n\mathbf{T(u_n)})$$

and, since the basis vectors $\mathbf{u_1}, \mathbf{u_2}, \ldots, \mathbf{u_k}$ are in $\mathcal{NS}(\mathbf{T})$,

$$\mathbf{w} = \mathbf{T(u)} = a_{k+1}\mathbf{T(u_{k+1})} + \ldots + a_n\mathbf{T(u_n)}.$$

This last equation shows that $\mathcal{B}_2 = \{\mathbf{T(u_{k+1})}, \ldots, \mathbf{T(u_n)}\}$ is a spanning set for the range of **T**. The proof will be completed if we can show that $\mathcal{B}_2$ is independent and hence is a basis for $\mathcal{R}(\mathbf{T})$. If

$$b_{k+1}\mathbf{T(u_{k+1})} + \ldots + b_n\mathbf{T(u_n)} = \mathbf{0},$$

then, by the linearity of **T**,

$$\mathbf{T}(b_{k+1}\mathbf{u_{k+1}} + \ldots + b_n\mathbf{u_n}) = \mathbf{0},$$

which shows that $\mathbf{y} = (b_{k+1}\mathbf{u_{k+1}} + \ldots + b_n\mathbf{u_n})$ is in $\mathcal{NS}(\mathbf{T})$. Therefore, since $\mathcal{B}_1$ is a basis for $\mathcal{NS}(\mathbf{T})$, there exist scalars c_i such that

$$\mathbf{y} = b_{k+1}\mathbf{u_{k+1}} + \ldots + b_n\mathbf{u_n} = c_1\mathbf{u_1} + \ldots + c_k\mathbf{u_k}.$$

Since $\mathcal{B}$ is a basis for $\mathcal{V}$, its elements must be independent; hence the last equation can hold only if all the coefficients are zero. Thus the spanning set $\mathcal{B}_2$ is independent and is therefore a basis for $\mathcal{R}(\mathbf{T})$. By counting the vectors in each of the bases we see that $\dim(\mathcal{NS}(\mathbf{T})) + \dim(\mathcal{R}(\mathbf{T})) = n = \dim(\mathcal{V})$ as claimed. $\square$

If we think of the dimension of a subspace as measuring the size of the subspace, then Theorem 4.4 says that if $\mathcal{NS}(\mathbf{T})$ becomes larger, then the range of **T** must become smaller.

We now consider an example of a linear transformation on a vector space whose elements are not matrices.

Example 2

Consider the linear transformation **D** on the polynomial subspace $\mathcal{P}_3 = \{a_0 + a_1x + a_2x^2 + a_3x^3\}$ of $\mathcal{P}$ defined by

$$\mathbf{D}(a_0 + a_1x + a_2x^2 + a_3x^3) = a_1 + 2a_2x + 3a_3x^2.*$$

The polynomials $p(x)$ for which $\mathbf{D}(p(x)) = 0$ are the constants. Thus the nullity of **D** is 1 and

$$\mathcal{NS}(\mathbf{D}) = \{a_0\} = \text{BASIS}\{1\} \subset \mathcal{P}_3.$$

---

*The calculus student should note that $\mathbf{D} = \dfrac{d}{dx}$.

The range of $\mathbf{D}$ is

$$\mathscr{R}(\mathbf{D}) = \{a_1 + 2a_2x + 3a_3x^2\}$$
$$= \text{BASIS}\{1 = \mathbf{D}(x),\ 2x = \mathbf{D}(x^2),\ 3x^2 = \mathbf{D}(x^3)\}$$
$$= \text{BASIS}\{1, x, x^2\},$$

so rank$(\mathbf{D}) = 3$. Note that rank$(\mathbf{D})$ + nullity$(\mathbf{D}) = 4 = \dim(\mathscr{P}_3)$, as predicted by Theorem 4.4. ■

We have already seen that the set of all solutions of the homogeneous linear problem $\mathbf{T}(\mathbf{x}) = \mathbf{0}$ is a subspace of the domain of $\mathbf{T}$. Several exercises have indicated that the set of all solutions of the nonhomogeneous linear problem $\mathbf{T}(\mathbf{x}) = \mathbf{w}$ is not a subspace of $\mathscr{V}$. The final theorem of this section describes the relationship between the solutions sets of the homogeneous and nonhomogeneous problems. This theorem may be interpreted geometrically as saying that the solution set of the nonhomogeneous problem is parallel to the subspace $\mathscr{NS}(\mathbf{T})$, or that the solution set of the nonhomogeneous problem is a translation of the subspace $\mathscr{NS}(\mathbf{T})$ (refer to Exercise 14 in Section 1.6 and Exercise 22 in Section 3.1). The diagram in Figure 4.4 illustrates the general situation.

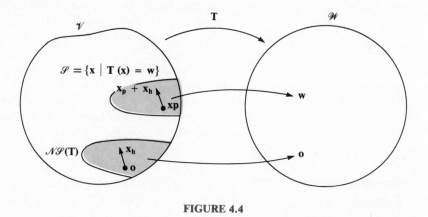

FIGURE 4.4

Theorem 4.5

Let $\mathbf{T}$ be a linear transformation from $\mathscr{V}$ to $\mathscr{W}$ and let $\mathbf{w}$ be a fixed nonzero vector in $\mathscr{W}$. Then every solution of the linear problem $\mathbf{T}(\mathbf{x}) = \mathbf{w}$ is of the form $\mathbf{x} = \mathbf{x_h} + \mathbf{x_p}$, where $\mathbf{x_h}$ is in $\mathscr{NS}(\mathbf{T})$ and $\mathbf{x_p}$ is a particular solution of the nonhomogeneous problem $\mathbf{T}(\mathbf{x}) = \mathbf{w}$. Equivalently,

$$\{\mathbf{x}\ \text{in}\ \mathscr{V}\ |\ \mathbf{T}(\mathbf{x}) = \mathbf{w}\} = \{\mathbf{x_h} + \mathbf{x_p}\ |\ \mathbf{x_h}\ \text{in}\ \mathscr{NS}(\mathbf{T})\} = \mathscr{NS}(\mathbf{T}) + \mathbf{x_p}.$$

Proof. If $\mathbf{x_p}$ is a particular solution of $\mathbf{T(x)} = \mathbf{w}$, then $\mathbf{T(x_p)} = \mathbf{w}$ and, from the linearity of $\mathbf{T}$, we have

$$\mathbf{T(x_h + x_p) = T(x_h) + T(x_p) = 0 + w = w}.$$

Thus any vector of the form $\mathbf{x_h + x_p}$ is a solution of $\mathbf{T(x)} = \mathbf{w}$.

If $\mathbf{y}$ is any other solution of $\mathbf{T(x)} = \mathbf{w}$ (that is, $\mathbf{T(y)} = \mathbf{w}$), then we can write

$$\mathbf{y = x_p + (y - x_p)};$$

the calculation

$$\mathbf{T(y - x_p) = T(y) - T(x_p) = w - w = 0}$$

shows that $\mathbf{x_h = y - x_p}$ is in $\mathscr{NS}(\mathbf{T})$. Thus, $\mathbf{y = x_p + x_h}$ as claimed. $\square$

The Gaussian elimination algorithm of Chapter 1 produces the general solution of the linear-equations problem and it is easy to write the solution in the form $X = X_p + X_h$.

Example 3

Express the general solution of the linear system

$$\begin{bmatrix} 1 & -2 & 1 & -4 \\ 1 & 3 & 7 & 2 \\ 1 & -12 & -11 & -16 \\ 2 & -14 & -10 & -20 \end{bmatrix} \begin{bmatrix} x_1 \\ x_2 \\ x_3 \\ x_4 \end{bmatrix} = \begin{bmatrix} 3 \\ 5 \\ -1 \\ 2 \end{bmatrix}$$

in the form $X = X_p + X_h$.

Since the coefficient matrix of this system is the matrix A of Example 1, we can use the reduction given there to see that the given system is equivalent to the triangular system

$$x_1 - 2x_2 + x_3 - 4x_4 = 3$$

$$5x_2 + 6x_3 + 6x_4 = 2.$$

If we set $x_3 = 5c_1$ and $x_4 = 5c_2$, we obtain the general solution

$$X = \begin{bmatrix} 19/5 - 17c_1 + 8c_2 \\ 2/5 - 6c_1 - 6c_2 \\ 5c_1 \\ 5c_2 \end{bmatrix}$$

$$= \begin{bmatrix} 19/5 \\ 2/5 \\ 0 \\ 0 \end{bmatrix} + \left(c_1 \begin{bmatrix} -17 \\ -6 \\ 5 \\ 0 \end{bmatrix} + c_2 \begin{bmatrix} 8 \\ -6 \\ 0 \\ 5 \end{bmatrix} \right) = X_p + X_h.$$

If we compare this solution with the general solution of $AX = 0$ given in Example 1, we see that the particular solution is

$$X_p = [19/5 \quad 2/5 \quad 0 \quad 0]^T$$

and that the general solution of the homogeneous system is

$$X_h = c_1 X_1 + c_2 X_2, \text{ where } X_1 = \begin{bmatrix} -17 \\ -6 \\ 5 \\ 0 \end{bmatrix} \text{ and } X_2 = \begin{bmatrix} 8 \\ -6 \\ 0 \\ 5 \end{bmatrix}$$

are the basis vectors for $\mathcal{NS}(T)$ found in Example 1. ■

Exercise Set 4.2

For range, read image

1. For each of the following linear transformations find $\mathcal{NS}(T)$, $\mathcal{R}(T)$, rank(T), and nullity(T).

   (a) $T\begin{bmatrix} x \\ y \end{bmatrix} = \begin{bmatrix} 1 & 2 \\ 2 & 4 \end{bmatrix}\begin{bmatrix} x \\ y \end{bmatrix}$

   (b) $D^2(a_0 + a_1x + a_2x^2 + a_3x^3) = 2a_2 + 6a_3x$

   (c) $T\begin{bmatrix} a & b \\ c & d \end{bmatrix} = \begin{bmatrix} a+b & 0 \\ 0 & c+d \end{bmatrix}$

   (d) $T[x \quad y \quad z] = [x + y + z \quad y + z \quad x + z]$

2. For each of the following linear transformations find a basis for the nullspace of T and a basis for the range of T.

   (a) $T(\mathbf{u}) = \begin{bmatrix} 1 & 3 & 4 \\ 3 & 4 & 7 \\ -2 & 2 & 0 \end{bmatrix}\mathbf{u}$ (b) $T(\mathbf{u}) = \begin{bmatrix} 2 & 0 & -3 & 1 \\ 2 & 1 & 1 & -4 \\ 2 & -1 & -11 & 12 \end{bmatrix}\mathbf{u}$

   (c) $T(\mathbf{u}) = \begin{bmatrix} 1 & 2 & 1 \\ 0 & 3 & 0 \\ 0 & 0 & 4 \end{bmatrix}\mathbf{u}$ (d) $T(\mathbf{u}) = \begin{bmatrix} 1 & 2 & 3 & 1 \\ 4 & 5 & 6 & 1 \\ 7 & 8 & 9 & 1 \end{bmatrix}\mathbf{u}$

3. (Calculus required) For each of the following linear transformations on $\mathcal{F}$ find a basis for $\mathcal{NS}(T)$.

   (a) $T(y) = \dfrac{dy}{dt} - 2y$ (b) $T(y) = \dfrac{d^2y}{dt^2}$

   (c) $T(y) = \dfrac{d^2y}{dt^2} + 4y$

4. For any linear transformation T from $\mathcal{V}$ to $\mathcal{W}$, show that if $\mathcal{S} = \{\mathbf{u}_1, \ldots, \mathbf{u}_m\}$ is a spanning set for $\mathcal{V}$, then $\{T(\mathbf{u}_1), T(\mathbf{u}_2), \ldots, T(\mathbf{u}_m)\}$ spans $\mathcal{R}(T)$.

5. Let T be a one-to-one linear transformation of $\mathcal{V}$ onto $\mathcal{W}$. Show that $\dim(\mathcal{V}) = \dim(\mathcal{W})$. What can you say about $\dim(\mathcal{V})$ and $\dim(\mathcal{W})$ if T is one-to-one but not onto $\mathcal{W}$? What can you say if T is neither one-to-one nor onto $\mathcal{W}$?

6. Let $\mathbf{T}$ be the linear transformation from $\mathcal{R}^{1\times 3}$ to $\mathcal{R}^{1\times 3}$ defined by
$\mathbf{T}[x_1 \quad x_2 \quad x_3] = [x_1 + x_2 + 2x_3 \quad 2x_1 + x_2 + x_3 \quad -x_1 - 2x_2 + 2x_3]$.
Find a basis for the range of $\mathbf{T}$.

7. Let $\mathbf{T}$ be a linear transformation from $\mathcal{V}$ to $\mathcal{W}$.
   (a) Given that $\dim(\mathcal{V}) = \dim(\mathcal{R}(\mathbf{T})) = 2$, find $\dim(\mathcal{NS}(\mathbf{T}))$.
   (b) Given that $\dim(\mathcal{NS}(\mathbf{T})) = \dim(\mathcal{R}(\mathbf{T})) = 2$, find $\dim(\mathcal{V})$.
   (c) Given that $\dim(\mathcal{V}) = 7$, $\dim(\mathcal{W}) = 5$, and $\text{rank}(\mathbf{T}) = 5$, find nullity$(\mathbf{T})$ and $\mathcal{R}(\mathbf{T})$.
   (d) Given that $\dim(\mathcal{V}) = 5$, $\dim(\mathcal{W}) = 1$, and $\mathbf{T} \neq 0$, find rank$(\mathbf{T})$ and nullity$(\mathbf{T})$.

8. Let $\mathbf{T}$ be a matrix transformation from $\mathcal{R}^n$ to $\mathcal{R}^n$ with matrix A. What condition on A would ensure that for any $\mathbf{v}$ in $\mathcal{R}^n$ there would exist a unique vector $\mathbf{u}$ in $\mathcal{R}^n$ such that $\mathbf{T}(\mathbf{u}) = \mathbf{v}$?

9. Let $\mathbf{T}$ be the transformation $\mathbf{T}(\mathbf{u}) = A\mathbf{u}$. What is the relationship between
$\mathcal{R}(A^T)$ and the rowspace of A?

10. Let $\mathbf{T}$ be a matrix transformation from $\mathcal{R}^3$ to $\mathcal{R}^3$ that has rank 2. Show that $\mathcal{NS}(\mathbf{T})$ is a line through the origin and that $\{\mathbf{u} \mid \mathbf{T}(\mathbf{u}) = \mathbf{w}\}$ is a line parallel to $\mathcal{NS}(\mathbf{T})$. What is the corresponding geometric interpretation if $\text{rank}(\mathbf{T}) = 1$?

11. Express the general solutions of the following linear systems in the form $X = X_p + X_h$, as in Example 3.

(a) $\begin{bmatrix} 1 & 5 & 0 & 0 & 5 \\ 0 & 0 & 1 & 0 & 3 \\ 0 & 0 & 0 & 1 & 4 \\ 0 & 0 & 0 & 0 & 0 \end{bmatrix} X = \begin{bmatrix} -1 \\ 1 \\ 2 \\ 0 \end{bmatrix}$
(b) $\begin{bmatrix} 1 & 3 & -2 & 4 \\ 0 & 2 & 3 & 0 \\ 0 & 0 & 0 & 3 \\ 0 & 0 & 0 & 0 \end{bmatrix} X = \begin{bmatrix} 3 \\ 4 \\ 5 \\ 0 \end{bmatrix}$

(c) $\begin{bmatrix} 1 & -1 & 2 & 3 \\ 2 & -1 & 0 & 2 \\ 4 & 1 & -11 & -1 \\ 1 & 2 & 3 & 8 \end{bmatrix} X = \begin{bmatrix} 1 \\ 1 \\ 1 \\ 1 \end{bmatrix}$
(d) $\begin{array}{l} x + 2y = 5 \\ 2x + 3y = 8 \\ x + 4y = 9 \end{array}$

†12. Use Theorem 4.4 and Exercise 8 of Section 3.5 to show that $\dim(\mathcal{RS}(A)) = \dim(\mathcal{CS}(A))$ for any matrix A.

4.3

The Algebra of Linear Transformations

Since the motivating example for the study of linear transformations is matrix multiplication and since there is a well-defined algebra of matrices, it is not surprising that there is also an algebra of linear transformations that generalizes matrix algebra. We begin by defining addition and scalar multiplication for linear transformations. Note that the definitions are like those given in Example 5 of Section 3.4 for the vector space of real-valued functions.

Definition 4.4

Let **S** and **T** be linear transformations from $\mathcal{V}$ to $\mathcal{W}$. The SUM **S** + **T** is the function from $\mathcal{V}$ to $\mathcal{W}$ defined by

$$(\mathbf{S} + \mathbf{T})(\mathbf{u}) = \mathbf{S}(\mathbf{u}) + \mathbf{T}(\mathbf{u})$$

and the SCALAR MULTIPLE $k\mathbf{T}$ is the function defined by

$$(k\mathbf{T})(\mathbf{u}) = k(\mathbf{T}(\mathbf{u})).$$

Note that for linear transformations, the sum and the scalar multiple are defined using the vector operations in the space $\mathcal{W}$ and the ranges of the linear transformations are all subspaces of $\mathcal{W}$.

It is easy to see that both **S** + **T** and $k\mathbf{T}$ are linear transformations from $\mathcal{V}$ to $\mathcal{W}$; for **S** + **T** the calculation is as follows:

$$(\mathbf{S} + \mathbf{T})(a\mathbf{u} + b\mathbf{v}) = \mathbf{S}(a\mathbf{u} + b\mathbf{v}) + \mathbf{T}(a\mathbf{u} + b\mathbf{v})$$

$$= \{a\mathbf{S}(\mathbf{u}) + b\mathbf{S}(\mathbf{v})\} + \{a\mathbf{T}(\mathbf{u}) + b\mathbf{T}(\mathbf{v})\}$$

$$= a\{\mathbf{S}(\mathbf{u}) + \mathbf{T}(\mathbf{u})\} + b\{\mathbf{S}(\mathbf{v}) + \mathbf{T}(\mathbf{v})\}$$

$$= a(\mathbf{S} + \mathbf{T})(\mathbf{u}) + b(\mathbf{S} + \mathbf{T})(\mathbf{v}).$$

Justify each step in the above calculation and then construct a similar argument to show that $k\mathbf{T}$ is linear.

Definition 4.5

The set of all linear transformations from $\mathcal{V}$ to $\mathcal{W}$ will be denoted by $\mathcal{L}(\mathcal{V}, \mathcal{W})$.

If **S** and **T** are matrix transformations, with $\mathbf{S}(\mathbf{u}) = A\mathbf{u}$ and $\mathbf{T}(\mathbf{u}) = B\mathbf{u}$, then

$$(\mathbf{S} + \mathbf{T})(\mathbf{u}) = \mathbf{S}(\mathbf{u}) + \mathbf{T}(\mathbf{u}) = A\mathbf{u} + B\mathbf{u} = (A + B)\mathbf{u}$$

and

$$(k\mathbf{T})(\mathbf{u}) = k(\mathbf{T}(\mathbf{u})) = k(B\mathbf{u}) = (kB)\mathbf{u}.$$

These calculations show that for linear transformations, the sum and scalar multiple are generalizations of familiar matrix operations. Thus, the following theorem is not at all surprising.

Theorem 4.6

$\mathcal{L}(\mathcal{V}, \mathcal{W})$, the set of all linear transformations from $\mathcal{V}$ to $\mathcal{W}$, is a real vector space with the operations in Definition 4.4.

The proof of this theorem is very easy, involving only routine calculations like those in Example 5 of Section 3.4. Recall that functions are equal if and only if their values are equal for every element of the domain. We will prove that addition of transformations is associative; the remaining details will be left to the exercises. Thus we let **T**, **S**, and **U** be elements of $\mathcal{L}(\mathcal{V}, \mathcal{W})$ and compute, for a typical vector **v** in $\mathcal{V}$,

$$((S + T) + U)(v) = (S + T)(v) + U(v) = (S(v) + T(v)) + U(v)$$

and

$$(S + (T + U))(v) = S(v) + (T + U)(v) = S(v) + (T(v) + U(v)).$$

The associativity of vector addition in $\mathcal{W}$ shows that the above expressions are equal for every vector $\mathbf{v}$ in $\mathcal{V}$, and hence that the functions $(S + T) + U$ and $S + (T + U)$ are equal. Therefore addition of linear transformations is associative, as claimed. □

The next question we consider is whether there is a way to define a product for linear transformations that is a generalization of matrix multiplication. The definition of multiplication of functions does not involve the product of the functional values; rather, it involves the familiar function of a function notion encountered in elementary algebra and calculus. For example, the function $f(x) = \sqrt{\sin x}$ can be viewed as the product of the sine function and the square-root function and could be evaluated on a pocket calculator by entering x, pushing the sin key, and then pushing the square-root key.

Definition 4.6

For $\mathbf{T}$ in $\mathcal{L}(\mathcal{V}, \mathcal{W})$ and $\mathbf{S}$ in $\mathcal{L}(\mathcal{W}, \mathcal{Z})$, the PRODUCT $\mathbf{ST}$ is the mapping from $\mathcal{V}$ to $\mathcal{Z}$ defined by

$$\mathbf{ST(u)} = \mathbf{S(T(u))}.$$

The simple diagrams in Figure 4.5 may be helpful in understanding Definition 4.6.

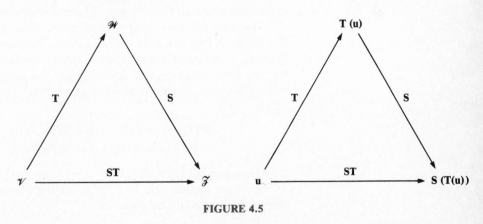

FIGURE 4.5

It is easy to see that the product of two linear transformations is another linear transformation; to verify this we compute, for arbitrary vectors $\mathbf{u}$ and $\mathbf{v}$ in $\mathcal{V}$ and arbitrary scalars a and b,

$$ST(a\mathbf{u} + b\mathbf{v}) = S(T(a\mathbf{u} + b\mathbf{v}))$$
$$= S(aT(\mathbf{u}) + bT(\mathbf{v}))$$
$$= aS(T(\mathbf{u})) + bS(T(\mathbf{v}))$$
$$= aST(\mathbf{u}) + bST(\mathbf{v}).$$

If **S** and **T** are matrix transformations with $S(\mathbf{u}) = A\mathbf{u}$ and $T(\mathbf{u}) = B\mathbf{u}$, then

$$ST(\mathbf{u}) = S(T(\mathbf{u})) = A(B\mathbf{u}) = (AB)\mathbf{u},$$

so that **ST** is the matrix transformation with matrix AB. Thus, the product introduced in Definition 4.6 is a generalization of matrix multiplication. This observation means that we cannot expect multiplication of transformations to satisfy properties not enjoyed by matrix multiplication. In particular, it must be the case that multiplication of transformations is not commutative, not always defined, and not always invertible.

The obvious way to extend the definition of the matrix inverse to more general transformations is to call a transformation $\mathbf{S} = \mathbf{T}^{-1}$ if **S** is such that $\mathbf{ST} = \mathbf{TS} = \mathbf{I}$, where **I** is the IDENTITY OPERATOR on $\mathcal{V}$ defined by $\mathbf{I}(\mathbf{x}) = \mathbf{x}$ for every **x** in $\mathcal{V}$ (recall Exercise 8 in Section 4.1). It is clear, from Figure 4.5, that this can occur only if $\mathcal{V} = \mathcal{W} = \mathcal{X}$; that is, only linear operators (transformations from $\mathcal{V}$ to $\mathcal{V}$) can have inverses. We will extend the familiar terminology associated with matrix inverses to the case of transformations; we will say that **T** is INVERTIBLE or NONSINGULAR if **T** has a multiplicative inverse. A necessary and sufficient condition that **T** be nonsingular is that **T** be both one-to-one and onto (see Exercise 12 for this section).

The multiplicative properties of transformations are very much like the properties of matrix multiplication. The proofs are usually less involved in the more general case of linear transformations, since we are free from subscripts and double sums. For example, to show that multiplication of transformations is associative we compute, for any vector **v** in $\mathcal{V}$,

$$S(TU)(\mathbf{v}) = S(TU(\mathbf{v})) = S(T(U(\mathbf{v}))) \quad \text{and}$$
$$(ST)U(\mathbf{v}) = ST(U(\mathbf{v})) = S(T(U(\mathbf{v}))).$$

It follows that $\mathbf{S(TU)} = \mathbf{(ST)U}$.

We have proved part 1 of the following theorem. Proofs of the other parts are left to the exercises.

Theorem 4.7

If **S**, **T**, and **U** are linear operators on $\mathcal{V}$, then

(1) $\mathbf{S(TU)} = \mathbf{(ST)U}$;
(2) $\mathbf{S(T + U)} = \mathbf{ST + SU}$;
(3) $\mathbf{(S + T)U} = \mathbf{SU + TU}$; and

(4) if $\dim(\mathcal{V}) < \infty$, then **T** is invertible if and only if **T** is one-to-one.

Our next example shows how these ideas can be applied to the linear operators studied in calculus.

Example 1

Let **D** be the differentiation operator $\mathbf{D}(y) = y' = dy/dx$. Compute $\mathbf{D}^2$, $(\mathbf{D} + 3)(\mathbf{D} - 5)$, and **DxD**.

Since $\mathbf{D}^2 y = \mathbf{D}(\mathbf{D}y) = \mathbf{D}(y') = y''$, we see that $\mathbf{D}^2$ is the second derivative operator.

$$(\mathbf{D} + 3)(\mathbf{D} - 5)y = (\mathbf{D} + 3)(y' - 5y) = \mathbf{D}(y' - 5y) + 3(y' - 5y)$$
$$= y'' - 5y' + 3y' - 15y = \mathbf{D}^2 y - 2\mathbf{D}y - 15y$$
$$= (\mathbf{D}^2 - 2\mathbf{D} - 15)y.$$

Therefore $(\mathbf{D} + 3)(\mathbf{D} - 5) = \mathbf{D}^2 - 2\mathbf{D} - 15$.

The operator **DxD** is to be viewed as the product of three linear operators: first, differentiation; second, multiplication by x; and third, another differentiation. Thus,

$$\mathbf{DxD}(y) = \mathbf{D}(\mathbf{x}(y')) = \mathbf{D}(xy') = xy'' + y' = (\mathbf{xD}^2 + \mathbf{D})(y).$$

Note that we have used the product rule for the derivative in the above calculation. Note also that this calculation shows that $\mathbf{DxD} \neq \mathbf{xD}^2$. ■

Exercise Set 4.3

1. Consider **T** and **U** in $\mathcal{L}(\mathcal{R}^2, \mathcal{R}^{1 \times 2})$ defined by

$$\mathbf{T}\begin{bmatrix} x \\ y \end{bmatrix} = [2x - 3y \quad 0] \quad \text{and} \quad \mathbf{U}\begin{bmatrix} x \\ y \end{bmatrix} = [3y \quad x + y].$$

Find: **(a)** $(\mathbf{T} + \mathbf{U})(\mathbf{e}_1)$; **(b)** $(3\mathbf{T} - 5\mathbf{U})(\mathbf{e}_2)$; and **(c)** $(2\mathbf{T} + \mathbf{U})\begin{bmatrix} 3 \\ 2 \end{bmatrix}$.

2. Consider **T** and **U** in $L(R^{2 \times 2}, R^{2 \times 2})$ defined by

$$\mathbf{T}(A) = A^T \quad \text{and} \quad \mathbf{U}\begin{bmatrix} a & b \\ c & d \end{bmatrix} = \begin{bmatrix} a + b & 0 \\ 0 & c + d \end{bmatrix}.$$

For $A = \begin{bmatrix} 1 & 2 \\ 3 & 4 \end{bmatrix}$, find: **(a)** $(\mathbf{U} + 2\mathbf{T})(A)$; **(b)** $\mathbf{T}^2(A)$; **(c)** $(\mathbf{UT})(A)$; **(d)** $(\mathbf{TU})(A)$; and **(e)** $(\mathbf{U}^2 + \mathbf{T}^2)(A)$.

3. Consider **T** in $\mathcal{L}(\mathcal{R}^3, \mathcal{R}^3)$ defined by $\mathbf{T}(\mathbf{u}) = A\mathbf{u}$ where

$$A = \begin{bmatrix} 2 & 1 & 1 \\ 2 & 3 & 2 \\ 1 & 1 & 2 \end{bmatrix}.$$

Show that $(\mathbf{T}^3 - 7\mathbf{T}^2 + 11\mathbf{T})\mathbf{u} = 5\mathbf{u}$ for every **u** in $\mathcal{R}^3$.

4. Consider $\mathbf{T}$ and $\mathbf{U}$ in $\mathscr{L}(\mathscr{P}, \mathscr{P})$ defined by

$$\mathbf{T}(a_0 + a_1x + a_2x^2 + \ldots + a_nx^n) = a_1 + 2a_2x + 3a_3x^2 + \ldots + na_nx^{n-1}$$

and

$$\mathbf{U}(a_0 + a_1x + a_2x^2 + \ldots + a_nx^n) = a_0x + \frac{a_1x^2}{2} + \frac{a_2x^3}{3} + \ldots + \frac{a_nx^{n+1}}{(n+1)}.$$

(a) Compute $\mathbf{TU}(x^2 + x + 5)$ and $\mathbf{UT}(x^2 + x + 5)$.

(b) Show that $\mathbf{TU} = \mathbf{I}$, but $\mathbf{UT} \neq \mathbf{I}$.

5. Let $\mathbf{D} = d/dx$ be the differentiation operator. Compute

(a) $(\mathbf{D} - 1)(\mathbf{D} + 1)$ (b) $(\mathbf{D}^2 + 5\mathbf{D} - 11)(\mathbf{D} + 2)$

(c) $(x\mathbf{D} + 3)(\mathbf{D}^2 + 3\mathbf{D} + 1)$ (d) $(x\mathbf{D} + 3)(e^x\mathbf{D} + \sin 3x)$

6. Let $\mathbf{T}_1$ and $\mathbf{T}_2$ be in $\mathscr{L}(\mathscr{V}, \mathscr{V})$. How do $\mathscr{NS}(\mathbf{T}_1\mathbf{T}_2)$, $\mathscr{NS}(\mathbf{T}_1)$, and $\mathscr{NS}(\mathbf{T}_2)$ compare? What if $\mathbf{T}_1\mathbf{T}_2 = \mathbf{T}_2\mathbf{T}_1$?

7. Verify that if $\mathbf{T}$ is in $\mathscr{L}(\mathscr{V}, \mathscr{W})$, then so is $k\mathbf{T}$.

8. Verify that multiplication of transformations distributes over addition of transformations (parts 2 and 3 of Theorem 4.7).

9. Let $\mathbf{T}$ be a matrix transformation from $\mathscr{R}^n$ to $\mathscr{R}^n$ with matrix A. What condition on A would ensure that for any $\mathbf{v}$ in $\mathscr{R}^n$ there would exist a unique vector $\mathbf{u}$ in $\mathscr{R}^n$ such that $\mathbf{T}(\mathbf{u}) = \mathbf{v}$?

10. Let $\mathscr{V}$ and $\mathscr{W}$ be finite dimensional vector spaces. Show that $\dim(\mathscr{L}(\mathscr{V}, \mathscr{W})) = \dim(\mathscr{V}) \times \dim(\mathscr{W})$.

(*Hint:* Let $\mathscr{B} = \{\mathbf{u}_1, \ldots, \mathbf{u}_n\}$ and $\mathscr{B}' = \{\mathbf{w}_1, \ldots, \mathbf{w}_m\}$ be bases for $\mathscr{V}$ and $\mathscr{W}$ respectively. Define $\mathbf{E}_{ij}$ in $\mathscr{L}(\mathscr{V}, \mathscr{W})$ by

$$\mathbf{E}_{ij}(\mathbf{u}_k) = \begin{cases} \mathbf{w}_i & \text{if } k = j \\ \mathbf{0} & \text{if } k \neq j. \end{cases}$$

Show that $\{\mathbf{E}_{ij} \mid i = 1, \ldots, n; j = 1, \ldots, m\}$ is a basis for $\mathscr{L}(\mathscr{V}, \mathscr{W})$.)

11. For $\mathbf{T}$ in $\mathscr{L}(\mathscr{V}, \mathscr{V})$ show that $\mathbf{T}^{-1}$ exists if and only if $\mathbf{T}$ is both one-to-one and onto. (*Hint:* If $\mathbf{T}(\mathbf{u}) = \mathbf{v}$, then define $\mathbf{T}^{-1}(\mathbf{v}) = \mathbf{u}$ and show that $\mathbf{T}^{-1}$ is linear.)

12. Let $\mathbf{T}$ be a linear operator on a finite dimensional space $\mathscr{V}$ for which

$$\{\mathbf{u}_1, \mathbf{u}_2, \ldots, \mathbf{u}_n\}$$

is a basis. Show that the following are equivalent:

(1) $\mathbf{T}^{-1}$ exists;

(2) if $\mathbf{T}(\mathbf{x}) = \mathbf{0}$, then $\mathbf{x} = \mathbf{0}$;

(3) $\mathbf{T}$ is one-to-one;

(4) $\{\mathbf{T}(\mathbf{u}_1), \mathbf{T}(\mathbf{u}_2), \ldots, \mathbf{T}(\mathbf{u}_n)\}$ is a basis for $\mathscr{V}$; and

(5) $\mathbf{T}$ is onto $\mathscr{V}$.

4.4   *Geometric Interpretation*

The vector spaces $\mathscr{R}^n$ hold a very prominent place in the development of linear algebra. Not only are they the motivating example for the study of more general vector spaces, but they are the range, of an arbitrary finite dimensional vector space $\mathscr{V}$, under the linear transformation that associates with every vector in $\mathcal{V}$ its coordinate matrix with respect to a fixed basis (see Example 4 in Section 4.1). For this reason we call the spaces $\mathcal{R}^n$ COORDINATE SPACES. Since the coordinate mapping is both one-to-one and onto we call the spaces $\mathcal{V}$ and $\mathcal{R}^n$ ISOMORPHIC, which means "same structure."

The matrix transformations, on the coordinate spaces $\mathcal{R}^n$, are more than just the motivating example for the study of linear transformations. We will now show that every linear transformation between coordinate spaces is a matrix transformation. To see this we consider **T**, any linear transformation from $\mathcal{R}^n$ to $\mathcal{R}^m$, and let

$$\mathcal{S} = \{\mathbf{e_1}, \mathbf{e_2}, \ldots, \mathbf{e_n}\}$$

be the standard basis for $\mathcal{R}^n$. For any vector $\mathbf{u} = [u_1 \quad \ldots \quad u_n]^T$ in $\mathcal{R}^n$ we can write

$$\mathbf{u} = u_1\mathbf{e_1} + u_2\mathbf{e_2} + \ldots + u_n\mathbf{e_n}$$

and then compute

$$\mathbf{T(u)} = u_1\mathbf{T(e_1)} + u_2\mathbf{T(e_2)} + \ldots + u_n\mathbf{T(e_n)}.$$

Since each $\mathbf{T(e_i)} = [a_{1i} \quad a_{2i} \quad \ldots \quad a_{mi}]^T$ is an $m \times 1$ matrix (an element of $\mathcal{R}^m$) we can write

$$\mathbf{T(u)} = \mathbf{T}\begin{bmatrix} u_1 \\ u_2 \\ \vdots \\ u_n \end{bmatrix} = [\mathbf{T(e_1)} \quad \mathbf{T(e_2)} \quad \ldots \quad \mathbf{T(e_n)}]\begin{bmatrix} u_1 \\ u_2 \\ \vdots \\ u_n \end{bmatrix}$$

$$= \begin{bmatrix} a_{11} & a_{12} & \cdots & a_{1n} \\ a_{21} & a_{22} & \cdots & a_{2n} \\ \vdots & \vdots & & \vdots \\ a_{m1} & a_{m2} & \cdots & a_{mn} \end{bmatrix}\begin{bmatrix} u_1 \\ u_2 \\ \vdots \\ u_n \end{bmatrix} = A\mathbf{u}.$$

Thus **T** is multiplication by the matrix A and **T** is a matrix transformation as claimed. We will call the matrix

$$A = [\mathbf{T(e_1)} \quad \mathbf{T(e_2)} \quad \ldots \quad \mathbf{T(e_n)}]$$

the STANDARD MATRIX OF **T**. We summarize this discussion in the following theorem.

Theorem 4.8

If **T** is a linear transformation from $\mathcal{R}^n$ to $\mathcal{R}^m$, then **T** is multiplication by the $m \times n$ matrix

$$A = [\mathbf{T(e_1)} \quad \mathbf{T(e_2)} \quad \ldots \quad \mathbf{T(e_n)}].$$

Example 1

For the linear operator on $\mathcal{R}^3$ defined by

$$\mathbf{T}\begin{bmatrix} x_1 \\ x_2 \\ x_3 \end{bmatrix} = \begin{bmatrix} 3x_1 + 4x_3 \\ 2x_2 - 5x_3 \\ -x_1 + 2x_2 \end{bmatrix}$$

we have

$$\mathbf{T}(\mathbf{e}_1) = \mathbf{T}\begin{bmatrix} 1 \\ 0 \\ 0 \end{bmatrix} = \begin{bmatrix} 3 \\ 0 \\ -1 \end{bmatrix}, \quad \mathbf{T}(\mathbf{e}_2) = \begin{bmatrix} 0 \\ 2 \\ 2 \end{bmatrix}, \quad \text{and} \quad \mathbf{T}(\mathbf{e}_3) = \begin{bmatrix} 4 \\ -5 \\ 0 \end{bmatrix},$$

so that the standard matrix of **T** is

$$A = [\mathbf{T}(\mathbf{e}_1) \quad \mathbf{T}(\mathbf{e}_2) \quad \mathbf{T}(\mathbf{e}_3)] = \begin{bmatrix} 3 & 0 & 4 \\ 0 & 2 & -5 \\ -1 & 2 & 0 \end{bmatrix}. \quad \blacksquare$$

Example 2

Find the standard matrix of the transformation **R** that rotates $\mathcal{R}^2$ through an angle θ (recall Exercise 14 in Section 4.1).

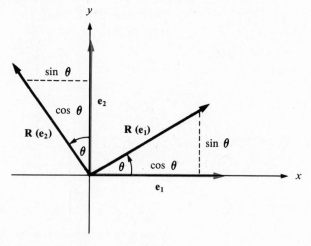

FIGURE 4.6

As we see from Figure 4.6,

$$\mathbf{R}(\mathbf{e}_1) = \mathbf{e}_1 \cos \theta + \mathbf{e}_2 \sin \theta$$

and

$$\mathbf{R}(\mathbf{e}_2) = -\mathbf{e}_1 \sin \theta + \mathbf{e}_2 \cos \theta,$$

so the standard matrix of **R** is $\begin{bmatrix} \cos \theta & -\sin \theta \\ \sin \theta & \cos \theta \end{bmatrix} = R.$

Note that the standard matrix of this rotation is an orthogonal matrix; that is, $R^T R = I.$ $\blacksquare$

We will devote the rest of this section to a geometric interpretation of linear operators on $\mathcal{R}^2$ and $\mathcal{R}^3$. The main tools in this study will be Theorems 4.2, 4.8, and the fact, proven in Theorem 1.9, that every nonsingular matrix can be written as a product of elementary matrices.

Since a linear operator on $\mathcal{R}^2$ maps points of the plane to points of the plane it will frequently be helpful to describe such a mapping by showing the image of the unit rectangle in the first quadrant (see Figure 4.7). Recall from Theorem 4.8 that **T** is completely determined by the images of the standard basis vectors; that is, by the vectors $\mathbf{T}(\mathbf{e}_1) = \mathbf{T}(1, 0)$ and $\mathbf{T}(\mathbf{e}_2) = \mathbf{T}(0, 1)$.

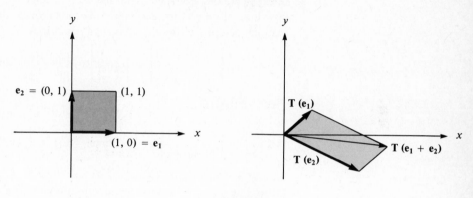

FIGURE 4.7

We begin by observing that a linear transformation on $\mathcal{R}^2$ or $\mathcal{R}^3$ always sends straight lines to straight lines and straight lines through the origin to straight lines through the origin. To see that this is the case we let P be any point on the line and let **u** be a vector parallel to the line. Then the line can be described as $\ell = \{\mathbf{OP} + k\mathbf{u} \,|\, k \text{ real}\}$ and the image of this line under **T** is the line $\mathbf{T}(\ell) = \{\mathbf{T}(\mathbf{OP}) + k\mathbf{T}(\mathbf{u}) \,|\, k \text{ real}\}$ which passes through the point $\mathbf{T}(\mathbf{OP})$ and is parallel to the vector $\mathbf{T}(\mathbf{u})$. See Figure 4.8.

FIGURE 4.8

Since the origin is a fixed point under any linear operator (that is, $\mathbf{T}(0,0) = (0,0)$), a line through the origin always maps into another line through the origin. Since two lines are parallel if and only if they are parallel to a common vector $\mathbf{u}$, it follows easily that parallel lines map into parallel lines and hence that parallelograms map into parallelograms. Thus, the image of the unit rectangle in Figure 4.7 must be a parallelogram.

From Theorem 1.9 we know that we can express any nonsingular matrix as a product of elementary matrices. Thus, from Section 4.3 it follows that any linear operator on $\mathcal{R}^2$ can be viewed as a product of linear operators associated with the 2×2 elementary matrices. Let us now look closely at the geometric interpretation of the matrix transformations determined by the elementary matrices

$$\begin{bmatrix} 0 & 1 \\ 1 & 0 \end{bmatrix}, \quad \begin{bmatrix} -1 & 0 \\ 0 & 1 \end{bmatrix}, \quad \begin{bmatrix} k & 0 \\ 0 & 1 \end{bmatrix}, \quad \begin{bmatrix} 1 & k \\ 0 & 1 \end{bmatrix}, \quad \text{and} \quad \begin{bmatrix} 1 & 0 \\ k & 1 \end{bmatrix}.$$

Reflections

If $\mathbf{T}\begin{bmatrix} x \\ y \end{bmatrix} = \begin{bmatrix} 0 & 1 \\ 1 & 0 \end{bmatrix}\begin{bmatrix} x \\ y \end{bmatrix} = \begin{bmatrix} y \\ x \end{bmatrix}$, then $\mathbf{T}(\mathbf{e}_1) = \mathbf{e}_2$ and $\mathbf{T}(\mathbf{e}_2) = \mathbf{e}_1$. Since

the line connecting (a, b) and (b, a) has slope -1, it is perpendicular to the line $y = x$, this transformation amounts to a REFLECTION IN THE LINE $y = x$. Figure 4.9 shows the image of a 1×2 rectangle under this transformation.

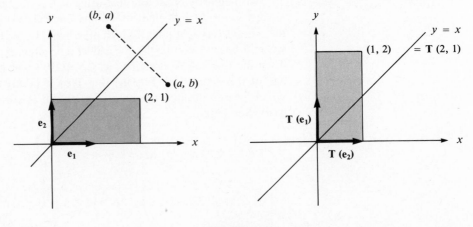

FIGURE 4.9

If $\mathbf{T}\begin{bmatrix} x \\ y \end{bmatrix} = \begin{bmatrix} -1 & 0 \\ 0 & 1 \end{bmatrix}\begin{bmatrix} x \\ y \end{bmatrix} = \begin{bmatrix} -x \\ y \end{bmatrix}$, then $\mathbf{T}(\mathbf{e}_1) = -\mathbf{e}_1$ and

$T(e_2) = e_2$ and T is a REFLECTION IN THE y-AXIS, as shown in Figure 4.10. Similarly, the transformation with matrix $\begin{bmatrix} 1 & 0 \\ 0 & -1 \end{bmatrix}$ represents a REFLECTION IN THE x-AXIS.

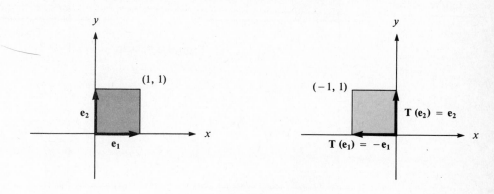

FIGURE 4.10

Expansion and Compressions

The matrix transformation with matrix $\begin{bmatrix} k & 0 \\ 0 & 1 \end{bmatrix}$, $k > 0$, sends e_1 to ke_1 and leaves e_2 fixed. If $k > 1$, this represents an EXPANSION IN THE x-DIRECTION BY A FACTOR OF k. If $k < 1$, this represents a CONTRACTION IN THE x-DIRECTION BY A FACTOR OF k. Figure 4.11 on page 236 illustrates these two cases. Similarly, the matrix $\begin{bmatrix} 1 & 0 \\ 0 & k \end{bmatrix}$, with $k > 0$, represents either an expansion or a contraction in the y-direction.

Shears

The matrix transformation with matrix $\begin{bmatrix} 1 & 0 \\ k & 1 \end{bmatrix}$ sends e_2 to e_2 and e_1 to $ke_2 + e_1$. We will call this transformation a SHEAR IN THE y-DIRECTION WITH FACTOR k. The effect on the unit rectangle is shown in Figure 4.12 on page 237.

Similarly, the matrix transformation with matrix $\begin{bmatrix} 1 & k \\ 0 & 1 \end{bmatrix}$ represents a SHEAR IN THE x-DIRECTION WITH FACTOR k.

All of the 2×2 elementary matrices have either been included in the preceding discussion or are simple products of the preceding ma-

FIGURE 4.11

trices. Thus we can describe the transformation associated with any nonsingular 2×2 matrix as a sequence of the simple transformations described above.

Example 3

Decompose the matrix transformation, with matrix $A = \begin{bmatrix} 1 & 2 \\ -1 & -4 \end{bmatrix}$

into a product of the elementary transformations (reflections, expansions and compressions, and shears) just described.

The row operations $(R_2 \leftarrow R_1 + R_2)$, $(R_2 \leftarrow -1/2R_2)$ and $(R_1 \leftarrow -2R_2 + R_1)$ reduce A to I, so we have, from Theorem 1.7,

$$\begin{bmatrix} 1 & -2 \\ 0 & 1 \end{bmatrix}\begin{bmatrix} 1 & 0 \\ 0 & -1/2 \end{bmatrix}\begin{bmatrix} 1 & 0 \\ 1 & 1 \end{bmatrix}A = I.$$

It follows that $A = \begin{bmatrix} 1 & 0 \\ 1 & 1 \end{bmatrix}^{-1}\begin{bmatrix} 1 & 0 \\ 0 & -1/2 \end{bmatrix}^{-1}\begin{bmatrix} 1 & -2 \\ 0 & 1 \end{bmatrix}^{-1}$

$$= \begin{bmatrix} 1 & 0 \\ -1 & 1 \end{bmatrix}\begin{bmatrix} 1 & 0 \\ 0 & -2 \end{bmatrix}\begin{bmatrix} 1 & 2 \\ 0 & 1 \end{bmatrix}$$

$$= \begin{bmatrix} 1 & 0 \\ -1 & 1 \end{bmatrix}\begin{bmatrix} 1 & 0 \\ 0 & 2 \end{bmatrix}\begin{bmatrix} 1 & 0 \\ 0 & -1 \end{bmatrix}\begin{bmatrix} 1 & 2 \\ 0 & 1 \end{bmatrix} = E_1 E_2 E_3 E_4.$$

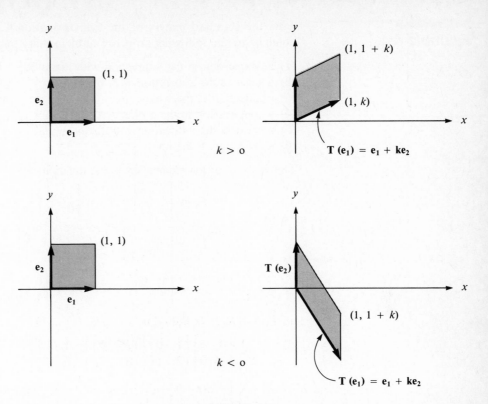

FIGURE 4.12

Note that we have written the Type I elementary matrix $\begin{bmatrix} 1 & 0 \\ 0 & -2 \end{bmatrix}$ as

the product $\begin{bmatrix} 1 & 0 \\ 0 & 2 \end{bmatrix} \begin{bmatrix} 1 & 0 \\ 0 & -1 \end{bmatrix}$, which represents a reflection followed

by an expansion. Thus, $\mathbf{T}$ can be viewed as the product of a shear in the x-direction (E_4), a reflection in the x-axis (E_3), an expansion in the y-direction (E_2) and a shear in the y-direction (E_1). The corners of the unit rectangle are successively transformed into

$$\begin{bmatrix} 1 \\ 0 \end{bmatrix} \longrightarrow \begin{bmatrix} 1 \\ 0 \end{bmatrix} \longrightarrow \begin{bmatrix} 1 \\ 0 \end{bmatrix} \longrightarrow \begin{bmatrix} 1 \\ 0 \end{bmatrix} \longrightarrow \begin{bmatrix} 1 \\ -1 \end{bmatrix} = \mathbf{T}\begin{bmatrix} 1 \\ 0 \end{bmatrix};$$

$$\begin{bmatrix} 0 \\ 1 \end{bmatrix} \longrightarrow \begin{bmatrix} 2 \\ 1 \end{bmatrix} \longrightarrow \begin{bmatrix} 2 \\ -1 \end{bmatrix} \longrightarrow \begin{bmatrix} 2 \\ -2 \end{bmatrix} \longrightarrow \begin{bmatrix} 2 \\ -4 \end{bmatrix} = \mathbf{T}\begin{bmatrix} 0 \\ 1 \end{bmatrix}; \text{ and}$$

$$\begin{bmatrix} 1 \\ 1 \end{bmatrix} \longrightarrow \begin{bmatrix} 3 \\ 1 \end{bmatrix} \longrightarrow \begin{bmatrix} 3 \\ -1 \end{bmatrix} \longrightarrow \begin{bmatrix} 3 \\ -2 \end{bmatrix} \longrightarrow \begin{bmatrix} 3 \\ -5 \end{bmatrix} = \mathbf{T}\begin{bmatrix} 1 \\ 1 \end{bmatrix}.$$

The successive transformations of the unit rectangle are shown in Figure 4.13 on page 239. ▄

Example 4

Find the standard matrix of the transformation $\mathbf{T}$, on $\mathcal{R}^2$, that is obtained by the following sequence of elementary transformations:

(1) an expansion in the y-direction with factor 3;
(2) a shear in the x-direction with factor -3;
(3) a reflection in the y-axis;
(4) a compression in the x-direction with factor $1/3$;
(5) a shear in the y-direction with factor 2; and
(6) a reflection in the line $y = x$.

The matrices of the elementary transformations are

$$E_1 = \begin{bmatrix} 1 & 0 \\ 0 & 3 \end{bmatrix}, \quad E_2 = \begin{bmatrix} 1 & -3 \\ 0 & 1 \end{bmatrix},$$

$$E_3 = \begin{bmatrix} -1 & 0 \\ 0 & 1 \end{bmatrix}, \quad E_4 = \begin{bmatrix} 1/3 & 0 \\ 0 & 1 \end{bmatrix},$$

$$E_5 = \begin{bmatrix} 1 & 0 \\ 2 & 1 \end{bmatrix}, \quad \text{and} \quad E_6 = \begin{bmatrix} 0 & 1 \\ 1 & 0 \end{bmatrix},$$

so $\mathbf{T(u)} = E_6 E_5 E_4 E_3 E_2 E_1 \mathbf{u}$

$$= \begin{bmatrix} 0 & 1 \\ 1 & 0 \end{bmatrix} \begin{bmatrix} 1 & 0 \\ 2 & 1 \end{bmatrix} \begin{bmatrix} 1/3 & 0 \\ 0 & 1 \end{bmatrix} \begin{bmatrix} -1 & 0 \\ 0 & 1 \end{bmatrix} \begin{bmatrix} 1 & -3 \\ 0 & 1 \end{bmatrix} \begin{bmatrix} 1 & 0 \\ 0 & 3 \end{bmatrix} \begin{bmatrix} x \\ y \end{bmatrix}$$

$$= \begin{bmatrix} -2/3 & 9 \\ -1/3 & 3 \end{bmatrix} \begin{bmatrix} x \\ y \end{bmatrix}.$$

Note that the product must be formed from right to left since the first transformation must go with the rightmost factor. ■

There is one other kind of transformation on $\mathcal{R}^2$ that is worth mentioning: the projections onto the coordinate axes. The transformation $\mathbf{P}_x = \begin{bmatrix} 1 & 0 \\ 0 & 0 \end{bmatrix} \begin{bmatrix} x \\ y \end{bmatrix} = \begin{bmatrix} x \\ 0 \end{bmatrix}$ is called a PROJECTION ONTO THE x-AXIS, while the transformation $\mathbf{P}_y = \begin{bmatrix} 0 & 0 \\ 0 & 1 \end{bmatrix} \begin{bmatrix} x \\ y \end{bmatrix} = \begin{bmatrix} 0 \\ y \end{bmatrix}$ is called a PROJECTION ONTO THE y-AXIS.

Note that these projections are not one-to-one mappings and hence are not invertible; their standard matrices are also singular. It is easy to see that $\mathcal{NS}(\mathbf{P}_x)$ is the y-axis while the x-axis is the nullspace of $\mathbf{P}_y$.

FIGURE 4.13

There are more general types of projections than those mentioned above. Suppose that W_1 is a subspace of $\mathcal{R}^n$ with a basis $\{u_1, \ldots, u_k\}$, which has been completed to a basis $\{u_1, \ldots, u_k, \ldots, u_n\}$ for $\mathcal{R}^n$. Let W_2 be the subspace of $\mathcal{R}^n$ spanned by $\{u_{k+1}, \ldots, u_n\}$ and consider $v = a_1 u_1 + \ldots + a_k u_k + \ldots + a_n u_n$, a typical vector in $\mathcal{R}^n$. The transformation $\mathbf{P}$ defined by

$$\mathbf{P}(v) = a_1 u_1 + a_2 u_2 + \ldots + a_k u_k$$

is called a PROJECTION ONTO W_1 PARALLEL TO W_2. Note that the range of $\mathbf{P}$ is W_1, that the nullspace of $\mathbf{P}$ is W_2, that $\mathbf{P}(x) = x$, for x in W_1, and that $\mathbf{P}^2(v) = \mathbf{P}(v)$ for every v in $\mathcal{V}$; that is, $\mathbf{P}^2 = \mathbf{P}$.

If $\{u_1, u_2, \ldots, u_n\}$ is an orthonormal basis for $\mathcal{R}^n$, then the subspace W_2 is the orthogonal complement of W_1 ($W_2 = W_1^\perp$) and $\mathbf{P}$ is called an ORTHOGONAL PROJECTION ONTO W_1.

Each 3×3 elementary matrix can also be interpreted geometrically, so the above results generalize easily to $\mathcal{R}^3$. For example,

$$\begin{bmatrix} 0 & 1 & 0 \\ 1 & 0 & 0 \\ 0 & 0 & 1 \end{bmatrix} \text{ is a reflection about the plane } y = x; \quad \begin{bmatrix} 1 & 0 & 0 \\ 0 & 1 & 0 \\ a & 0 & 1 \end{bmatrix} \text{ is a shear}$$

in the z direction, of the xz-plane, with factor a; and $\begin{bmatrix} 1 & 0 & 0 \\ 0 & 3 & 0 \\ 0 & 0 & 1 \end{bmatrix}$ is an

expansion in the y-direction with a factor of 3.

The rotation and projection matrices also generalize to $\mathcal{R}^3$:

$$\begin{bmatrix} \cos\theta & 0 & -\sin\theta \\ 0 & 1 & 0 \\ \sin\theta & 0 & \cos\theta \end{bmatrix} \text{ is a rotation in the } xz\text{-plane and } \begin{bmatrix} 1 & 0 & 0 \\ 0 & 0 & 0 \\ 0 & 0 & 1 \end{bmatrix} \text{ is}$$

a projection onto the xz-plane. We leave the details of the three-dimensional case to the exercises. The ideas introduced in this section have many applications to computer graphics.

Exercise Set 4.4

1. Find the standard matrix of the linear operator $\mathbf{T}$ on $\mathcal{R}^2$ given the following information.

(a) $\mathbf{T}\begin{bmatrix} 1 \\ 3 \end{bmatrix} = \begin{bmatrix} 2 \\ 6 \end{bmatrix}$ and $\mathbf{T}\begin{bmatrix} 1 \\ -3 \end{bmatrix} = \begin{bmatrix} 3 \\ -9 \end{bmatrix}$

$\left[Hint:\ e_1 = \dfrac{1}{2}\left(\begin{bmatrix} 1 \\ 3 \end{bmatrix} + \begin{bmatrix} 1 \\ -3 \end{bmatrix} \right) \right]$

(b) $T\begin{bmatrix} 1 \\ 1 \end{bmatrix} = \begin{bmatrix} 1 \\ 1 \end{bmatrix}$ and $T\begin{bmatrix} 2 \\ 3 \end{bmatrix} = \begin{bmatrix} 5 \\ 5 \end{bmatrix}$

(c) $T\begin{bmatrix} 1 \\ 2 \end{bmatrix} = \begin{bmatrix} 3 \\ 6 \end{bmatrix}$ and $T\begin{bmatrix} 1 \\ 1 \end{bmatrix} = \begin{bmatrix} -2 \\ -2 \end{bmatrix}$

2. Find the standard matrix of the linear operator T on $\mathcal{R}^3$ for which

$$T\begin{bmatrix} 1 \\ 0 \\ 0 \end{bmatrix} = \begin{bmatrix} 2 \\ 3 \\ 1 \end{bmatrix}, \quad T\begin{bmatrix} 1 \\ 1 \\ 0 \end{bmatrix} = \begin{bmatrix} 0 \\ -3 \\ 8 \end{bmatrix}, \quad \text{and} \quad T\begin{bmatrix} 1 \\ 1 \\ 1 \end{bmatrix} = \begin{bmatrix} 7 \\ 5 \\ 3 \end{bmatrix}.$$

3. Show that a product of two rotations in $\mathcal{R}^2$ is another rotation in $\mathcal{R}^2$.

4. Show that every linear transformation T from $\mathcal{R}^{1 \times n}$ to $\mathcal{R}^{1 \times m}$ is right multiplication by a suitable matrix A; that is,

$$T(\mathbf{u}) = \mathbf{u}A \quad \text{for every } \mathbf{u} \text{ in } \mathcal{R}^{1 \times n}.$$

In each of the following problems a transformation T on $\mathcal{R}^2$ is described as a sequence of simple transformations. In each case find the standard matrix of T and the image of the unit rectangle under T.

5. T is **(a)** a compression by a factor of $1/4$ in the x-direction;
   (b) a shear by a factor of 2 in the y-direction;
   (c) a shear by a factor of -3 in the x-direction; and
   (d) a reflection in the line $y = x$.

6. T is **(a)** a reflection in the y-axis;
   (b) an expansion by a factor of 3 in the y-direction;
   (c) a shear by a factor of 1 in the x-direction; and
   (d) a rotation through 45 degrees.

7. T is **(a)** a compression by a factor of $1/3$ in both directions;
   (b) a shear by a factor of -3 in the x-direction;
   (c) a shear by a factor of 2 in the y-direction; and
   (d) a reflection in the line $y = x$.

8. T is **(a)** an expansion by a factor of 2 in the x-direction;
   (b) a reflection in the x-axis;
   (c) a shear by a factor of $-1/2$ in the y-direction; and
   (d) a shear by a factor of 5 in the x-direction.

In each of the following problems describe the given matrix transformation on $\mathcal{R}^2$ as a product of expansions, contractions, reflections, and shears.

9. $T(\mathbf{u}) = \begin{bmatrix} 4 & -3 \\ 3 & -2 \end{bmatrix} \mathbf{u}$ 10. $T(\mathbf{u}) = \begin{bmatrix} 1 & 4 \\ 2 & 9 \end{bmatrix} \mathbf{u}$

11. $T(\mathbf{u}) = \begin{bmatrix} \cos\theta & -\sin\theta \\ \sin\theta & \cos\theta \end{bmatrix} \mathbf{u}$ 12. $T(\mathbf{u}) = \begin{bmatrix} 3 & 1 \\ 6 & 3 \end{bmatrix} \mathbf{u}$

In each of the following problems a transformation $\mathbf{T}$ on $\mathcal{R}^3$ is described as a sequence of elementary transformations. Find the standard matrix of $\mathbf{T}$. In each case find the image of the unit cube under $\mathbf{T}$.

13. $\mathbf{T}$ is **(a)** an expansion by a factor of 2 in the z-direction;

  (b) a shear by a factor of 3 in the z-direction of the xz-plane;

  (c) a reflection in the plane $x = y$; and

  (d) a projection onto the yz-plane.

14. $\mathbf{T}$ is **(a)** a rotation through 30 degrees in the xy-plane;

  (b) an expansion by a factor 2 in the y-direction;

  (c) a reflection in the xy-plane; and

  (d) a shear by a factor of 3 in the y-direction of the yz-plane.

In Exercises 15 and 16 describe the given matrix transformation on $\mathcal{R}^3$ as a product of elementary transformations.

15. $\mathbf{T(u)} = \begin{bmatrix} -1 & 1 & 3 \\ 2 & 1 & 1 \\ 4 & 2 & 3 \end{bmatrix}\mathbf{u}$ 16. $\mathbf{T(u)} = \begin{bmatrix} 1 & -1 & -1 \\ 5 & -9 & -13 \\ 3 & 1 & -3 \end{bmatrix}\mathbf{u}$

17. Determine a rotation $R(\theta) = \begin{bmatrix} \cos\theta & -\sin\theta \\ \sin\theta & \cos\theta \end{bmatrix}$ so that

$$R(\theta)\begin{bmatrix} a & b \\ c & d \end{bmatrix} = \begin{bmatrix} a' & b' \\ 0 & d' \end{bmatrix}.$$

18. Determine a rotation R in the yz-plane so that $RA = R\begin{bmatrix} a & b & c \\ d & e & f \\ g & h & i \end{bmatrix} = $

$\begin{bmatrix} a & b & c \\ d' & e' & f' \\ 0 & h' & i' \end{bmatrix}.$

Show that RAR^T has a zero in position $3, 1$.

19. Show that every 2×2 orthogonal matrix is either a rotation or a reflection in a coordinate axis followed by a rotation.

20. Determine a sequence of three plane rotations R_1, R_2, and R_3 so that

$$R_3R_2R_1A = R_3R_2R_1\begin{bmatrix} a & b & c \\ d & e & f \\ g & h & i \end{bmatrix} = \begin{bmatrix} * & * & * \\ 0 & * & * \\ 0 & 0 & * \end{bmatrix} = T$$

  (a) Show how this result can be used to find the QR factorization of a matrix A (see Exercise 23 of Section 3.7).

  (b) Show that if A is an orthogonal matrix, then T is a diagonal matrix.

  (c) Use part (b) to show that any 3×3 orthogonal matrix is a product of plane rotations and reflections.

21. **(a)** Show that each reflection matrix has eigenvalues 1 and -1.

  (b) Show that the matrices representing plane shears have two eigenvalues equal to 1, but only a one-dimensional subspace of eigenvectors.

   (c) Show that a plane rotation has no real eigenvalues.

   (d) Interpret each of statements (a) through (c) geometrically.

22. Let $\mathbf{w}$ be a unit vector in $\mathcal{R}^n$ and let $H = I - 2\mathbf{w}\mathbf{w}^T$.

   (a) Show that $H\mathbf{w} = -\mathbf{w}$.

   (b) Show that if $\mathbf{x}$ is orthogonal to $\mathbf{w}$, then $H\mathbf{x} = \mathbf{x}$.

   (c) Show that H can be interpreted as a reflection in the subspace orthogonal to $\mathbf{w}$.

23. Find the orthogonal projection of $\mathbf{u} = [1 \quad 2 \quad 3]^T$ onto the subspace of $\mathcal{R}^3$ spanned by $\{\mathbf{v_1} = [1 \quad 5 \quad 2]^T, \mathbf{v_2} = [7 \quad -1 \quad -1]^T\}$.

24. Consider the linear operator on $\mathcal{R}^2$ defined by $\mathbf{T}\begin{bmatrix} x \\ y \end{bmatrix} = \begin{bmatrix} a & b \\ c & d \end{bmatrix} \begin{bmatrix} x \\ y \end{bmatrix}$.

   (a) Show that the image of the unit square is a parallelogram.

   (b) Show that the area of the parallelogram in part (a) is $|\det(A)|$. (*Hint:* Recall Example 4 in Section 2.3.)

   (c) Let $\mathscr{S}$ be any plane area containing the origin and let $\mathbf{T}(\mathscr{S})$ be its image under $\mathbf{T}$. Show that the area of $\mathbf{T}(\mathscr{S})$ is equal to the area of $\mathscr{S}$ multiplied by $|\det(A)|$.

   (d) If $\mathscr{S}$ is the unit circle and $A = \begin{bmatrix} a & 0 \\ 0 & d \end{bmatrix}$, find the area of $\mathbf{T}(\mathscr{S})$.

   Describe $\mathbf{T}(\mathscr{S})$ geometrically.

   (e) What is the image of the unit circle under an arbitrary linear operator?

4.5   *Matrices and Linear Transformations*

In the last section we saw that every linear transformation between coordinate spaces can be described as multiplication by a suitable matrix. This is equivalent to saying that the vector space $\mathscr{L}(\mathcal{R}^n, \mathcal{R}^m)$, of all linear transformations from $\mathcal{R}^n$ to $\mathcal{R}^m$, is really just $\mathcal{R}^{m \times n}$, the vector space of all $m \times n$ matrices; that is,

$$\mathscr{L}(\mathcal{R}^n, \mathcal{R}^m) = \mathcal{R}^{m \times n}.$$

In this section we will see that there is also a close connection between matrix multiplication and linear transformations between arbitrary finite dimensional vector spaces.

   Let $\mathbf{T}$ be any linear transformation from the vector space $\mathcal{V}$ to the vector space $\mathcal{W}$. If we choose a basis $\mathscr{B} = \{\mathbf{v_1}, \mathbf{v_2}, \ldots, \mathbf{v_n}\}$ for $\mathcal{V}$, then we have shown that the coordinate mapping $[\quad]_{\mathscr{B}}$, which sends the vector $\mathbf{u}$ in $\mathcal{V}$ to its $\mathscr{B}$-coordinate matrix $[\mathbf{u}]_{\mathscr{B}}$ in $\mathcal{R}^n$, is a one-to-one linear mapping. Similarly, if we choose a basis $\mathscr{B}' = \{\mathbf{w_1}, \mathbf{w_2}, \ldots, \mathbf{w_m}\}$ for $\mathcal{W}$, then the coordinate mapping $[\quad]_{\mathscr{B}'}$, which sends $\mathbf{w}$ in $\mathcal{W}$ to $[\mathbf{w}]_{\mathscr{B}'}$ in $\mathcal{R}^m$, is also one-to-one and linear. The diagrams in Figure 4.14 illustrate the various transformations.

FIGURE 4.14

What is the relationship between $[\mathbf{T}(\mathbf{u})]_{\mathcal{B}'}$ and $[\mathbf{u}]_{\mathcal{B}}$? It certainly seems reasonable that $\mathbf{T}$ will "induce" a linear transformation $\mathbf{T}'$ from $\mathcal{R}^n$ to $\mathcal{R}^m$ such that $\mathbf{T}'([\mathbf{u}]_{\mathcal{B}}) = [\mathbf{T}(\mathbf{u})]_{\mathcal{B}'}$. According to Theorem 4.8, such a transformation must be a matrix transformation. Thus, we seek an $m \times n$ matrix A such that

$$A[\mathbf{u}]_{\mathcal{B}} = [\mathbf{T}(\mathbf{u})]_{\mathcal{B}'} \text{ for all } \mathbf{u} \text{ in } \mathcal{V}.$$

If we could find such a matrix, then the evaluation of $\mathbf{T}(\mathbf{u})$ could be accomplished by the following three steps, all of which could be easily handled on a digital computer.

(1) Find the coordinate matrix $[\mathbf{u}]_{\mathcal{B}}$; this normally involves solving a linear system.
(2) Multiply $[\mathbf{u}]_{\mathcal{B}}$ by A to find $[\mathbf{T}(\mathbf{u})]_{\mathcal{B}'} = A[\mathbf{u}]_{\mathcal{B}}$.
(3) Find $\mathbf{T}(\mathbf{u})$ from its coordinate matrix $[\mathbf{T}(\mathbf{u})]_{\mathcal{B}'}$.

The technique for finding a matrix A which satisfies

$$A[\mathbf{u}]_{\mathcal{B}} = [\mathbf{T}(\mathbf{u})]_{\mathcal{B}'} \tag{4.3}$$

for all $\mathbf{u}$ in $\mathcal{V}$, is like that used to find the change of basis matrix in Section 3.9. Since Equation 4.3 must hold for all vectors $\mathbf{u}$ in $\mathcal{V}$, it must hold for the vectors $\mathbf{v}_i$ of the basis $\mathcal{B}$. For $\mathbf{u} = \mathbf{v}_1$ we have

$$[\mathbf{v}_1]_{\mathcal{B}} = \mathbf{e}_1 = [1 \quad 0 \quad 0 \quad \cdots \quad 0]^T,$$

so that Equation 4.3 yields

$$[\mathbf{T}(\mathbf{v}_1)]_{\mathcal{B}'} = A[\mathbf{v}_1]_{\mathcal{B}} = A\mathbf{e}_1 = \begin{bmatrix} a_{11} \\ a_{21} \\ \vdots \\ a_{m1} \end{bmatrix} = \text{Col}_1(A).$$

Thus, the first column of A is the $\mathcal{B}'$-coordinate matrix of the image of the first basis vector of $\mathcal{B}$.

Similarly, for $\mathbf{u} = \mathbf{v}_i$ we have $[\mathbf{v}_i]_{\mathcal{B}} = \mathbf{e}_i$, so from Equation 4.3

$$[\mathbf{T}(\mathbf{v}_i)]_{\mathcal{B}'} = A[\mathbf{v}_i]_{\mathcal{B}} = A\mathbf{e}_i = \text{Col}_i(A) = \begin{bmatrix} a_{1i} \\ a_{2i} \\ \vdots \\ a_{mi} \end{bmatrix}.$$

This result shows that the matrix A of Equation 4.3, *if it exists,* is unique and

$$A = \begin{bmatrix} [\mathbf{T}(\mathbf{v}_1)]_{\mathcal{B}'} & [\mathbf{T}(\mathbf{v}_2)]_{\mathcal{B}'} & \cdots & [\mathbf{T}(\mathbf{v}_n)]_{\mathcal{B}'} \end{bmatrix}.$$

We now need to show that (4.3) holds for an arbitrary vector $\mathbf{x}$ in $\mathcal{V}$. Since the $\mathbf{v}_i$ form a basis for $\mathcal{V}$, there exists unique scalars b_i such that

$$\mathbf{x} = b_1\mathbf{v}_1 + b_2\mathbf{v}_2 + \ldots + b_n\mathbf{v}_n.$$

Since $\mathbf{T}$ is linear we have

$$\mathbf{T}(\mathbf{x}) = b_1\mathbf{T}(\mathbf{v}_1) + b_2\mathbf{T}(\mathbf{v}_2) + \ldots + b_n\mathbf{T}(\mathbf{v}_n)$$

and, since the coordinate mapping $[\quad]_{\mathcal{B}'}$ is also linear, we see that

$$[\mathbf{T}(\mathbf{x})]_{\mathcal{B}'} = b_1[\mathbf{T}(\mathbf{v}_1)]_{\mathcal{B}'} + b_2[\mathbf{T}(\mathbf{v}_2)]_{\mathcal{B}'} + \ldots + b_n[\mathbf{T}(\mathbf{v}_n)]_{\mathcal{B}'}$$

$$= \begin{bmatrix} [\mathbf{T}(\mathbf{v}_1)]_{\mathcal{B}'} & [\mathbf{T}(\mathbf{v}_2)]_{\mathcal{B}'} & \cdots & [\mathbf{T}(\mathbf{v}_n)]_{\mathcal{B}'} \end{bmatrix} \begin{bmatrix} b_1 \\ b_2 \\ \vdots \\ b_n \end{bmatrix} = A[\mathbf{x}]_{\mathcal{B}}.$$

We summarize this argument with our next theorem.

Theorem 4.9

Let $\mathcal{V}$ and $\mathcal{W}$ be finite dimensional vector spaces and let

$$\mathcal{B} = \{\mathbf{v}_1, \ldots \mathbf{v}_n\} \quad \text{and} \quad \mathcal{B}' = \{\mathbf{w}_1, \ldots, \mathbf{w}_m\}$$

be bases for $\mathcal{V}$ and $\mathcal{W}$ respectively. If $\mathbf{T}$ is any linear transformation from $\mathcal{V}$ to $\mathcal{W}$, then

$$A = \begin{bmatrix} [\mathbf{T}(\mathbf{v}_1)]_{\mathcal{B}'} & [\mathbf{T}(\mathbf{v}_2)]_{\mathcal{B}'} & \cdots & [\mathbf{T}(\mathbf{v}_n)]_{\mathcal{B}'} \end{bmatrix}$$

is the unique $m \times n$ matrix satisfying

$$[\mathbf{T}(\mathbf{u})]_{\mathcal{B}'} = A[\mathbf{u}]_{\mathcal{B}} \quad \text{for every } \mathbf{u} \text{ in } \mathcal{V}.$$

The matrix A of Theorem 4.9 is called the MATRIX REPRESENTATIVE OF $\mathbf{T}$ with respect to the bases $\mathcal{B}$ and $\mathcal{B}'$. This matrix will be denoted by $[T]_{\mathcal{B}'\mathcal{B}}$ when a compact notation is desired. In this notation, Equation 4.3 becomes

$$[T(u)]_{\mathcal{B}'} = [T]_{\mathcal{B}'\mathcal{B}}[u]_{\mathcal{B}}.$$

In the important case where $\mathcal{V} = \mathcal{W}$, the matrix representative is a square matrix. In this case it is natural to take $\mathcal{B}' = \mathcal{B}$ and Equation 4.3 becomes

$$[T(u)]_{\mathcal{B}} = [T]_{\mathcal{B}\mathcal{B}}[u]_{\mathcal{B}}.$$

If $\mathcal{V} = \mathcal{W} = \mathcal{R}^n$ and the basis is the standard basis $\mathcal{S}$, then the matrix representative $[T]_{\mathcal{S}\mathcal{S}}$ is the standard matrix of the transformation which was introduced in Section 4.4. In this case finding the coordinate vectors is easy because $[v]_{\mathcal{S}} = v$ for every v in $\mathcal{V}$. In general, finding the matrix representative $[T]_{\mathcal{B}'\mathcal{B}}$ involves the following steps.

---

(1) Find $T(v_i)$ for each v_i in the basis $\mathcal{B}$.
(2) Find $[T(v_i)]_{\mathcal{B}'}$ for each v_i in V (this normally involves solving a set of linear systems, all with the same coefficient matrix).
(3) Write down $[T]_{\mathcal{B}'\mathcal{B}}$, using the fact that its columns are the coordinate matrices from step 2.

In general, the second step will require the bulk of the computational effort.

Example 1

Let T be the linear operator on $\mathcal{R}^2$ defined by

$$T(u) = Au = \begin{bmatrix} 2 & 1 \\ -2 & 5 \end{bmatrix} \begin{bmatrix} x \\ y \end{bmatrix}.$$

Find the matrix representative of T with respect to the basis

$$\mathcal{B} = \left\{ u_1 = \begin{bmatrix} 1 \\ 2 \end{bmatrix}, u_2 = \begin{bmatrix} 1 \\ 1 \end{bmatrix} \right\}.$$

The matrix A is both the standard matrix of T and the matrix representative relative to the standard basis for $\mathcal{R}^2$. In order to find the matrix representative, relative to the $\mathcal{B}$ basis, we first need to compute

$$T(u_1) = Au_1 = \begin{bmatrix} 2 & 1 \\ -2 & 5 \end{bmatrix} \begin{bmatrix} 1 \\ 2 \end{bmatrix} = \begin{bmatrix} 4 \\ 8 \end{bmatrix}$$

and

$$T(u_2) = Au_2 = \begin{bmatrix} 2 & 1 \\ -2 & 5 \end{bmatrix} \begin{bmatrix} 1 \\ 1 \end{bmatrix} = \begin{bmatrix} 3 \\ 3 \end{bmatrix}.$$

The second step in finding $[\mathbf{T}]_{\mathcal{BB}}$ is to find the $\mathcal{B}$-coordinate matrices of $\mathbf{T}(\mathbf{u}_1)$ and $\mathbf{T}(\mathbf{u}_2)$. In order to find the $\mathcal{B}$-coordinate matrix of $\mathbf{T}(\mathbf{u}_1)$ we need to find scalars a and b so that

$$\mathbf{T}(\mathbf{u}_1) = \begin{bmatrix} 4 \\ 8 \end{bmatrix} = a\mathbf{u}_1 + b\mathbf{u}_2 = a\begin{bmatrix} 1 \\ 2 \end{bmatrix} + b\begin{bmatrix} 1 \\ 1 \end{bmatrix}.$$

The solution of this system is obvious; it is $a = 4$ and $b = 0$ so that $\mathbf{T}(\mathbf{u}_1) = 4\mathbf{u}_1 + 0\mathbf{u}_2 = 4\mathbf{u}_1$. Similarly, we see that $\mathbf{T}(\mathbf{u}_2) = 0\mathbf{u}_1 + 3\mathbf{u}_2 = 3\mathbf{u}_2$. Thus, $[\mathbf{T}(\mathbf{u}_1)]_{\mathcal{B}} = \begin{bmatrix} 4 \\ 0 \end{bmatrix}$ and $[\mathbf{T}(\mathbf{u}_2)]_{\mathcal{B}} = \begin{bmatrix} 0 \\ 3 \end{bmatrix}$. These coordinate matrices are the columns of the matrix that represents $\mathbf{T}$ with respect to the $\mathcal{B}$ basis. That matrix is $[\mathbf{T}]_{\mathcal{BB}} = \begin{bmatrix} 4 & 0 \\ 0 & 3 \end{bmatrix}$. Note that this is a much simpler matrix than the matrix A that was originally used to define the transformation.

We can use the factorization

$$\begin{bmatrix} 4 & 0 \\ 0 & 3 \end{bmatrix} = \begin{bmatrix} 4 & 0 \\ 0 & 1 \end{bmatrix}\begin{bmatrix} 1 & 0 \\ 0 & 3 \end{bmatrix}$$

to observe that $\mathbf{T}$ is the product of two expansions: by a factor of 3 in the $\mathbf{u}_2$ direction and by a factor of 4 in the $\mathbf{u}_1$ direction. Figure 4.15 shows the image of a 2×1 rectangle under this transformation. ■

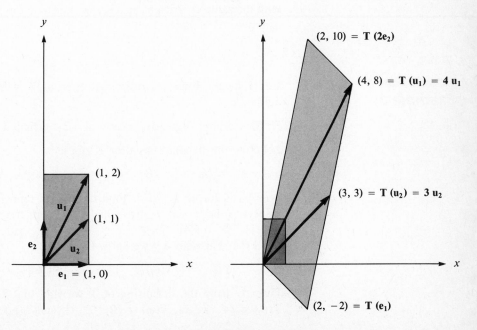

FIGURE 4.15

Example 2

Let $\mathbf{D}$ be the linear operator on $\mathcal{P}_3$ defined by

$$\mathbf{D}(a_0 + a_1x + a_2x^2 + a_3x^3) = a_1 + 2a_2x + 3a_3x^2.$$

Find the matrix representative of $\mathbf{D}$ with respect to the basis $\mathcal{B} = \{1, x, x^2, x^3\}$ of $\mathcal{P}_3$ and use this matrix to find $\mathbf{D}(5 - 7x + 11x^2 + 4x^3)$.

Direct calculation shows that $\mathbf{D}(1) = 0$, $\mathbf{D}(x) = 1$, $\mathbf{D}(x^2) = 2x$, and $\mathbf{D}(x^3) = 3x^2$, so that

$$[\mathbf{D}]_{\mathcal{B}\mathcal{B}} = [[0]_{\mathcal{B}} \quad [1]_{\mathcal{B}} \quad [2x]_{\mathcal{B}} \quad [3x^2]_{\mathcal{B}}] = \begin{bmatrix} 0 & 1 & 0 & 0 \\ 0 & 0 & 2 & 0 \\ 0 & 0 & 0 & 3 \\ 0 & 0 & 0 & 0 \end{bmatrix}.$$

To find $\mathbf{D}(p(x))$ for $p(x) = 5 - 7x + 11x^2 + 4x^3$, we first note that $[p(x)]_{\mathcal{B}} = [5 \quad -7 \quad 11 \quad 4]^T$ and then compute

$$[\mathbf{D}]_{\mathcal{B}}[p(x)]_{\mathcal{B}} = \begin{bmatrix} 0 & 1 & 0 & 0 \\ 0 & 0 & 2 & 0 \\ 0 & 0 & 0 & 3 \\ 0 & 0 & 0 & 0 \end{bmatrix} \begin{bmatrix} 5 \\ -7 \\ 11 \\ 4 \end{bmatrix} = \begin{bmatrix} -7 \\ 22 \\ 12 \\ 0 \end{bmatrix} = [\mathbf{D}(p(x))]_{\mathcal{B}},$$

from which it follows that $\mathbf{D}(p(x)) = -7 + 22x + 12x^2$. ■

The hardest step in finding the matrix representative of a linear transformation is finding the coordinates of the images of the basis vectors. This has been easy in the first two examples, but in general it requires the solution of a set of systems of linear equations, all with the same coefficient matrix. We illustrate this process in our next example.

Example 3

Let $\mathbf{T}$ be the linear transformation from $\mathcal{V} = \mathcal{R}^{1\times4}$ to $\mathcal{W} = \mathcal{R}^{1\times3}$ defined by

$$\mathbf{T}[x \quad y \quad z \quad w] = [3x + 4y - w \quad 2y + 5w \quad x + y + z + w].$$

Let $\mathcal{S}$ be the standard basis for $\mathcal{V}$ and let

$$\mathcal{B}' = \{\mathbf{u}_1 = [1 \quad 3 \quad 1], \mathbf{u}_2 = [1 \quad 1 \quad 0], \mathbf{u}_3 = [0 \quad 3 \quad 1]\}$$

be a basis for $\mathcal{W} = \mathcal{R}^{1\times3}$. Find the matrix representative of $\mathbf{T}$ with respect to the $\mathcal{S}$ and $\mathcal{B}'$ bases and use this matrix to compute $\mathbf{T}(\mathbf{v})$ if $\mathbf{v} = [1 \quad 2 \quad 3 \quad 4]$.

From Theorem 4.9 we know that

$$[\mathbf{T}]_{\mathcal{B}'\mathcal{S}} = [[\mathbf{T}(\mathbf{e}_1)]_{\mathcal{B}'} \quad [\mathbf{T}(\mathbf{e}_2)]_{\mathcal{B}'} \quad [\mathbf{T}(\mathbf{e}_3)]_{\mathcal{B}'} \quad [\mathbf{T}(\mathbf{e}_4)]_{\mathcal{B}'}].$$

Directly from the definition of $\mathbf{T}$ we find that $\mathbf{T}(\mathbf{e}_1) = [3 \quad 0 \quad 1]$, $\mathbf{T}(\mathbf{e}_2) = [4 \quad 2 \quad 1)$, $\mathbf{T}(\mathbf{e}_3) = [0 \quad 0 \quad 1]$, and $\mathbf{T}(\mathbf{e}_4) = [-1 \quad 5 \quad 1]$. In order to find $[\mathbf{T}(\mathbf{e}_1)]_{\mathscr{B}'}$ we need to find scalars x_i so that

$$\begin{aligned}[3 \quad 0 \quad 1] &= x_1\mathbf{u}_1 + x_2\mathbf{u}_2 + x_3\mathbf{u}_3 \\ &= x_1[1 \quad 3 \quad 1] + x_2[1 \quad 1 \quad 0] + x_3[0 \quad 3 \quad 1].\end{aligned}$$

This system is equivalent to the system

$$Ax = \begin{bmatrix} 1 & 1 & 0 \\ 3 & 1 & 3 \\ 1 & 0 & 1 \end{bmatrix}\begin{bmatrix} x_1 \\ x_2 \\ x_3 \end{bmatrix} = \begin{bmatrix} 3 \\ 0 \\ 1 \end{bmatrix} = \mathbf{T}(\mathbf{e}_1)^T.$$

To find the $\mathscr{B}'$-coordinates of the other $\mathbf{T}(\mathbf{e}_i)$ we need to solve the linear systems $AX = \mathbf{T}(\mathbf{e}_i)^T$; $i = 2, 3, 4$. These four systems can be solved by the single row reduction

$$\begin{bmatrix} 1 & 1 & 0 & 3 & 4 & 0 & -1 \\ 3 & 1 & 3 & 0 & 2 & 0 & 5 \\ 1 & 0 & 1 & 1 & 1 & 1 & 1 \end{bmatrix} \longrightarrow$$

$$\begin{bmatrix} 1 & 1 & 0 & 3 & 4 & 0 & -1 \\ 0 & -2 & 3 & -9 & -10 & 0 & 8 \\ 0 & -1 & 1 & -2 & -3 & 1 & 2 \end{bmatrix} \longrightarrow$$

$$\begin{bmatrix} 1 & 0 & 0 & 0 & 0 & 0 & 0 \\ 0 & 0 & 1 & -5 & -4 & -2 & 4 \\ 0 & 1 & -1 & 2 & 3 & -1 & -2 \end{bmatrix} \longrightarrow$$

$$\begin{bmatrix} 1 & 0 & 0 & 6 & 5 & 3 & -3 \\ 0 & 1 & 0 & -3 & -1 & -3 & 2 \\ 0 & 0 & 1 & -5 & -4 & -2 & 4 \end{bmatrix}.$$

From the above reduction it follows that $[\mathbf{T}(\mathbf{e}_1)]_{\mathscr{B}'} = [\,6 \quad -3 \quad -5\,]^T$, $[\mathbf{T}(\mathbf{e}_2)]_{\mathscr{B}'} = [\,5 \quad -1 \quad -4\,]^T$, and so forth, and that the matrix representative of $\mathbf{T}$ with respect to the given bases is

$$[\mathbf{T}]_{\mathscr{B}'\mathscr{S}} = \begin{bmatrix} 6 & 5 & 3 & -3 \\ -3 & -1 & -3 & 2 \\ -5 & -4 & -2 & 4 \end{bmatrix}.$$

In order to find $\mathbf{T}(\mathbf{v})$ we observe that $[\mathbf{v}]_{\mathscr{S}} = \mathbf{v}^T$ and compute

$$[\mathbf{T}(\mathbf{v})]_{\mathscr{S}} = [\mathbf{T}]_{\mathscr{B}'\mathscr{S}}[\mathbf{v}]_{\mathscr{S}} = \begin{bmatrix} 6 & 5 & 3 & -3 \\ -3 & -1 & -3 & 2 \\ -5 & -4 & -2 & 4 \end{bmatrix}\begin{bmatrix} 1 \\ 2 \\ 3 \\ 4 \end{bmatrix} = \begin{bmatrix} 13 \\ -6 \\ -3 \end{bmatrix}$$

Thus, $\mathbf{T}(\mathbf{v}) = 13\mathbf{u}_1 - 6\mathbf{u}_2 - 3\mathbf{u}_3$

$$\begin{aligned} &= 13[1 \quad 3 \quad 1] - 6[1 \quad 1 \quad 0] - 3[0 \quad 3 \quad 1] \\ &= [7 \quad 24 \quad 10], \end{aligned}$$

which agrees with the value of $\mathbf{T}(\mathbf{v})$ obtained by direct evaluation. ■

Exercise Set 4.5

1. Let $A = \begin{bmatrix} 3 & -2 & 1 & 0 \\ 1 & 6 & 2 & 1 \\ -3 & 0 & 7 & 1 \end{bmatrix}$ be the matrix representative of **T** in

   $\mathscr{L}(\mathscr{R}^4, \mathscr{R}^3)$ with respect to the bases $\mathscr{B} = \{v_1, v_2, v_3, v_4\}$ and
   $\mathscr{B}' = \{w_1, w_2, w_3\}$, where

   $$v_1 = \begin{bmatrix} 0 \\ 1 \\ 1 \\ 1 \end{bmatrix}, \quad v_2 = \begin{bmatrix} 2 \\ 1 \\ -1 \\ -1 \end{bmatrix}, \quad v_3 = \begin{bmatrix} 1 \\ 4 \\ -1 \\ 2 \end{bmatrix}, \quad v_4 = \begin{bmatrix} 6 \\ 9 \\ 4 \\ 2 \end{bmatrix},$$

   $$w_1 = \begin{bmatrix} 0 \\ 8 \\ 8 \end{bmatrix}, \quad w_2 = \begin{bmatrix} -7 \\ 8 \\ 1 \end{bmatrix}, \quad \text{and} \quad w_3 = \begin{bmatrix} -6 \\ 9 \\ 1 \end{bmatrix},$$

   (a) find $[T(v_1)]_{\mathscr{B}'}$, $[T(v_2)]_{\mathscr{B}'}$, $[T(v_3)]_{\mathscr{B}'}$, and $[T(v_4)]_{\mathscr{B}'}$;
   (b) find $T(v_1)$, $T(v_2)$, $T(v_3)$, and $T(v_4)$; and
   (c) find $T \begin{bmatrix} 2 \\ 2 \\ 0 \\ 0 \end{bmatrix}$.

2. Let **T** be the linear transformation from $\mathscr{R}^{1 \times 4}$ to $\mathscr{R}^{1 \times 3}$ defined by
   $T[x_1 \quad x_2 \quad x_3 \quad x_4] = [x_1 + 2x_2 - x_3 \quad 2x_1 - 5x_2 \quad 7x_1 - 3x_2]$. Find
   the matrix representative of **T** relative to the standard bases of $\mathscr{R}^{1 \times 4}$
   and $\mathscr{R}^{1 \times 3}$.

3. Let **U** be the linear transformation from $\mathscr{R}^3$ to $\mathscr{R}^{1 \times 5}$ defined by
   $U[x_1 \quad x_2 \quad x_3]^T = [x_1 + x_2 \quad 0 \quad x_2 + x_3 \quad 0 \quad x_3 + x_1]$. Consider
   $u_1 = [1 \quad 1 \quad 1]^T$, $u_2 = [1 \quad 1 \quad 0]^T$, and $u_3 = [0 \quad 1 \quad 1]^T$ as the basis
   for $\mathscr{R}^3$ and the standard basis for $\mathscr{R}^{1 \times 5}$. Find the matrix representative
   of **U** with respect to these bases.

4. Define a linear operator **T** on $\mathscr{R}^{1 \times 4}$ by
   $$T[x_1 \quad x_2 \quad x_3 \quad x_4] = [x_1 + 2x_2 \quad x_1 - 3x_3 \quad x_1 + 4x_4 \quad x_2 + x_3].$$
   Find the matrix representative of **T** with respect to the basis
   $$\mathscr{B} = \{w_1 = [1 \quad 2 \quad 0 \quad 3], w_2 = [0 \quad 1 \quad -1 \quad 0],$$
   $$w_3 = [0 \quad 0 \quad 1 \quad 2], w_4 = [0 \quad 0 \quad 0 \quad 1]\}.$$

5. Let **D** be the linear operator on $\mathscr{P}_3$ considered in Example 2. Find the
   matrix representative of **D** with respect to the basis
   $$\mathscr{B} = \{-3 + x^2, -3 + x, 7 - x - x^2, x^3\}.$$

6. Let **T** be a linear operator on $\mathscr{R}^{1 \times 3}$ which is represented, relative to the
   standard basis, by
   $$A = \begin{bmatrix} 7 & -8 & -8 \\ 9 & -16 & -18 \\ -5 & 11 & 13 \end{bmatrix}.$$

The vectors $\mathbf{u_1} = [0 \quad -1 \quad 1]$, $\mathbf{u_2} = [1 \quad 3 \quad -2]$, and $\mathbf{u_3} = [2 \quad 0 \quad 1]$ form a basis for $\mathcal{R}^{1\times 3}$. Find the matrix representative of $\mathbf{T}$ with respect to this basis.

7. Let $\mathbf{I}$ in $\mathcal{L}(\mathcal{V}, \mathcal{V})$ be the identity operator on $\mathcal{V}$ defined by $\mathbf{I}(\mathbf{u}) = \mathbf{u}$, for all $\mathbf{u}$ in $\mathcal{V}$. Show that the matrix representative of $\mathbf{I}$ is the identity matrix I, no matter which basis is chosen for $\mathcal{V}$.

8. Let $\mathbf{T}$ be a linear operator on $\mathcal{R}^{1\times 3}$ that is represented, relative to

   the standard basis, by $A = \begin{bmatrix} 2 & 1 & 1 \\ 2 & 3 & 2 \\ 1 & 1 & 2 \end{bmatrix}$. Find $[\mathbf{T}]_{\mathcal{B}\mathcal{B}}$ where

   $\mathcal{B} = \{\mathbf{u_1} = [1 \quad -2 \quad 2], \mathbf{u_2} = [-2 \quad 1 \quad 2], \mathbf{u_3} = [2 \quad 2 \quad 1]\}$.

9. Show that the $\mathcal{B}$ to $\mathcal{B}'$ change-of-basis matrix in $\mathcal{R}^n$ is a matrix representative of the identity operator $\mathbf{I}$.

10. Let $\mathbf{T}$ be a linear operator on $\mathcal{V}$. Call a subspace $\mathcal{W}$ of $\mathcal{V}$ $\mathbf{T}$-INVARIANT if $\mathbf{T}(u)$ is in $\mathcal{W}$ for every $\mathbf{u}$ in $\mathcal{W}$. Show that $\mathcal{NS}(\mathbf{T})$ and $\mathcal{E} = \{\mathbf{u} \mid \mathbf{T}(\mathbf{u}) = 2\mathbf{u}\}$ are $\mathbf{T}$-invariant subspaces of $\mathcal{V}$.

11. Find the matrix representative of the transposition operator $\mathbf{T}$ $(\mathbf{T}(A) = A^T)$ on $\mathcal{R}^{2\times 2}$ relative to the basis

$$\mathcal{S} = \left\{ \begin{bmatrix} 1 & 0 \\ 0 & 0 \end{bmatrix}, \begin{bmatrix} 0 & 1 \\ 0 & 0 \end{bmatrix}, \begin{bmatrix} 0 & 0 \\ 1 & 0 \end{bmatrix}, \begin{bmatrix} 0 & 0 \\ 0 & 1 \end{bmatrix} \right\} \text{ for } \mathcal{R}^{2\times 2}.$$

12. Let $\mathcal{B}_1$ be a basis for the vector space $\mathcal{V}_1$, $\mathcal{B}_2$ a basis for the vector space $\mathcal{V}_2$ and $\mathcal{B}_3$ a basis for the vector space $\mathcal{V}_3$. Consider $\mathbf{T}_1$ in $\mathcal{L}(\mathcal{V}_1, \mathcal{V}_2)$ and $\mathbf{T}_2$ in $\mathcal{L}(\mathcal{V}_2, \mathcal{V}_3)$ so that $\mathbf{T}_2\mathbf{T}_1$ is in $\mathcal{L}(\mathcal{V}_1, \mathcal{V}_3)$. Show that
$$[\mathbf{T}_2\mathbf{T}_1]_{\mathcal{B}_3\mathcal{B}_1} = [\mathbf{T}_2]_{\mathcal{B}_3\mathcal{B}_2}[\mathbf{T}_1]_{\mathcal{B}_2\mathcal{B}_1}.$$

13. Let $\mathbf{T}$ be the matrix transformation on $\mathcal{R}^3$ whose standard matrix is

$$C = \begin{bmatrix} 0 & 1 & 0 \\ 0 & 0 & 1 \\ 15 & -23 & 9 \end{bmatrix}.$$

Find the matrix representative of $\mathbf{T}$ relative to the basis

$$\mathcal{B} = \left\{ \mathbf{u_1} = \begin{bmatrix} 1 \\ 5 \\ 25 \end{bmatrix}, \mathbf{u_2} = \begin{bmatrix} 1 \\ 3 \\ 9 \end{bmatrix}, \mathbf{u_3} = \begin{bmatrix} 1 \\ 1 \\ 1 \end{bmatrix} \right\}.$$

14. Explain why $[\mathbf{T}]_{\mathcal{B}'\mathcal{B}}$ is a better notation for the matrix representative than $[\mathbf{T}]_{\mathcal{B}\mathcal{B}'}$.

15. Let $\mathcal{V}$ and $\mathcal{W}$ be finite dimensional vector spaces with bases $\mathcal{B}$ and $\mathcal{B}'$ respectively; let $\dim(\mathcal{V}) = n$ and $\dim(\mathcal{W}) = m$. Consider the mapping which sends a linear transformation $\mathbf{T}$ to its matrix representative $[\mathbf{T}]_{\mathcal{B}'\mathcal{B}}$. Show that this mapping is a one-to-one linear transformation of $\mathcal{L}(\mathcal{V}, \mathcal{W})$ onto $\mathcal{R}^{m\times n}$.

Chapter 4

Review Exercises

1. Let $\mathbf{T}$ be a linear transformation from $\mathcal{V}$ to $\mathcal{W}$. Complete the following from the list on the right.

   (a) Domain of $\mathbf{T}$ = _____   **1.** $\mathcal{W}$ **2.** $\mathcal{V}$ **3.** $\dim(\mathcal{V})$ **4.** $\mathcal{RS}(A)$
   (b) $\mathcal{NS}(\mathbf{T})$ = _____   **5.** $\mathcal{RS}(\mathbf{T})$ **6.** $\{\mathbf{x} \mid \mathbf{T}(\mathbf{x}) = 2\mathbf{x}\}$
   (c) Range of $\mathbf{T}$ = _____   **7.** $\{\mathbf{T}(\mathbf{u}) \mid \mathbf{u} \text{ in } \mathcal{V}\}$ **8.** $\{\mathbf{u} \mid \mathbf{T}(\mathbf{u}) = \mathbf{0}\}$
   (d) $\text{rank}(\mathbf{T})$ = _____   **9.** $\dim(\mathcal{NS}(\mathbf{T}))$ **10.** $\dim(\mathcal{R}(\mathbf{T}))$
   (e) $\text{nullity}(\mathbf{T})$ = _____   **11.** $\dim\mathcal{V} - \text{rank}\mathbf{T}$ **12.** $\dim\mathcal{V} + \text{rank}\mathbf{T}$

2. Consider the function $\mathbf{T}$ from $\mathcal{R}^3$ to $\mathcal{R}^{2\times 2}$ defined by

$$\mathbf{T}\begin{bmatrix} x \\ y \\ z \end{bmatrix} = \begin{bmatrix} x + y & 3 + z \\ 0 & 2x + 3z \end{bmatrix}.$$

   Determine if $\mathbf{T}$ is linear. Explain your reasoning.

3. Given $\mathbf{T}$, a function from $\mathcal{R}^{2\times 2}$ to $\mathcal{R}$ defined by $\mathbf{T}(A) = \det(A^T)$, determine whether $\mathbf{T}$ is a linear transformation. Explain carefully.

4. For $\mathbf{T}$ in $\mathcal{L}(\mathcal{V}, \mathcal{W})$, prove that $\mathcal{NS}(\mathbf{T})$ is a subspace of $\mathcal{V}$.

5. Is the transformation $\mathbf{T}\begin{bmatrix} x_1 \\ x_2 \\ x_3 \end{bmatrix} = \begin{bmatrix} 1 + x_1 \\ x_1 x_2 \\ x_1 + x_2 + x_3{}^3 \end{bmatrix}$ linear?

6. Let $\mathbf{T}$ be a linear transformation from $\mathcal{R}^2$ to $\mathcal{R}^{1\times 3}$ such that

   $\mathbf{T}(\mathbf{e}_1) = \begin{bmatrix} 3 & 5 & 1 \end{bmatrix}$ and $\mathbf{T}(\mathbf{e}_2) = \begin{bmatrix} -1 & 4 & 3 \end{bmatrix}$. Find $\mathbf{T}(7\mathbf{e}_1 - 2\mathbf{e}_2)$, $\mathbf{T}\begin{bmatrix} 4 \\ 2 \end{bmatrix}$,

   and $\mathbf{T}\begin{bmatrix} 2 \\ -5 \end{bmatrix}$.

7. For the linear operator $\mathbf{Q}$ on $\mathcal{P}_3$ defined by

$$\mathbf{Q}(a + bx + cx^2 + dx^3) = b + cx^2,$$

   find $\mathcal{NS}(\mathbf{Q})$, $\mathcal{R}(\mathbf{Q})$, $\text{rank}(\mathbf{Q})$, and $\text{nullity}(\mathbf{Q})$.

8. Let $\mathbf{T}$ in $\mathcal{L}(\mathcal{R}^4, \mathcal{R}^3)$ be defined by $\mathbf{T}(\mathbf{u}) = \begin{bmatrix} 1 & 2 & 3 & 1 \\ 4 & 5 & 6 & 1 \\ 7 & 8 & 9 & 1 \end{bmatrix}\mathbf{u}$.

   (a) Find a basis for $\mathcal{NS}(\mathbf{T})$.
   (b) Find the rank of $\mathbf{T}$.
   (c) Find a basis for the range of $\mathbf{T}$.

9. Let $\mathbf{T}$ be a linear operator on $\mathcal{R}^2$ which is described geometrically by the following sequence of steps. Find the standard matrix of $\mathbf{T}$.

   (1) Compression by a factor of $1/3$ in the x-direction
   (2) A shear, by a factor of -3, in the x-direction
   (3) A shear, by a factor of 2, in the y-direction
   (4) A reflection in the line $y = x$

10. Let **T** be a linear operator on $\mathfrak{R}^2$ which is described geometrically by the following sequence of steps. Find the standard matrix of **T**.

    (1) An expansion by a factor of 3 in the y-direction
    (2) A shear by a factor of 2 in the y-direction
    (3) A reflection in the line $y = x$
    (4) A rotation through an angle of 45 degrees

5

Similar Matrices and the Eigenvalue Problem

In this chapter we will introduce and study the relationship of similarity between matrices. We will show how this important relationship arises naturally in the study of the matrix representatives of a linear operator and provides geometric motivation for the further study of eigenvalues and eigenvectors. Some of the special topics we will consider include the eigenvalue problem for symmetric matrices, the special case of orthogonal similarity, and similarity to a triangular matrix. The chapter concludes with a discussion of the power method for computing eigenvalues and eigenvectors on a computer.

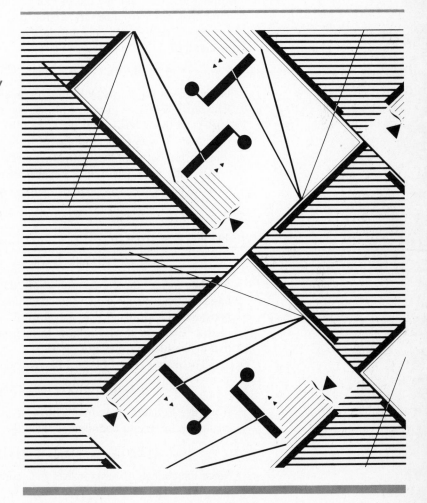

5.1

The Change-of-Basis Problem

In Chapter 4 you had several opportunities to observe that the matrix representative of a linear operator $\mathbf{T}$ in $\mathscr{L}(\mathcal{V}, \mathcal{V})$ depends on the choice of the basis for the vector space $\mathcal{V}$. In general, we have seen that changing the basis of the vector space changes the matrix representative; it would be naive to expect otherwise. In this section we will investigate the relationship between the various matrix representatives of the linear operator $\mathbf{T}$. This investigation will lead us to the study of similar matrices and to a closer look at eigenvalues and eigenvectors, a topic that was introduced in Section 2.4.

Let $\mathbf{T}$ be a linear operator on the finite dimensional vector space $\mathcal{V}$ and let

$$\mathscr{B} = \{\mathbf{u}_1, \mathbf{u}_2, \ldots, \mathbf{u}_n\} \quad \text{and} \quad \mathscr{B}' = \{\mathbf{v}_1, \mathbf{v}_2, \ldots, \mathbf{v}_n\}$$

be two bases for $\mathcal{V}$; we will think of $\mathscr{B}$ as the *old basis* and $\mathscr{B}'$ as the *new basis*. Let $A = [\mathbf{T}]_{\mathscr{B}} = [\mathbf{T}]_{old}$ be the matrix that represents $\mathbf{T}$ with respect to the B basis, and let $A' = [\mathbf{T}]_{\mathscr{B}'} = [\mathbf{T}]_{new}$ be the matrix that represents $\mathbf{T}$ with respect to the $\mathscr{B}'$ basis. We will now investigate the relationship between the matrices A and A'.

Recall, from Theorem 3.22, that there is a unique $\mathscr{B}'$ to $\mathscr{B}$ (new to old) change-of-basis matrix P satisfying

$$P[\mathbf{w}]_{\mathscr{B}'} = [\mathbf{w}]_{\mathscr{B}} \text{ for any vector } \mathbf{w} \text{ in } \mathcal{V}, \tag{5.1}$$

and that P is given by

$$P = \left[[\mathbf{v}_1]_{\mathscr{B}} \quad [\mathbf{v}_2]_{\mathscr{B}} \quad \cdots \quad [\mathbf{v}_n]_{\mathscr{B}} \right].$$

In other words, the columns of P are the old coordinates of the new basis vectors. We have also shown that the change-of-basis matrix P is nonsingular and that P^{-1} is the $\mathscr{B}$ to $\mathscr{B}'$ (old to new) change-of-basis matrix. Exercise 15 shows how to interpret P as a matrix representative of the identity operator on $\mathcal{V}$.

The matrix $A = [\mathbf{T}]_{\mathscr{B}} = [\mathbf{T}]_{old}$ satisfies the relation

$$[\mathbf{T}(\mathbf{x})]_{\mathscr{B}} = A[\mathbf{x}]_{\mathscr{B}} \text{ for all vectors } \mathbf{x} \text{ in } \mathcal{V}.$$

If we use Equation 5.1, first with $\mathbf{w} = \mathbf{x}$ and then with $\mathbf{w} = \mathbf{T}(\mathbf{x})$, to change to $\mathscr{B}'$ (new) coordinates, then the last equation becomes

$$P[\mathbf{T}(\mathbf{x})]_{\mathscr{B}'} = [\mathbf{T}(\mathbf{x})]_{\mathscr{B}} = A[\mathbf{x}]_{\mathscr{B}} = A(P[\mathbf{x}]_{\mathscr{B}'}).$$

Since the change-of-basis matrix P is nonsingular, it follows that

$$[\mathbf{T}(\mathbf{x})]_{\mathscr{B}'} = (P^{-1}AP)[\mathbf{x}]_{\mathscr{B}'}.$$

This last relation holds for all $\mathbf{x}$ in $\mathcal{V}$, so $P^{-1}AP = A' = [\mathbf{T}]_{new} = [\mathbf{T}]_{\mathscr{B}'}$ is the matrix representative of $\mathbf{T}$ with respect to the $\mathscr{B}'$ basis. This discussion is summarized in the following theorem.

Theorem 5.1

Let **T** be a linear operator on the finite dimensional vector space $\mathcal{V}$, whose matrix representative with respect to the $\mathcal{B}$ basis is A and whose matrix representative with respect to the $\mathcal{B}'$ basis is A'. If P is the $\mathcal{B}'$ to $\mathcal{B}$ change-of-basis matrix, then

$$A' = P^{-1}AP \quad \text{and} \quad A = PA'P^{-1}. \tag{5.2}$$

Theorem 5.1 provides motivation for the study of the important matrix relation introduced in the next definition.

Definition 5.1

The $n \times n$ matrices A and B are said to be SIMILAR ($A \approx B$) if there exists a nonsingular matrix Q such that $Q^{-1}AQ = B$.

We can now rephrase Theorem 5.1 to say that the different matrix representatives of a linear operator are similar matrices. The converse is also true, so we have the following result that provides geometric motivation for the study of similarity.

Theorem 5.2

Two $n \times n$ matrices are similar if and only if they represent the same linear operator relative to different bases for some n-dimensional vector space $\mathcal{V}$.

Note that Theorem 5.2 does not specify the space $\mathcal{V}$. If the matrices are $n \times n$, then any n-dimensional space will do; in particular we could take $\mathcal{V} = \mathcal{R}^n$.

Similarity, although a less restrictive relation than equality, shares three important properties with the relations of equality and of row equivalence: similarity is *reflexive* ($A \approx A$), *symmetric* (if $A \approx B$, then $B \approx A$), and *transitive* (if $A \approx B$ and $B \approx C$, then $A \approx C$) (see Exercise 10 of this section).

It is easy to become confused when applying Theorems 5.1 or 5.2; is P the $\mathcal{B}$ to $\mathcal{B}'$ change-of-basis matrix or the $\mathcal{B}'$ to $\mathcal{B}$ change-of-basis matrix? The key to keeping it all straight is to remember Theorem 5.2, note that the *rightmost matrix is the first to have its effect felt*, and then decide what that effect must be. For example,

$[\mathbf{T(x)}]_{\mathcal{B}'} = P^{-1}AP[\mathbf{x}]_{\mathcal{B}'}$, where

$$P[\mathbf{x}]_{\mathcal{B}'} = [\mathbf{x}]_{\mathcal{B}} \qquad (\mathcal{B}' \text{ to } \mathcal{B} \text{ coordinate change});$$

$$A[\mathbf{x}]_{\mathcal{B}} = [\mathbf{T(x)}]_{\mathcal{B}} \qquad (A \text{ represents } \mathbf{T} - \mathcal{B} \text{ basis}); \text{ and}$$

$$P^{-1}[\mathbf{T(x)}]_{\mathcal{B}} = [\mathbf{T(x)}]_{\mathcal{B}} \qquad (\mathcal{B} \text{ to } \mathcal{B}' \text{ coordinate change}).$$

It may also be helpful to write the first equation of (5.2) as

$$A' \qquad = \qquad P^{-1} \qquad A \qquad P$$
[new matrix of **T**] = [old to new] [old matrix of **T**] [new to old].
$\qquad$ [$\mathcal{B}'$ basis] $\qquad$ [$\mathcal{B}$ to $\mathcal{B}'$] $\qquad$ [$\mathcal{B}$ basis] $\qquad$ [$\mathcal{B}'$ to $\mathcal{B}$]

Remember to read the last line from right to left.

Example 1

Let **T** be the linear operator on $\mathcal{R}^2$ defined by

$$\mathbf{T(u)} = A\mathbf{u} = \begin{bmatrix} 1 & 2 \\ 3 & 4 \end{bmatrix} \begin{bmatrix} u_1 \\ u_2 \end{bmatrix}.$$

The matrix representative of **T** with respect to the standard basis of $\mathcal{R}^2$ is $A = [\mathbf{T}]_{\mathcal{S}}$. Use Theorem 5.1 to find $A' = [\mathbf{T}]_{\mathcal{B}'}$, which is the matrix representative of **T** with respect to the basis

$$\mathcal{B}' = \left\{ \mathbf{v}_1 = \begin{bmatrix} 1 \\ 2 \end{bmatrix}, \mathbf{v}_2 = \begin{bmatrix} 2 \\ 3 \end{bmatrix} \right\}.$$

From Theorem 5.1 we know that A and A' are similar matrices:

$$A' = [\mathbf{T}]_{\mathcal{B}'} = P^{-1}[\mathbf{T}]_{\mathcal{S}}P = P^{-1}AP,$$

where P must be the $\mathcal{B}'$ to $\mathcal{S}$ (new to old) change-of-basis matrix. Thus, P must satisfy

$$P[\mathbf{x}]_{\mathcal{B}'} = [\mathbf{x}]_{\mathcal{S}} \text{ for all } \mathbf{x} \text{ in } \mathcal{R}^2.$$

From Equation 5.1 we have, since $[\mathbf{x}]_{\mathcal{S}} = \mathbf{x}$ for all $\mathbf{x}$ in $\mathcal{R}^2$,

$$P = \begin{bmatrix} [\mathbf{v}_1]_{\mathcal{S}} & [\mathbf{v}_2]_{\mathcal{S}} \end{bmatrix} = \begin{bmatrix} 1 & 2 \\ 2 & 3 \end{bmatrix}.$$

$$\text{Thus, } A' = \begin{bmatrix} 1 & 2 \\ 2 & 3 \end{bmatrix}^{-1} \begin{bmatrix} 1 & 2 \\ 3 & 4 \end{bmatrix} \begin{bmatrix} 1 & 2 \\ 2 & 3 \end{bmatrix} = \begin{bmatrix} -3 & 2 \\ 2 & -1 \end{bmatrix} \begin{bmatrix} 5 & 8 \\ 11 & 18 \end{bmatrix}$$

$$= \begin{bmatrix} 7 & 12 \\ -1 & -2 \end{bmatrix}$$

is the matrix representative of **T** with respect to the $\mathcal{B}'$ basis. From this last matrix we see that we must have

$$\mathbf{T(v_1)} = 7\mathbf{v}_1 - \mathbf{v}_2 \quad \text{and} \quad \mathbf{T(v_2)} = 12\mathbf{v}_1 - 2\mathbf{v}_2.$$

These last relations can be checked directly; for $\mathbf{v}_1$ we have

$$\mathbf{T(v_1)} = \begin{bmatrix} 1 & 2 \\ 3 & 4 \end{bmatrix} \begin{bmatrix} 1 \\ 2 \end{bmatrix} = \begin{bmatrix} 5 \\ 11 \end{bmatrix} = 7 \begin{bmatrix} 1 \\ 2 \end{bmatrix} - \begin{bmatrix} 2 \\ 3 \end{bmatrix} = 7\mathbf{v}_1 - \mathbf{v}_2. \quad \blacksquare$$

In general, finding the change-of-basis matrix of Theorem 5.1 involves solving a set of linear systems. The solution of these systems is easy if the old basis is the standard basis as in Example 1. Our next example illustrates the more general case.

Example 2

Let **T** be the linear operator on $\mathcal{R}^3$ whose matrix representative with respect to the old basis,

$$\mathcal{B} = \left\{ \mathbf{u}_1 = \begin{bmatrix} 1 \\ 1 \\ 1 \end{bmatrix}, \mathbf{u}_2 = \begin{bmatrix} 1 \\ 0 \\ 1 \end{bmatrix}, \mathbf{u}_3 = \begin{bmatrix} 1 \\ 1 \\ 0 \end{bmatrix} \right\},$$

is $A = [\mathbf{T}]_{\mathscr{B}} = \begin{bmatrix} 11 & -21 & 6 \\ 3 & -5 & 2 \\ -2 & 5 & 1 \end{bmatrix}$.

Find the matrix representative of $\mathbf{T}$ with respect to the new basis

$$\mathscr{B}' = \left\{ \mathbf{v_1} = \begin{bmatrix} -3 \\ -2 \\ -4 \end{bmatrix}, \mathbf{v_2} = \begin{bmatrix} 6 \\ 4 \\ 7 \end{bmatrix}, \mathbf{v_3} = \begin{bmatrix} 7 \\ 5 \\ 8 \end{bmatrix} \right\}.$$

From Theorem 5.2 we know that $[\mathbf{T}]_{\mathscr{B}'}$ is similar to A; thus there exists a nonsingular matrix Q such that

$$[\mathbf{T}]_{\mathscr{B}'} = Q^{-1}[\mathbf{T}]_{\mathscr{B}}Q = Q^{-1}AQ.$$

It follows that Q is the $\mathscr{B}'$ to $\mathscr{B}$ (new to old) change-of-basis matrix

$$Q = \begin{bmatrix} [\mathbf{v_1}]_{\mathscr{B}} & [\mathbf{v_2}]_{\mathscr{B}} & [\mathbf{v_3}]_{\mathscr{B}} \end{bmatrix}.$$

In order to find the matrix Q we need to find the $\mathscr{B}$-coordinates of the $\mathbf{v_i}$. The three vector equations

$$x_1\mathbf{u_1} + x_2\mathbf{u_2} + x_3\mathbf{u_3} = \mathbf{v_i}; \quad i = 1, 2, 3$$

lead easily to the three systems

$$\begin{bmatrix} 1 & 1 & 1 \\ 1 & 0 & 1 \\ 1 & 1 & 0 \end{bmatrix}\begin{bmatrix} x_1 \\ x_2 \\ x_3 \end{bmatrix} = \begin{bmatrix} -3 \\ -2 \\ -4 \end{bmatrix}, \begin{bmatrix} 6 \\ 4 \\ 7 \end{bmatrix}, \begin{bmatrix} 7 \\ 5 \\ 8 \end{bmatrix}.$$

We can solve these three systems by the single row reduction

$$\begin{bmatrix} 1 & 1 & 1 & -3 & 6 & 7 \\ 1 & 0 & 1 & -2 & 4 & 5 \\ 1 & 1 & 0 & -4 & 7 & 8 \end{bmatrix} \longrightarrow$$

$$\begin{bmatrix} 1 & 1 & 1 & -3 & 6 & 7 \\ 0 & -1 & 0 & 1 & -2 & -2 \\ 0 & 0 & -1 & -1 & 1 & 1 \end{bmatrix} \longrightarrow$$

$$\begin{bmatrix} 1 & 0 & 0 & -3 & 5 & 6 \\ 0 & 1 & 0 & -1 & 2 & 2 \\ 0 & 0 & 1 & 1 & -1 & -1 \end{bmatrix}.$$

It follows that $\mathbf{v_1} = -3\mathbf{u_1} - \mathbf{u_2} + \mathbf{u_3}$, $\mathbf{v_2} = 5\mathbf{u_1} + 2\mathbf{u_2} - \mathbf{u_3}$, and that $\mathbf{v_3} = 6\mathbf{u_1} + 2\mathbf{u_2} - \mathbf{u_3}$.

Thus, the $\mathscr{B}'$ to $\mathscr{B}$ change-of-basis matrix is $Q = \begin{bmatrix} -3 & 5 & 6 \\ -1 & 2 & 2 \\ 1 & -1 & -1 \end{bmatrix}$.

In Section 1.7 we saw that the most efficient way to compute $Q^{-1}AQ$ was to compute AQ and then use the row reduction

$$[Q \,|\, AQ] \longrightarrow [I \,|\, Q^{-1}AQ].$$

In this case $AQ = \begin{bmatrix} -6 & 7 & 18 \\ -2 & 3 & 6 \\ 2 & -1 & -3 \end{bmatrix}$ and the row reduction

$$\begin{bmatrix} -3 & 5 & 6 & -6 & 7 & 18 \\ -1 & 2 & 2 & -2 & 3 & 6 \\ 1 & -1 & -1 & 2 & -1 & -3 \end{bmatrix} \longrightarrow$$

$$\begin{bmatrix} 0 & 2 & 3 & 0 & 4 & 9 \\ 0 & 1 & 1 & 0 & 2 & 3 \\ 1 & -1 & -1 & 2 & -1 & -3 \end{bmatrix} \longrightarrow$$

$$\begin{bmatrix} 0 & 0 & 1 & 0 & 0 & 3 \\ 0 & 1 & 1 & 0 & 2 & 3 \\ 1 & 0 & 0 & 2 & 1 & 0 \end{bmatrix} \longrightarrow \begin{bmatrix} 1 & 0 & 0 & 2 & 1 & 0 \\ 0 & 1 & 0 & 0 & 2 & 0 \\ 0 & 0 & 1 & 0 & 0 & 3 \end{bmatrix}$$

shows that $[\mathbf{T}]_{\mathscr{B}'} = Q^{-1}AQ = \begin{bmatrix} 2 & 1 & 0 \\ 0 & 2 & 0 \\ 0 & 0 & 3 \end{bmatrix} = [\mathbf{T}]_{new}$.

From the last matrix we see that

$$\mathbf{T}(\mathbf{v}_1) = 2\mathbf{v}_1, \quad \mathbf{T}(\mathbf{v}_2) = \mathbf{v}_1 + 2\mathbf{v}_2, \quad \text{and} \quad \mathbf{T}(\mathbf{v}_3) = 3\mathbf{v}_3.$$

It follows, as in Section 4.4, that $\mathbf{T}$ can be viewed geometrically as the product of expansions by a factor of 2 in the $\mathbf{v}_1$ and $\mathbf{v}_2$ directions, a shear in the $\mathbf{v}_1$ direction of the $\mathbf{v}_1\mathbf{v}_2$ plane with factor $1/2$, and an expansion by a factor of 3 in the $\mathbf{v}_3$ direction.   ■

Example 3

Let $\mathbf{T}$ be the linear operator on $\mathscr{R}^{1 \times 3}$ whose matrix representative, relative to the standard basis, is

$$A = [\mathbf{T}]_{\mathscr{S}} = \begin{bmatrix} 17 & 12 & 18 \\ -16 & -9 & -24 \\ -5 & -4 & -4 \end{bmatrix}.$$

Find the matrix representative of $\mathbf{T}$ with respect to the basis

$$\mathscr{E} = \{\mathbf{x}_1 = [10 \quad -8 \quad -3], \mathbf{x}_2 = [-3 \quad 3 \quad 1], \mathbf{x}_3 = [-3 \quad 2 \quad 1]\}.$$

From Equation 5.2 we have

$$C = [\mathbf{T}]_{\mathscr{E}} = P^{-1}[\mathbf{T}]_{\mathscr{S}}P = P^{-1}AP,$$

where P is the $\mathscr{E}$ to $\mathscr{S}$ change-of-basis matrix given by

$$P = [[\mathbf{x}_1]_{\mathscr{S}} \quad [\mathbf{x}_2]_{\mathscr{S}} \quad [\mathbf{x}_3]_{\mathscr{S}}] = \begin{bmatrix} 10 & -3 & -3 \\ -8 & 3 & 2 \\ -3 & 1 & 1 \end{bmatrix}.$$

It follows that

$$
C = \begin{bmatrix} 10 & -3 & -3 \\ -8 & 3 & 2 \\ -3 & 1 & 1 \end{bmatrix}^{-1} \begin{bmatrix} 17 & 12 & 18 \\ -16 & -9 & -24 \\ -5 & -4 & -4 \end{bmatrix} \begin{bmatrix} 10 & -3 & -3 \\ -8 & 3 & 2 \\ -3 & 1 & 1 \end{bmatrix}
$$

$$
= \begin{bmatrix} 10 & -3 & -3 \\ -8 & 3 & 2 \\ -3 & 1 & 1 \end{bmatrix}^{-1} \begin{bmatrix} 20 & 3 & -9 \\ -16 & -3 & 6 \\ -6 & -1 & 3 \end{bmatrix} = \begin{bmatrix} 2 & 0 & 0 \\ 0 & -1 & 0 \\ 0 & 0 & 3 \end{bmatrix}
$$

and we have

$$
\mathbf{T(x_1)} = 2\mathbf{x_1}, \quad \mathbf{T(x_2)} = -\mathbf{x_2}, \quad \text{and} \quad \mathbf{T(x_3)} = 3\mathbf{x_3}. \quad \blacksquare
$$

It would certainly be easier to study the linear operator **T** of Example 3 relative to the $\mathscr{E}$ basis rather than relative to the standard basis. In particular we could describe **T** as the product of two expansions (by a factor of 2 in the $\mathbf{x_1}$-direction and by a factor of 3 in the $\mathbf{x_3}$-direction) and a reflection in the $\mathbf{x_1 x_3}$-plane.

The last example suggests some questions about linear operators.

Is it always possible to find a basis for $\mathscr{V}$ relative to which a given linear operator is represented by a diagonal matrix? Equivalently, can we find, for any square matrix A, a nonsingular matrix P such that $P^{-1}AP$ is a diagonal matrix? If so, how? If not, when? If not, how close can we come to a diagonal representative?

Exercise Set 5.1

1. Let $\mathscr{B} = \{\mathbf{u_1}, \mathbf{u_2}, \mathbf{u_3}, \mathbf{u_4}\}$ and $\mathscr{B}' = \{\mathbf{v_1}, \mathbf{v_2}, \mathbf{v_3}, \mathbf{v_4}\}$ be two bases for a four-dimensional vector space $\mathscr{V}$. Define a linear operator **T** by

$$
\mathbf{T(u_1)} = 2\mathbf{u_1} - 3\mathbf{u_2} + \mathbf{u_3} \qquad \mathbf{T(u_2)} = 4\mathbf{u_1} - 5\mathbf{u_4}
$$
$$
\mathbf{T(u_3)} = \mathbf{u_1} + 4\mathbf{u_3} \qquad \qquad \mathbf{T(u_4)} = 5\mathbf{u_1} + \mathbf{u_2} - \mathbf{u_4}.
$$

   (a) Find $[\mathbf{T}]_{\mathscr{B}}$.

   (b) If $\mathbf{v_1} = \mathbf{u_1} - 2\mathbf{u_2} + \mathbf{u_3} + \mathbf{u_4}$, $\mathbf{v_2} = \mathbf{u_2} + \mathbf{u_3} - 3\mathbf{u_4}$, $\mathbf{v_3} = \mathbf{u_3} + 5\mathbf{u_4}$, and $\mathbf{v_4} = \mathbf{u_4}$, find P so that $P[\mathbf{w}]_{\mathscr{B}'} = [\mathbf{w}]_{\mathscr{B}}$ for all **w** in $\mathscr{V}$.

   (c) Find $[\mathbf{T}]_{\mathscr{B}'}$.

   (d) Express each $\mathbf{T(v_i)}$ as a linear combination of the $\mathbf{v_j}$.

2. Let $\mathscr{V}$ be an arbitrary three-dimensional real vector space and let **T** be a linear operator on $\mathscr{V}$. You are given two bases for $\mathscr{V}$,

$$
\mathscr{B} = \{\mathbf{v_1}, \mathbf{v_2}, \mathbf{v_3}\} \quad \text{and} \quad \mathscr{B}' = \{\mathbf{w_1}, \mathbf{w_2}, \mathbf{w_3}\},
$$

and the following information.

$$T(v_1) = v_1 - 3v_3 \qquad w_1 = 2v_1 - v_2 \qquad v_1 = w_1 + w_2$$
$$T(v_2) = 2v_2 + 5v_3 \qquad w_2 = -v_1 + v_2 \qquad v_2 = w_1 + 2w_2$$
$$T(v_3) = 2v_1 + 7v_2 + v_3 \qquad w_3 = -v_1 + v_3 \qquad v_3 = w_1 + w_2 + w_3$$

(a) Find $[T]_{\mathcal{B}}$, which is the matrix representative of T with respect to the $\mathcal{B}$ basis.

(b) Find a nonsingular matrix P such that $[T]_{\mathcal{B}'} = P^{-1}[T]_{\mathcal{B}}P$.

(c) Compute $[T]_{\mathcal{B}'}$.

3. Let T be the linear operator on $\mathcal{R}^2$ defined by

$$T(v) = \begin{bmatrix} 2 & -1 \\ 4 & 3 \end{bmatrix} v.$$

(a) Use Theorem 5.1 to find the matrix representative of T with respect to the basis

$$\mathcal{B} = \{u_1 = [4 \quad 3]^T, u_2 = [3 \quad 2]^T\}.$$

(b) Find $[T]_{\mathcal{B}}$ by finding $T(u_1)$ and $T(u_2)$ directly.

4. Let T be the linear operator on $\mathcal{R}^2$ whose matrix representative relative to the standard basis is $[T]_{\mathcal{G}} = \begin{bmatrix} 3 & 2 \\ -1 & 0 \end{bmatrix}$.

(a) Use Theorem 5.1 to find the matrix representative of T with respect to the basis

$$\mathcal{B} = \{u_1 = [1 \quad -1]^T, u_2 = [2 \quad -1]^T\}.$$

(b) Find $[T]_{\mathcal{B}}$ by finding $T(u_1)$ and $T(u_2)$ directly.

5. Let T be the linear operator on $\mathcal{R}^2$ whose matrix representative relative to the basis

$$\mathcal{B} = \left\{ u_1 = \begin{bmatrix} 1 \\ 2 \end{bmatrix}, u_2 = \begin{bmatrix} 2 \\ 3 \end{bmatrix} \right\} \text{ is } [T]_{\mathcal{B}} = \begin{bmatrix} 2 & 1 \\ 0 & 2 \end{bmatrix}.$$

Find the matrix representative of T with respect to the basis

$$\mathcal{B}' = \left\{ v_1 = \begin{bmatrix} 2 \\ 1 \end{bmatrix}, v_2 = \begin{bmatrix} 1 \\ 0 \end{bmatrix} \right\}.$$

6. Consider the basis $\mathcal{B} = \{u_1 = [1 \quad 2 \quad 3], u_2 = [2 \quad 4 \quad 5], u_3 = [3 \quad 5 \quad 6]\}$

and let T be the linear operator on $R^{1 \times 3}$ for which $[T]_{\mathcal{B}} = \begin{bmatrix} 1 & 3 & 3 \\ 1 & 4 & 3 \\ 1 & 3 & 4 \end{bmatrix}$.

Find the matrix representative of T with respect to the standard basis.

7. Let T be a linear operator on $\mathcal{R}^3$ whose standard matrix is $\begin{bmatrix} 2 & 1 & 1 \\ 2 & 3 & 2 \\ 1 & 1 & 2 \end{bmatrix}$.

Find the matrix representative of T with respect to the basis

$$\mathcal{B} = \left\{ u_1 = \begin{bmatrix} 1 \\ 2 \\ 1 \end{bmatrix}, u_2 = \begin{bmatrix} -1 \\ 1 \\ 0 \end{bmatrix}, u_3 = \begin{bmatrix} 1 \\ 0 \\ -1 \end{bmatrix} \right\}$$

and describe T geometrically.

8. Let **T** be the linear operator on $\mathcal{R}^3$ whose standard matrix is
$\begin{bmatrix} 0 & 1 & 0 \\ 0 & 0 & 1 \\ 8 & -12 & 6 \end{bmatrix}$. Find the matrix representative of **T** with respect to
the basis $\mathcal{B} = \{[1 \quad 2 \quad 4]^T, [0 \quad 1 \quad 4]^T, [0 \quad 0 \quad 1]^T\}$.

9. Let **T** be a linear operator on $\mathcal{R}^4$ whose standard matrix is
$$[\mathbf{T}]_\mathcal{S} = \begin{bmatrix} 6 & 0 & -3 & 0 \\ 4 & 2 & -1 & -1 \\ 3 & 1 & 1 & -1 \\ 6 & 2 & 0 & -1 \end{bmatrix}.$$

Find the matrix representative of **T** with respect to the basis
$\mathcal{B} = \{[1 \quad 1 \quad 1 \quad 2]^T, [3 \quad 3 \quad 4 \quad 8]^T, [3 \quad 4 \quad 3 \quad 6]^T, [0 \quad 1 \quad 0 \quad 1]^T\}$.
Can you interpret **T** geometrically?

10. Verify that similarity is a symmetric, reflexive, and transitive relation on the set of all $n \times n$ matrices.

†11. Show that similar matrices have the same rank, nullity, trace, and determinant.

12. Show that similar matrices have the same characteristic polynomial; that is, show that if $B = P^{-1}AP$, then $\det(\lambda I - B) = \det(\lambda I - A)$.

13. (a) Show that if $B = P^{-1}AP$, then $B^k = P^{-1}A^kP$.

(b) Let $p(x) = a_0 + a_1 x + a_2 x^2 + \ldots + a_n x^n$ be any polynomial and define $p(A) = a_0 I + a_1 A + a_2 A^2 + \ldots + a_n A^n$. Show that if A and B are similar, then so are $p(A)$ and $p(B)$.

14. Show that if A and B are similar, then A^T and B^T are also similar. Moreover, if A is nonsingular, B is also nonsingular and A^{-1} and B^{-1} are similar.

15. (a) Let $\mathcal{B}$ and $\mathcal{B}'$ be two different bases for the vector space $\mathcal{V}$ and let $P = [\mathbf{I}]_{\mathcal{B}'\mathcal{B}}$ be the matrix that represents the identity operator on V with respect to the bases $\mathcal{B}$ and $\mathcal{B}'$ (recall Section 4.5). Show that P is the $\mathcal{B}$ to $\mathcal{B}'$ change-of-basis matrix.

(b) Use Exercise 12 in Section 4.5 to show that
$$P^{-1}[\mathbf{T}]_{\mathcal{B}'}P = [\mathbf{I}]_{\mathcal{B}\mathcal{B}'}[\mathbf{T}]_{\mathcal{B}'}[\mathbf{I}]_{\mathcal{B}'\mathcal{B}} = [\mathbf{ITI}]_{\mathcal{B}\mathcal{B}} = [\mathbf{T}]_{\mathcal{B}}.$$
This point of view, and the diagram below, can be useful tools for remembering how to construct the $\mathcal{B}$ to $\mathcal{B}'$ change-of-basis matrix P.

16. Use the result of Exercise 15 and the diagram below to describe the change-of-basis theorem.

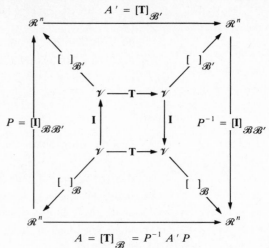

17. Let **T** be a linear transformation from $\mathcal{V}$ to $\mathcal{W} \neq \mathcal{V}$. Let $\mathcal{B}$ and $\mathcal{B}'$ be two bases for $\mathcal{V}$ and let $\mathcal{C}$ and $\mathcal{C}'$ be two bases for $\mathcal{W}$. Construct a diagram like that in Exercise 16 and then determine the relationship between the matrix representatives $[\mathbf{T}]_{\mathcal{C}\mathcal{B}}$ and $[\mathbf{T}]_{\mathcal{C}'\mathcal{B}'}$.

5.2

Diagonalization — Eigenvalues and Eigenvectors

Eigenvalues and eigenvectors were first introduced in Section 2.4 as an application of determinants and of homogeneous systems. Section 2.4 was listed as optional in Chapter 2. If you omitted it then, it is now time to study it carefully. If you included Section 2.4 when you studied Chapter 2, review the basic definitions, theorems, and computational procedures of Section 2.4 at this time. In this section we will provide further motivation for the study of eigenvalues and eigenvectors as well as a good bit of additional information about them.

It is useful to give a slightly more general definition of eigenvalues and eigenvectors than that given in Chapter 2. Since we can now view square matrices as special kinds of linear operators, it is natural to extend these notions to linear operators on arbitrary vector spaces.

Definition 5.2

Let **T** be any linear operator on the vector space $\mathcal{V}$ and let λ be a scalar. A nonzero vector **v** in $\mathcal{V}$ such that $\mathbf{T}(\mathbf{v}) = \lambda\mathbf{v}$ is called an EIGENVECTOR of **T**; the scalar λ is called the EIGENVALUE of **T**, associated with the eigenvector **v**.

Theorem 4.8 tells us that if $\mathcal{V} = \mathcal{R}^n$, then $\mathbf{T}$ is a matrix transformation, $\mathbf{T(v)} = A\mathbf{v}$, so an eigenvector of $\mathbf{T}$ must satisfy $\mathbf{T(v)} = A\mathbf{v} = \lambda\mathbf{v}$. Thus, $\mathbf{v}$ is an eigenvector of the matrix A associated with the eigenvalue λ, and we see that Definition 5.2 is a generalization of Definition 2.5.

In order to provide motivation for the study of eigenvectors and eigenvalues, let us look again at the change-of-basis problem considered in the last section. Let $\mathbf{T}$ be any linear operator on $\mathcal{V}$ and consider the two bases

$$\mathcal{B} = \{\mathbf{u_1}, \ldots, \mathbf{u_n}\} \quad \text{and} \quad \mathcal{B}' = \{\mathbf{v_1}, \ldots, \mathbf{v_n}\}.$$

We know that if $A = [\mathbf{T}]_{\mathcal{B}}$ and P is the $\mathcal{B}'$ to $\mathcal{B}$ change-of-basis matrix, then $[\mathbf{T}]_{\mathcal{B}'} = P^{-1}AP$ is the matrix representative of $\mathbf{T}$ with respect to the $\mathcal{B}'$ basis.

Suppose now that $[\mathbf{T}]_{\mathcal{B}'}$ is a diagonal matrix:

$$[\mathbf{T}]_{\mathcal{B}'} = P^{-1}AP = \begin{bmatrix} \lambda_1 & & & \\ & \lambda_2 & & \\ & & \ddots & \\ & & & \lambda_n \end{bmatrix} = \mathrm{Dg}\{\lambda_1, \lambda_2, \ldots, \lambda_n\}.$$

It follows, from the way that the matrix representative is constructed, that

$$\mathbf{T(v_1)} = \lambda_1\mathbf{v_1}, \mathbf{T(v_2)} = \lambda_2\mathbf{v_2}, \ldots, \mathbf{T(v_n)} = \lambda_n\mathbf{v_n};$$

that is, the new basis $\mathcal{B}'$ consists of eigenvectors of the linear operator $\mathbf{T}$ and the diagonal elements of $[\mathbf{T}]_{\mathcal{B}'}$ are the eigenvalues of $\mathbf{T}$.

From the basic "eigenrelation" $\mathbf{T(v_i)} = \lambda_i\mathbf{v_i}$ we obtain, by taking coordinates with respect to the $\mathcal{B}$ basis,

$$[\mathbf{T(v_i)}]_{\mathcal{B}} = [\lambda_i\mathbf{v_i}]_{\mathcal{B}},$$

which, by Theorem 4.9, is equivalent to

$$[\mathbf{T}]_{\mathcal{B}}[\mathbf{v_i}]_{\mathcal{B}} = \lambda_i[\mathbf{v_i}]_{\mathcal{B}}$$

or

$$A[\mathbf{v_i}]_{\mathcal{B}} = \lambda_i[\mathbf{v_i}]_{\mathcal{B}}.$$

Since the change-of-basis matrix P is given by

$$P = [[\mathbf{v_1}]_{\mathcal{B}} \quad \cdots \quad [\mathbf{v_n}]_{\mathcal{B}}],$$

it follows that

$$A\mathrm{Col}_i(P) = \lambda_i\mathrm{Col}_i(P).$$

Thus we see that the columns of the diagonalizing matrix P are eigenvectors of the matrix A.

We summarize the discussion above with two equivalent theorems. The second theorem is nothing but the matrix version of the first.

Theorem 5.3

Let $\mathbf{T}$ be a linear operator on an n-dimensional vector space $\mathcal{V}$. If there exists a basis $\mathcal{B}' = \{\mathbf{v}_1, \mathbf{v}_2, \ldots, \mathbf{v}_n\}$ for $\mathcal{V}$, relative to which $\mathbf{T}$ is represented by the diagonal matrix $D = Dg\{\lambda_1, \lambda_2, \ldots, \lambda_n\}$, then each of the basis vectors of $\mathcal{B}'$ is an eigenvector of $\mathbf{T}$; that is, $\mathbf{T}(\mathbf{v_i}) = \lambda_i \mathbf{v_i}$; $i = 1, 2, \ldots, n$.

Theorem 5.4

If the $n \times n$ matrix A is similar to a diagonal matrix

$$P^{-1}AP = \begin{bmatrix} \lambda_1 & & & \\ & \lambda_2 & & \\ & & \cdot & \\ & & & \cdot \\ & & & & \lambda_n \end{bmatrix} = Dg\{\lambda_1, \lambda_2, \ldots, \lambda_n\},$$

then the columns of P are a linearly independent set of eigenvectors for A; that is

$$A\mathrm{Col}_i(P) = \lambda_i \mathrm{Col}_i(P), \quad i = 1, 2, \ldots, n.$$

In our first example we will review the computational techniques for finding eigenvalues and eigenvectors. These techniques were introduced in Section 2.4.

Example 1

Let $\mathbf{T}$ be the matrix transformation on $\mathcal{R}^3$ whose standard matrix is

$$A = \begin{bmatrix} 2 & 1 & 1 \\ 2 & 3 & 2 \\ 1 & 1 & 2 \end{bmatrix}; \quad \text{that is, } \mathbf{T}(\mathbf{v}) = A\mathbf{v}.$$

Find the eigenvalues and eigenvectors of $\mathbf{T}$.

Since λ is an eigenvalue of A if and only if $\det(\lambda I - A) = 0$, we first compute $\det(\lambda I - A)$, the characteristic polynomial of A.

$$c(\lambda) = \det(\lambda I - A) = \det \begin{bmatrix} \lambda - 2 & -1 & -1 \\ -2 & \lambda - 3 & -2 \\ -1 & -1 & \lambda - 2 \end{bmatrix}$$

$$= (\lambda - 2) \det \begin{bmatrix} \lambda - 3 & -2 \\ -1 & \lambda - 2 \end{bmatrix} + \det \begin{bmatrix} -2 & -2 \\ -1 & \lambda - 2 \end{bmatrix}$$

$$- \det \begin{bmatrix} -2 & \lambda - 3 \\ -1 & -1 \end{bmatrix}$$

$$= (\lambda - 2)[(\lambda - 3)(\lambda - 2) - 2] + [-2(\lambda - 2) - 2]$$
$$- [2 + (\lambda - 3)]$$
$$= \lambda^3 - 7\lambda^2 + 11\lambda - 5$$

To find the eigenvalues of A, we need to find the roots of the characteristic equation $c(\lambda) = 0$. The only rational possibilities are the factors of -5; that is, ± 1 and ± 5. Direct evaluation (synthetic division is an efficient way to do the evaluations) leads quickly to the factorization

$$c(\lambda) = \lambda^3 - 7\lambda^2 + 11\lambda - 5 = (\lambda - 5)(\lambda - 1)(\lambda - 1),$$

so the eigenvalues of $\mathbf{T}$ are $\lambda_1 = 5$ and $\lambda_2 = \lambda_3 = 1$. In order to find an eigenvector for $\lambda_1 = 5$ we need to solve the homogeneous system $(\lambda_1 I - A)X = 0$. We see from the row reduction

$$[5I - A \,|\, 0] = \begin{bmatrix} 3 & -1 & -1 & 0 \\ -2 & 2 & -2 & 0 \\ -1 & -1 & 3 & 0 \end{bmatrix} \longrightarrow \begin{bmatrix} 0 & 4 & -8 & 0 \\ 0 & -4 & 8 & 0 \\ 1 & 1 & -3 & 0 \end{bmatrix} \longrightarrow$$

$$\begin{bmatrix} 1 & 1 & -3 & 0 \\ 0 & 1 & -2 & 0 \\ 0 & 0 & 0 & 0 \end{bmatrix} \longrightarrow \begin{bmatrix} 1 & 0 & -1 & 0 \\ 0 & 1 & -2 & 0 \\ 0 & 0 & 0 & 0 \end{bmatrix}$$

that the general solution of $(5I - A)X = 0$ is $X = c_1 \begin{bmatrix} 1 \\ 2 \\ 1 \end{bmatrix} = c_1 \mathbf{v}_1$. To

find the eigenvectors for $\lambda = 1$ we solve the homogeneous system $(I - A)X = 0$ via the row reduction

$$[I - A \,|\, 0] = \begin{bmatrix} -1 & -1 & -1 & 0 \\ -2 & -2 & -2 & 0 \\ -1 & -1 & -1 & 0 \end{bmatrix} \longrightarrow \begin{bmatrix} 1 & 1 & 1 & 0 \\ 0 & 0 & 0 & 0 \\ 0 & 0 & 0 & 0 \end{bmatrix}.$$

The general solution of $(I - A)X = 0$ is thus $X = c_2 \begin{bmatrix} -1 \\ 1 \\ 0 \end{bmatrix} + c_3 \begin{bmatrix} -1 \\ 0 \\ 1 \end{bmatrix}$.

In this case we see that the set of all eigenvalues of $\mathbf{T}$ (or of A), corresponding to the repeated eigenvalue $\lambda = 1$, is the two-dimensional subspace of $\mathcal{R}^3$ spanned by the eigenvectors

$$\mathbf{v}_2 = \begin{bmatrix} -1 \\ 1 \\ 0 \end{bmatrix} \quad \text{and} \quad \mathbf{v}_3 = \begin{bmatrix} -1 \\ 0 \\ 1 \end{bmatrix}. \quad \blacksquare$$

Since the operator equation $\mathbf{T}(\mathbf{v}) = \lambda\mathbf{v}$ is equivalent to the equation $(\lambda \mathbf{I} - \mathbf{T})(\mathbf{v}) = \mathbf{0}$, we see that the set of all eigenvectors of $\mathbf{T}$, associated with the eigenvalue λ, is the subspace $\mathcal{NS}(\lambda \mathbf{I} - \mathbf{T})$ of $\mathcal{V}$. These subspaces will be called the EIGENSPACES of $\mathbf{T}$. Note that $\mathcal{NS}(\mathbf{T}) = \mathcal{NS}(-\mathbf{T})$ is the eigenspace associated with the eigenvalue $\lambda = 0$; thus $\mathbf{T}$ is singular if and only if $\lambda = 0$ is an eigenvalue of $\mathbf{T}$.

Example 2

Students who have studied calculus will observe that the familiar differentiation formulas

$$\frac{d}{dt}(e^{at}) = ae^{at} \quad \text{and} \quad \frac{d^2}{dt^2}(\sin at) = -a^2 \sin at$$

say that the exponential function is an eigenvector of the derivative operator and that the sine function is an eigenvector of the second derivative operator. What are the corresponding eigenvalues? ■

A very important property of eigenvectors, which was not included in Section 2.4, is contained in the next theorem.

Theorem 5.5

Eigenvectors associated with distinct eigenvalues of $\mathbf{T}$ are linearly independent.

Proof.   Let $\lambda_1, \lambda_2, \ldots, \lambda_k$ be distinct eigenvalues of $\mathbf{T}$ and let $\mathbf{v}_1, \mathbf{v}_2, \ldots, \mathbf{v}_k$ be the associated eigenvectors; that is, $\mathbf{T}(\mathbf{v}_i) = \lambda_i \mathbf{v}_i$ and each $\mathbf{v}_i \neq \mathbf{0}$.

If $\{\mathbf{v}_1, \mathbf{v}_2, \ldots, \mathbf{v}_k\}$ is a linearly dependent set, then there must be a first vector $\mathbf{v}_s$ in the list such that $\{\mathbf{v}_1, \ldots, \mathbf{v}_{s-1}\}$ is independent and $\{\mathbf{v}_1, \ldots, \mathbf{v}_s\}$ is dependent. Thus, $\mathbf{v}_s$ is in SPAN$\{\mathbf{v}_1, \ldots, \mathbf{v}_{s-1}\}$ and there exist scalars a_i such that

$$\mathbf{v}_s = a_1\mathbf{v}_1 + a_2\mathbf{v}_2 + \ldots + a_{s-1}\mathbf{v}_{s-1}. \tag{5.3}$$

Taking the image of Equation 5.3 under $\mathbf{T}$ yields

$$\mathbf{T}(\mathbf{v}_s) = a_1\mathbf{T}(\mathbf{v}_1) + a_2\mathbf{T}(\mathbf{v}_2) + \ldots + a_{s-1}\mathbf{T}(\mathbf{v}_{s-1}),$$

which, using the fact that the $\mathbf{v}_i$ are eigenvectors of $\mathbf{T}$, becomes

$$\lambda_s\mathbf{v}_s = a_1\lambda_1\mathbf{v}_1 + a_2\lambda_2\mathbf{v}_2 + \ldots + a_{s-1}\lambda_{s-1}\mathbf{v}_{s-1}. \tag{5.4}$$

Multiplying Equation 5.3 by λ_s yields the additional relation

$$\lambda_s\mathbf{v}_s = a_1\lambda_s\mathbf{v}_1 + a_2\lambda_s\mathbf{v}_2 + \ldots + a_{s-1}\lambda_s\mathbf{v}_{s-1}. \tag{5.5}$$

Now, subtracting Equation 5.5 from Equation 5.4 yields the equation

$$\mathbf{0} = a_1(\lambda_1 - \lambda_s)\mathbf{v}_1 + a_2(\lambda_2 - \lambda_s)\mathbf{v}_2 + \ldots + a_{s-1}(\lambda_{s-1} - \lambda_s)\mathbf{v}_{s-1}.$$

Since the set $\{\mathbf{v}_1, \mathbf{v}_2, \ldots, \mathbf{v}_{s-1}\}$ is linearly independent, each of the coefficients in the last equation must vanish. Thus, each $a_i(\lambda_i - \lambda_s) = 0$

and, since $\lambda_i \neq \lambda_s$ for $i = 1, 2, \ldots, s - 1$, it follows that each $a_i = 0$ and hence, from Equation 5.3, that $v_s = 0$. This is a contradiction that follows from the stated assumption that the set $\{v_1, v_2, \ldots, v_k\}$ is dependent. That assumption must be incorrect and so it follows that the set $\{v_1, v_2, \ldots, v_k\}$ is linearly independent. $\square$

Theorem 5.5 allows us to conclude, in the case of distinct eigenvalues, that the matrix P, whose columns are eigenvectors of the matrix A, must be nonsingular and its columns form a basis for $\mathcal{R}^n$. When the matrix A has repeated eigenvalues it may or may not be possible to find a set of eigenvectors large enough to diagonalize A.

Example 3

Find a nonsingular matrix P such that $P^{-1}AP$ is diagonal, where

$$A = \begin{bmatrix} 2 & 1 & 1 \\ 2 & 3 & 2 \\ 1 & 1 & 2 \end{bmatrix} \text{ is the matrix of Example 1.}$$

By Theorem 5.4 the columns of P must be eigenvectors of A. Using the eigenvectors found in Example 1 we see that a suitable matrix P is

$$P = \begin{bmatrix} 1 & -1 & -1 \\ 2 & 1 & 0 \\ 1 & 0 & 1 \end{bmatrix} = \begin{bmatrix} v_1 & v_2 & v_3 \end{bmatrix}.$$

For this matrix P we must have, as a consequence of Theorem 5.4,

$$P^{-1}AP = \begin{bmatrix} \lambda_1 & 0 & 0 \\ 0 & \lambda_2 & 0 \\ 0 & 0 & \lambda_3 \end{bmatrix} = \begin{bmatrix} 5 & 0 & 0 \\ 0 & 1 & 0 \\ 0 & 0 & 1 \end{bmatrix}. \quad \blacksquare$$

The diagonalization technique illustrated in this example is certain to produce a suitable (nonsingular) diagonalizing matrix P if the eigenvalues of A are distinct. If the $n \times n$ matrix A has repeated eigenvalues, as in this example, then it may happen that A does not have enough linearly independent eigenvectors to make a basis for $\mathcal{R}^n$. We illustrate this case in our next example.

Example 4

Diagonalize $J = \begin{bmatrix} 2 & 1 & 0 \\ 0 & 2 & 1 \\ 0 & 0 & 2 \end{bmatrix}$.

If $P^{-1}JP$ is to be a diagonal matrix, then the columns of P must be three linearly independent eigenvectors of J. The characteristic polynomial of J is

$$\det(\lambda I - J) = \det\begin{bmatrix} \lambda - 2 & -1 & 0 \\ 0 & \lambda - 2 & -1 \\ 0 & 0 & \lambda - 2 \end{bmatrix} = (\lambda - 2)^3,$$

so $\lambda = 2$ is a repeated eigenvalue of J. The system $(2I - J)X = 0$

is $\begin{bmatrix} 0 & -1 & 0 \\ 0 & 0 & -1 \\ 0 & 0 & 0 \end{bmatrix}\begin{bmatrix} x_1 \\ x_2 \\ x_3 \end{bmatrix} = \begin{bmatrix} 0 \\ 0 \\ 0 \end{bmatrix}$, for which the general solution is

$X = c_1\begin{bmatrix} 1 \\ 0 \\ 0 \end{bmatrix}$. Thus, the eigenspace of J, for $\lambda = 2$, is one-dimensional

and we cannot find three linearly independent eigenvectors. Hence, J is not similar to a diagonal matrix.   ■

We summarize this discussion in the final theorem of this section.

Theorem 5.6

An $n \times n$ matrix A is similar to a diagonal matrix if and only if A has n linearly independent eigenvectors. If the eigenvalues of A are distinct, then A is similar to a diagonal matrix.

In our next example we show how the results of this section can be used to uncouple the equations in a system of first-order linear differential equations.

Example 5

Let $Y(t) = [y_1(t) \quad y_2(t) \quad \dots \quad y_n(t)]^T$ be an $n \times 1$ matrix of unknown functions that satisfy the differential equations

$$y_1'(t) = a_{11}y_1(t) + a_{12}y_2(t) + \dots + a_{1n}y_n(t)$$
$$y_2'(t) = a_{21}y_1(t) + a_{22}y_2(t) + \dots + a_{2n}y_n(t)$$
$$\vdots \qquad \vdots \qquad \vdots \qquad \vdots$$
$$y_n'(t) = a_{n1}y_1(t) + a_{n2}y_2(t) + \dots + a_{nn}y_n(t)$$

as well as the initial conditions $y_i(0) = c_i$, $i = 1, 2, \dots, n$.

We can write these equations as the single matrix equation

$$Y'(t) = AY(t) \quad \text{with initial condition} \quad Y(0) = C.$$

The solution of this system is easy if the coefficient matrix A is diagonal; it is known from calculus that the solution of $y'(t) = ay(t)$; $y(0) = c$ is the exponential function $y = ce^{at}$. If A is a diagonal matrix, then we say that the equations of the system are uncoupled. If the equations are coupled (that is, if A is not a diagonal matrix) we can try to uncouple them by making a linear substitution of the form $Y(t) = QZ(t)$, where Q is an $n \times n$ nonsingular matrix. It follows that $Y'(t) = QZ'(t)$, so the system $Y' = AY$ reduces to $QZ'(t) = AQZ(t)$ or

$Z'(t) = Q^{-1}AQZ(t)$ with $Z(0) = Q^{-1}C$. If Q is chosen so that $Q^{-1}AQ = Dg\{\lambda_1, \lambda_2, \ldots, \lambda_n\}$ is a diagonal matrix, then we have reduced the problem to that of solving the simple equations

$$z_i'(t) = \lambda_i z_i(t); \quad i = 1, 2, \ldots, n.$$

The solutions are $z_i(t) = z_i(0)e^{\lambda_i t}$, so that the solution of the original system is

$$Y(t) = QZ(t) = Q\begin{bmatrix} z_1(0)e^{\lambda_1 t} \\ z_2(0)e^{\lambda_2 t} \\ \vdots \\ z_n(0)e^{\lambda_n t} \end{bmatrix}.$$

If $\lambda = a + bi$ and $\overline{\lambda} = a - bi$ are a pair of complex eigenvalues of A, then you can use the Euler formula, $e^{i\theta} = \cos\theta + i\sin\theta$, to write $e^{\lambda t} = e^{at}(\cos bt + i\sin bt)$ and $e^{\overline{\lambda} t} = e^{at}(\cos bt - i\sin bt)$. The solution of the system will involve the two real functions $e^{at}\cos bt$ and $e^{at}\sin bt$.* ■

Exercise Set 5.2

For each of the following matrices find the characteristic equation, the eigenvalues, and, if possible, a nonsingular diagonalizing matrix.

1. $\begin{bmatrix} 5 & 1 \\ 4 & 8 \end{bmatrix}$ **2.** $\begin{bmatrix} 0 & 1 \\ -4 & 4 \end{bmatrix}$ **3.** $\begin{bmatrix} 1 & 3 \\ -1 & 5 \end{bmatrix}$ **4.** $\begin{bmatrix} 1 & 7 \\ 3 & -4 \end{bmatrix}$

5. $\begin{bmatrix} 3 & -1 & 1 \\ -1 & 3 & -1 \\ 1 & -1 & 3 \end{bmatrix}$ **6.** $\begin{bmatrix} 5 & 0 & 1 \\ -2 & 2 & 0 \\ -2 & 0 & 2 \end{bmatrix}$ **7.** $\begin{bmatrix} 0 & 1 & 0 \\ 0 & 0 & 1 \\ 20 & -24 & 9 \end{bmatrix}$

8. $\begin{bmatrix} 5 & 6 & 2 \\ 0 & -1 & -8 \\ 1 & 0 & -2 \end{bmatrix}$ **9.** $\begin{bmatrix} 0 & 0 & 4 \\ 1 & 0 & 1 \\ 0 & 1 & -4 \end{bmatrix}$ **10.** $\begin{bmatrix} 2 & -1 & 0 \\ -1 & 2 & -1 \\ 0 & -1 & 2 \end{bmatrix}$

11. $\begin{bmatrix} 0 & 1 & 0 & 0 \\ 0 & 0 & 1 & 0 \\ 0 & 0 & 0 & 1 \\ -36 & 0 & 13 & 0 \end{bmatrix}$ **12.** $\begin{bmatrix} 1 & 1 & 1 & 1 \\ 1 & 1 & 1 & 1 \\ 1 & 1 & 1 & 1 \\ 1 & 1 & 1 & 1 \end{bmatrix}$ **13.** $\begin{bmatrix} 2 & -1 & 0 & 0 \\ -1 & 2 & -1 & 0 \\ 0 & -1 & 2 & -1 \\ 0 & 0 & -1 & 2 \end{bmatrix}$

The following matrices have some complex eigenvalues. Find them and then find a complex diagonalizing matrix.

14. $\begin{bmatrix} 2 & 3 \\ -3 & 2 \end{bmatrix}$ **15.** $\begin{bmatrix} 0 & 1 & 0 \\ 0 & 0 & 1 \\ 12 & -4 & 3 \end{bmatrix}$ **16.** $\begin{bmatrix} \cos\theta & -\sin\theta \\ \sin\theta & \cos\theta \end{bmatrix}$

---

*For more details on applications to systems of linear differential equations you may consult Cullen, C. G. *Linear Algebra and Differential Equations*. Boston: Prindle, Weber & Schmidt, 1979.

17. (a) Determine how the eigenvalues and eigenvectors of A^k are related to those of A.

(b) Let $p(x)$ be any polynomial. Determine how the eigenvalues and eigenvectors of $p(A)$ are related to those of A.

18. Show that the eigenvectors of $A = \begin{bmatrix} 1 & 0 & 1 \\ 0 & 1 & 2 \\ 1 & 2 & 5 \end{bmatrix}$ are mutually orthogonal.

19. Show that if A is similar to a diagonal matrix, then A and A^T are similar.

20. How can one define the characteristic equation of a linear operator **T**?

21. The COMPANION MATRIX of the polynomial

$$p(x) = x^n + a_{n-1}x^{n-1} + \ldots + a_2x^2 + a_1x + a_0$$

is the $n \times n$ matrix $C = \begin{bmatrix} 0 & 1 & 0 & \cdots & 0 \\ 0 & 0 & 1 & \cdots & 0 \\ \vdots & \vdots & \vdots & & \vdots \\ 0 & 0 & 0 & \cdots & 1 \\ -a_0 & -a_1 & -a_2 & \cdots & -a_{n-1} \end{bmatrix}$.

Show that the characteristic polynomial of A is $p(\lambda)$. (*Hint:* Use the elementary column operations $C_1 \leftarrow \lambda C_2 + C_1$, $C_1 \leftarrow \lambda^2 C_3 + C_1$, $C_1 \leftarrow \lambda^3 C_4 + C_1, \ldots, C_1 \leftarrow \lambda^{n-1}C_n + C_1$).

22. Let A be an $n \times n$ matrix, set $B_0 = I$, and let, for $k = 1, 2, 3, \ldots, n$,

$$A_k = AB_{k-1}$$
$$c_k = -(1/k)\, \text{tr}(A_k)$$
$$B_k = A_k + c_k I.$$

We can show that the characteristic equation of A is

$$c(\lambda) = \lambda^n + c_1\lambda^{n-1} + \ldots + c_{n-1}\lambda + c_n.$$

This is called the SOURIAU-FRAME method and can be used to compute the characteristic equation on a digital computer. Show that this method requires about n^4 multiplications and use this method to compute the characteristic equations of the matrices in parts (a) and (b).

(a) $\begin{bmatrix} 2 & -2 & 3 \\ 1 & 1 & 1 \\ 1 & 3 & -1 \end{bmatrix}$

(b) $\begin{bmatrix} 2 & 1 & -1 & 2 \\ 1 & 3 & 2 & -3 \\ -1 & 2 & 1 & -1 \\ 2 & -3 & -1 & 4 \end{bmatrix}$

(c) Write a computer program to find the characteristic equation of an $n \times n$ matrix.

23. A matrix A is said to be TRIDIAGONAL if $a_{ij} = 0$ for $j \neq i - 1, i, i + 1$. The matrices of Exercises 10 and 13 are tridiagonal.

(a) Write down a general 5×5 tridiagonal matrix.

(b) Show that the characteristic polynomial of an $n \times n$ tridiagonal matrix is the polynomial $c_n(\lambda)$, where

$$c_0(\lambda) = 1, \quad c_1(\lambda) = \lambda - a_{11}$$

and, for $i = 2, 3, \ldots, n$,

$$c_i(\lambda) = (\lambda - a_{ii})c_{i-1}(\lambda) - a_{i(i-1)}a_{(i-1)i}c_{i-2}(\lambda).$$

(*Hint:* Expand $\det(\lambda I - A)$ by the last column.)

(c) Use this recursion to find the characteristic equations of the matrices in Exercises 10 and 13.

(d) For the matrix of Exercise 13 use the above recursion to evaluate $c_4(3)$, $c_4(2)$, and $c_4(5)$.

†**24.** Show that the eigenvalues of $A = \begin{bmatrix} a & b \\ b & c \end{bmatrix}$ are real.

25. Use the results of Exercise 3 to solve

$$Y'(t) = \begin{bmatrix} 1 & 3 \\ -1 & 5 \end{bmatrix} Y(t), Y(0) = \begin{bmatrix} 1 \\ 4 \end{bmatrix}.$$

26. Use the results of Exercise 6 to solve

$$Y'(t) = \begin{bmatrix} 5 & 0 & 1 \\ -2 & 2 & 0 \\ -2 & 0 & 2 \end{bmatrix} Y(t), Y(0) = \begin{bmatrix} 2 \\ -3 \\ 5 \end{bmatrix}.$$

5.3 *Orthogonal Similarity and Symmetric Matrices*

The eigenvalue problem for an $n \times n$ real matrix A is complicated by the following facts.

(1) Some, or perhaps all, of the eigenvalues may be complex.
(2) Complex eigenvalues will lead to complex eigenvectors.*
Although the theory is pretty much the same in the complex case, the computations are noticeably more difficult.
(3) Repeated eigenvalues may mean that the matrix is not diagonalizable; this is the case if there are not enough eigenvectors, as illustrated in Example 4 in Section 5.2.

Let $\lambda = a + bi$ be a complex eigenvalue of the real $n \times n$ matrix A and let X be the corresponding complex eigenvector of A. It is easy to see that we can write $X = X_1 + X_2i$, where X_1 and X_2 are real $n \times 1$ matrices. We now have

$$AX = A(X_1 + X_2i) = (a + bi)(X_1 + X_2i)$$

or

$$AX_1 + AX_2i = (aX_1 - bX_2) + (bX_1 + aX_2)i.$$

---

*Exercises 1–4 of this section provide a brief review of complex numbers.

By equating real and imaginary parts in the above expression we obtain

$$AX_1 = aX_1 - bX_2 \quad \text{and} \quad AX_2 = bX_1 + aX_2 .$$

It then follows that the complex vector $\overline{X} = X_1 - X_2i$ satisfies

$$A\overline{X} = A(X_1 - X_2i) = AX_1 - AX_2i = (aX_1 - bX_2) - (bX_1 + aX_2)i$$
$$= (a - bi)(X_1 - X_2i) = \overline{\lambda}\,\overline{X} ,$$

where $\overline{\lambda} = a - bi$ is the *complex conjugate* of $\lambda = a + bi$ and $\overline{X} = X_1 - X_2i$ is the *complex conjugate* of the vector $X = X_1 + X_2i$. Thus, complex eigenvalues of real matrices must occur in conjugate pairs and the eigenvectors also occur in conjugate pairs; that is, if $AX = \lambda X$, then we also have the conjugate relation $A\overline{X} = \overline{\lambda}\,\overline{X}$.

The symmetric matrices, which arise frequently in applications, are a very important type of matrix with many interesting properties.

Definition 5.3

An $n \times n$ real matrix A is said to be SYMMETRIC if $A^T = A$, that is, if $a_{ij} = a_{ji}$ for all i and j.

The matrices

$$A = \begin{bmatrix} 1 & -2 & 3 \\ -2 & 0 & 5 \\ 3 & 5 & 6 \end{bmatrix}, \quad B = \begin{bmatrix} 2 & 3 \\ 3 & 0 \end{bmatrix}, \quad \text{and}$$

$$C = \begin{bmatrix} 2 & 1 & -1 & 4 \\ 1 & 3 & 2 & -3 \\ -1 & 2 & 1 & -1 \\ 4 & -3 & -1 & 4 \end{bmatrix}$$

are all examples of symmetric matrices. Note that elements that are symmetrically placed with respect to the main diagonal are equal; for example,

$$a_{23} = a_{32} = 5, \quad b_{12} = b_{21} = 3, \quad c_{14} = c_{41} = 4, \quad \text{and so forth} .$$

If λ is an eigenvalue of a symmetric matrix A and X is the associated eigenvector, then, if λ is complex, we have the two relations

$$AX = \lambda X \quad \text{and} \quad A\overline{X} = \overline{\lambda}\,\overline{X} ,$$

from which we obtain the two equations

$$\overline{X}^T AX = \overline{X}^T(AX) = \overline{X}^T(\lambda X) = \lambda \overline{X}^T X$$

and

$$\overline{X}^T AX = (\overline{X}^T A)X = (A\overline{X})^T X = (\overline{\lambda}\,\overline{X})^T X = \overline{\lambda}\,\overline{X}^T X .$$

It follows that

$$\lambda \overline{X}^T X = \overline{\lambda}\,\overline{X}^T X \quad \text{or} \quad (\lambda - \overline{\lambda})\overline{X}^T X = 0 .$$

Since $X^TX > 0$ by Exercise 4, we see that $\lambda = \overline{\lambda}$; that is, λ is a real number. We state this important property of symmetric matrices as our next theorem.

Theorem 5.7

The eigenvalues of a symmetric matrix are all real.

Another useful property of symmetric matrices will be demonstrated next. Some laboratory evidence in support of this theorem was provided in Exercise 18 of Section 5.2.

Theorem 5.8

Eigenvectors of a symmetric matrix A, associated with distinct eigenvalues, must be orthogonal.

Proof.   Suppose that $AX_1 = \lambda_1 X_1$ and $AX_2 = \lambda_2 X_2$, where $\lambda_1 \neq \lambda_2$. By direct calculation we have

$$X_2{}^T(AX_1) = X_2{}^T(\lambda_1 X_1) = \lambda_1(X_2{}^TX_1)$$

and

$$(X_2{}^TA)X_1 = (AX_2)^TX_1 = (\lambda_2 X_2)^TX_1 = \lambda_2(X_2{}^TX_1)\,.$$

Since $\lambda_1 \neq \lambda_2$ it follows that $X_2{}^TX_1 = (X_2 \cdot X_1) = 0$, and hence that the eigenvectors X_1 and X_2 are orthogonal as claimed. $\square$

If the $n \times n$ matrix A is symmetric, with distinct eigenvalues, then Theorem 5.8 says that the eigenvectors of A form a basis for $\mathscr{R}^n$, whose vectors are mutually orthogonal. If the eigenvectors are scaled so that they are unit vectors, then they will form an orthonormal basis for $\mathscr{R}^n$. In this case the matrix P, whose columns are the orthonormal eigenvectors of A, will be an orthogonal matrix ($P^TP = I$), and we will have

$$P^TAP = P^{-1}AP = D\,,$$

where $D = Dg\{\lambda_1, \ldots, \lambda_n\}$ is the $n \times n$ diagonal matrix whose diagonal entries are the eigenvalues of A.

Definition 5.4

The matrices A and B are said to be ORTHOGONALLY SIMILAR if there exists an orthogonal matrix P such that $P^TAP = B$.

Example 1

Find an orthogonal matrix P such that P^TAP is diagonal if

$$A = \begin{bmatrix} 5 & -2 \\ -2 & 8 \end{bmatrix}.$$

The characteristic polynomial of A is

$$\det(\lambda I - A) = \det\begin{bmatrix} \lambda - 5 & 2 \\ 2 & \lambda - 8 \end{bmatrix} = \lambda^2 - 13\lambda + 36$$

$$= (\lambda - 9)(\lambda - 4).$$ To find an eigenvector for $\lambda = 9$, we solve the homogeneous system $(9I - A)X = 0$ using the row reduction

$$\begin{bmatrix} 4 & 2 & 0 \\ 2 & 1 & 0 \end{bmatrix} \longrightarrow \begin{bmatrix} 2 & 1 & 0 \\ 0 & 0 & 0 \end{bmatrix}.$$

Thus, $X_1 = \begin{bmatrix} 1 \\ -2 \end{bmatrix}$ is an eigenvector for $\lambda = 9$. Similarly, the reduction

$$\begin{bmatrix} -1 & 2 & 0 \\ 2 & -4 & 0 \end{bmatrix} \longrightarrow \begin{bmatrix} 1 & -2 & 0 \\ 0 & 0 & 0 \end{bmatrix}$$ leads to the eigenvector $X_2 = \begin{bmatrix} 2 \\ 1 \end{bmatrix}$

for $\lambda_2 = 4$. Note that $X_1 \cdot X_2 = 0$, as guaranteed by Theorem 5.8. In order to change the basis $\{X_1, X_2\}$ into an orthonormal basis for $\mathcal{R}^2$ we need only divide each of the basis vectors by its length to obtain the orthonormal basis

$$\left\{ X_1' = \frac{1}{\sqrt{5}}\begin{bmatrix} 1 \\ -2 \end{bmatrix}, X_2' = \frac{1}{\sqrt{5}}\begin{bmatrix} 2 \\ 1 \end{bmatrix} \right\}.$$

It now follows that the orthogonal matrix $P = \dfrac{1}{\sqrt{5}}\begin{bmatrix} 1 & 2 \\ -2 & 1 \end{bmatrix}$ is such that $P^T A P = \begin{bmatrix} 9 & 0 \\ 0 & 4 \end{bmatrix}$. ■

We have previously observed that the change-of-basis matrix between two orthonormal bases of $\mathcal{R}^n$ is an orthogonal matrix (see Theorem 3.23). If $\mathcal{B}$ and $\mathcal{B}'$ are two orthonormal bases for $\mathcal{R}^n$ and $\mathbf{T}$ is a linear operator on $\mathcal{R}^n$, then, from Theorem 5.1, we have

$$[\mathbf{T}]_{\mathcal{B}'} = P^{-1}[\mathbf{T}]_{\mathcal{B}} P = P^T[\mathbf{T}]_{\mathcal{B}} P.$$

This observation proves our next theorem.

Theorem 5.9

Orthogonally similar matrices represent the same linear operator with respect to different orthonormal bases of $\mathcal{R}^n$.

If the matrix A is orthogonally similar to a diagonal matrix D, $P^T A P = D$, then

$$A = PDP^T \quad \text{and} \quad A^T = (P^T)^T D^T P^T = PDP^T = A,$$

so A is symmetric.

In Section 5.4 we will show that any symmetric matrix is orthogonally similar to a diagonal matrix; that is, there are always enough eigenvectors. Thus, the following theorem is true.

Theorem 5.10

A real matrix A is orthogonally similar to a diagonal matrix if and only if A is symmetric.

To find the orthogonal diagonalizing matrix of Theorem 5.10 one first finds bases for the eigenspaces of A and then uses the Gram-Schmidt process to find orthonormal bases for the eigenspaces of dimension greater than one. The one-dimensional eigenspaces are treated as in Example 1.

Example 2

Find an orthogonal matrix P such that $P^T A P$ is a diagonal matrix if

$$A = \begin{bmatrix} 2 & -1 & 1 \\ -1 & 2 & -1 \\ 1 & -1 & 2 \end{bmatrix}.$$

The characteristic polynomial of A is $\det(\lambda I - A) =$

$$\det \begin{bmatrix} \lambda - 2 & 1 & -1 \\ 1 & \lambda - 2 & 1 \\ -1 & 1 & \lambda - 2 \end{bmatrix} = \lambda^3 - 6\lambda^2 + 9\lambda - 4 = (\lambda - 1)^2(\lambda - 4),$$

so the eigenvalues of A are $\lambda_1 = \lambda_2 = 1$ and $\lambda_3 = 4$. In order to find the eigenspace $\mathscr{E}_1 = \{X \mid AX = X\}$ we solve the homogeneous system $(I - A)X = 0$ using the row reduction

$$\begin{bmatrix} -1 & 1 & -1 & 0 \\ 1 & -1 & 1 & 0 \\ -1 & 1 & -1 & 0 \end{bmatrix} \longrightarrow \begin{bmatrix} 1 & -1 & 1 & 0 \\ 0 & 0 & 0 & 0 \\ 0 & 0 & 0 & 0 \end{bmatrix}.$$

Thus, $\mathscr{E}_1 = \{X \mid x_1 = x_2 - x_3\} = \left\{ c_1 \begin{bmatrix} 1 \\ 1 \\ 0 \end{bmatrix} + c_2 \begin{bmatrix} -1 \\ 0 \\ 1 \end{bmatrix} \right\}$, so

$X_1 = [1 \quad 1 \quad 0]^T$ and $X_2 = [-1 \quad 0 \quad 1]^T$ form a basis for the eigenspace $\mathscr{E}_1$. Note that X_1 and X_2 are not orthogonal vectors. The eigenspace $\mathscr{E}_4 = \{X \mid AX = 4X\}$ is determined by solving the system $(4I - A)X = 0$. From the row reduction

$$\begin{bmatrix} 2 & 1 & -1 & 0 \\ 1 & 2 & 1 & 0 \\ -1 & 1 & 2 & 0 \end{bmatrix} \longrightarrow \begin{bmatrix} 0 & -3 & -3 & 0 \\ 1 & 2 & 1 & 0 \\ 0 & 3 & 3 & 0 \end{bmatrix} \longrightarrow$$

$$\begin{bmatrix} 0 & 1 & 1 & 0 \\ 1 & 2 & 1 & 0 \\ 0 & 0 & 0 & 0 \end{bmatrix} \longrightarrow \begin{bmatrix} 0 & 1 & 1 & 0 \\ 1 & 0 & -1 & 0 \\ 0 & 0 & 0 & 0 \end{bmatrix}$$

we see that $\mathscr{E}_4$ is the one-dimensional subspace of $\mathscr{R}^3$ spanned by $X_3 = [1, -1, 1]^T$. Note that X_3 is orthogonal to both X_1 and X_2, as guaranteed by Theorem 5.8, and hence to any vector in the subspace $\mathscr{E}_1$.

We now use the Gram-Schmidt process of Section 3.8 to find an orthonormal basis for $\mathscr{E}_1$. We therefore need to determine a scalar a so that the vector

$$X_2' = aX_1 + X_2,$$

of $\mathscr{E}_1$, is orthogonal to X_1. From the equation

$$0 = X_1 \cdot X_2' = aX_1 \cdot X_1 + X_1 \cdot X_2 = 2a - 1$$

we see that we must take $a = 1/2$ and that

$$X_2' = \frac{1}{2}X_1 + X_2 = \frac{1}{2}[-1 \quad 1 \quad 2]^T.$$

is orthogonal to X_1.

We now have the mutually orthogonal set of eigenvectors X_1, X_2', and X_3. Normalizing these vectors yields the following orthonormal set of eigenvectors of A, which form an orthonormal basis for $\mathscr{R}^3$:

$$Y_1 = \frac{1}{\sqrt{2}}\begin{bmatrix} 1 \\ 1 \\ 0 \end{bmatrix}, \quad Y_2 = \frac{1}{\sqrt{6}}\begin{bmatrix} -1 \\ 1 \\ 2 \end{bmatrix}, \quad \text{and} \quad Y_3 = \frac{1}{\sqrt{3}}\begin{bmatrix} 1 \\ -1 \\ 1 \end{bmatrix}.$$

It follows that

$$P = [Y_1 \quad Y_2 \quad Y_3] = \frac{1}{\sqrt{6}}\begin{bmatrix} \sqrt{3} & -1 & \sqrt{2} \\ \sqrt{3} & 1 & \sqrt{2} \\ 0 & 2 & \sqrt{2} \end{bmatrix}$$

is an orthogonal matrix ($P^T P = I$) such that

$$P^T A P = \begin{bmatrix} 1 & 0 & 0 \\ 0 & 1 & 0 \\ 0 & 0 & 4 \end{bmatrix}. \quad \blacksquare$$

Exercise Set 5.3

†1. For the complex numbers $z = 3 + 2i$ and $w = 4 - 5i$, compute the following values.

(a) $z + w$ (b) $2z - 3w$ (c) $z + \overline{w}$ (d) $\overline{z} + \overline{w}$ (e) $\overline{zw}$

(f) $\overline{z}\,\overline{w}$ (g) $\overline{z}z$ (h) $\begin{bmatrix} \overline{z} \\ \overline{w} \end{bmatrix}^T \begin{bmatrix} z \\ w \end{bmatrix}$

†**2.** If $z = a + bi$ and $w = c + di$ are complex numbers show that statements (a) through (e) are true.

(a) $\overline{\overline{z}} = z$ (b) $\overline{z + w} = \overline{z} + \overline{w}$ (c) $\overline{zw} = \overline{z}\,\overline{w}$

(d) $\overline{z}z = a^2 + b^2 \geq 0$

(e) $z = a + bi$ is real ($b = 0$) if and only if $\overline{z} = z$, and z is pure imaginary ($a = 0$) if and only if $\overline{z} = -z$.

(f) Show that $\mathscr{C}$, the set of all complex numbers, is a two-dimensional real vector space.

†**3.** Let A and B be matrices with complex entries.

(a) Show that there exists unique real matrices A_1 and A_2 such that $A = A_1 + A_2 i$.

(b) Let $\overline{A}$ be the matrix obtained from A by taking the complex conjugate of each entry of A and let $A^H = \overline{A}^T$ be the HERMITIAN CONJUGATE of A. Show that $\overline{AB} = \overline{A}\,\overline{B}$ and that $(AB)^H = B^H A^H$.

(c) Show that conjugation and transposition are commutative, that is, $\overline{A^T} = \overline{A}^T$.

†**4.** Let X be an $n \times 1$ matrix with complex entries. Show that
$$\overline{X}^T X = X^H X = \overline{x}_1 x_1 + \overline{x}_2 x_2 + \ldots + \overline{x}_n x_n \geq 0$$
and hence that $X^H X = 0$ if and only if $X = 0$.

5. Find the eigenvalues and eigenvectors of the following matrices and verify directly that the eigenvectors are orthogonal.

(a) $\begin{bmatrix} 3 & 1 \\ 1 & 3 \end{bmatrix}$ (b) $\begin{bmatrix} 4 & 2 & 2 \\ 2 & 3 & 0 \\ 2 & 0 & 5 \end{bmatrix}$

For each of the following matrices, find an orthogonal diagonalizing matrix.

6. $\begin{bmatrix} 3 & 1 \\ 1 & 3 \end{bmatrix}$ **7.** $\begin{bmatrix} 16 & -12 \\ -12 & 9 \end{bmatrix}$ **8.** $\begin{bmatrix} 2 & -1 \\ -1 & 2 \end{bmatrix}$

9. $\begin{bmatrix} 4 & 2 & 2 \\ 2 & 3 & 0 \\ 2 & 0 & 5 \end{bmatrix}$ **10.** $\begin{bmatrix} 3 & 4 & 2 \\ 4 & 3 & 2 \\ 2 & 2 & 0 \end{bmatrix}$ **11.** $\begin{bmatrix} 2 & 0 & 0 \\ 0 & -2 & 2 \\ 0 & 2 & 1 \end{bmatrix}$

12. $\begin{bmatrix} 2 & 1 & 1 \\ 1 & 1 & 0 \\ 1 & 0 & 1 \end{bmatrix}$ **13.** $\begin{bmatrix} 3 & 1 & 0 & 0 \\ 1 & 3 & 0 & 0 \\ 0 & 0 & 16 & -12 \\ 0 & 0 & -12 & 9 \end{bmatrix}$ **14.** $\begin{bmatrix} 4 & 0 & -1 & 1 \\ 0 & 3 & 0 & 0 \\ -1 & 0 & 4 & -1 \\ 1 & 0 & -1 & 4 \end{bmatrix}$

15. Show that if A is symmetric, then so is $P^T A P$.

16. Let A be any $m \times n$ matrix.

(a) Show that $A^T A$ is symmetric with positive diagonal elements.

(b) Show that $\mathscr{NS}(A^T A) = \mathscr{NS}(A)$ and hence that $\operatorname{rank}(A^T A) = \operatorname{rank}(A)$.

17. Show that the eigenvalues of $A = \begin{bmatrix} 0 & b \\ -b & 0 \end{bmatrix}$ are pure imaginary.

18. A real matrix A is SKEW SYMMETRIC if $A^T = -A$.

(a) Show that the diagonal elements of a skew symmetric matrix are zero.

    (b) Show that the eigenvalues of a skew symmetric matrix are pure imaginary.

19. Show that any $n \times n$ real matrix can be written, in a unique way, as the sum of a symmetric matrix and a skew symmetric matrix.

20. An $n \times n$ complex matrix A is called HERMITIAN if $A^H = A$.

    (a) Show that the diagonal elements of a Hermitian matrix are real.

    (b) Show that the eigenvalues of a Hermitian matrix are real.

21. Let A be an $n \times n$ symmetric matrix whose eigenvalues are all positive. Show that $X^T A X > 0$ for all $X \neq 0$ in $\mathcal{R}^n$. (*Hint:* Express X as a linear combination of eigenvectors of A.)

22. **(a)** Assume that A is similar to a diagonal matrix with $\lambda_1, \ldots, \lambda_k$ as distinct eigenvalues. Show that there exist matrices E_i such that
$$A = \lambda_1 E_1 + \lambda_2 E_2 + \ldots + \lambda_k E_k,$$
where $E_i + E_2 + \ldots + E_k = I$, $E_i E_j = 0$ for $i \neq j$, and $E_i^2 = E_i$. (*Hint:* First consider the case where A is a diagonal matrix.)

    (b) Use this decomposition to compute A^r and $\sqrt{A}$.

    (c) How many different square roots of A are there?

23. Prove that symmetric matrices with the same characteristic polynomial must be similar. Is this true for nonsymmetric matrices?

24. Show that an orthogonal triangular matrix with positive diagonal entries must be I. Use this fact to show that a factorization $A = QR$, where Q is orthogonal and R is upper triangular with positive diagonal entries, must be unique.

5.4   *Schur's Theorem**

Theorem 5.10, which has not as yet been proven completely, asserts that every symmetric matrix is orthogonally similar to a diagonal matrix. If A is not symmetric, then it cannot be orthogonally similar to a diagonal matrix (refer to Exercise 19). The following theorem describes the simplest matrix that is orthogonally similar to a general matrix A under the additional assumption that all the eigenvalues of A are real. The proof is a bit involved, but we include it because it can be used as the basis for a computer algorithm to do the reduction.

Theorem 5.11   SCHUR'S THEOREM: If A is an $n \times n$ matrix, all of whose eigenvalues are real, then A is orthogonally similar to an upper triangular matrix T; that is, there exists an orthogonal matrix P such that

---

*This section is optional.

$$P^T A P = T = \begin{bmatrix} \lambda_1 & t_{12} & t_{13} & \cdot & \cdot & \cdot & t_{1n} \\ 0 & \lambda_2 & t_{23} & \cdot & \cdot & \cdot & t_{2n} \\ 0 & 0 & \lambda_3 & \cdot & \cdot & \cdot & t_{3n} \\ \vdots & \vdots & \vdots & & & & \vdots \\ 0 & 0 & 0 & \cdot & \cdot & \cdot & \lambda_n \end{bmatrix}.$$

Proof. The proof is by mathematical induction on n (a loop in the computer program). Since the assertion is obvious for $n = 1$, we need to show that if the theorem is true for $n = r - 1$, then it is also true for $n = r$. It will then follow that the theorem is true for all n.

Let A be an $r \times r$ matrix with real eigenvalues and let λ_1 be an eigenvalue of A associated with the eigenvector X_1. We will assume that X_1 has been scaled so that it is a unit vector; thus we have

$$AX_1 = \lambda_1 X_1 \quad \text{and} \quad X_1^T X_1 = X_1 \cdot X_1 = 1.$$

Using either Theorem 3.20 or the Gram-Schmidt orthogonalization process, construct an orthogonal matrix P_1 whose first column is X_1. Let $P_1 = [X_1 \ X_2 \ \ldots \ X_r]$ and compute

$$P_1^T A P_1 = \begin{bmatrix} X_1^T \\ X_2^T \\ \vdots \\ X_r^T \end{bmatrix} [AX_1 \ AX_2 \ \cdots \ AX_r]$$

$$= \begin{bmatrix} X_1^T \\ X_2^T \\ \vdots \\ X_r^T \end{bmatrix} [\lambda_1 X_1 \ AX_2 \ \cdots \ AX_r]$$

$$= \begin{bmatrix} \lambda_1 X_1^T X_1 & X_1^T AX_2 & \cdots & X_1^T AX_r \\ \hline \lambda_1 X_2^T X_1 & & & \\ \vdots & & A_2 & \\ \lambda_1 X_r^T X_1 & & & \end{bmatrix} = \begin{bmatrix} \lambda_1 & * \\ 0 & A_2 \end{bmatrix}.$$

Note that the subdiagonal elements in the first column are 0 because $X_j^T X_1 = 0$ for $j \neq 1$, by the orthogonality of P_1, and that the matrix A_2 is an $(r - 1) \times (r - 1)$ matrix whose eigenvalues are also eigenvalues of A and, hence, real. Since we have assumed that the theorem is true for $n = r - 1$, we can assert the existence of an $(r - 1) \times (r - 1)$ orthogonal matrix P_2 such that

$$P_2^T A_2 P_2 = \begin{bmatrix} \lambda_2 & & & & * \\ 0 & \lambda_3 & & & \\ \vdots & & \ddots & & \\ \vdots & & & & \\ 0 & 0 & & 0 & \lambda_r \end{bmatrix} = T_2$$

is an upper triangular matrix.

Now the $r \times r$ matrix $P_3 = \begin{bmatrix} 1 & 0 \\ 0 & P_2 \end{bmatrix}$ is also orthogonal and we have

$$(P_1 P_3)^T A (P_1 P_3) = P_3^T P_1^T A P_1 P_3 = \begin{bmatrix} 1 & 0 \\ 0 & P_2^T \end{bmatrix} \begin{bmatrix} \lambda_1 & * \\ 0 & A_1 \end{bmatrix} \begin{bmatrix} 1 & 0 \\ 0 & P_2 \end{bmatrix}$$

$$= \begin{bmatrix} \lambda_1 & * \\ 0 & P_2^T A_2 P_2 \end{bmatrix} = \begin{bmatrix} \lambda_1 & * \\ 0 & T_2 \end{bmatrix} = T.$$

Thus $P = P_1 P_3$ is an orthogonal matrix such that

$$P^T A P = T = \begin{bmatrix} \lambda_1 & t_{12} & t_{13} & \cdots & t_{1n} \\ 0 & \lambda_2 & t_{23} & \cdots & t_{2n} \\ 0 & 0 & \lambda_3 & \cdots & t_{3n} \\ \vdots & \vdots & \vdots & & \vdots \\ 0 & 0 & 0 & \cdots & \lambda_n \end{bmatrix}$$

is the desired triangular matrix.   □

An immediate consequence of Schur's theorem is the unproven part of Theorem 5.10. If A is symmetric, then, by Theorem 5.7, the eigenvalues of A are real and from Theorem 5.11 we have an orthogonal matrix P such that $P^T A P = U$ is an upper triangular matrix. The calculation

$$U^T = (P^T A P)^T = P^T A^T (P^T)^T = P^T A P = U$$

shows that U is symmetric as well as upper triangular; this combination is possible only if U is a diagonal matrix.

We have shown that every symmetric matrix is orthogonally similar to a diagonal matrix, thus completing the proof of Theorem 5.10.

Example 1

Determine if $A = \begin{bmatrix} -2 & 1 & 0 \\ -4 & 0 & 3 \\ 0 & 1 & -2 \end{bmatrix}$ is orthogonally similar to a triangular matrix, and if it is, find the transforming matrix.

The characteristic polynomial of A is

$$\det(\lambda I - A) = \lambda^3 + 4\lambda^2 + 5\lambda + 2 = (\lambda + 2)(\lambda + 1)^2,$$

so the eigenvalues of A are $\lambda_1 = -2$ and $\lambda_2 = \lambda_3 = -1$. Since the eigenvalues of A are all real, Theorem 5.11 tells us that A is orthogonally similar to a triangular matrix. A unit eigenvector for $\lambda_1 = -2$ is $X_1 = [3 \quad 0 \quad 4]^T/5$ and, using Theorem 3.20, we construct the orthogonal matrix

$$
P_1 = \left[\begin{array}{c|cc}
3/5 & 0 & 4/5 \\
\hline
0 & & \\
4/5 & \begin{array}{cc} -1 \\ 1-3/5 \end{array} \begin{bmatrix} 0 & 0 \\ 0 & 16/25 \end{bmatrix} + I
\end{array}\right] = \frac{1}{5}\begin{bmatrix} 3 & 0 & 4 \\ 0 & 5 & 0 \\ 4 & 0 & -3 \end{bmatrix},
$$

whose first column is X_1. Direct multiplication yields

$$
P_1^T A P_1 = \begin{bmatrix} -2 & 7/5 & 0 \\ 0 & 0 & -5 \\ 0 & 1/5 & -2 \end{bmatrix} = \begin{bmatrix} \lambda_1 & * \\ 0 & A_2 \end{bmatrix}.
$$

The eigenvalues of A_2 are $\lambda_2 = \lambda_3 = -1$ and a unit eigenvector of

$$
A_2 \text{ is } X_2 = \frac{1}{\sqrt{26}}\begin{bmatrix} 5 \\ 1 \end{bmatrix}. \text{ Now } P_2 = \frac{1}{\sqrt{26}}\begin{bmatrix} 5 & 1 \\ 1 & -5 \end{bmatrix}
$$

is a 2×2 orthogonal matrix whose first column is X_2. Moreover,

$$
P_2^T A_2 P_2 = \begin{bmatrix} -1 & 338/65 \\ 0 & -1 \end{bmatrix} \text{ is upper triangular}.
$$

$$
\text{Thus, } P = P_1\begin{bmatrix} 1 & 0 \\ 0 & P_2 \end{bmatrix} = \frac{1}{5\sqrt{26}}\begin{bmatrix} 3\sqrt{26} & 4 & -20 \\ 0 & 25 & 5 \\ 4\sqrt{26} & -3 & 15 \end{bmatrix}
$$

is such that $P^T A P$ is upper triangular. Direct calculation yields

$$
P^T A P = \begin{bmatrix} 1 & 0 \\ 0 & P_2^T \end{bmatrix}\begin{bmatrix} -2 & 7/5 & 0 \\ 0 & 0 & -5 \\ 0 & 1/5 & -2 \end{bmatrix}\begin{bmatrix} 1 & 0 \\ 0 & P_2 \end{bmatrix}
$$

$$
= \begin{bmatrix} -2 & 7/\sqrt{26} & 7/5\sqrt{26} \\ 0 & -1 & 338/65 \\ 0 & 0 & -1 \end{bmatrix}. \quad \blacksquare
$$

The unpleasant arithmetic encountered in Example 1 is typical of the computations encountered in the orthogonal triangularization algorithm based on the proof of Schur's theorem. For this reason it is best to do the arithmetic by machine, using either a good pocket calculator or a computer program like MATALG.

Many applications involve polynomial expressions in a matrix A.

If $q(x) = a_n x^n + a_{n-1}x^{n-1} + \ldots + a_2 x^2 + a_1 x + a_0$

is any polynomial and A is any square matrix, then we define

$$q(A) = a_n A^n + a_{n-1} A^{n-1} + \ldots + a_2 A^2 + a_1 A + a_0 I.$$

We saw in Exercise 11 of Section 5.1 that if A and B are similar, then so are $q(A)$ and $q(B)$; that is, if $B = P^{-1}AP$, then $q(B) = P^{-1}q(A)P$.

If A is a diagonal matrix, $A = Dg\{\lambda_1, \lambda_2, \ldots, \lambda_n\}$, then it is easy to see that

$$q(A) = Dg\{q(\lambda_1), q(\lambda_2), \ldots, q(\lambda_n)\}.$$

Similarly, if A is upper triangular; that is, if

$$A = \begin{bmatrix} \lambda_1 & & & & \\ & \lambda_2 & & * & \\ & & \ddots & & \\ & 0 & & \ddots & \\ & & & & \lambda_n \end{bmatrix}, \text{ then}$$

$$q(A) = \begin{bmatrix} q(\lambda_1) & & & & \\ & q(\lambda_2) & & * & \\ & & \ddots & & \\ & 0 & & \ddots & \\ & & & & q(\lambda_n) \end{bmatrix}.$$

is also an upper triangular matrix (refer to Exercise 12).

If $c(x) = \det(xI - A)$ is the characteristic polynomial of A, then $c(\lambda_i) = 0$ for each eigenvalue of A. If A is diagonalizable, with

$$P^{-1}AP = D = Dg\{\lambda_1, \lambda_2, \ldots, \lambda_n\},$$

then

$$c(D) = Dg\{c(\lambda_1), c(\lambda_2), \ldots, c(\lambda_n)\} = 0$$

and

$$c(A) = P^{-1}c(D)P = P^{-1}0P = 0.$$

We have proven, for the case of a diagonalizable matrix A, the following famous theorem. The proof for the nondiagonalizable case will be omitted.

Theorem 5.12

CAYLEY-HAMILTON THEOREM: If A is any $n \times n$ matrix and

$$c(\lambda) = \lambda^n + a_{n-1}\lambda^{n-1} + \ldots + a_2\lambda^2 + a_1\lambda + a_0$$

is the characteristic polynomial of A, then $c(A) = 0$; that is,

$$A^n + a_{n-1}A^{n-1} + \ldots + a_2 A^2 + a_1 A + a_0 I = 0.$$

Example 2

Verify the Cayley-Hamilton theorem for the matrix

$$A = \begin{bmatrix} 2 & 1 & 1 \\ -1 & 2 & -1 \\ -1 & 1 & 3 \end{bmatrix}.$$

The characteristic polynomial of A is

$$\det(\lambda I - A) = \lambda^3 - 7\lambda^2 + 19\lambda - 19,$$

which has two complex roots and one irrational real root. We compute

$$A^2 = \begin{bmatrix} 2 & 5 & 4 \\ -3 & 2 & -6 \\ -6 & 4 & 7 \end{bmatrix} \quad \text{and} \quad A^3 = \begin{bmatrix} -5 & 16 & 9 \\ -2 & -5 & -23 \\ -23 & 9 & 11 \end{bmatrix}$$

and then

$$A^3 - 7A^2 + 19A - 19I = \begin{bmatrix} 0 & 0 & 0 \\ 0 & 0 & 0 \\ 0 & 0 & 0 \end{bmatrix}.$$

Note that we can now write $A^3 = 7A^2 - 19A + 19I$ so that

$$A^4 = AA^3 = 7A^3 - 19A^2 + 19A$$

$$= 7(7A^2 - 19A + 19I) - 19A^2 + 19A$$

$$= 49A^2 - 133A + 133I - 19A^2 + 19A = 30A^2 - 114A + 133I$$

and

$$A^5 = AA^4 = 30A^3 - 114A^2 + 133A$$

$$= 30(7A^2 - 19A + 19I) - 114A^2 + 133A$$

$$= 96A^2 - 437A + 570I.$$

Clearly we could continue the above process, thus expressing any power of A, and hence any polynomial in A, as a polynomial of degree 2 in A. We can also write $A^3 - 7A^2 - 19A = -19I$ or $(-1/19)(A^2 - 7A - 19I)A = I$ so that $A^{-1} = (-1/19)(A^2 - 7A - 19I)$ is also a second-degree polynomial in A. ■

The arguments used in Example 2 can be applied to any matrix to yield the following useful corollary of the Cayley-Hamilton theorem.

Theorem 5.13

For any $n \times n$ matrix A the following statements are true.

(1) If A^{-1} exists, it is a polynomial in A of degree less than n.
(2) A^k $(k \geq n)$ is a polynomial in A of degree less than n.
(3) For any polynomial $p(x)$, there exists a polynomial $r(x)$ of degree less than n such that $p(A) = r(A)$.

Exercise Set 5.4

For each of the following matrices, determine if it is orthogonally similar to a triangular matrix, and if it is, find an orthogonal triangularizing matrix.

1. $\begin{bmatrix} 5 & 1 \\ 4 & 8 \end{bmatrix}$ 2. $\begin{bmatrix} 3 & 2 \\ -2 & 4 \end{bmatrix}$ 3. $\begin{bmatrix} -2 & 1 & 0 \\ -4 & 0 & 3 \\ 0 & 1 & 2 \end{bmatrix}$

4. $\begin{bmatrix} 0 & 1 & 0 \\ 0 & 0 & 1 \\ 26 & -21 & 6 \end{bmatrix}$ 5. $\begin{bmatrix} 2 & 1 & 1 \\ 2 & 3 & 2 \\ 1 & 1 & 2 \end{bmatrix}$ 6. $\begin{bmatrix} \cos\theta & 0 & -\sin\theta \\ 0 & 1 & 0 \\ \sin\theta & 0 & \cos\theta \end{bmatrix}$

7. Write a computer subroutine that uses Theorem 3.20 to construct an orthogonal matrix whose first column is a scalar multiple of a given vector.

8. Show that if A and B are orthogonal matrices, then so are AB and

$$\begin{bmatrix} A & 0 \\ 0 & B \end{bmatrix}.$$

9. Let $\mathbf{w}$ be a unit vector in $\mathcal{R}^n$ and define $H = I - 2\mathbf{w}\mathbf{w}^T$.

   (a) Show that H is both symmetric and orthogonal.

   (b) Compute $H\mathbf{w}$.

   (c) Show that if $\mathbf{u}$ and $\mathbf{w}$ are orthogonal, then $H\mathbf{u} = \mathbf{u}$.

   (d) Use parts (b) and (c) to show that, for $n = 3$, H is a reflection in the plane perpendicular to $\mathbf{w}$. For this reason these matrices are called ELEMENTARY REFLECTORS.

   (e) Show that the orthogonal matrix of Theorem 3.20 is an elementary reflector with $\mathbf{w} = \dfrac{-1}{\sqrt{2(1 - x_1)}}\begin{bmatrix} x_1 - 1 \\ Y \end{bmatrix}$.

10. Assuming the existence of a subroutine EIGEN(n, A, X, r) which will compute one eigenvalue $\lambda = r$ and the corresponding eigenvector X, write a computer program that will triangularize an $n \times n$ matrix.

11. A real matrix is called NORMAL if $A^T A = AA^T$.

   (a) Show that if A is normal, then so is $P^T AP$ for any orthogonal P.

   (b) Show that a normal triangular matrix must be diagonal.

   (c) Show, as a corollary of Schur's theorem, that every normal matrix with real eigenvalues is orthogonally similar to a diagonal matrix and hence must be symmetric.

   (d) Show that any orthogonal matrix is also a normal matrix.

†12. Let T be an upper triangular matrix and let $p(x)$ be any polynomial.

   (a) Show that the diagonal elements of T^k are $(t_{ii})^k$.

   (b) Show that the diagonal elements of $p(T)$ are $p(t_{ii})$.

13. Verify the Cayley-Hamilton theorem for each of the matrices in Exercises 1–5.

14. For each of the matrices in Exercises 1–5, express A^{-1} as a polynomial in A.

15. If the polynomial $p(x)$ is divided by the polynomial $c(x)$, the result is a quotient $q(x)$ and a remainder $r(x)$ that is either zero or of degree lower than $c(x)$. These polynomials must satisfy

$$p(x) = q(x)c(x) + r(x).$$

Show that $p(A) = r(A)$ if $c(x)$ is the characteristic polynomial of A.

16. Let $p(x) = x^5 - 2x^4 + x^2 - 7x + 2$ and use the result of Exercise 15 to evaluate $p(A)$ for each of the matrices in Exercises 1–5.

17. Show that the inverse of a symmetric matrix must be symmetric.

18. Find J^k for $J = \begin{bmatrix} \lambda & 1 & 0 \\ 0 & \lambda & 1 \\ 0 & 0 & \lambda \end{bmatrix}$.

†**19.** Show that if a matrix A is orthogonally similar to a diagonal matrix, then A must be symmetric.

20. An $n \times n$ complex matrix U is called UNITARY if $U^H U = I$.

(a) Show that a real unitary matrix is orthogonal.

(b) Generalize Theorem 3.20 to the complex case. Two complex vectors X and Y are said to be orthogonal if $X^H Y = 0$.

(c) Generalize Schur's theorem to the case where A might have some complex eigenvalues.

5.5   *The Power Method**

The methods described in the text for computing eigenvalues and eigenvectors, although they are important for understanding the concepts involved, do not provide good algorithms for use with a modern digital computer. In this section we will describe one method that works well on a computer.

In most cases numerical algorithms are designed to produce a sequence of approximations to the solution of a problem rather than an exact solution. The generated sequence should converge to the exact solution of the problem, but the machine is incapable of carrying the process to the limit; the best we can do is accept one of the terms of the approximating sequence as being "good enough." The error introduced by this approximation is called the TRUNCATION ERROR.

A simple example of such a sequence of approximations is

$$x_0 = 1$$

$$x_{n+1} = \frac{2}{3} x_n + \frac{a}{3x_n^2}, \quad n = 0, 1, 2, \ldots,$$

*This section is optional.

which comes from using Newton's method (recall Exercise 24 of Section 2.4), for the function $f(x) = x^3 - a$, to compute $\sqrt[3]{a}$. For $a = 10$ the terms of this sequence are as follows, correct to six significant figures:

$$x_1 = 4$$

$$x_2 = 2.875$$

$$x_3 = 2.31994$$

$$x_4 = 2.16596$$

$$x_5 = 2.15449$$

$$x_6 = 2.15443$$

$$x_7 = 2.15443 .$$

If we accept 2.15443 as a "good enough" approximation, then we could check the accuracy of our approximation by computing the residual

$$10 - (2.15443)^3 = 10 - 9.999934693 = 6.5307 \times 10^{-5},$$

which indicates that our approximation is slightly in error. Now it can be shown that the sequence just listed converges to $\sqrt[3]{10}$, but we have stopped short of the limit and taken an approximation which is correct to six significant figures. If we use the very accurate approximation $\sqrt[3]{10} = 2.154434692$, we see that the truncation error in our approximation is

$$|\sqrt[3]{10} - 2.15443|, \text{ which is about } .000004692 = 4.692 \times 10^{-6}.$$

For more general algorithms the truncation error can only be approximated since the exact solution is unknown.

A second kind of error that is always present in machine calculations is ROUND-OFF ERROR. This error comes from the fact that the machine represents every number as a decimal with a finite number of significant figures. Thus the machine can only approximate numbers like $1/3 = 3.333333333\ldots$, $\sqrt{2} = 1.414213562\ldots$, and $\pi = 3.141592654\ldots$. After each calculation the machine either rounds or chops the result to the number of places that is proper for that machine.

A well-conceived algorithm should attempt to control both round-off error and truncation error. Detailed error analysis and the design of efficient algorithms is the subject matter of a branch of mathematics called numerical analysis. Readers who are interested in scientific computing should explore this area at the earliest possible opportunity.

If one has a matrix A and a nonzero vector X, then one can check to see if X is an eigenvector of A by comparing the entries of X and AX. If X is not an eigenvector, then it can be shown that AX is, in some sense, "closer" to an eigenvector than was X. This is the basis for the POWER METHOD that we will now describe.

Suppose that the eigenvalues of A are $\{\lambda_1, \lambda_2, \ldots, \lambda_n\}$ and that λ_1 is such that $|\lambda_1| > |\lambda_i|, i = 2, 3, \ldots, n$. In this case we say that λ_1 is the DOMINANT EIGENVALUE of A.

The power method will normally produce approximations to the eigenvector associated with the dominant eigenvalue of A. The basic algorithm is described here:

POWER METHOD

Y_0 arbitrary nonzero vector
For $i = 0, 1, 2, 3, \ldots$ (5.6)
    $Y_{i+1} = AY_i$
(Y_n approaches a dominant eigenvector of A)

Example 1

First solution: Use five iterations of the power method to estimate the dominant eigenvalue and eigenvector for the matrix $A = \begin{bmatrix} 1 & 2 \\ 2 & 3 \end{bmatrix}$.

If we take $Y_0 = \begin{bmatrix} 1 \\ 1 \end{bmatrix}$, then we have

$$Y_1 = AY_0 = \begin{bmatrix} 3 \\ 5 \end{bmatrix}, \quad Y_2 = AY_1 = \begin{bmatrix} 13 \\ 21 \end{bmatrix}, \quad Y_3 = AY_2 = \begin{bmatrix} 55 \\ 89 \end{bmatrix},$$

$$Y_4 = AY_3 = \begin{bmatrix} 233 \\ 377 \end{bmatrix}, \quad \text{and } Y_5 = AY_4 = \begin{bmatrix} 987 \\ 1597 \end{bmatrix}.$$

The sequence of Y_i is clearly not converging. Observe, however, that $987/233 = 4.23605$ and that $1597/377 = 4.23607$. Thus, $Y_5 = AY_4$ is approximately $(4.2361)Y_4$. We conclude that $\lambda = 4.2361$ is an approximation to an eigenvalue of A and that Y_4 is an approximation to the associated eigenvector. Since any scalar multiple of an eigenvector is also an eigenvector we can scale Y_4 to make the numbers smaller. Thus

$$X_1 = (1/233)Y_4 = \begin{bmatrix} 1.000 \\ 1.618 \end{bmatrix} \text{ is also an approximate eigenvector.}$$

For this simple problem it is easy to check the above calculations. The characteristic polynomial of A is $\lambda^2 - 4\lambda - 1$ and the roots, computed by the quadratic formula, are $(4 \pm \sqrt{20})/2$. The largest root, correct to ten places, is $\lambda_1 = 4.236067977$, so we see that five iterations of the power method have produced an approximation that is correct to five significant figures. This degree of accuracy would be adequate for most applications. If more accurate results are needed, then one can compute more terms of the sequence. ▬

As one can see, even for the simple matrix of Example 1 the elements of the Y_i may grow quickly. Since a scalar multiple of an eigenvector is still an eigenvector, it is advisable to scale the Y_i in order to keep the entries small. The scaling can be done in a number of ways; when working by hand it is best to do the scaling so that the largest entry of each Y_i is 1. When writing a program for a computer, it is best to scale by $\|Y\|$, thus turning each Y_i into a unit vector. In either case the sequence of scale factors approaches the absolute value of the dominant eigenvalue (refer to Exercise 6). If the dominant eigenvalue is negative, the signs of the eigenvector approximations will alternate.

POWER METHOD WITH SCALING

Y_0 arbitrary nonzero vector
For $i = 0, 1, 2, \ldots$

$$Y_{i+1}' = AY_i$$

$$s_{i+1} = \|Y_{i+1}'\| \quad \text{or} \quad s_{i+1} = \max_j \{|(Y_{i+1}')_j|\} \qquad (5.7)$$

$$Y_{i+1} = (1/s_{i+1})Y_{i+1}'$$

(Y_n approaches a dominant eigenvector of A and s_n approaches the absolute value of the dominant eigenvalue of A)

Example 1

Second solution: We will use the second scaling method of Method 5.7 to obtain an alternate solution of Example 1.

$$Y_0 = \begin{bmatrix} 1 \\ 1 \end{bmatrix}$$

$$Y_1' = AY_0 = \begin{bmatrix} 3 \\ 5 \end{bmatrix} \qquad s_1 = 5 \qquad Y_1 = \frac{Y_1'}{s_1} = \begin{bmatrix} 0.6 \\ 1.0 \end{bmatrix}$$

$$Y_2' = AY_1 = \begin{bmatrix} 2.6 \\ 4.2 \end{bmatrix} \qquad s_2 = 4.2 \qquad Y_2 = \frac{Y_2'}{s_2} = \begin{bmatrix} 0.619 \\ 1.000 \end{bmatrix}$$

$$Y_3' = AY_2 = \begin{bmatrix} 2.619 \\ 4.238 \end{bmatrix} \qquad s_3 = 4.238 \qquad Y_3 = \begin{bmatrix} 0.61798 \\ 1.00000 \end{bmatrix}$$

$$Y_4' = AY_3 = \begin{bmatrix} 2.61798 \\ 4.23596 \end{bmatrix} \quad s_4 = 4.23596 \quad Y_4 = \begin{bmatrix} 0.618036 \\ 1.000000 \end{bmatrix}$$

$$Y_5' = AY_4 = \begin{bmatrix} 2.618036 \\ 4.236074 \end{bmatrix} \quad s_5 = 4.236074 \quad Y_5 = \begin{bmatrix} 0.6180335 \\ 1.0000000 \end{bmatrix}$$

Therefore, a dominant eigenvector is approximately $X_1 = Y_5$ and the dominant eigenvalue is about $\lambda_1 = 4.23607$. Note that the eigenvector obtained here is an approximate scalar multiple of the one obtained in the first solution. ■

Example 2

Use the power method, with the first scaling method of 5.7, to compute the dominant eigenvector and corresponding eigenvalue for

$A = \begin{bmatrix} 1 & 1 & 3 \\ 1 & -2 & 1 \\ 3 & 1 & 3 \end{bmatrix}$. We take $Y_0 = \begin{bmatrix} 1 \\ 1 \\ 1 \end{bmatrix}$ and, using a programmable

pocket calculator, compute the following sequence of unit vectors.

$$Y_1' = AY_0 = \begin{bmatrix} 5 \\ 0 \\ 7 \end{bmatrix}; \quad s_1 = \sqrt{74} = 8.6023; \quad Y_1 = \frac{Y_1'}{s_1} = \begin{bmatrix} .5812 \\ 0 \\ .8137 \end{bmatrix}$$

$$Y_2' = \begin{bmatrix} 3.02244 \\ 1.39497 \\ 4.18491 \end{bmatrix}; \quad s_2 = 5.34739; \quad Y_2 = \begin{bmatrix} .565217 \\ .260870 \\ .782609 \end{bmatrix}$$

$$Y_3' = \begin{bmatrix} 3.17391 \\ .82609 \\ 4.30435 \end{bmatrix}; \quad s_3 = 4.1143; \quad Y_3 = \begin{bmatrix} .58652 \\ .15266 \\ .79542 \end{bmatrix}$$

$$Y_4' = \begin{bmatrix} 3.1254307 \\ 1.0766265 \\ 4.2984715 \end{bmatrix}; \quad s_4 = 5.42257; \quad Y_4 = \begin{bmatrix} .5763741 \\ .1985453 \\ .7926996 \end{bmatrix}$$

$$Y_5' = \begin{bmatrix} 3.1530182 \\ .9719831 \\ 4.3057664 \end{bmatrix}; \quad s_5 = 5.42456; \quad Y_5 = \begin{bmatrix} .58124818 \\ .12918177 \\ .79340967 \end{bmatrix}$$
$$\vdots \qquad\qquad\qquad \vdots \qquad\qquad\qquad \vdots$$

Continuing this iteration, we obtain, after thirty steps,

$$Y_{29} = Y_{30} = \begin{bmatrix} .5797702392 \\ .1849535588 \\ .7935103344 \end{bmatrix} \quad \text{and} \quad s_{29} = s_{30} = 5.425002162$$

as very accurate estimates of the dominant eigenvector and eigenvalue of the matrix A. ■

To understand how the power method works, we assume that A is diagonalizable and that $\{X_1, X_2, \ldots, X_n\}$ is a basis of $\mathcal{R}^n$ consisting of eigenvectors of A, with $AX_i = \lambda_i X_i$, $i = 1, 2, \ldots, n$.

The starting vector Y_0 in Method 5.6 can be written as a linear combination of the above basis vectors:

$$Y_0 = a_1 X_1 + a_2 X_2 + \ldots + a_n X_n.$$

If we repeatedly multiply this last equation by A we obtain

$$Y_1 = AY_0 = a_1 \lambda_1 X_1 + a_2 \lambda_2 X_2 + \ldots + a_n \lambda_n X_n$$

$$Y_2 = AY_1 = a_1 \lambda_1^2 X_1 + a_2 \lambda_2^2 X_2 + \ldots + a_n \lambda_n^2 X_2$$

$$\vdots$$

$$Y_k = aY_{k-1} = a_1 \lambda_1^k X_1 + a_2 \lambda_2^k X_k + \ldots + a_n \lambda_n^k X_k$$

$$= \lambda_1^k \left(a_1 X_1 + a_2 \left(\frac{\lambda_2}{\lambda_1}\right)^k X_2 + \ldots + a_n \left(\frac{\lambda_n}{\lambda_1}\right)^k X_n \right). \qquad (5.8)$$

Since λ_1 is the dominant eigenvalue of A, we have, for $i \neq 1$, $|\lambda_i / \lambda_1| < 1$, so that $(\lambda_i / \lambda_1)^k$ gets small as k gets large. Thus we see that Y_k approaches a scalar multiple of the dominant eigenvector X_1.

Equation 5.8 is the key to understanding what happens if A does not have a dominant eigenvalue ($\lambda_1 = \pm \lambda_2$) or if the dominant eigenvalue is complex. We leave the details, and some experimentation with the exceptional cases, to the exercises.

Many applications require only the dominant eigenvalue. If one needs more eigenvalues, then one can use the technique introduced in the proof of Schur's theorem to transform the $n \times n$ matrix A into

$\begin{bmatrix} \lambda_1 & * \\ 0 & A_2 \end{bmatrix}$ where A_2 is an $(n - 1) \times (n - 1)$ matrix whose eigenvalues

are also eigenvalues of A. The power method can then be used to obtain the dominant eigenvalue of A_2 and hence a second eigenvalue of A. In theory, one could use this procedure to find all the eigenvalues of A. In practice, the truncation errors accumulate so that only the first few eigenvalues are found accurately.

There are very powerful methods designed to accurately compute all of the eigenvalues and eigenvectors of an $n \times n$ matrix A. One such method is the QR-algorithm. This algorithm has been implemented in a very fine software package called EISPACK. The details of this method are too advanced for this text, although EISPACK can be used as a "black box" without understanding the details of the method.

Exercise Set 5.5

Use the power method, with the second scaling method of 5.7, to compute the dominant eigenvalue and dominant eigenvector of the matrices in Exercises 1–4.

1. $\begin{bmatrix} 0 & 1 \\ -7 & -6 \end{bmatrix}$
2. $\begin{bmatrix} 1 & 1 & 3 \\ 1 & -2 & 1 \\ 3 & 1 & 3 \end{bmatrix}$
3. $\begin{bmatrix} -3 & 5 & -2 \\ -6 & 8 & -2 \\ -4 & 4 & 1 \end{bmatrix}$
4. $\begin{bmatrix} 0 & 3 & 0 \\ 0 & 0 & 3 \\ 5 & 2 & 2 \end{bmatrix}$

5. Consider $A = \begin{bmatrix} 6 & -4 \\ 8 & -6 \end{bmatrix}$.

   (a) Find the characteristic polynomial and show that the eigenvalues are $\lambda_1 = 2$ and $\lambda_2 = -2$. Find eigenvectors for each of these eigenvalues.

   (b) Take $Y_0 = [2 \quad 1]^T$ and compute eight terms of the sequence Y_i, using the second scaling method of Method 5.7. Is this sequence converging?

   (c) What is the sequence $Y_0, Y_2, Y_4, Y_6, \ldots$ converging to? What is the sequence $Y_1, Y_3, Y_5, Y_7, \ldots$ converging to?

   (d) Show that in general the sequence $Y_0, Y_2, Y_4, \ldots$ can be used to compute the dominant eigenvalue and dominant eigenvector of the $n \times n$ matrix A^2, provided that the dominant eigenvalue is real.

   (e) Take $A = \begin{bmatrix} -2 & -1 & 4 \\ 2 & 1 & -2 \\ -1 & -1 & 3 \end{bmatrix}$ and $Y_0 = \begin{bmatrix} 3 \\ 1 \\ 2 \end{bmatrix}$ and compute six terms of the sequence defined by Method 5.6. What do your calculations show?

†6. Show that if Y is an eigenvector of A and $\|Y\| = 1$, then $\|AY\|$ is the absolute value of the corresponding eigenvalue.

7. Consider $A = \begin{bmatrix} 2 & 4 \\ -4 & 3 \end{bmatrix}$.

   (a) Find the characteristic polynomial and the eigenvalues of A.

   (b) Take $Y_0 = [1 \quad 1]^T$ and compute eight terms of the sequence Y_i using the second scaling method of Method 5.7.

   (c) Let $Z_7 = AY_6$ and $Z_8 = AZ_7$ and then compute $Z_8 - 5Z_7 + 22Y_6$. Does this procedure suggest a way of estimating the dominant complex eigenvalue of this matrix?

8. Suppose that the eigenvalues of A are $\{\lambda_1, \lambda_2, \ldots, \lambda_n\}$. Show that the eigenvalues of $A - kI$ are $\lambda_i - k$, $i = 1, 2, \ldots, n$. What is the dominant eigenvalue of $A - kI$ if k is an approximation for the dominant eigenvalue of A? Describe how this observation can be used to compute a second eigenvalue and eigenvector of A.

9. Let A be a symmetric matrix with eigenvalues $\lambda_1 > \lambda_2 > \ldots > \lambda_n$, and let X_1 be a unit eigenvector associated with the dominant eigenvalue λ_1. Let $X_1, X_2, \ldots, X_n$ be a full set of eigenvectors for A. Show that the matrix $A_1 = A - \lambda_1 X_1 X_1^T$ has eigenvalues $0, \lambda_2, \lambda_3, \ldots, \lambda_n$ and eigenvectors $X_1, X_2, \ldots, X_n$. Describe how this could be used, in conjunction

with the power method, to find all the eigenvectors and all the eigenvalues of A.

10. Write a computer subroutine EIGEN(n, A, X, r) that will take a given
matrix A and a given first approximation X and return an approximate
eigenvector X and the corresponding eigenvalue r. Stop the iteration
when $\|Y_{k+1} - Y_k\| \leq .0001$ or when $k = 50$, whichever comes first.

11. Combine the result of Exercise 10 and the program from Exercise 10 of
Section 5.4 to obtain a program which will compute all the real eigenvalues of an $n \times n$ matrix A. Test your program on some of the matrices
of this section.

Chapter 5   Review Exercises

1. Find the characteristic polynomial of $B = \begin{bmatrix} -1 & 1 & 3 \\ 2 & 1 & 1 \\ 4 & 2 & 3 \end{bmatrix}$.

2. For $A = \begin{bmatrix} 3 & -6 \\ -2 & 2 \end{bmatrix}$, find P such that $P^{-1}AP$ is diagonal.

3. Given that the characteristic polynomial of $A = \begin{bmatrix} 3 & -1 & 1 \\ -1 & 3 & -1 \\ 1 & -1 & 3 \end{bmatrix}$ is

$(\lambda - 2)^2(\lambda - 5)$,

(a) find a basis for the two-dimensional subspace of all eigenvectors of
A for $\lambda = 2$; and

(b) find an orthonormal basis for this subspace.

4. Let **T** be a linear operator on $\mathcal{R}^2$ whose matrix representative with
respect to the standard basis is $[\mathbf{T}]_{\mathcal{S}} = \begin{bmatrix} 1 & 1 \\ -2 & 4 \end{bmatrix}$.

(a) Find the matrix representative $[\mathbf{T}]_{\mathcal{B}}$ if
$$\mathcal{B} = \left\{ \mathbf{v}_1 = \begin{bmatrix} 1 \\ 1 \end{bmatrix}, \mathbf{v}_2 = \begin{bmatrix} 1 \\ 2 \end{bmatrix} \right\}.$$

(b) Find $\mathbf{T}(\mathbf{v}_1)$ and $\mathbf{T}(\mathbf{v}_2)$.

5. Given that the eigenvalues of $B = \begin{bmatrix} 2 & 1 & 1 \\ 2 & 3 & 2 \\ 1 & 1 & 2 \end{bmatrix}$ are 5, 1, and 1, find a

matrix P so that $P^{-1}AP$ is a diagonal matrix.

6. Let **T** be a linear operator on $\mathcal{R}^{1\times3}$ defined by $\mathbf{T}[x_1 \ \ x_2 \ \ x_3] =$
$[3x_1 - 5x_2 + 7x_3 \ \ \ 3x_2 - 4x_3 \ \ \ x_1 + x_2 + x_3]$.

(a) Find the matrix A that represents **T** with respect to the basis
$$\mathcal{B} = \{\mathbf{u}_1 = [1 \ \ 3 \ \ 2], \mathbf{u}_2 = [0 \ \ 1 \ \ 2], \mathbf{u}_3 = [0 \ \ 2 \ \ 3]\}.$$

(b) Use the matrix from part (a) to find $[\mathbf{T}(\mathbf{x})]_{\mathcal{B}}$ if
$\mathbf{x} = \mathbf{u}_1 + 2\mathbf{u}_2 - 3\mathbf{u}_3$.

(c) Let $\mathscr{S}$ be the standard basis for $\mathscr{R}^{1\times3}$. Let A' be the matrix which represents **T** with respect to $\mathscr{S}$. Find a matrix P such that $A = P^{-1}A'P$.

7. Find the characteristic equation and the eigenvalues of

$$A = \begin{bmatrix} 3 & -4 & 3 \\ -4 & 3 & 0 \\ 3 & 0 & 3 \end{bmatrix}.$$

8. For $A = \begin{bmatrix} 2 & 6 \\ 2 & 1 \end{bmatrix}$, find P such that $P^{-1}AP$ is diagonal.

9. Let $A = \begin{bmatrix} 1 & 3 & 0 \\ 4 & 2 & 1 \\ 2 & 7 & -2 \\ 0 & -1 & 5 \end{bmatrix}$ be the matrix representative of the linear

transformation **T** with respect to the basis

$$\mathscr{B} = \left\{ \mathbf{u_1} = \begin{bmatrix} 1 \\ 0 \\ 1 \end{bmatrix}, \mathbf{u_2} = \begin{bmatrix} 1 \\ 2 \\ 0 \end{bmatrix}, \mathbf{u_3} = \begin{bmatrix} 1 \\ 1 \\ 1 \end{bmatrix} \right\} \text{ for } \mathscr{R}^3$$

and the basis

$$\mathscr{B}' = \left\{ \begin{bmatrix} 1 & 1 \\ 1 & 1 \end{bmatrix}, \begin{bmatrix} 1 & 1 \\ 1 & 0 \end{bmatrix}, \begin{bmatrix} 1 & 1 \\ 0 & 0 \end{bmatrix}, \begin{bmatrix} 1 & 0 \\ 0 & 0 \end{bmatrix} \right\} \text{ for } \mathscr{R}^{2\times2}.$$

(a) Find $[\mathbf{T(u_1)}]_{\mathscr{B}'}$.

(b) Find $\mathbf{T(u_2)}$.

(c) Find $\mathbf{T}\begin{bmatrix} 6 \\ 7 \\ 4 \end{bmatrix}$.

10. Given that the characteristic equation of $D = \begin{bmatrix} 1 & 4 & -3 \\ 4 & 0 & -4 \\ -3 & -4 & 1 \end{bmatrix}$ is

$$\lambda^3 - 2\lambda^2 - 40\lambda - 64,$$

(a) find the eigenvalues of D;

(b) find a nonsingular matrix P such that $P^{-1}DP$ is a diagonal matrix; and

(c) find an orthogonal matrix Q such that Q^TDQ is a triangular matrix.

11. Let $\mathscr{V}$ be an arbitrary three-dimensional real vector space and let **T** be a linear operator on $\mathscr{V}$. You are given

$$\mathscr{B} = \{\mathbf{v_1, v_2, v_3}\} \quad \text{and} \quad \mathscr{B}' = \{\mathbf{w_1, w_2, w_3}\}$$

as two bases for V, and the following information.

$$\begin{array}{ll} \mathbf{T(v_1)} = \mathbf{v_1} + 5\mathbf{v_3} & \mathbf{w_1} = \mathbf{v_1} + \mathbf{v_2} + \mathbf{v_3} \\ \mathbf{T(v_2)} = 2\mathbf{v_2} - 5\mathbf{v_3} & \mathbf{w_2} = -\mathbf{v_2} + \mathbf{v_3} \\ \mathbf{T(v_3)} = 2\mathbf{v_1} - \mathbf{v_2} + \mathbf{v_3} & \mathbf{w_3} = 2\mathbf{v_1} - \mathbf{v_2} + 4\mathbf{v_3} \end{array}$$

$$\mathbf{v_1} = -3\mathbf{w_1} - 5\mathbf{w_2} + 2\mathbf{w_3}$$
$$\mathbf{v_2} = 2\mathbf{w_1} + 2\mathbf{w_2} - \mathbf{w_3}$$
$$\mathbf{v_3} = 2\mathbf{w_1} + 3\mathbf{w_2} - \mathbf{w_3}$$

(a) Find $[\mathbf{T}]_{\mathscr{B}}$, the matrix representative of $\mathbf{T}$ with respect to the $\mathscr{B}$ basis.

(b) Find a nonsingular matrix P such that $[\mathbf{T}]_{\mathscr{B}'} = P^{-1}[\mathbf{T}]_{\mathscr{B}} P$.

12. Find an orthogonal matrix R so that $R^T A R$ is upper triangular if

$$A = \begin{bmatrix} 3 & 1 \\ -1 & 1 \end{bmatrix}.$$

13. For $A = \begin{bmatrix} 3 & 4 & -3 \\ 4 & 2 & -4 \\ -3 & -4 & 3 \end{bmatrix}$, the characteristic equation is $\lambda^3 - 8\lambda^2 - 20\lambda$.

Find an orthogonal matrix Q so that $Q^T A Q$ is a diagonal matrix.

14. Let $A = \begin{bmatrix} 5 & 2 & 3 \\ -4 & 6 & 2 \end{bmatrix}$ be the matrix representative of the linear transformation $\mathbf{T}: \mathscr{R}^3 \to \mathscr{R}^2$, with respect to the basis

$$\mathscr{B} = \left\{ \mathbf{u}_1 = \begin{bmatrix} 1 \\ 1 \\ 1 \end{bmatrix}, \mathbf{u}_2 = \begin{bmatrix} 1 \\ -1 \\ -1 \end{bmatrix}, \mathbf{u}_3 = \begin{bmatrix} 4 \\ -1 \\ 2 \end{bmatrix} \right\} \text{ for } \mathscr{R}^3$$

and the basis $\mathscr{B}' = \left\{ \mathbf{v}_1 = \begin{bmatrix} 2 \\ 3 \end{bmatrix}, \mathbf{v}_2 = \begin{bmatrix} 3 \\ 5 \end{bmatrix} \right\}$ for $\mathscr{R}^2$.

(a) Find $[\mathbf{T}(\mathbf{u}_2)]_{\mathscr{B}'}$ (b) Find $\mathbf{T}(\mathbf{u}_3)$ (c) Find $\mathbf{T}\begin{bmatrix} 6 \\ -1 \\ 2 \end{bmatrix}$

6

Linear Programming

In this chapter we will see how to use the row reduction techniques of Chapter 1 to solve a class of optimization problems that occur frequently in a wide variety of applications to business, economics, and political and social situations. We will first discuss a simple example from a geometric point of view and then develop the simplex method for solving such problems.

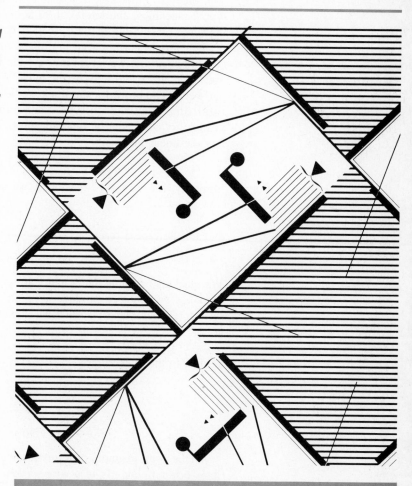

6.1   The Geometric Point of View

A certain process is capable of producing two items: type A and type B. The profit on each type A product is \$30 and the profit on each type B product is \$36. Manufacturing each product requires processing by three different machines. Each type A product requires 1 hour on machine 1, 1 hour on machine 2, and 3 hours on machine 3. Each type B product requires 2 hours on machine 1, 1 hour on machine 2, and 1 hour on machine 3. There is a total of 70 hours per week available on type 1 machines, a total of 40 hours per week available on type 2 machines, and a total of 90 hours per week available on type 3 machines. Assuming that all products can be sold, how many of each product should be produced in order to maximize the profit?

Suppose that we let x represent the number of type A items produced per week and let y be the number of type B products generated per week. The profit for the week would then be $P = 30x + 36y$. The total time required on type 1 machines is then $x + 2y$ and the availability constraint on type 1 machines means that x and y must satisfy the linear inequality

$$x + 2y \leq 70 .$$

Similarly, for the type 2 machines, we must have

$$x + y \leq 40 ,$$

and for the type 3 machines

$$3x + y \leq 90 .$$

We must also assume that x and y are greater than or equal to 0, since it makes no sense to produce a negative number of items. Thus we wish to choose x and y so that we maximize the objective function

$$P = 30x + 36y \tag{6.1}$$

subject to the constraints

$$
\begin{aligned}
x + 2y &\leq 70 & x &\geq 0 \\
x + y &\leq 40 & y &\geq 0 . \\
3x + y &\leq 90 & & \tag{6.2}
\end{aligned}
$$

This is an example of a *linear programming problem;* such problems occur frequently in a wide variety of business and scientific applications. We say that it is a linear problem because the objective function (see Equation 6.1) is a linear function. Here, "programming" refers to planning, not to computer programming. We will be able to solve this problem in a geometric way that will give insight into how we should proceed with more complicated problems of this type.

The set of all points (x, y) that satisfy the constraints (given in System 6.2) is called the FEASIBLE SET for the problem; its elements are called FEASIBLE SOLUTIONS. To see what the feasible set looks like we first observe that a straight line $ax + by = c$ divides the plane into two half planes, as shown in Figure 6.1.

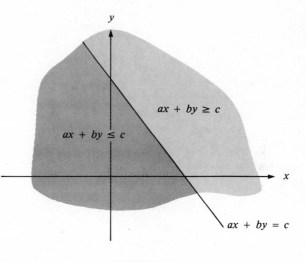

FIGURE 6.1

Thus, each of the linear inequalities in System 6.2 determines a half plane and the feasible set, determined by System 6.2, is the intersection of the five half planes shown in Figure 6.2 on the following page. Thus the feasible set determined by System 6.2 is the polygonal region shown in Figure 6.3 on the following page. The vertices of the polygon are determined by finding the points of intersections of the lines, two at a time.

We want to find the maximum value of $P = 30x + 36y$ for (x, y) in the feasible set. Since the feasible set contains an infinite number of points, it is clearly impossible to check the value of P for each point in the feasible set. We can gain some further insight into where the maximum occurs by interpreting Equation 6.1 geometrically.

For any fixed value of P, the objective function (Equation 6.1) represents a straight line with slope $-5/6$. Thus, we can interpret Equation 6.1 as a one-parameter family of parallel lines with P as the parameter. Several of these lines are shown in Figure 6.4 on page 301. Note that P increases as the lines move to the right of the origin.

FIGURE 6.2

FIGURE 6.3

FIGURE 6.4

If we superimpose Figure 6.4 on Figure 6.3 we obtain the diagram in Figure 6.5. Solving the optimization problem posed at the start of this section amounts to finding the rightmost line in Figure 6.4 that intersects the feasible set in Figure 6.3. Figure 6.5 shows that the maximum occurs for the line through the vertex $(10, 30)$, for which the value of P is 1380. It should be clear geometrically that the maximum value must occur at a vertex of the feasible set; at such a vertex at least two of the inequalities of System 6.2 must actually be equalities.

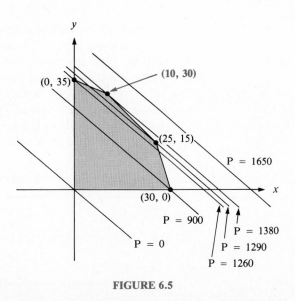

FIGURE 6.5

For more general problems of this type this observation is still valid; that is, *the optimal solution occurs at one of the vertices of the feasible set*. If there are three unknowns in the problem, then the feasible set is a polyhedron that is the intersection of several half spaces of R^3. If there are n unknowns, then the feasible set will be the intersection of several half spaces of R^n. For $n > 3$ it will be impossible to proceed geometrically, so we must devise algebraic methods to handle the problem.

The key to success in the algebraic treatment of the problem is to devise an efficient way of searching through the vertices of the feasible set for the optimal solution. This search involves finding the vertices and then deciding which of the vertices to check and in what order to check them. If the number of variables and the number of constraints are large, then the number of vertices of the feasible set may be so large that it is impractical to either find or check all of the vertices of the feasible set.

It is easier to deal with linear equations than with linear inequalities, so we begin by changing the constraints in System 6.2 into equalities. We make this change by introducing new variables into the inequalities. Thus, the inequality $x + 2y \leq 70$ is equivalent to the equation $x + 2y + u = 70$ with the restriction that $u \geq 0$. We will call variables like u SLACK VARIABLES; in many applications they represent the amount of unused resources. If we introduce slack variables into each of the inequalities of System 6.2, then we have the linear system

$$\begin{aligned} x + 2y + u \quad\quad\quad &= 70 \\ x + y \quad\quad + v \quad\quad &= 40 \\ 3x + y \quad\quad\quad\quad + w &= 90, \end{aligned} \qquad (6.3)$$

in which all the variables are required to be positive.

The linear system 6.3 has three equations in five unknowns and is of rank 3, so any two of the unknowns can be assigned values arbitrarily and then the other three values will be uniquely determined. A solution obtained by setting any two of the variables equal to 0 is called a BASIC SOLUTION of System 6.3. For example, if we set $x = y = 0$, then it follows at once that $u = 70$, $v = 40$, and $w = 90$. For this problem there are a total of ten basic solutions that are listed here:

	1	2	3	4	5	6	7	8	9	10
x	0	0	0	0	70	40	30	10	26	25
y	0	35	40	90	0	0	0	30	22	15
u	70	0	-10	-110	0	30	40	0	0	15
v	40	5	0	-50	-30	0	10	0	-10	0
w	90	55	50	0	-120	-30	0	30	0	0.

Of these ten basic solutions, only solutions 1, 2, 7, 8, and 10 are feasible solutions; the others fail to be feasible because at least one variable is negative. These feasible solutions are called BASIC FEASIBLE SOLUTIONS; in any basic feasible solution the BASIC VARIABLES are those that are not set equal to 0. For solution 8 the basic variables are x, y, and w. There is a correspondence between the basic feasible solutions of System 6.3 and the vertices of the feasible set shown in Figure 6.3; for each slack variable set equal to 0, the corresponding inequality in System 6.2 becomes an equality and the points that satisfy the equation are thus on the boundary of the feasible set. For example, in solution 10 the slack variables v and w are set equal to 0 so that the last two inequalities of System 6.2 become equalities and solution 10 corresponds to the intersection of the lines $x + y = 40$ and $3x + y = 90$. That intersection is the point $V_3 = (25, 15)$ of Figure 6.3. Similarly, solution 1 corresponds to the origin, solution 2 to $V_1 = (0, 35)$, solution 7 to $V_4 = (30, 0)$, and solution 8 to $V_2 = (10, 30)$.

The augmented matrix of System 6.3 is

$$\begin{bmatrix} 1 & 2 & 1 & 0 & 0 & 70 \\ 1 & 1 & 0 & 1 & 0 & 40 \\ 3 & 1 & 0 & 0 & 1 & 90 \end{bmatrix}. \tag{6.4}$$

Upon examination of Matrix 6.4 we see that the columns associated with the variables u, v, and w are columns of the identity matrix I. This situation makes it very easy to solve for u, v, and w if x and y are set equal to 0; this leads to the basic feasible solution 1 above.

Any matrix that is row equivalent to Matrix 6.4 is the augmented matrix of a linear system equivalent to System 6.3. We can find other basic feasible solutions by using elementary row operations to change Matrix 6.4 so that other columns become columns of I. For example, the row operations $R_2 \leftarrow -R_1 + R_2$ and $R_3 \leftarrow -3R_1 + R_3$ will change column 1 of Matrix 6.4 into the first column of I and will not change columns 4 and 5. This leads to the augmented matrix

$$\begin{bmatrix} 1 & 2 & 1 & 0 & 0 & 70 \\ 0 & -1 & -1 & 1 & 0 & -30 \\ 0 & -5 & -3 & 0 & 1 & -120 \end{bmatrix}.$$

If we now set $y = u = 0$, we find that $x = 70$, $y = -30$, and $z = -120$. This is solution 5 above, which is a basic solution of System 6.3 but not a basic feasible solution.

Similarly, the operations $R_1 \leftarrow 1/2R_1$, $R_2 \leftarrow -R_1 + R_2$, and $R_3 \leftarrow -R_1 + R_3$ on Matrix 6.4 lead to the augmented matrix

$$\begin{bmatrix} 1/2 & 1 & 1/2 & 0 & 0 & 35 \\ 1/2 & 0 & -1/2 & 1 & 0 & 5 \\ 5/2 & 0 & -1/2 & 0 & 1 & 55 \end{bmatrix}. \tag{6.5}$$

If we set $x = u = 0$ we obtain $y = 35$, $v = 5$, and $w = 55$. This is the second basic feasible solution listed in the table on page 302. It corresponds to the vertex V_1 of the feasible set shown in Figure 6.3.

For the original optimization problem (6.1), (6.2), which is equivalent to the optimization problem (6.1), (6.3), the easiest basic feasible solution to find is the first one in the table on page 302; that is, $[x \quad y \quad u \quad v \quad w] = [0 \quad 0 \quad 70 \quad 40 \quad 90]$. For this basic feasible solution the value of the objective function is $P = 30x + 36y = 0$, which is clearly not optimal. In order to increase P we need to increase one of the nonbasic variables; that is, we need to increase either x or y. Since a unit change in y will increase P more than a unit change in x, let us increase y while holding x fixed at $x = 0$. To see how much we can increase y, we look at the equations of System 6.3, with $x = 0$. These equations are

$$2y + u = 70 \qquad u = 70 - 2y$$
$$y + v = 40 \quad \text{or} \quad v = 40 - y$$
$$y + w = 90 \qquad w = 90 - y.$$

The restriction that $u \geq 0$ means that $y \leq 35$; $v \geq 0$ requires $y \leq 40$; and $w \geq 0$ requires $y \leq 90$. If $y \leq 35$, then u, v, and w will all be nonnegative. If we take $y = 35$, then $u = 0$, $v = 5$, and $w = 35$. This amounts to changing the set of basic variables from $\{u, v, w\}$ to $\{y, v, w\}$ and the new basic feasible solution is $[x \quad y \quad u \quad v \quad w] = [0 \quad 35 \quad 0 \quad 5 \quad 55]$, which corresponds to the vertex V_1 of the feasible set shown in Figure 6.3 on page 300.

The value of the objective function for this feasible solution is

$$P = 30x + 36y = 1260.$$

We have seen that System 6.3 is equivalent to the system whose augmented matrix is Matrix 6.5 and that this new system leads easily to the above basic feasible solution. The equations of that system are

$$\frac{x}{2} + y + \frac{u}{2} \qquad\qquad = 35$$

$$\frac{x}{2} \qquad - \frac{u}{2} + v \qquad = 5$$

$$\frac{5x}{2} \qquad - \frac{u}{2} \qquad + w = 55. \qquad\qquad (6.6)$$

If we now use the first equation of System 6.6 to eliminate y from Equation 6.1, thus expressing P in terms of the current nonbasic variables, we have

$$P = 30x + 36y = 30x + 36\left(35 - \frac{x}{2} - \frac{u}{2}\right) = 1260 + 12x - 18u.$$

In order to further increase P we need to increase x while holding $u = 0$, since increasing u will decrease P. To see how much we can increase x we look at the equations of System 6.6 which, with $u = 0$, reduce to

$$\frac{x}{2} + y = 35 \qquad y = 35 - \frac{x}{2}$$

$$\frac{x}{2} + v = 5 \quad \text{or} \quad v = 5 - \frac{x}{2}.$$

$$\frac{5x}{2} + w = 55 \qquad w = 55 - \frac{5x}{2}$$

Now $y \geq 0$ requires that $x \leq 70$, and $v \geq 0$ requires that $x \leq 10$, while $w \geq 0$ requires that $x \leq 22$. If we take $x = 10$, then $y = 30$, $v = 0$, and $w = 30$ are all nonnegative. We have again changed the set of basic variables, from $\{y, v, w\}$ to $\{x, y, w\}$, and the associated basic feasible solution is $[x \quad y \quad u \quad v \quad w] = [10 \quad 30 \quad 0 \quad 0 \quad 30]$, which corresponds to the vertex V_2 of the feasible set shown in Figure 6.3 on page 300. The value of the objective function for this feasible solution is

$$P = 1260 + 12x - 18u = 1380.$$

To see if any further increase is possible we need to express P in terms of the new nonbasic variables, and in order to do this we need to modify Matrix 6.5 so that the first column changes into the second column of I. The elementary row operations

$$R_2 \leftarrow 2R_2, R_1 \leftarrow \frac{-R_2}{2} + R_1, \quad \text{and} \quad R_3 \leftarrow \frac{-5R_2}{2} + R_3$$

change Matrix 6.5 to the matrix

$$\begin{bmatrix} 0 & 1 & 1 & -1 & 0 & 30 \\ 1 & 0 & -1 & 2 & 0 & 10 \\ 0 & 0 & 2 & -5 & 1 & 30 \end{bmatrix},$$

which is the augmented matrix of a system equivalent to System 6.3

$$y + u - v = 30$$

$$x - u + 2v = 10$$

$$2u - 5v + w = 30.$$

We can use the second equation of this system to eliminate x from the last expression for P and thus express P in terms of the current nonbasic variables:

$$P = 1260 + 12(10 - 2v + u) - 18u = 1380 - 24v - 6u.$$

Since any increase in the nonbasic variables u and v will result in a decrease in P we conclude that we have found the maximum P and the corresponding optimal production schedule for our original problem. Note that this is the same optimal solution that we obtained earlier by a geometric argument.

Exercise Set 6.1

1. Sketch the region in the xy-plane determined by the inequalities
$$3x + 2y \le 46 \qquad x \ge 0$$
$$3x - 4y \le 16 \qquad 0 \le y \le 8$$

2. Sketch the region in the xy-plane determined by the inequalities
$$2x + 15y \le 300$$
$$16x + 5y \le 330 \qquad x \ge 0$$
$$10x + 15y \ge 150 \qquad y \ge 0.$$
$$2x - 3y \le 34$$

3. Sketch the region in $\mathcal{R}^2$ determined by the inequalities
$$5x_1 - 6x_2 \le 38$$
$$x_1 - x_2 \ge -9 \qquad x_1 \ge 0$$
$$10x_1 + 7x_2 \ge 114 \qquad x_2 \ge 0.$$

4. Sketch the region in $\mathcal{R}^2$ determined by the inequalities
$$x_1 + x_2 \le 8$$
$$3x_1 - x_2 \le 12$$
$$-x_1 + 3x_2 \le 4.$$

5. Sketch the region in $\mathcal{R}^3$ determined by the inequalities
$$6x + 5y \le 60$$
$$6x - 13y + 6z \le 60 \qquad x \ge 0$$
$$-78x + 25y + 30z \le 300 \qquad y \ge 0$$
$$42x - 19y + 42z \le 636 \qquad z \ge 0.$$

6. For the region described in Exercise 1, find the indicated values.
   (a) Maximum value of $z_1 = 24x + 16y$
   (b) Minimum value of z_1
   (c) Maximum value of $z_2 = 12y + 30x$
   (d) Minimum value of z_2

7. For the region described in Exercise 2, find the indicated values.
   (a) Maximum value of $z_1 = 20x + 25y$
   (b) Minimum value of z_1
   (c) Maximum value of $z_2 = 20x - 15y$
   (d) Minimum value of z_2

8. For the region described in Exercise 3, find the indicated values.
   (a) Maximum value of $z = 8x_1 + 6x_2$
   (b) Minimum value of z

9. For the region described in Exercise 4, find the indicated values.
   (a) Maximum value of $z = 3x_1 + 5x_2$
   (b) Minimum value of z

10. For the region described in Exercise 5, find the indicated values.
    (a) Maximum value of $P = 10x + 12y + 15z$
    (b) Maximum value of $R = 18x + 32y - 17z$

11. Change the system of linear inequalities in Exercise 1 into a system of linear equations by adding slack variables. List all the basic solutions of this system and then find all the basic feasible solutions. Match the basic feasible solutions with the vertices of the region in Exercise 1.

12. Change the system of linear inequalities in Exercise 5 into a system of linear equations by adding slack variables.
    (a) How many basic solutions does this system have?
    (b) Find all basic solutions for which $x = z = 0$. Are any of them basic feasible solutions?

13. A truck farmer plans to grow corn, green beans, and zucchini. He has 15 acres of land available and $600 for seed and fertilizer. He has 80 hours per week of labor available to work the crops. The labor and capital requirements as well as the estimated profit for each crop are given below.

	Labor (hours)	Capital ($)	Profit ($)
Corn/acre	5	30	45
Green beans/acre	8	20	60
Zucchini/acre	3	15	20

How much of each crop should be planted in order to maximize the profit to the farmer?

14. The farmer in Exercise 13 has just received word that corn prices are going down while bean and zucchini prices are going up. The revised estimates are that corn will produce $40 profit per acre and beans will produce $70 profit per acre, while zucchini will produce $30 profit per acre. How should the farmer change his planting plans in order to take advantage of this change in the market?

15. An investment analyst must make decisions about the investment of a $600,000 trust fund. He can invest in common stocks, which pay a 4-percent dividend and are expected to have a 10-percent capital gain; he can invest in corporate bonds, which pay 12-percent interest and are expected to have a 1-percent capital gain; or he can invest in

government bonds, which pay a 10-percent dividend but have no potential for capital growth. His company restricts trust-fund investments so that no more than 60 percent can be invested in stocks, no more than 30 percent can be invested in corporate bonds, and at least 20 percent must be invested in government bonds.

(a) How should the funds be invested to generate the maximum amount of dividends and interest?

(b) The owner of the trust fund wants to use $40,000 of the income and will reinvest any additional income. How should the fund be invested to generate the maximum capital growth while meeting the income requirements?

6.2   *Different Types of Linear Programming Problems*

In the last section we considered a simple example of a special type of linear programming problem. In this section we will describe the general linear programming problem, show how to change the general problem into standard form, and consider some additional examples. We will also show how to use matrix notation to describe the standard linear-programming problem.

The GENERAL LINEAR PROGRAMMING PROBLEM, in the n variables $x_1, x_2, \ldots, x_n$, requires that we either maximize or minimize a linear objective function such as

$$z = c_1 x_1 + c_2 x_2 + \ldots + c_n x_n \tag{6.7}$$

subject to certain linear constraints of the form

$$a_{11} x_1 + a_{12} x_2 + \ldots + a_{1n} x_n \ (\leq, \geq, =) \ b_1$$
$$a_{21} x_1 + a_{22} x_2 + \ldots + a_{2n} x_n \ (\leq, \geq, =) \ b_2$$
$$\vdots \qquad \vdots \qquad \qquad \vdots \qquad \qquad \vdots$$
$$a_{m1} x_1 + a_{m2} x_2 + \ldots + a_{mn} x_n \ (\leq, \geq, =) \ b_m, \tag{6.8}$$

where in each equation in System 6.8 one and only one of the symbols $(\leq)$, $(\geq)$, and $(=)$ appear. The variables may, or may not, be required to be nonnegative.

We will say that the above problem is a STANDARD LINEAR PROGRAMMING PROBLEM if it is a maximization problem in which each of the constraints is a $(\leq)$ inequality and the variables are all constrained to be nonnegative. The problem considered in Section 6.1 was a standard linear programming problem. A general linear programming problem may fail to be in standard form for any of the following reasons.

(1) It is a minimization problem.
(2) Some of the constraints are either ($\geq$) or ($=$) constraints.
(3) Some variables are not required to be nonnegative.

Example 1

The problem of finding the minimum value of $z = 3x_1 - 2x_2 + 4x_3$ subject to the constraints

$$2x_1 + 3x_2 - 3x_3 \leq 4 \qquad x_1 \geq 0$$
$$3x_1 + 2x_2 + 2x_3 \leq 6 \qquad x_2 \geq 0$$
$$-x_1 + 3x_2 + 2x_3 \geq -7 \qquad x_3 \geq 0$$

is not a standard linear programming problem for two reasons: it is a minimization problem and the third constraint is a ($\geq$) inequality. ▪

Example 2

The problem of finding the maximum of $z = 5x_1 - 6x_2 + x_3$ subject to the constraints

$$x_1 + x_2 + x_3 \geq 1$$
$$x_1 + 2x_2 + 3x_3 \leq 6 \qquad x_1 \geq 0$$
$$x_1 + 3x_2 \qquad = 7 \qquad x_2 \geq 0$$

fails to be in standard form because the first and third constraints are not ($\leq$) inequalities and because x_3 is not constrained to be nonnegative. ▪

As we will now show, it is easy to change any linear programming problem into an equivalent standard linear programming problem.

If the given linear programming problem is a minimization problem, then we can change it into a maximization problem using the observation that

$$\min(z) = -\max(-z),$$

where z is any real-valued function. Thus, to minimize the objective function we can maximize its negative and then change the sign. Since the inequality $a \leq b$ is equivalent to the inequality $-a \geq -b$, changing a ($\geq$) constraint to a ($\leq$) constraint involves only changing the direction of the inequality by multiplying by -1.

Example 3

We can change the linear programming problem in Example 1 to a standard linear programming problem by multiplying the third inequality by -1 and by changing the signs of the coefficients in the objective function. The equivalent standard linear programming problem is then the following.

Maximize $w = -3x_1 + 2x_2 - 4x_3$ subject to the constraints

$$2x_1 + 3x_2 - 3x_3 \leq 4 \qquad x_1 \geq 0$$

$$3x_1 + 2x_2 + 2x_3 \leq 6 \qquad x_2 \geq 0$$

$$x_1 - 3x_2 - 2x_3 \leq 7 \qquad x_3 \geq 0. \quad \blacksquare$$

Since the equality $x = a$ is equivalent to the two inequalities $x \leq a$ and $-x \leq -a$ $(x \geq a)$, we can replace an $(=)$ constraint with two $(\leq)$ constraints.

Moreover, since any number, positive or negative, can be written as the difference of two positive numbers, we can replace any unconstrained variable x_j by the difference

$$x_j = x_{j+} - x_{j-}$$

where x_{j+} and x_{j-} are constrained to be nonnegative.

Example 4

We can change the nonstandard linear programming problem of Example 2 to a standard linear programming problem by multiplying the first inequality by -1, replacing x_3 by the difference $x_{3+} - x_{3-}$, and replacing the third constraint with two inequalities. The equivalent standard linear programming problem is then the following.

Maximize $z = 5x_1 - 6x_2 + x_{3+} - x_{3-}$ subject to the constraints

$$-x_1 - x_2 - x_{3+} + x_{3-} \leq -1 \qquad x_1 \geq 0$$

$$x_1 + 2x_2 + 3x_{3+} - 3x_{3-} \leq 6 \qquad x_2 \geq 0$$

$$x_1 + 3x_2 \qquad\qquad\quad \leq 7 \qquad x_{3+} \geq 0$$

$$-x_1 - 3x_2 \qquad\qquad\quad \leq -7 \qquad x_{3-} \geq 0. \quad \blacksquare$$

In order to use matrix notation to describe linear programming problems we need to introduce inequalities for matrices.

Definition 6.1

If A and B are matrices of the same size, say m by n, then $A \leq B$ means that $a_{ij} \leq b_{ij}$ for all relevant i and j. $A \geq B$, $A > B$, and $A < B$ are all defined in terms of their components in the obvious way.

Using Definition 6.1 we can describe the standard linear programming problem as follows: given an $n \times 1$ matrix C, an $m \times 1$ matrix B, and an $m \times n$ matrix A, we seek an $n \times 1$ matrix X, that is, an element of $\mathcal{R}^n$ that maximizes $z = C^T X$ subject to the constraints $AX \leq B$ and $X \geq 0$.

The FEASIBLE SET for this problem is the subset of $\mathcal{R}^n$ satisfying the constraints; that is,

$$\mathcal{FS} = \{X \mid AX \leq B \text{ and } X \geq 0\}.$$

We will say that the feasible set $\mathscr{FS}$ is BOUNDED if there exists a scalar r such that, for every X in $\mathscr{FS}$, $\|X\| = \sqrt{X^T X} \leq r$. In $\mathscr{R}^2$ this means that the feasible set is contained in a circle of radius r with its center at the origin. If $\mathscr{FS}$ is not bounded we will say it is UNBOUNDED. Examples of both bounded and unbounded feasible sets were contained in the exercises of Section 6.1.

The basic theoretical result about the standard linear programming problem is contained in the following theorem. The proof will be omitted.

Theorem 6.1

For the standard linear programming problem

$$\text{maximize } z = C^T X \text{ subject to } AX \leq B \text{ and } X \geq 0$$

either

(1) there is a unique solution;
(2) no solution exists; or
(3) there are an infinite number of solutions.

The geometric interpretation of the problem, given in Section 6.1, makes this result clear if the number of variables is two or three. The no-solution case can occur only if the feasible set is unbounded, but an unbounded feasible set does not necessarily mean that there is no optimal solution. If there are solutions, then they can be interpreted as vertices or extreme points of the feasible set. The case of an infinite number of solutions occurs when all the points on some "edge" or "face" of the feasible set are optimal solutions.

As we saw in Section 6.1, we can change the inequalities of a standard linear programming problem into equalities by introducing slack variables. For the standard linear programming problem,

$$\text{maximize } z = C^T X \text{ subject to } AX \leq B \text{ and } X \geq 0,$$

we introduce slack variables $s_1, s_2, \ldots, s_m$ satisfying

$$
\begin{aligned}
a_{11}x_1 + \ldots + a_{1n}x_n + s_1 \qquad\qquad &= b_1 \\
a_{21}x_1 + \ldots + a_{2n}x_n \qquad + s_2 \qquad &= b_2 \\
\vdots \qquad\qquad \vdots \qquad\qquad \vdots \\
a_{m1}x_1 + \ldots + a_{mn}x_n \qquad\qquad + s_m &= b_m
\end{aligned}
$$

and the additional constraints $s_1 \geq 0, s_2 \geq 0, \ldots, s_m \geq 0$. This last system can be written in matrix form as

$$[A\,|\,I]\begin{bmatrix} X \\ S \end{bmatrix} = B; \quad X \text{ and } S = [s_1 \quad s_2 \quad \cdots \quad s_m]^T \geq 0.$$

The objective function can also be written in terms of the slack variables as

$$z = c_1x_1 + \ldots + c_nx_n + 0s_1 + \ldots + 0s_m = [C^T \quad 0]\begin{bmatrix} X \\ S \end{bmatrix}.$$

A linear programming problem that requires the maximization of the objective function, in which all the constraints are (=) constraints and all the variables are constrained to be nonnegative, is said to be in CANONICAL FORM. By adding slack variables we can turn any standard linear-programming problem into a canonical linear programming problem. The solution techniques to be introduced in the next section require that the linear programming problem be in canonical form.

Example 5

The linear programming problem of Example 3 is equivalent to the following canonical problem.

Maximize $w = -3x_1 + 2x_2 - 4x_3$ subject to the constraints

$$2x_1 + 3x_2 - 3x_3 + s_1 \qquad\qquad = 4 \qquad x_1 \geq 0 \quad s_1 \geq 0$$
$$3x_1 + 2x_2 + 2x_3 \qquad + s_2 \qquad = 6 \qquad x_2 \geq 0 \quad s_2 \geq 0$$
$$x_1 - 3x_2 - 2x_3 \qquad\qquad + s_3 = 7 \qquad x_3 \geq 0 \quad s_3 \geq 0.$$

This problem can be written in matrix form:

$$\text{maximize } w = [-3 \quad 2 \quad -4 \quad 0 \quad 0 \quad 0]\begin{bmatrix} x_1 \\ x_2 \\ x_3 \\ s_1 \\ s_2 \\ s_3 \end{bmatrix} = [C^T \quad 0]\begin{bmatrix} X \\ S \end{bmatrix}$$

satisfying

$$[A \,|\, I]\begin{bmatrix} X \\ S \end{bmatrix} = \begin{bmatrix} 2 & 3 & -3 & 1 & 0 & 0 \\ 3 & 2 & 2 & 0 & 1 & 0 \\ 1 & -3 & -2 & 0 & 0 & 1 \end{bmatrix}\begin{bmatrix} X \\ S \end{bmatrix} = B = \begin{bmatrix} 4 \\ 6 \\ 7 \end{bmatrix}$$

and $\begin{bmatrix} X \\ S \end{bmatrix} \geq 0$. ■

Let us now consider the canonical linear programming problem:

$$\text{maximize } z = C^TX \text{ subject to } AX = B \text{ and } X \geq 0,$$

where A is m by $(m + n)$. We can find an element of the feasible set for this problem by assigning nonnegative values to n of the variables and then computing the values of the remaining variables from the system $AX = B$. If the values turn out to be nonnegative, then we have found a feasible solution.

If n of the variables are set to 0, then we will call the resulting solution a BASIC SOLUTION; the variables set to 0 are called the NON-BASIC VARIABLES, and the variables that are computed are called the BASIC VARIABLES. If a basic solution is also a member of the feasible set, then it is called a BASIC FEASIBLE SOLUTION.

The basic feasible solutions are of importance because they correspond to the vertices of the feasible set. The following theorem is the basis for the computational procedure that will be introduced in the next section.

Theorem 6.2

If the canonical linear-programming problem

$$\text{maximize } z = C^T X \text{ subject to } AX = B \text{ and } X \geq 0$$

has an optimal solution, then there is a basic feasible solution that is optimal.

Exercise Set 6.2

For Exercises 1–5, find an equivalent problem in standard form and rewrite the problem in matrix form.

1. Maximize $z = 2x + 5y$ subject to
$$\begin{array}{l} 2x + y \geq 2 \\ x + y \leq 8 \\ x + y \geq 3 \end{array} \quad \begin{array}{l} x \geq 0 \\ y \geq 0. \end{array}$$

2. Minimize $P = 2x + 3y + z$ subject to
$$\begin{array}{l} x + y + z \leq 12 \\ x + 3y + 3z = 24 \\ 3x + 6y + 4z \leq 90 \end{array} \quad \begin{array}{l} x \geq 0 \\ z \geq 0. \end{array}$$

3. Minimize $C = 3x - 4y + 6z$ subject to
$$\begin{array}{l} x + y + z = 1 \\ 2x + 3y + 5z \leq 4 \\ 4x + 3y + z \leq 2 \end{array} \quad x \geq 0.$$

4. Minimize $P = 6x + y$ subject to
$$\begin{array}{l} x + y \leq 8 \\ 2x - y \geq -2 \\ 2x + y \leq 2 \end{array} \quad \begin{array}{l} x \geq 0 \\ y \geq 0. \end{array}$$

5. Minimize $C = x + 3y + 2u + 2v$ subject to
$$\begin{array}{l} x + y = 150 \\ u + v = 150 \\ x + u \leq 200 \\ y + v \leq 250 \end{array} \quad \begin{array}{l} x \geq 0 \\ y \geq 0 \\ u \geq 0 \\ v \geq 0. \end{array}$$

6. For each of the problems in Exercises 1–5, find an equivalent problem in canonical form.

7. Find all basic solutions and all basic feasible solutions of
$$\begin{array}{l} 2x + y + s_1 = 6 \\ x + y + s_2 = 2. \end{array}$$

8. For the standard linear programming problem

   maximize $z = 24x + 16y$ subject to
   $$\begin{array}{ll} 3x + 2y \le 46 & x \ge 0 \\ 3x - 4y \le 16 & 0 \le y \le 10, \end{array}$$

   find the equivalent canonical problem. Find all basic solutions of the canonical restraint system and identify which of the basic solutions is also feasible.

9. Verify or find a counterexample for each of the following assertions about matrix inequalities.

   (a) $A \ge B$ implies $-A \le -B$.

   (b) For any two matrices A and B either $A > B$, $A < B$, or $A = B$.

   (c) $A \ge 0$ and $B \ge C$ implies $AB \ge AC$.

   (d) $A \ge B$ and $C \ge D$ implies $A + C \ge B + D$.

10. A subset $\mathcal{S}$ of $\mathcal{R}^2$ or $\mathcal{R}^3$ is CONVEX if whenever P and Q are any two points of $\mathcal{S}$, then the line segment joining P and Q lies entirely in $\mathcal{S}$.

    (a) Which of the following subsets of $\mathcal{R}^2$ are convex?

    (b) For a subset $\mathcal{S}$ of $\mathcal{R}^2$ or $\mathcal{R}^3$ show that $\mathcal{S}$ is convex if and only if $tx + (1 - t)y$ is in $\mathcal{S}$ for every x and y in $\mathcal{S}$ and every real number t such that $0 \le t \le 1$.

    Define a convex set in $\mathcal{R}^n$ by generalizing part (b); $\mathcal{S} \subseteq \mathcal{R}^n$ is convex if and only if $tx + (1 - t)y$ is in $\mathcal{S}$ for every x and y in $\mathcal{S}$ and for every real number t such that $0 \le t \le 1$.

    (c) Show that the half space of $\mathcal{R}^n$ defined by
    $$a_1x_1 + a_2x_2 + \ldots + a_nx_n \le b$$
    is a convex set in $\mathcal{R}^n$.

    (d) Show that the intersection of convex sets is a convex set.

    (e) Show that the feasible set of a linear-programming problem is a convex set.

11. A truck-rental company has two models of refrigerated trucks available for rent. The old model has 2 cubic yards of refrigerated space, 4 cubic yards of unrefrigerated space, and costs $.35 per mile to rent. The new model has 3 cubic yards of refrigerated space, 3 cubic yards of unrefrigerated space, and costs $.50 per mile to rent. A shipper needs 90 cubic

yards of refrigerated space and 120 cubic yards of unrefrigerated space. How many of each type of truck should he rent to keep his shipping costs to a minimum? Find a linear programming problem that models this situation.

12. A nutritionist wants to make an egg and milk mixture that contains at least 480 calories, at least 30 grams of protein, and at least 1600 units of vitamin A. Each large egg contains 80 calories, 6 grams of protein, 580 units of vitamin A, and costs $.08. Each quart of whole milk contains 640 calories, 40 grams of protein, 1360 units of vitamin A, and costs $.60. How should the two ingredients be combined so that the cost of the resulting mixture is minimized? Find a linear programming problem that models this situation. Solve the problem geometrically.

13. A local brewery makes four products: premium beer, premium ale, regular beer, and regular ale. The current stock of ingredients consists of 480 pounds of corn, 160 ounces of hops, and 1190 pounds of malt. There are 280 hours of labor available to process these ingredients. Each barrel of premium beer requires 5 pounds of corn, 4 ounces of hops, 40 pounds of malt, 14 hours of labor, and will generate a profit of $23. Each barrel of regular beer uses 10 pounds of corn, 4 ounces of hops, 35 pounds of malt, 7 hours of labor, and generates a profit of $15. Each barrel of premium ale uses 15 pounds of corn, 2 ounces of hops, 22 pounds of malt, 14 hours of labor, and generates a profit of $22. Each barrel of regular ale uses 17 pounds of corn, 2 ounces of hops, 22 pounds of malt, 14 hours of labor, and generates a profit of $14. How many barrels of each product should be made in order to maximize profit? Find a canonical linear programming problem to describe this situation.

6.3   *The Simplex Method*

In this section we will describe the *simplex method* for solving the canonical linear programming problem. The *simplex method* was developed by George Dantzig in 1947. This method, which uses relatively simple techniques, has been very successful in providing solutions to a wide class of linear programming problems arising in applications. The basic ideas of the method were illustrated in the algebraic solution of the simple production-planning problem discussed in detail in Section 6.1.

We will describe the method for solving a standard linear programming problem, or equivalently for solving the canonical linear programming problem that arises from a standard linear programming problem when slack variables are added to each of the constraints. The method starts from the basic feasible solution in which the slack variables are the basic variables, and at each step moves from the current basic feasible solution to another basic feasible solution for which the value of the objective function is larger. At each step the set of basic variables is modified by bringing in one new basic variable (the ENTERING VARIABLE) and deleting one of the old basic variables (the DEPARTING VARIABLE). At each step the constraint system is modified by elementary row operations to make the computation of the new basic feasible solution easy and to express the objective function in terms of the new nonbasic variables.

To deal with the computations we will introduce the SIMPLEX TABLEAU; it is essentially the augmented matrix of the constraint system with an additional row representing the objective function. Consider again the canonical form for the motivating problem of Section 6.1:

maximize $z = 30x + 36y$ subject to the constraints

$$x + 2y + u \qquad\qquad = 70$$
$$x + \ y \qquad + v \qquad = 40$$
$$3x + \ y \qquad\qquad + w = 90$$

$$x \geq 0$$
$$y \geq 0.$$

The initial tableau for this problem is

		x	y	u	v	w	z	
Current	u	1	2	1	0	0	0	70
basic	v	1	1	0	1	0	0	40
variables	w	3	1	0	0	1	0	90
		-30	-36	0	0	0	1	0 ← objective row .

(6.9)

The first row and first column of the tableau are labels; the labels in the first column specify the current basic variables. The center part of the tableau is the augmented matrix of the constraint system, with an extra column for the variable z. The last row represents the objective function written in the form

$$-30x - 36y - 0u - 0v - 0w + z = 0.$$

In general, the canonical linear programming problem, maximize $z = C_1^T X + u_1$ subject to $A_1 X = B_1$ and $X \geq 0$, would lead to the tableau

	A_1	0	B_1
	$-C_1^T$	1	u_1 .

If $B_1 \geq 0$ and the matrix A_1 has the identity matrix I as an $m \times m$ submatrix, then we can easily obtain a basic feasible solution by taking as the basic variables those variables corresponding to the columns of A_1

which are columns of I. Note that Tableau 6.9 has these properties; the *simplex method* is designed to preserve these properties at each step. Moreover, if the objective row has zeros in the columns associated with the basic variables, then the value of the objective function at the basic feasible solution is u_1.

For example, the *simplex tableau*

	x_1	x_2	x_3	x_4	x_5	x_6	z	
x_3	0	5	1	−4	0	9	0	6
x_1	1	−2	0	3	0	−5	0	2
x_5	0	3	0	6	1	4	0	7
	0	−7	0	8	0	10	1	150

(6.10)

makes it easy to compute the basic feasible solution associated with the basic variables $\{x_3, x_1, x_5\}$. That solution, which is obtained by setting $x_2 = x_4 = x_6 = 0$, is

$$X = [2 \quad 0 \quad 6 \quad 0 \quad 7 \quad 0]^T.$$

From the last row of the tableau we see that the objective function, expressed in terms of the nonbasic variables, is

$$z = 7x_2 - 8x_4 - 10x_6 + 150,$$

and that the value of the objective function for the current basic feasible solution is $z = 150$. Note that we could increase the value of the objective function by increasing the nonbasic variable x_2 (from its current value of 0 to a positive value), but increasing x_4 or x_6 would decrease the value of the objective function.

This argument extends easily to show that *a basic feasible solution is optimal if and only if there are no negative entries in the objective row of the associated tableau*.

In going from one tableau to the next, the choice of the entering variable is made so as to increase the objective function as much as possible. *Thus we choose, as the entering variable, the variable with the most negative element in the objective row; in case of a tie choose either.* For our motivating problem we see, from Tableau 6.9, that y should be the entering variable for the first main step.

The choice of the departing variable is dictated by the requirement that the new basic solution should be a basic feasible solution; that is, it must be a nonnegative basic solution.

For Tableau 6.9, y is to become a basic variable, but x will still be nonbasic at the next step. Thus $x = 0$ for the new basic feasible solution and the other variables must satisfy the constraint equations

$$2y + u = 70, \quad y + v = 40, \quad \text{and} \quad y + w = 90.$$

It follows that $u = 70 - 2y \geq 0$ if and only if $y \leq 35 = 70/2$. Similarly, $v = 40 - y \geq 0$ if and only if $y \leq 40$, and $w = 90 - y \geq 0$ if and only if $y \leq 90$. All (u, v, and w) will be nonnegative if $y \leq 35$. Thus, the maximum possible value for the entering variable is $y = 35$. This value will force $u = 0$ so that u will be the departing basic variable.

For Tableau 6.10, x_2 is to become a basic variable but x_4 and x_6 will remain nonbasic. Thus, for the new basic feasible solution, $x_4 = x_6 = 0$. The equations of the constraint system then reduce to

$$5x_2 + x_3 = 6, \quad x_1 - 2x_2 = 2, \quad \text{and} \quad 3x_2 + x_5 = 7.$$

We therefore see that $x_3 \geq 0$ if and only if $x_2 \leq 6/5$, $x_1 \geq 0$ if and only if $x_2 \geq -1$, and $x_5 \geq 0$ if and only if $x_2 \leq 7/3$. All (x_3, x_1, and x_5) will be nonnegative if $x_2 \leq 6/5$. Thus, the maximum possible value for the entering variable is $x_2 = 6/5$; this forces $x_3 = 0$ so that x_3 is the departing variable. Note that the negative entry in the 2, 2 position of Tableau 6.10 leads to an equation that imposes no additional restriction on x_2; it is already required to be nonnegative.

In general we divide the entries in the last column of the tableau by the corresponding positive entries in the column associated with the entering variable. We will call these ratios the FEASIBILITY RATIOS for the basic variables. *We choose as the departing variable the basic variable with the smallest nonnegative feasibility ratio; in case of a tie choose either.* It follows that the new basic solution will be nonnegative and hence a basic feasible solution.

For Tableau 6.9 the feasibility ratios are as follows:

$$\text{for } u: \frac{70}{2} = 35, \quad \text{for } v: \frac{40}{1} = 40, \quad \text{and} \quad \text{for } w: \frac{90}{1} = 90.$$

Thus, u should be chosen as the departing variable.

For Tableau 6.10 the feasibility ratios are as follows:

$$\text{for } x_3: \frac{6}{5} = 1.2, \quad \text{and} \quad \text{for } x_5: \frac{7}{3} = 2.33.$$

Thus, x_3 should be chosen as the departing variable. Note that we do not compute a feasibility ratio for x_1 since the entry in the pivot column corresponding to x_1 is negative.

Once the new set of basic variables has been determined, we need to modify the tableau in order to find the new basic feasible solution and a new description of the objective function. We do this modification by using elementary row operations to change the column associated with the entering variable (the PIVOT COLUMN) into the column of the identity matrix I that was associated with the departing variable in the old tableau. In the linear programming literature, this modification of the tableau is called PIVOTING.

THE SIMPLEX METHOD

(1) Select the entering variable (pivot column) by choosing a column with the most negative entry in the objective row.

(2) Select the departing variable (pivot row) by choosing a row that has a positive entry in the pivot column and for which the feasibility ratio is minimum.

(3) Use elementary row operations to change the element at the intersection of the pivot row and the pivot column (the pivot) to 1 and to change all other entries in the pivot column to 0.

(4) Stop if all the entries in the objective row are nonnegative; otherwise go to step 1.

Example 1

Use the simplex method to solve the motivating example discussed in Section 6.1. The initial tableau for this problem was given in Tableau 6.9. It is

We have already decided that the entering variable should be y and that the departing variable should be u. The pivot row and the pivot column have been highlighted and marked with small arrows. The pivot is the entry 2 in the $1, 2$ position of the tableau. The reduction described in step 3 of the simplex method can be accomplished by the following sequence of elementary row operations:

$$R_1 \leftarrow \left(\frac{1}{2}\right)R_1, \quad R \leftarrow -R_1 + R_2, \quad R_3 \leftarrow -R_1 + R_3,$$

$$\text{and} \quad R_4 \leftarrow 36R_1 + R_4.$$

The new tableau and the associated basic feasible solution are

	x	y	u	v	w	z	
y	1/2	1	1/2	0	0	0	35
v	1/2	0	−1/2	1	0	0	5
w	5/2	0	−1/2	0	1	0	55
	−12	0	18	0	0	1	1260

$$X = \begin{bmatrix} x \\ y \\ u \\ v \\ w \end{bmatrix} = \begin{bmatrix} 0 \\ 35 \\ 0 \\ 5 \\ 55 \end{bmatrix}.$$

This solution is not optimal because there is one negative entry (-12) in the objective row. Thus the entering variable for the next stage is x and the pivot column is the first column. The feasibility ratios are as follows.

$$y: \frac{35}{1/2} = 70 \qquad v: \frac{5}{1/2} = 10 \qquad w: \frac{55}{5/2} = 22$$

We therefore choose v as the departing variable. Thus the pivot is the $1/2$ in the 2, 1 position of the tableau. We now modify the tableau using the elementary row operations

$$R_2 \leftarrow 2R_2, \quad R_1 \leftarrow \frac{-R_2}{2} + R_1, \quad R_3 \leftarrow \frac{-5R_2}{2} + R_3,$$

$$\text{and} \quad R_4 \leftarrow 12R_2 + R_4.$$

The modified tableau and the associated basic feasible solution are

	x	y	u	v	w	z	
y	0	1	1	−1	0	0	30
x	1	0	−1	2	0	0	10
w	0	0	2	−5	1	0	30
	0	0	6	24	0	1	1380

$$\text{and} \quad X = \begin{bmatrix} x \\ y \\ u \\ v \\ w \end{bmatrix} = \begin{bmatrix} 10 \\ 30 \\ 0 \\ 0 \\ 30 \end{bmatrix}.$$

Since there are no negative elements in the objective row, we conclude that we have found the optimal solution. The maximum value of z is 1380. ■

Example 2

Use the simplex method to solve the following standard linear programming problem.

$$\text{Maximize } z = 8x_1 + 9x_2 + 5x_3 \text{ subject to}$$

$$\begin{aligned} x_1 + x_2 + 2x_3 &\leq 2 \\ 3x_1 + 4x_2 + 6x_3 &\leq 5 \quad \text{and} \\ 7x_1 + 7x_2 + 4x_3 &\leq 10 \end{aligned} \qquad \begin{aligned} x_1 &\geq 0 \\ x_2 &\geq 0 . \\ x_3 &\geq 0 \end{aligned}$$

If we add slack variables (x_4, x_5, x_6) to each of the inequalities to change the problem into canonical form, we obtain the initial tableau and initial feasible solution given below.

$$
\begin{array}{c|ccccccc|c}
 & x_1 & x_2 & x_3 & x_4 & x_5 & x_6 & z & \\
\hline
x_4 & 1 & 1 & 2 & 1 & 0 & 0 & 0 & 2 \\
x_5 & 3 & 4 & 6 & 0 & 1 & 0 & 0 & 5 \\
x_6 & 7 & 7 & 4 & 0 & 0 & 1 & 0 & 10 \\
\hline
 & -8 & -9 & -5 & 0 & 0 & 0 & 1 & 0
\end{array}
\qquad
X = \begin{bmatrix} 0 \\ 0 \\ 0 \\ 2 \\ 5 \\ 10 \end{bmatrix}.
$$

The most negative entry in the objective row is the -9 in the x_2 column, so we select x_2 as the entering variable and the second column as the pivot column. The feasibility ratios are $2/1 = 2$, $5/4 = 1.25$, and $10/7 = 1.42$, so we select x_5 as the departing variable. The pivot is thus the entry 4 in the 2, 2 position of the tableau. The pivoting operation is accomplished by the elementary row operations

$$
R_2 \leftarrow \frac{R_2}{4}, \quad R_1 \leftarrow -R_2 + R_1, \quad R_3 \leftarrow -7R_2 + R_3, \quad \text{and}
$$

$$
R_4 \leftarrow 9R_2 + R_4 .
$$

The modified tableau is then

$$
\begin{array}{c|ccccccc|c}
 & x_1 & x_2 & x_3 & x_4 & x_5 & x_6 & z & \\
\hline
x_4 & 1/4 & 0 & 2/4 & 1 & -1/4 & 0 & 0 & 3/4 \\
x_2 & 3/4 & 1 & 6/4 & 0 & 1/4 & 0 & 0 & 5/4 \\
x_6 & 7/4 & 0 & -26/4 & 0 & -7/4 & 1 & 0 & 5/4 \\
\hline
 & -5/4 & 0 & 17/2 & 0 & 9/4 & 0 & 1 & 45/4,
\end{array}
$$

from which we conclude that the value of z for the basic feasible solution $X = [0 \quad 5/4 \quad 0 \quad 3/4 \quad 0 \quad 5/4]^T$ is $45/4$. Since there are still negative entries in the objective row, this solution is not optimal.

For the next step of the simplex method the entering variable will be x_1. The feasibility ratios are then $(3/4)/(1/4) = 3$, $(5/4)/(3/4) = 1.67$, and $(5/4)/(7/4) = .714$, so we must select x_6 as the departing variable. The pivot is then the 3, 1 entry of the tableau, and the operations

$$
R_3 \leftarrow \frac{4R_3}{7}, \quad R_2 \leftarrow \frac{-3}{4}R_3 + R_2, \quad R_1 \leftarrow \frac{-1}{4}R_3 + R_1, \quad \text{and}
$$

$$
R_4 \leftarrow \frac{5}{4}R_3 + R_4
$$

lead to the modified tableau

	x_1	x_2	x_3	x_4	x_5	x_6	z	
x_4	0	0	10/7	1	0	$-1/7$	0	4/7
x_2	0	1	30/7	0	1	$-3/7$	0	5/7
x_1	1	0	$-26/7$	0	-1	4/7	0	5/7
	0	0	27/7	0	1	5/7	1	85/7,

from which we conclude that the optimal solution of the problem is $X = [5/7 \quad 5/7 \quad 0 \quad 4/7 \quad 0 \quad 0]^T$, for which $z = 85/7$. ▬

If there is a tie in the selection process for the entering variable (see step 1), then any one of the tied variables may be chosen. Let us agree to always choose the variable with the smallest subscript as the entering variable.

Ties in the choice of the departing variable require more attention. If two basic variables have the same minimum nonnegative feasibility ratio, then we can choose either as the departing variable, but the variable not chosen will be a basic variable with value zero at the next step. At the subsequent step this variable will have a feasibility ratio of zero and is the natural choice for the departing variable if the corresponding entry in the pivot column is positive. It is easy to see that the next pivoting step will not increase the value of the objective function. For this reason such a problem is said to be DEGENERATE. With degenerate problems there is a slim possibility that the algorithm will enter an infinite cycle and hence fail to find the optimal solution. We will leave a detailed discussion of the cycling phenomenon to more advanced texts, noting only that it rarely occurs in practical problems.

Example 3

Use the simplex method to solve the maximization problem in Exercise 10(a) of Section 6.1.

We set $x = x_1$, $y = x_2$, and $z = x_3$ and denote the slack variables by x_4, x_5, x_6, and x_7. The initial tableau is then

	x_1	x_2	x_3	x_4	x_5	x_6	x_7	P	
x_4	6	5	0	1	0	0	0	0	60
x_5	6	-13	6	0	1	0	0	0	60
x_6	-78	25	30	0	0	1	0	0	300
x_7	42	-19	42	0	0	0	1	0	636
	-10	-12	-15	0	0	0	0	1	0.

We choose x_3 as the entering variable and compute the feasibility ratios $60/6 = 10$ for x_5, $300/30 = 10$ for x_6, and $636/42 = 15.14$ for x_7. Note that there is a tie in the choice of the departing variable. We choose x_5 and pivot on the 2, 3 position of the tableau. This step yields

	x_1	x_2	x_3	x_4	x_5	x_6	x_7	P	
x_4	6	5	0	1	0	0	0	0	60
x_3	1	$-13/6$	1	0	1/6	0	0	0	10
←— x_6	-108	90	0	0	-5	1	0	0	0
x_7	0	72	0	0	-7	0	1	0	216
	5	$-89/2$	0	0	5/2	0	0	1	150.

Note that the tie for departing variable at step 1 has resulted in a value of 0 for the basic variable x_6. Thus this is an example of a **degenerate** problem. We now choose x_2 as the entering variable and compute the feasibility ratios

$$60/5 = 12 \text{ for } x_4, \quad 0/90 = 0 \text{ for } x_6, \quad \text{and} \quad 216/72 = 3 \text{ for } x_7.$$

We choose x_6 as the departing variable and pivot on the 3, 2 position of the tableau to obtain the new tableau

	x_1	x_2	x_3	x_4	x_5	x_6	x_7	P	
x_4	12	0	0	1	5/18	$-1/18$	0	0	60
x_3	$-8/5$	0	1	0	5/108	13/540	0	0	10
x_2	$-6/5$	1	0	0	$-1/18$	1/90	0	0	0
←— x_7	432/5	0	0	0	-3	$-4/5$	1	0	216
	$-242/5$	0	0	0	1/36	89/180	0	1	150.

Note that the current value of the objective function (150) did not increase at the last step. This is because the problem is degenerate and the departing basic variable had value 0. We now choose x_1 as the entering variable and compute the feasibility ratios

$$\frac{60}{12} = 5 \text{ for } x_4 \quad \text{and} \quad \frac{216}{432/5} = 2.5 \text{ for } x_7.$$

Note that we do not compute the feasibility ratio for x_3 and x_2 since these variables have negative entries in the pivot column. You should check that choosing either of these as the departing variable would result in a nonfeasible basic solution. We choose x_7 as the departing variable and pivot on the 4, 1 position of the tableau to obtain

$$\downarrow$$

	x_1	x_2	x_3	x_4	x_5	x_6	x_7	P	
← x_4	0	0	0	1	25/36	1/18	−5/36	0	30
x_3	0	0	1	0	−1/108	1/108	1/54	0	14
x_2	0	1	0	0	−7/72	0	1/72	0	3
x_1	1	0	0	0	−5/144	−1/108	5/432	0	5/2
	0	0	0	0	−119/72	5/108	121/216	1	271 .

Note that the value of the objective function increased at the last step
and that all of the current basic variables have positive values. This
solution is still not optimal, so we choose x_5 as the entering variable
and note that only x_4 has a positive value in the pivot column. Thus
we choose x_4 as the departing variable and pivot on the $1,5$ position of
the tableau to obtain

	x_1	x_2	x_3	x_4	x_5	x_6	x_7	P	
x_5	0	0	0	36/25	1	2/25	−1/5	0	216/5
x_3	0	0	1	1/75	0	1/100	1/60	0	72/5
x_2	0	1	0	7/50	0	7/900	−1/180	0	36/5
x_1	1	0	0	1/20	0	−7/1080	1/216	0	4
	0	0	0	119/50	0	241/1350	31/135	1	1712/5 .

Since all the entries in the objective row are now nonnegative, this
tableau represents the optimal solution. The maximum value of 342.4
occurs at the point $(x_1, x_2, x_3) = (4, 7.2, 14.4)$. ■

Exercise Set 6.3

*Given the following simplex tableaux identify the current basic variables,
find the associated basic feasible solution and its objective value, choose
the entering and departing variable, find the tableau associated with the
new set of basic variables, and find the new basic feasible solution and its
objective value.*

1.

	x	y	u	v	w	z	
	3	2	2	1	0	0	10
	7	4	1	0	1	0	3
	−5	−8	−9	0	0	1	0

2.

	x_1	x_2	x_3	x_4	x_5	z	
	5	1	6	0	0	0	10
	3	0	3	1	0	0	15
	−10	0	2	0	1	0	20
	−9	0	2	0	0	1	80

3.

x_1	x_2	x_3	x_4	x_5	x_6	x_7	z	
1	4	1	0	1	0	3	0	12
0	2	3	0	2	1	−2	0	6
0	−5	5	1	3	0	4	0	10
0	−8	−8	0	4	0	−6	1	120

4.

x_1	x_2	x_3	x_4	x_5	x_6	x_7	z	
0	−1	2	0	1	4	0	0	12
0	2	5	1	0	0	2	0	10
1	3	1	0	0	7	3	0	7
0	−15	−20	0	0	8	3	1	325

Use the simplex method to solve each of the following linear programming problems.

5. Maximize $z = 2x + 3y$ subject to $\begin{aligned} x + y &\le 160 \\ 3x + 2y &\le 240 \end{aligned}$ $\quad \begin{aligned} x &\ge 0 \\ y &\ge 0 \end{aligned}$.

6. Maximize $z = 40x_1 + 60x_2$ subject to $\begin{bmatrix} 2 & 1 \\ 1 & 1 \\ 1 & 3 \end{bmatrix}\begin{bmatrix} x_1 \\ x_2 \end{bmatrix} \le \begin{bmatrix} 70 \\ 40 \\ 90 \end{bmatrix}$, $\begin{bmatrix} x_1 \\ x_2 \end{bmatrix} \ge 0$.

7. Maximize $w = x + 2y$ subject to $\begin{bmatrix} -2 & 1 \\ 1 & -1 \end{bmatrix}\begin{bmatrix} x \\ y \end{bmatrix} \le \begin{bmatrix} 2 \\ 4 \end{bmatrix}$, $\begin{bmatrix} x \\ y \end{bmatrix} \ge 0$.

8. Maximize $P = 200x + 300y + 100z$ subject to
$$\begin{aligned} x + 2y + 2z &\le 50 \\ x + y + 3z &\le 40 \\ x + y + z &\le 30 \end{aligned} \qquad \begin{aligned} x &\ge 0 \\ y &\ge 0 \\ z &\ge 0 \end{aligned}.$$

9. Maximize $w = 2x_1 + 5x_2 + 4x_3 - 15$ subject to
$$\begin{aligned} x_1 + 2x_2 + x_3 &\le 4 \\ x_1 + 2x_2 + 2x_3 &\le 6 \end{aligned} \qquad x_i \ge 0, \quad i = 1, 2, 3.$$

10. Maximize $M = 5x_1 + 4x_2$ subject to $\begin{aligned} 4x_1 + 2x_2 &\le 80{,}000 \\ x_1 + 3x_2 &\le 30{,}000 \end{aligned} \quad \begin{aligned} x_1 &\ge 0 \\ x_2 &\ge 0 \end{aligned}$.

11. Maximize $P = 23x_1 + 16x_2 + 19x_3 + 14x_4$ subject to
$$\begin{bmatrix} 1 & 1 & 1 & 1 \\ 4 & 3 & 4 & 3 \\ 1 & 0 & 1 & 0 \\ 0 & 1 & 0 & 1 \end{bmatrix}\begin{bmatrix} x_1 \\ x_2 \\ x_3 \\ x_4 \end{bmatrix} \le \begin{bmatrix} 30 \\ 200 \\ 20 \\ 20 \end{bmatrix} \qquad \begin{bmatrix} x_1 \\ x_2 \\ x_3 \\ x_4 \end{bmatrix} \ge 0.$$

12. Maximize $P = 10x + 12y + 15z + 100$ subject to
$$\begin{aligned} 2x + 4y + 5z &\le 10{,}000 \\ 3x + 3y + 4z &\le 12{,}000 \end{aligned} \qquad x, y, z \ge 0.$$

13. Exercise 13 of Section 6.1 (see page 307).

14. Exercise 10(a) of Section 6.1 (see page 307).

15. Exercise 10(b) of Section 6.1 (see page 307).

16. Exercise 13 of Section 6.2 (see page 315).

†17. The following simplex tableau is not in a form for which a basic feasible solution is obvious. Show that $X = [0 \quad 6 \quad 3 \quad 0]^T$ is a basic feasible solution, make the necessary adjustments in the tableau, and then complete the solution.

x_1	x_2	x_3	x_4	z	
2	1	1	0	0	9
1	1	0	−1	0	6
−4	−5	0	0	1	0

†18. The following simplex tableau is not in a form for which a basic feasible solution is obvious. Show that $X = [2 \quad 2 \quad 5 \quad 0]^T$ is a basic feasible solution, make the necessary adjustments in the tableau, and then complete the solution.

x_1	x_2	x_3	x_4	z	
1	1	1	0	0	9
1	2	0	−1	0	6
0	1	0	0	0	2
−5	−4	0	0	1	0

19. Consider this canonical linear programming problem:

$$\text{Maximize } z = C^T X \text{ subject to } AX = B \quad \text{and}$$
$$X \geq 0 \text{ where } A \text{ is } m \text{ by } m + n.$$

Show that this problem is not degenerate if every m by m submatrix of $[A \,|\, B]$ is nonsingular.

6.4   *Refinements of the Simplex Method*

The basic simplex method, as described in Section 6.3, will not work for all linear-programming problems. Difficulties occur if an initial basic feasible solution is not obvious. This is the case if the original problem has (=) constraints or (≥) constraints with positive right-hand sides.

Consider this problem:

$$\text{maximize } z = 40x_1 + 30x_2 \text{ subject to}$$

$$x_1 + 2x_2 \geq 6$$
$$2x_1 + x_2 \geq 4 \qquad x_1 \geq 0$$
$$x_1 + x_2 \leq 5 \qquad x_2 \geq 0. \tag{6.10}$$
$$2x_1 + x_2 \leq 8$$

Since this is a two-dimensional problem we can easily sketch the feasible set determined by the constraints in System 6.10. This set is shown in Figure 6.6.

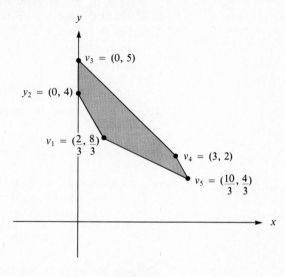

FIGURE 6.6

If we add nonnegative slack variables to the constraints of System 6.10 we obtain the system

$$
\begin{aligned}
x_1 + 2x_2 - x_3 \quad\quad\quad\quad &= 6 \\
2x_1 + x_2 \quad - x_4 \quad\quad\quad &= 4 \\
x_1 + x_2 \quad\quad + x_5 \quad &= 5 \\
2x_1 + x_2 \quad\quad\quad + x_6 &= 8
\end{aligned}
\qquad \text{all } x_i \geq 0 .
$$
(6.11)

Note that the ($\geq$) inequalities in System 6.10 lead to slack variables with negative coefficients in System 6.11. A basic solution of System 6.11 is readily available:

$$x_1 = x_2 = 0, \quad x_3 = -6, \quad x_4 = -4, \quad x_5 = 5, \quad \text{and} \quad x_6 = 8 .$$

It is not a feasible solution since x_3 and x_4 are negative. Suppose we modify System 6.11 as follows:

$$
\begin{aligned}
x_1 + 2x_2 - x_3 \quad\quad\quad + x_7 \quad &= 6 \\
2x_1 + x_2 \quad - x_4 \quad\quad\quad + x_8 &= 4 \\
x_1 + x_2 \quad\quad + x_5 \quad\quad &= 5 \\
2x_1 + x_2 \quad\quad\quad + x_6 \quad &= 8
\end{aligned}
\qquad \text{all } x_i \geq 0 .
$$
(6.12)

We will call x_7 and x_8 ARTIFICIAL VARIABLES. Note that a basic feasible solution of System 6.12 is readily available:

$$x_1 = x_2 = x_3 = x_4 = 0, \quad x_5 = 5, \quad x_6 = 8, \quad x_7 = 6, \quad \text{and} \quad x_8 = 4 \, .$$

Note also that any basic feasible solution of System 6.12, for which $x_7 = x_8 = 0$, would yield a basic feasible solution of System 6.11; it would also yield a maximum value of 0 for the auxiliary objective function $P = -x_7 - x_8$. Thus we could try to find a basic feasible solution for System 6.11 by finding an optimal solution to the auxiliary problem,

maximize $P = -x_7 - x_8$ satisfying System 6.12 .

The simplex tableau for this problem is

	x_1	x_2	x_3	x_4	x_5	x_6	x_7	x_8	P	
x_7	1	2	−1	0	0	0	1	0	0	6
x_8	2	1	0	−1	0	0	0	1	0	4
x_5	1	1	0	0	1	0	0	0	0	5
x_6	2	1	0	0	0	1	0	0	0	8
	0	0	0	0	0	0	1	1	1	0,

which is not quite in standard form since the objective row has non-zero entries in the columns associated with the basic variables x_7 and x_8. The elementary row operations

$$R_5 \leftarrow -R_1 + R_5 \quad \text{and} \quad R_5 \leftarrow -R_2 + R_5$$

lead to the tableau

	x_1	x_2	x_3	x_4	x_5	x_6	x_7	x_8	P	
x_7	1	2	−1	0	0	0	1	0	0	6
x_8	2	1	0	−1	0	0	0	1	0	4
x_5	1	1	0	0	1	0	0	0	0	5
x_6	2	1	0	0	0	1	0	0	0	8
	−3	−3	1	1	0	0	0	0	1	−10,

(6.13)

which is suitable for use by the simplex algorithm.

The simplex method, applied to Tableau 6.13, leads to the tableau

	x_1	x_2	x_3	x_4	x_5	x_6	x_7	x_8	P	
x_2	0	1	−2/3	1/3	0	0	2/3	−1/3	0	8/3
x_1	1	0	1/3	−2/3	0	0	−1/3	2/3	0	2/3
x_5	0	0	1/3	1/3	1	0	−1/3	−1/3	0	5/3
x_6	0	0	0	1	0	1	0	−1	0	4
	0	0	0	0	0	0	1	1	1	0

(6.14)

from which we conclude that

$$x_1 = \frac{2}{3}, \quad x_2 = \frac{8}{3}, \quad x_3 = x_4 = 0, \quad x_5 = \frac{5}{3}, \quad x_6 = 4, \quad \text{and}$$

$$x_7 = x_8 = 0$$

is a basic feasible solution of System 6.12 and an optimal solution of the auxiliary problem. Since $x_7 = x_8 = 0$ we can also conclude that

$$x_1 = \frac{2}{3}, \quad x_2 = \frac{8}{3}, \quad x_3 = x_4 = 0, \quad x_5 = \frac{5}{3}, \quad \text{and} \quad x_6 = 4$$

is a basic feasible solution of System 6.11. Note that this solution corresponds to the vertex V_1 of the feasible set shown in Figure 6.6 on page 327.

If we delete the columns of Tableau 6.14 associated with the artificial variables (x_7 and x_8), we obtain a constraint system equivalent to System 6.11 that is associated with the preceding basic feasible solution.

If we now change the objective row in Tableau 6.14, with columns 7 and 8 deleted, to correspond to the original objective function, we obtain the tableau

	x_1	x_2	x_3	x_4	x_5	x_6	z	
x_2	0	1	$-2/3$	$1/3$	0	0	0	$8/3$
x_1	1	0	$1/3$	$-2/3$	0	0	0	$2/3$
x_5	0	0	$1/3$	$1/3$	1	0	0	$5/3$
x_6	0	0	0	1	0	1	0	4
	-40	-30	0	0	0	0	1	0 .

Note that the objective row of this tableau has nonzero entries in the columns associated with the basic variables x_1 and x_2. The elementary row operations

$$R_5 \leftarrow 40R_2 + R_5 \quad \text{and} \quad R_5 \leftarrow 30R_1 + R_5$$

change the last tableau to

	x_1	x_2	x_3	x_4	x_5	x_6	z	
x_2	0	1	$-2/3$	$1/3$	0	0	0	$8/3$
x_1	1	0	$1/3$	$-2/3$	0	0	0	$2/3$
x_5	0	0	$1/3$	$1/3$	1	0	0	$5/3$
x_6	0	0	0	1	0	1	0	4
	0	0	$-20/3$	$-50/3$	0	0	1	$320/3$.

(6.15)

If we apply the simplex method to Tableau 6.15 we are led to the tableau

	x_1	x_2	x_3	x_4	x_5	x_6	z	
x_2	0	1	0	0	2	-1	0	2
x_1	1	0	0	0	-1	1	0	3
x_3	0	0	1	0	3	-1	0	1
x_4	0	0	0	1	0	1	0	4
	0	0	0	0	20	10	1	180,

from which we see that the maximum value of z is 180 for $x_1 = 3$ and $x_2 = 2$; this is the vertex V_4 shown in Figure 6.6 on page 327.

The method we have illustrated above is called the TWO-PHASE METHOD. It is widely used in computer software for solving linear programming problems.

TWO-PHASE METHOD

(1) Change the original problem into a canonical linear-programming problem by adding slack variables to the inequality constraints. (The slack variables have a coefficient of -1 for a ($\geq$) inequality.)

(2) Add artificial variables to those constraints that do not have a slack variable with a positive coefficient.

(3) Set up the simplex tableau for the constraints in step 2 and the objective function $M = -(\text{sum of the artificial variables})$.

(4) Use row operations to modify the tableau so that the objective row has zeros in the columns associated with the initial basic variables.

(5) Use the simplex method to solve the auxiliary problem.

(6) If the auxiliary problem has a solution with all the artificial variables equal to zero, then construct a basic feasible solution of the problem by deleting the artificial variables from the solution of the auxiliary problem. If the solution of the auxiliary problem has nonzero artificial variables, then the original problem has no feasible solution.

(7) Modify the final tableau of the auxiliary problem by
   (a) deleting the columns associated with the artificial variables; and
   (b) changing the objective row to correspond to the objective function of the original problem.

(8) Use row operations to modify the tableau so that the objective row has zeros in the columns associated with the current basic variables.

(9) Use the simplex method to solve the original problem.

Exercise Set 6.4

1. Maximize $P = 2x + 3z$ subject to $\begin{aligned} x + 3y + z &= 5 \\ x + 2y + 7z &\leq 4 \end{aligned}$ $x, y, z \geq 0$.

2. (a) Maximize $z = 40x + 80y$ subject to $\begin{aligned} x + 2y &\geq 4 \\ 3x + 5y &\leq 30 \\ -2x + 3y &\leq 9 \\ 3x - y &\leq 12 \end{aligned}$ $\begin{aligned} x &\geq 0 \\ y &\geq 0 \end{aligned}$.

   (b) Minimize z subject to the constraints in part (a).

3. Maximize $P = 3x_1 + 2x_2 - x_3 + 4x_4$ subject to

   $$\begin{aligned} 7x_2 + x_3 + 2x_4 &\geq 3 \\ 2x_1 + x_2 + 3x_3 - 4x_4 &\leq 7 \\ x_1 + x_2 + 2x_3 \phantom{{}- 4x_4} &= 9 \end{aligned}$$ all $x_i \geq 0$.

4. Solve the minimization problem of Exercise 11 of Section 6.2 on page 314.

5. Solve the minimization problem of Exercise 12 of Section 6.2 on page 315.

6. Solve the brewery problem of Exercise 13 of Section 6.2 (on page 315) with the additional constraint that at least 5 barrels of premium ale must be produced.

7. Solve the transportation problem (recall Section 1.3) whose data graph is shown below.

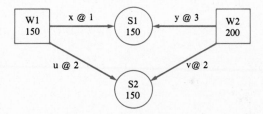

8. Solve the transportation problem whose data graph is shown below.

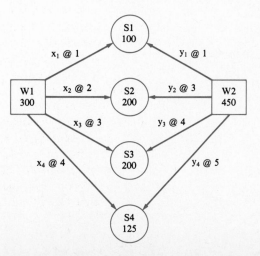

7

Selected Applications

In this chapter we will present four applications of linear algebra. Each section of this chapter is essentially independent of the others. Those topics from Chapters 1 through 5 that are required for each section are indicated at the beginning of each section.

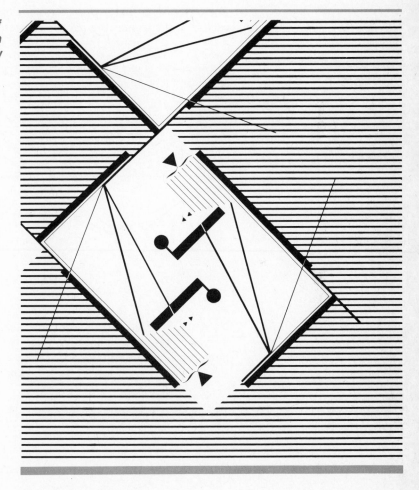

7.1   *Graph Theory*

Requires: Chapter 1

Graph theory is a branch of mathematics that is used to analyze relationships among elements of a group. The group may consist of people, cities, objects, teams, or the like. Graph theory has important applications to many disciplines, including business, sociology, biology, engineering, medicine, and communications.

Definition 7.1

A GRAPH G is a finite set of vertices along with a finite set of edges, each of which connects a pair of vertices.

Example 1

Consider the graph in Figure 7.1 that represents the roads connecting seven different towns.

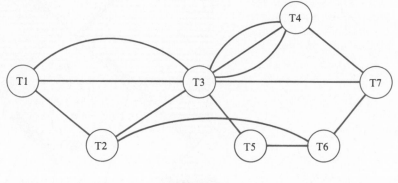

FIGURE 7.1

Note that there are two edges connecting T1 and T3, indicating that there are two roads connecting these two towns. Similarly, there are three roads connecting T3 and T4, one road connecting T5 and T6, and no direct route from T2 to T5.   ■

Example 2

The graph in Figure 7.2 represents which members of a group of eight people are acquainted.

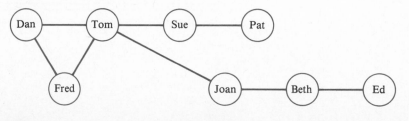

FIGURE 7.2

From this graph we see that Ed knows only Beth, Sue knows Tom and Pat, Fred and Dan do not know any of the women, and Pat does not know any of the men. ■■

Many applications assign a direction to the edges of a graph; such graphs are called DIRECTED GRAPHS or DIGRAPHS.

Example 3

Consider the digraph in Figure 7.3 that represents the routes of a small commuter airline serving five cities.

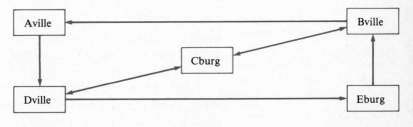

FIGURE 7.3

The arrows on the edges indicate the directions in which the planes fly. There is direct round-trip service from Cburg to Bville, but only one-way service from Bville to Aville. To get from Aville to Bville it is necessary to visit either Cburg or Eburg. ■■

Example 4

The digraph in Figure 7.4 represents the results of an incomplete round-robin tennis tournament involving eight players.

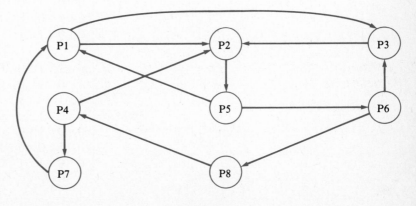

FIGURE 7.4

The existence of an edge between any two players means that those two players have competed. The arrow indicates the direction of dominance; that is, it points toward the loser. Thus, P1 has defeated P2 and P3, but has lost to P5 and P7, while P8 has defeated P4 but lost to P6. ▄▄

In many applications of graph theory the graphs are very large and it is not easy to draw the desired conclusions by looking at the graph. Matrix theory can be useful in dealing with such graphs. We associate with a given graph two matrices that we now define formally.

Definition 7.2

Let G be a graph with n vertices. The MATRIX of G is the $n \times n$ matrix $M(G)$ whose i,jth entry is the number of edges connecting vertex i to vertex j. The ADJACENCY MATRIX of G is the $n \times n$ matrix $A(G)$ whose i,jth entry is 1 if and only if there is an edge containing vertex i and vertex j. All other entries of $A(G)$ are 0. If G is a digraph, then the adjacency matrix has a 1 in position i,j if and only if there is an edge from vertex i to vertex j.

Note that for an undirected graph $M(G) = A(G)$ if there are no multiple edges and both $M(G)$ and $A(G)$ are always symmetric. $A(G)$ can be obtained from $M(G)$ by changing all nonzero entries to 1.

Example 5

For the graph of Example 1,

$$M(G) = \begin{bmatrix} 0 & 1 & 2 & 0 & 0 & 0 & 0 \\ 1 & 0 & 1 & 0 & 0 & 1 & 0 \\ 2 & 1 & 0 & 3 & 1 & 0 & 1 \\ 0 & 0 & 3 & 0 & 0 & 0 & 1 \\ 0 & 0 & 1 & 0 & 0 & 1 & 0 \\ 0 & 1 & 0 & 0 & 1 & 0 & 1 \\ 0 & 0 & 1 & 1 & 0 & 1 & 0 \end{bmatrix} \quad \text{and}$$

$$A(G) = \begin{bmatrix} 0 & 1 & 1 & 0 & 0 & 0 & 0 \\ 1 & 0 & 1 & 0 & 0 & 1 & 0 \\ 1 & 1 & 0 & 1 & 1 & 0 & 1 \\ 0 & 0 & 1 & 0 & 0 & 0 & 1 \\ 0 & 0 & 1 & 0 & 0 & 1 & 0 \\ 0 & 1 & 0 & 0 & 1 & 0 & 1 \\ 0 & 0 & 1 & 1 & 0 & 1 & 0 \end{bmatrix}.$$

For the graph of Example 2,

$$M(G) = A(G) = \begin{array}{c} \\ \text{Dan} \\ \text{Tom} \\ \text{Fred} \\ \text{Sue} \\ \text{Joan} \\ \text{Beth} \\ \text{Ed} \\ \text{Pat} \end{array} \begin{array}{cccccccc} \text{D} & \text{T} & \text{F} & \text{S} & \text{J} & \text{B} & \text{E} & \text{P} \\ \left[\begin{array}{cccccccc} 0 & 1 & 1 & 0 & 0 & 0 & 0 & 0 \\ 1 & 0 & 1 & 1 & 1 & 0 & 0 & 0 \\ 1 & 1 & 0 & 0 & 0 & 0 & 0 & 0 \\ 0 & 1 & 0 & 0 & 0 & 0 & 0 & 1 \\ 0 & 1 & 0 & 0 & 0 & 1 & 0 & 1 \\ 0 & 0 & 0 & 0 & 1 & 0 & 1 & 0 \\ 0 & 0 & 0 & 0 & 0 & 1 & 0 & 0 \\ 0 & 0 & 0 & 1 & 1 & 0 & 0 & 0 \end{array}\right] \end{array}.$$

For the digraph of Example 3 the adjacency matrix is

$$A(G) = \begin{bmatrix} 0 & 0 & 0 & 1 & 0 \\ 1 & 0 & 1 & 0 & 0 \\ 0 & 1 & 0 & 1 & 0 \\ 0 & 0 & 1 & 0 & 1 \\ 0 & 1 & 0 & 0 & 0 \end{bmatrix}.$$

For the digraph of Example 4 the adjacency matrix is

$$A(G) = \begin{bmatrix} 0 & 1 & 1 & 0 & 0 & 0 & 0 & 0 \\ 0 & 0 & 0 & 0 & 1 & 0 & 0 & 0 \\ 0 & 1 & 0 & 0 & 0 & 0 & 0 & 0 \\ 0 & 1 & 0 & 0 & 0 & 0 & 1 & 0 \\ 1 & 0 & 0 & 0 & 0 & 1 & 0 & 0 \\ 0 & 0 & 1 & 0 & 0 & 0 & 0 & 1 \\ 1 & 0 & 0 & 0 & 0 & 0 & 0 & 0 \\ 0 & 0 & 0 & 1 & 0 & 0 & 0 & 0 \end{bmatrix}. \quad \blacksquare$$

We will say that the vertices V_i and V_j of the graph G are CON-NECTED if there is a sequence of edges connecting V_i and V_j. In the graph of Example 1 there is a two-step path (T1 $\rightarrow$ T3, T3 $\rightarrow$ T5) connecting T1 and T5. There is also a three-step path connecting these two vertices.

If A is the adjacency matrix of the graph G, then the entries of A^2 will tell us about the two-step paths in the graph G. To see how this works we suppose that the entry in the i,j position of A^2 is nonzero; that is,

$$[A^2]_{ij} = a_{i1}a_{1j} + a_{i2}a_{2j} + \ldots + a_{in}a_{nj} \neq 0. \qquad (7.1)$$

There must be at least one term, say $a_{ik}a_{kj}$, in this sum that is nonzero. In G this means that there is an edge from V_i to V_k and also an edge from V_k to V_j. Thus there is a two-step path from V_i to V_j. If $[A^2]_{ij} = r \neq 0$, then there must be r nonzero terms in Equation 7.1 and hence r

two-step paths from V_i to V_j. This result extends easily to any power of A, so we have the following theorem.

Theorem 7.1

If $A(G)$ is the adjacency matrix of the graph G and the i, j entry of $[A(G)]^s = r$, then there are r paths of length s connecting vertex i to vertex j.

If the graph is undirected and has multiple edges, then a similar argument establishes the next theorem.

Theorem 7.2

If the i, j entry of $[M(G)]^s = r$, then in G there are r paths of length s connecting vertex i to vertex j.

Example 6

For the matrix $M = M(G)$ from Examples 4 and 1 we use the MATALG program to compute

$$M^2 = \begin{bmatrix} 5 & 2 & 1 & 6 & 2 & 1 & 2 \\ 2 & 3 & 2 & 3 & 2 & 0 & 2 \\ 1 & 2 & 16 & 1 & 0 & 3 & 3 \\ 6 & 3 & 1 & 10 & 3 & 1 & 3 \\ 2 & 2 & 0 & 3 & 2 & 0 & 2 \\ 1 & 0 & 3 & 1 & 0 & 3 & 0 \\ 2 & 2 & 3 & 3 & 2 & 0 & 3 \end{bmatrix} \quad \text{and}$$

$$M^3 = \begin{bmatrix} 4 & 7 & 34 & 5 & 2 & 6 & 8 \\ 7 & 4 & 20 & 8 & 2 & 7 & 5 \\ 34 & 20 & 10 & 51 & 19 & 5 & 20 \\ 5 & 8 & 51 & 6 & 2 & 9 & 12 \\ 2 & 2 & 19 & 2 & 0 & 6 & 3 \\ 6 & 7 & 5 & 9 & 6 & 0 & 7 \\ 8 & 5 & 20 & 12 & 3 & 7 & 6 \end{bmatrix}.$$

Since the $1, 7$ entries of M^2 and M^3 are 2 and 8 respectively, Theorem 7.2 tells us that there are only two paths of length 2 connecting T1 and T7 while there are eight paths of length 3 connecting T1 and T7. Can you find them? Because the $4, 5$ entry of M^3 is 2 there should be precisely two paths of length 3 connecting T4 and T5; this can be verified by direct inspection of the graph.

Because every *off-diagonal entry* (an entry not positioned on the diagonal of a matrix) of M^3 is nonzero it follows that every town is connected to every other town by at least one path of length 3. From

M^2 we see that the only pairs not connected by a path of length 2 are (T2, T6), (T3, T5), (T5, T6), and (T6, T7); all of these are connected by paths of length 1. Thus, every pair of towns is connected by a path of length 1 or 2 and this is equivalent to the fact that the matrix $M + M^2$ has no zero entries. ∎

Definition 7.3

A digraph is STRONGLY CONNECTED if and only if there is a path connecting any pair of vertices.

If the graph contained n vertices then it would never be necessary to consider a path of length greater than $n - 1$. Thus the graph would be strongly connected if and only if the matrix

$$A(G) + A(G)^2 + \ldots + A(G)^{n-1} = D$$

had no zero entries. The calculations in Example 6 show that the graph of Example 1 is strongly connected.

Example 7

Who should be declared the winner of the incomplete tournament described in Example 4, assuming that no further matches can be played?

At first glance it would appear that P4, P5, and P6 are tied with a record of two wins and one loss. Let us break the tie by considering the quality of the wins. Clearly a win over a strong player is more impressive than a win over a weak player. We will consider Pi to have second-order dominance over Pj if Pi beat someone who beat Pj, that is, if there is a path of length 2 from Pi to Pj in the graph of Example 4. The matrix of second-order dominance would be $A(G)^2$ and the sum $A(G) + A(G)^2$ would represent both the first- and second-order dominance. This matrix is

$$A(G) + A(G)^2 = \begin{bmatrix} 0 & 1 & 1 & 0 & 1 & 0 & 0 & 0 \\ 1 & 0 & 0 & 0 & 1 & 1 & 0 & 0 \\ 0 & 1 & 0 & 0 & 1 & 0 & 0 & 0 \\ 1 & 1 & 0 & 0 & 1 & 0 & 1 & 0 \\ 1 & 1 & 2 & 2 & 0 & 1 & 0 & 1 \\ 0 & 1 & 1 & 1 & 0 & 0 & 0 & 1 \\ 1 & 1 & 1 & 0 & 0 & 0 & 0 & 0 \\ 0 & 1 & 0 & 1 & 0 & 0 & 1 & 0 \end{bmatrix}.$$

The sum of the entries in row i indicates the number of players over which the ith player has either first- or second-order dominance. It is now clear that P5 is the strongest player and should be declared the winner. ∎

Exercise Set 7.1

1. Find the matrix $M(G)$ for the following graph.

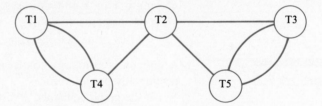

Find the number of paths of length 2 connecting T1 and T3. Find the number of paths of length 3 connecting T4 and T3.

2. Find the adjacency matrix $A(G)$ for the following digraphs.

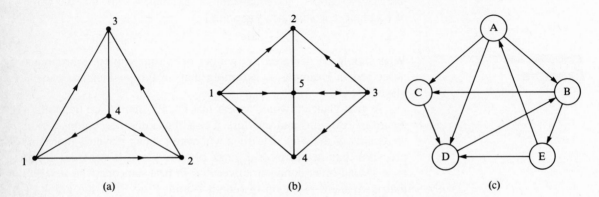

(a)   (b)   (c)

3. Find a directed graph whose adjacency matrix is as shown.

(a) $\begin{bmatrix} 0 & 0 & 1 & 1 & 0 \\ 0 & 0 & 0 & 1 & 1 \\ 0 & 1 & 0 & 0 & 0 \\ 1 & 0 & 0 & 0 & 0 \\ 0 & 0 & 1 & 0 & 0 \end{bmatrix}$ (b) $\begin{bmatrix} 0 & 1 & 0 & 0 & 0 \\ 0 & 0 & 1 & 0 & 0 \\ 0 & 0 & 0 & 1 & 0 \\ 0 & 0 & 0 & 0 & 1 \\ 1 & 0 & 0 & 0 & 0 \end{bmatrix}$

(c) $\begin{bmatrix} 0 & 0 & 0 & 1 & 0 & 0 \\ 1 & 0 & 1 & 1 & 0 & 0 \\ 0 & 0 & 0 & 1 & 0 & 1 \\ 0 & 0 & 1 & 0 & 0 & 1 \\ 0 & 0 & 0 & 0 & 0 & 0 \\ 1 & 1 & 0 & 0 & 1 & 0 \end{bmatrix}$

4. Which of the digraphs in Exercises 2 and 3 are strongly connected?

5. Make a reasonable definition of equivalence for two digraphs and then determine how the adjacency matrices of two equivalent graphs are related.

6. A CLIQUE in a digraph is a subset of the vertices of the graph satisfying the following three conditions.

1. The subset contains at least three vertices.

2. For each pair of vertices in the subset there is a two-way path connecting them.

3. No other vertices can be added to the subset and still satisfy property 2.

(a) Find a clique in the graph of Exercise 2(b).

(b) Show that the ith vertex is a member of a clique if and only if the entry in the i, i position of S^3 is not 0, where S is defined by $s_{ij} = a_{ij}a_{ji}$, and $A = A(G)$.

7- A group of six individuals have been interacting for some time and the following influence pattern has been observed:

Adam influences Fred and Bill;

Dan is influenced by Ed;

Dan influences Charlie and Bill;

Bill and Fred influence Charlie;

Fred is influenced by Dan;

Charlie influences Adam and Ed; and

Ed is influenced by Adam, Fred, and Bill.

(a) Represent the above information by a digraph and find its adjacency matrix.

(b) Who influences the most people? Who is influenced by the most people?

(c) By considering second-order influence, determine who is the best choice for leader of the group.

8. Suppose that the Big Ten football season ended in a tie between Michigan and Iowa. Use graph theory to determine which team will be the conference's best representative in the Rose Bowl; consider first- and second-order dominance. The records of the teams are given in the table below, where + means a victory and 0 means the teams did not play.

	I L L	I N D	I O W	M I C	M S U	M I N	N W U	O S U	P U R	W I S	W-L
Illinois (ILL)	0	+	−	−	+	0	+	+	−	+	5–3
Indiana (IND)	−	0	0	−	−	−	+	−	−	−	1–7
Iowa (IOW)	+	0	0	+	+	+	+	−	+	+	7–1
Michigan (MIC)	+	+	−	0	+	+	0	+	+	+	7–1
Michigan State (MSU)	−	+	−	−	0	+	+	0	+	+	5–3
Minnesota (MIN)	+	−	−	−	−	0	+	−	+	+	4–4
Northwestern (NWU)	−	−	−	0	−	−	0	−	−	−	0–8
Ohio State (OSU)	−	+	+	−	0	+	+	+	+	−	6–2
Purdue (PUR)	+	+	−	−	−	−	+	−	0	0	3–5
Wisconsin (WIS)	−	+	−	−	−	−	−	+	0	0	2–6

7.2

Least Squares Approximations

Requires: Chapters 1 and 3

A very common problem in applications is to find an analytic expression that represents a collection of experimental data. We know that two points determine a straight line, that three points determine a parabola, that four points determine a cubic polynomial, and so on. Suppose that we were given the six data points graphed in Figure 7.5. The relationship between x and y seems to be almost linear, so it would not seem reasonable to fit a fifth-degree polynomial, as in Figure 7.5(b), to this data. A much more reasonable approach is to find a linear function $y = a + bx$ that best represents the given data; such a line is drawn in Figure 7.5(c).

(a)

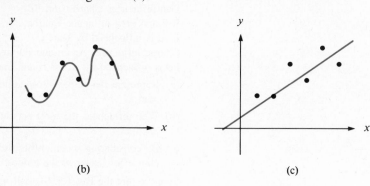

(b)   (c)

FIGURE 7.5

Let us try to find the straight line $y = a + bx$ that best fits the data points

$$(x_1, y_1), (x_2, y_2), \ldots, (x_k, y_k).$$

Ideally we would like each of the data points to lie on the line; that is, we would want a and b to satisfy each of the k equations

$$a + bx_i = y_i, \quad i = 1, 2, \ldots, k. \tag{7.2}$$

This system of k equations in 2 unknowns is overdetermined in the sense that there are more equations than unknowns. We have seen in Chapter 1 that such a system is almost certain to be inconsistent. Instead of insisting that each data point lie on the line, let us try to find a line that is close to each of the data points.

We will measure the deviation of a given data point from the line $y = a + bx$ by the difference

$$d_i = y_i - (a + bx_i) \quad i = 1, 2, \ldots, k, \tag{7.3}$$

which represents the vertical distance from the point to the line, as shown in Figure 7.6.

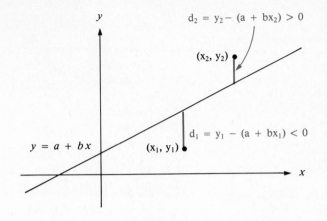

FIGURE 7.6

If we define the matrices

$$A = \begin{bmatrix} 1 & x_1 \\ 1 & x_2 \\ \vdots & \vdots \\ 1 & x_k \end{bmatrix}, \quad X = \begin{bmatrix} a \\ b \end{bmatrix}, \quad Y = \begin{bmatrix} y_1 \\ y_2 \\ \vdots \\ y_k \end{bmatrix}, \quad \text{and} \quad D = \begin{bmatrix} d_1 \\ d_2 \\ \vdots \\ d_k \end{bmatrix},$$

then we can write (inconsistent) System 7.2 as $AX = Y$ and Equation 7.3 as the matrix equation

$$Y - AX = D.$$

We will measure the deviation of the data points from the line by the norm of the vector D in $\mathcal{R}^k$. Thus, we wish to choose X so that

$$\|Y - AX\|^2 = \|D\|^2 = \sum_{i=1}^{k} d_i^2$$

is as small as possible. The resulting vector X_0 in $\mathcal{R}^2$ is called the LEAST SQUARES SOLUTION to the inconsistent system $AX = Y$.

Note that the vector AX is an element of the column space of the matrix A. Thus, we wish to choose X_0 so that AX_0 is the element of $\mathcal{CS}(A)$ that is closest to Y. The diagram in Figure 7.7 on the following page illustrates the situation for the case $k = 3$. Note that we want to choose X_0 so that AX_0 is the orthogonal projection of Y onto $\mathcal{CS}(A)$ and $D_0 = Y - AX_0$ is orthogonal to the column space of A.

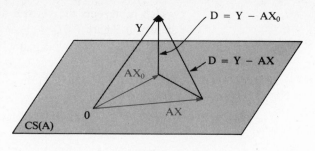

FIGURE 7.7

If $k > 3$, then the preceding geometric argument is not valid, but it is still true that we wish to choose X_0 so that $D_0 = Y - AX_0$ is orthogonal to $\mathscr{CS}(A)$. To see that this is true in general we consider

$$D = Y - AX = Y - AX_0 + AX_0 - AX$$
$$= (Y - AX_0) + A(X - X_0).$$

If $Y - AX_0$ is orthogonal to $\mathscr{CS}(A)$, then by the generalization of the Pythagorean theorem to $\mathscr{R}^k$, given in Exercise 31 of Section 3.8, it follows that

$$\|D\|^2 = \|Y - AX_0\|^2 + \|AX_0 - AX\|^2 = \|D_0\|^2 + \|AX_0 - AX\|^2.$$

Thus $\|D_0\| \leq \|D\|$ for every choice of X.

If X is to be chosen so that $Y - AX$ is orthogonal to $CS(A)$, then $Y - AX$ must be orthogonal to each column of the matrix A. This is equivalent to the matrix condition

$$A^T(Y - AX) = 0 \quad \text{or} \quad A^TAX = A^TY.$$

The last equation above is called the NORMAL EQUATION for the least squares problem; for the case of finding the best-fitting straight line it is a system of two linear equations in two unknowns.

Note that the above argument does not require that A have only two columns; it applies to any overdetermined system $AX = Y$. Note also that rank (A^TA) = rank(A), by Exercise 9 of Section 3.8, so that A^TA will be nonsingular if and only if the columns of A are linearly independent. We summarize this discussion in the following theorem.

Theorem 7.3

Let A be a $m \times n$ matrix with $m > n$. The least squares solution of the linear system $AX = Y$ is an exact solution of the normal equations

$$A^TAX = A^TY. \tag{7.4}$$

If A has rank n, then the $n \times n$ matrix A^TA is nonsingular and the unique solution of the normal equations is

$$X = (A^TA)^{-1}A^TY.$$

Example 1

Find the straight line that best fits the following four data points: $(1, 3)$, $(2, 2)$, $(3, 5)$, and $(4, 4)$.

The matrices associated with this problem are

$$A = \begin{bmatrix} 1 & 1 \\ 1 & 2 \\ 1 & 3 \\ 1 & 4 \end{bmatrix}, \quad Y = \begin{bmatrix} 3 \\ 2 \\ 5 \\ 4 \end{bmatrix}, \quad X = \begin{bmatrix} a \\ b \end{bmatrix}, \quad A^T A = \begin{bmatrix} 4 & 10 \\ 10 & 30 \end{bmatrix}, \quad \text{and}$$

$$A^T Y = \begin{bmatrix} 14 \\ 38 \end{bmatrix}.$$

Thus Normal Equations 7.4, in matrix form, are

$$\begin{bmatrix} 4 & 10 \\ 10 & 30 \end{bmatrix} \begin{bmatrix} a \\ b \end{bmatrix} = \begin{bmatrix} 14 \\ 38 \end{bmatrix}$$

and the solution, obtained by Gaussian elimination, is $a = 2$ and $b = 3/5$. The best-fitting line, $y = 2 + 3x/5$, and the data points are shown in Figure 7.8.

FIGURE 7.8

In the following tables we show the deviations for the best-fitting line and also for the line $y = 1 + x$, which also fits the data reasonably well.

Data		Best-Fitting Line		Line $y = 1 + x$	
x	y	y	d^2	y	d^2
1	3	2.6	.16	2	1
2	2	3.2	1.44	3	1
3	5	3.8	1.44	4	1
4	4	4.4	.16	5	1
			3.20		4

Note that the sum of the squares of the deviations for the least squares line is 3.20, which is less than the 4 that is the sum of the squares of the deviations for the line $y = 1 + x$. ■

If we want to fit a higher degree polynomial to a given set of data points, the basic procedure is the same; the only thing that changes is the number of columns of the matrix A. Suppose we want to fit a polynomial of degree 3 to the data

$$(x_1, y_1), (x_2, y_2), \ldots, (x_k, y_k).$$

If we let $y = a + bx + cx^2 + dx^3$ be the desired cubic polynomial, we are led to the linear system

$$
\begin{aligned}
a + bx_1 + cx_1^{\,2} + dx_1^{\,3} &= y_1 \\
a + bx_2 + cx_2^{\,2} + dx_2^{\,3} &= y_2 \\
&\;\;\vdots \\
a + bx_k + cx_k^{\,2} + dx_k^{\,3} &= y_k.
\end{aligned}
$$

If $k > 4$, this system is overdetermined and we do not expect that Gaussian elimination will produce an exact solution. In this case we will seek a least squares solution. In matrix notation this system is

$$
AX = \begin{bmatrix}
1 & x_1 & x_1^{\,2} & x_1^{\,3} \\
1 & x_2 & x_2^{\,2} & x_2^{\,3} \\
\cdot & \cdot & \cdot & \cdot \\
1 & x_k & x_k^{\,2} & x_k^{\,3}
\end{bmatrix}
\begin{bmatrix}
a \\ b \\ c \\ d
\end{bmatrix}
=
\begin{bmatrix}
y_1 \\ y_2 \\ \vdots \\ y_k
\end{bmatrix}
= Y.
$$

Note that every 4×4 submatrix of A is a Vandermonde matrix and is hence nonsingular by Exercise 13 of Section 2.3. Therefore A is of rank 4 provided only that the x_i are distinct. By Theorem 7.3 the least squares solution of this system is the exact solution of the normal equations

$$A^T A X = A^T Y.$$

Example 2

Find the parabola $y = a + bx + cx^2$ that best fits the five data points $(0, 4), (1, 1), (2, 4), (3, 6), (4, 8)$.

Attempting to fit $y = a + bx + cx^2$ to the given data leads to the following overdetermined system

$$
AX = \begin{bmatrix}
1 & 0 & 0 \\
1 & 1 & 1 \\
1 & 2 & 4 \\
1 & 3 & 9 \\
1 & 4 & 16
\end{bmatrix}
\begin{bmatrix}
a \\ b \\ c
\end{bmatrix}
=
\begin{bmatrix}
4 \\ 1 \\ 4 \\ 6 \\ 8
\end{bmatrix}
= Y,
$$

for which the normal equations are

$$A^TAX = \begin{bmatrix} 5 & 10 & 30 \\ 10 & 30 & 100 \\ 30 & 100 & 354 \end{bmatrix} \begin{bmatrix} a \\ b \\ c \end{bmatrix} = \begin{bmatrix} 23 \\ 59 \\ 199 \end{bmatrix} = A^TY.$$

The solution of the normal equations is easily found to be

$$X = \begin{bmatrix} 230/70 \\ -89/70 \\ 45/70 \end{bmatrix} = \begin{bmatrix} 3.29 \\ -1.27 \\ 0.64 \end{bmatrix},$$

so the best-fitting parabola is $y = (230 - 89x + 45x^2)/70$. For this parabola the sum of the squares of the deviations is 4.64. The data points and the least squares parabola are shown in Figure 7.9. ■

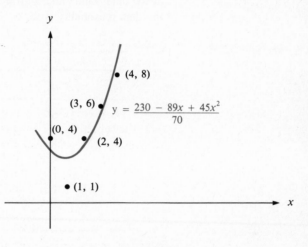

FIGURE 7.9

The basic procedure we have illustrated above can also be used to fit functions of several variables to given data. We illustrate this in our next example.

Example 3

Find the plane $z = a + bx + cy$ that best fits the data

$(1, 1, 5),\quad (2, 1, 4),\quad (1, 2, 8),\quad (0, 1, 4),\quad (1, 0, 2),\quad$ and $\quad (2, 2, 6)$.

If the given points are to lie on the plane $z = a + bx + cy$, then the following equations must be satisfied:

$$\begin{array}{rrrrr} a & + & b & + & c & = 5 \\ a & + & 2b & + & c & = 4 \\ a & + & b & + & 2c & = 8 \\ a & & & + & c & = 4 \\ a & + & b & & & = 2 \\ a & + & 2b & + & 2c & = 6 \end{array} \quad \text{or} \quad \begin{bmatrix} 1 & 1 & 1 \\ 1 & 2 & 1 \\ 1 & 1 & 2 \\ 1 & 0 & 1 \\ 1 & 1 & 0 \\ 1 & 2 & 2 \end{bmatrix} \begin{bmatrix} a \\ b \\ c \end{bmatrix} = \begin{bmatrix} 5 \\ 4 \\ 8 \\ 4 \\ 2 \\ 6 \end{bmatrix}.$$

This system is overdetermined and does not have an exact solution. The best solution, in the least squares sense, is the solution of the normal equations

$$A^TAX = \begin{bmatrix} 6 & 7 & 7 \\ 7 & 11 & 9 \\ 7 & 9 & 11 \end{bmatrix} \begin{bmatrix} a \\ b \\ c \end{bmatrix} = \begin{bmatrix} 29 \\ 35 \\ 41 \end{bmatrix},$$

for which the exact solution is $a = 96/44$, $b = -16/44$, and $c = 116/44$. Thus the best fitting plane is $z = (96 - 16x + 116y)/44$, for which the sum of the squares of the errors is 2.35. The following tables compare the value of this function with the values of z at the data points and with the plane $z = 2 + x + 2y$, which also fits the data reasonably well.

Data			LS-plane		$z = 2 + x + 2y$	
x	y	z	z	d^2	z	d^2
1	1	5	4.45	.2975	5	1
2	1	4	4.09	.0080	6	4
1	2	8	7.09	.8260	7	1
0	1	4	4.82	.6400	3	1
1	0	2	1.86	.0198	3	1
2	2	6	6.73	.5290	8	4
				2.350		12

The procedure just illustrated seems quite straightforward, but there is a computational problem that should be mentioned briefly. The difficulty is that the normal equations are notoriously ill-conditioned; that is, small changes in the data can lead to large changes in the solution. If the computations are being done on a modern digital computer this is certainly cause for concern and caution. A thorough discussion of ill-conditioning is beyond the scope of this book, but we should at least mention the problem and indicate how to improve the computational procedure.

The key ingredient is commonly called the QR factorization of the $m \times n$ coefficient matrix A. In this factorization the $m \times n$ matrix Q is to have orthonormal columns, so that $Q^TQ = I$, and the $n \times n$ matrix R is to be upper triangular and nonsingular. As we indicated in Exercise 26 of Section 3.8, finding the matrices Q and R is equivalent to the Gram-Schmidt process for the column space of A. In practice, Q is usually found as a product of plane rotations (see Exercises 17 and

20 of Section 4.4) or as a product of elementary reflectors (see Exercise 9 of Section 5.4 and Exercise 9 of this section).

If we have the QR factorization of the matrix A, then we can use it to simplify Normal Equations 7.4 as follows:

$$A^TAX = A^TY \longrightarrow (QR)^T(QR)X = (QR)^TY \longrightarrow$$

$$R^T(Q^TQ)RX = R^TQ^TY \longrightarrow R^TRX = R^TQ^TY \longrightarrow RX = Q^TY.$$

Note that the final step used the fact that R, and hence R^T, is nonsingular if A has rank n. Note also that the final equation, $RX = Q^TY$, can be obtained directly from $AX = Y$ by substituting QR for A and then multiplying by Q^T.

Example 4

For the 6×3 matrix A of Example 3 we have the following MATALG-generated QR factorization: $A = QR$ with

$$Q = \begin{bmatrix} .40825 & -.099015 & -.073127 \\ .40825 & .49507 & -.25595 \\ .40825 & -.099015 & .54845 \\ .40825 & -.69310 & .10969 \\ .40825 & .099015 & -.69471 \\ .40825 & .49507 & .36564 \end{bmatrix}$$

$$\text{and } R \begin{bmatrix} 2.4495 & 2.8577 & 2.8577 \\ 0 & 1.6833 & 0.49507 \\ 0 & 0 & 1.6088 \end{bmatrix}$$

and the system $RX = Q^TY$ is

$$\begin{bmatrix} 2.4495 & 2.8577 & 2.8577 \\ 0 & 1.6833 & 0.49507 \\ 0 & 0 & 1.6088 \end{bmatrix} \begin{bmatrix} a \\ b \\ c \end{bmatrix} = \begin{bmatrix} 11.839 \\ .69310 \\ 4.2414 \end{bmatrix}$$

The solution of this system is $a = 2.1818$, $b = -.36364$, and $c = 2.6364$, which agrees with the solution obtained in Example 3 by solving Normal Equations 7.4 exactly. ■

Exercise Set 7.2

1. Find the straight line that best fits the following sets of data points, and compare the sum of the squares of the deviations from this line and the line determined by the first two points.

   (a) $(1, 2)$, $(2, 4)$, $(3, 8)$

   (b) $(1, 4)$, $(2, 9)$, $(3, 10)$, $(4, 15)$, $(5, 16)$, $(6, 21)$

   (c) $(1, 4.2)$, $(2, 4.7)$, $(3, 4.9)$, $(4, 5.6)$

   (d) $(40, 472)$, $(45, 457)$, $(50, 442)$, $(55, 423)$, $(60, 411)$

2. Find the parabola that best fits the following sets of data points.
    (a) $(-2, 45)$, $(0, 10)$, $(2, -16)$, $(4, -30)$
    (b) $(1, 1)$, $(2, 0)$, $(3, 1)$, $(4, -2)$, $(5, -5)$
    (c) $(3, 0)$, $(4, -10)$, $(6, -48)$, $(7, -76)$

3. Find the plane that best fits the following sets of data points.
    (a) $(1, 1, 1)$, $(0, 2, 1)$, $(1, 2, 0)$, $(0, 4, 0)$
    (b) $(1, 2, -1)$, $(2, 3, -3)$, $(3, 2, 5)$, $(2, 1, 8)$, $(2, 2, 3)$
    (c) Compare the sum of the squares of the deviations for the plane in part (b) and the plane $z = 3x - 5y + 6$.

4. Show that System 7.2 is consistent if and only if all the data points lie on a straight line. Show that A has rank 2 unless all the data points lie on a vertical line.

5. Solve both of the following 2×2 systems exactly.

$$x + 3y = 4 \qquad\qquad x + 3y = 4$$
$$1.001x + 3y = 4.001 \qquad 1.001x + 3y = 4.004$$

Interpret each of the equations as a straight line and explain why the small change in the right-hand side generates such a large change in the answer. This is an example of an ill-conditioned system.

6. Find the QR factorization for each of the coefficient matrices in Exercises 1(a), 2(a), and 3(a) and use it to solve the normal equations.

7. In physics, Hooke's law asserts that there is a linear relationship, $s = a + hf$, between the length s of a spring and the force f applied to it. The following data were collected in an elementary physics laboratory; use it to estimate the natural length (a) of the spring and the Hooke's law constant (h) for the given spring.

f (pounds)	.5	1	2	4
s (feet)	4	8	12	24

8. Let $\mathcal{W}$ be an n-dimensional subspace of $\mathcal{R}^m$ and let $\{\mathbf{u_1}, \ldots, \mathbf{u_n}\}$ be an orthonormal basis for $\mathcal{W}$. Extend this basis to an orthonormal basis $\{\mathbf{u_1}, \ldots, \mathbf{u_n}, \ldots \mathbf{u_m}\}$ for $\mathcal{R}^m$. Let

$$\mathbf{v} = a_1\mathbf{u_1} + \ldots + a_n\mathbf{u_n} + \ldots + a_m\mathbf{u_m}$$

be an arbitrary vector in $\mathcal{R}^m$. Consider any vector

$$\mathbf{w} = b_1\mathbf{u_1} + \ldots + b_n\mathbf{u_n}$$

in $\mathcal{W}$. Compute $\|\mathbf{v} - \mathbf{w}\|$ and determine when $\mathbf{w}$ is closest to $\mathbf{v}$.

9. Let $\mathbf{w}$ be a unit vector in R^n and define $H = I - 2\mathbf{w}\mathbf{w}^T$.
    (a) Show that H is both orthogonal and symmetric.
    (b) Show that $H\mathbf{w} = -\mathbf{w}$ and that if $\mathbf{u} \cdot \mathbf{w} = 0$, then $H\mathbf{u} = \mathbf{u}$. Show that these results can be interpreted geometrically to mean that H is a reflection in the subspace orthogonal to $\mathbf{w}$. Hence, these matrices are called ELEMENTARY REFLECTORS.
    (c) Show that if $\|\mathbf{u}\| = \|\mathbf{v}\|$ and $\mathbf{w} = (\mathbf{u} - \mathbf{v})/\|\mathbf{u} - \mathbf{v}\|$, then $H\mathbf{u} = \mathbf{v}$.
    (d) For $\mathbf{u} = [3 \quad 4 \quad 12]^T$, find $\mathbf{w}$ such that $H\mathbf{u} = [13 \quad 0 \quad 0]^T$.
    (e) If A is $m \times n$ with $m > n$, show that there exist elementary reflectors $H_1, \ldots, H_{n-1}$ such that

$$H_{n-1} \ldots H_2 H_1 A = \begin{bmatrix} R \\ 0 \end{bmatrix}, \quad \text{where } R \text{ is } n \times n \text{ and triangular.}$$

(f) Show how the result of part (e) can be used to find the QR factorization of the matrix A.

7.3   *Quadratic Forms*

Requires: Sections 4.5 and 5.4

In this section we will show how matrix techniques are useful in the study of quadratic forms. Quadratic forms are of importance in geometry, statistics, physics, engineering, number theory, and other areas of mathematics.

In general, a QUADRATIC FORM in the n variables $x_1, x_2, \ldots, x_n$ is a second-degree polynomial of the form

$$\begin{aligned}
Q = {} & a_{11}x_1^2 + a_{12}x_1x_2 + \cdots + a_{1n}x_1x_n \\
{} & + a_{21}x_2x_1 + a_{22}x_2^2 + \cdots + a_{2n}x_2x_n \\
& \quad \vdots \qquad\qquad \vdots \qquad\qquad\quad \vdots \\
{} & + a_{n1}x_nx_1 + a_{n2}x_2x_n + \cdots + a_{nn}x_n^2 \\
= {} & \sum_{j=1}^{n} \left(\sum_{i=1}^{n} a_{ij}x_ix_j \right).
\end{aligned}$$

It is much easier to deal with quadratic forms in matrix notation than in the above double-sum notation. If we assume, without loss of generality, that $a_{ij} = a_{ji}$ (if not we can write $a_{ij}x_ix_j + a_{ji}x_jx_i = bx_ix_j + bx_jx_i$, where $b = [a_{ij} + a_{ji}]/2$), then we can write Q as

$$Q = [x_1 \quad x_2 \quad \ldots \quad x_n] A \begin{bmatrix} x_1 \\ x_2 \\ \vdots \\ x_n \end{bmatrix} = X^T A X,$$

where X is a vector from $\mathcal{R}^n$. Thus we can view Q as defining a function from $\mathcal{R}^n$ to $\mathcal{R}$, that is, a real-valued function on $\mathcal{R}^n$.

Some examples of quadratic forms and their matrix representations are

$$Q_1 = x_1^2 + 2x_1x_2 + 3x_2^2 = [x_1 \quad x_2] \begin{bmatrix} 1 & 1 \\ 1 & 3 \end{bmatrix} \begin{bmatrix} x_1 \\ x_2 \end{bmatrix};$$

$$Q_2 = x_1^2 - 5x_2^2 + 3x_3^2 = [x_1 \quad x_2 \quad x_3] \begin{bmatrix} 1 & 0 & 0 \\ 0 & -5 & 0 \\ 0 & 0 & 3 \end{bmatrix} \begin{bmatrix} x_1 \\ x_2 \\ x_3 \end{bmatrix}; \quad \text{and}$$

$$Q_3 = 3x_1^2 - 5x_1x_2 + 4x_2^2 + 3x_1x_3 + 8x_2x_3 + x_3^2 + 10x_1x_4 + 12x_3x_4$$

$$= [x_1 \quad x_2 \quad x_3 \quad x_4] \begin{bmatrix} 3 & -5/2 & 3/2 & 5 \\ -5/2 & 4 & 4 & 0 \\ 3/2 & 4 & 1 & 6 \\ 5 & 0 & 6 & 0 \end{bmatrix} \begin{bmatrix} x_1 \\ x_2 \\ x_3 \\ x_4 \end{bmatrix}.$$

You have most likely encountered quadratic forms in two variables in your algebra and calculus studies. There you considered the problem of graphing the equation

$$ax^2 + 2bxy + cy^2 = k^2 \tag{7.5}$$

This problem is relatively easy if $b = 0$. Recall that the graph of

$$ax^2 + cy^2 = k^2$$

is an ellipse if a and c are both positive and a hyperbola if ac is negative (see Figure 7.10). If $k = 0$, the hyperbola degenerates to two parallel lines and the ellipse shrinks to a single point. The easy case ($b = 0$) corresponds to the case where the matrix of the quadratic form is a diagonal matrix.

To identify and sketch the more general curve given in Equation 7.5 we will use the eigenvalue theory for symmetric matrices that was developed in Sections 5.3 and 5.4.

From Theorem 5.10 we know that for any 2×2 symmetric matrix A there exists an orthogonal matrix P such that P^TAP is a diagonal matrix with the eigenvalues of A on the diagonal;

$$P^TAP = \begin{bmatrix} \lambda_1 & 0 \\ 0 & \lambda_2 \end{bmatrix}. \tag{7.6}$$

Moreover, the columns of P are an orthonormal set of eigenvectors for the matrix A.

By Exercise 19 of Section 4.4 we know that every 2×2 orthogonal matrix is of one of the following forms:

$$P_1 = \begin{bmatrix} \cos\theta & -\sin\theta \\ \sin\theta & \cos\theta \end{bmatrix} \quad \text{or}$$

$$P_2 = \begin{bmatrix} \cos\theta & \sin\theta \\ \sin\theta & -\cos\theta \end{bmatrix} = \begin{bmatrix} \cos\theta & -\sin\theta \\ \sin\theta & \cos\theta \end{bmatrix} \begin{bmatrix} 1 & 0 \\ 0 & -1 \end{bmatrix}.$$

It was shown in Section 4.4 that P_1 represents a rotation through an angle θ and that P_2 represents a reflection in the x-axis followed by a rotation through an angle θ. Note that $\det P_1 = 1$ and $\det P_2 = -1$. It is easy to see that the matrix P of Equation 7.6 can always be chosen to be a rotation.

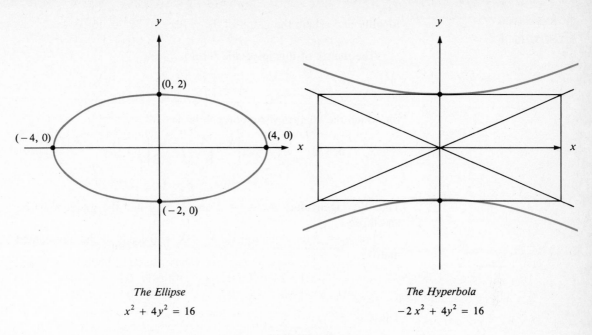

The Ellipse

$x^2 + 4y^2 = 16$

The Hyperbola

$-2x^2 + 4y^2 = 16$

FIGURE 7.10

If we now make the coordinate change

$$X = \begin{bmatrix} x \\ y \end{bmatrix} = PX' = P\begin{bmatrix} x' \\ y' \end{bmatrix},$$

where P is the matrix of Equation 7.6, we have

$$Q = X^T A X = (PX')^T A (PX) = X'^T (P^T AP) X' = \lambda_1 x'^2 + \lambda_2 y'^2,$$

and the troublesome cross-product term has been eliminated. Since

$$P = \begin{bmatrix} \cos\theta & -\sin\theta \\ \sin\theta & \cos\theta \end{bmatrix}$$

for some θ, the substitution $X = PX'$ amounts to a rotation of the co-ordinate axes through an angle θ. If we let $\mathbf{u}_1$ and $\mathbf{u}_2$ be the new basis vectors, then P is the change-of-basis matrix from the new basis $\{\mathbf{u}_1, \mathbf{u}_2\}$ to the old basis $\mathcal{S} = \{\mathbf{e}_1, \mathbf{e}_2\}$. Thus, from Theorem 3.22

$$P = \begin{bmatrix} [\mathbf{u}_1]_{\mathcal{S}} & [\mathbf{u}_2]_{\mathcal{S}} \end{bmatrix} = \begin{bmatrix} \mathbf{u}_1 & \mathbf{u}_2 \end{bmatrix},$$

and we see that the eigenvectors of A are vectors in the direction of the new coordinate axes.

Example 1

Identify and sketch the graph of $3x^2 + 2xy + 3y^2 = 16$.

The matrix of this quadratic form is

$$A = \begin{bmatrix} 3 & 1 \\ 1 & 3 \end{bmatrix},$$

for which the characteristic polynomial is

$$\det(\lambda I - A) = \det\begin{bmatrix} \lambda - 3 & -1 \\ -1 & \lambda - 3 \end{bmatrix}$$

$$= \lambda^2 - 6\lambda + 8 = (\lambda - 2)(\lambda - 4).$$

Thus the eigenvalues are $\lambda_1 = 2$ and $\lambda_2 = 4$ and the graph must be an ellipse.

For $\lambda_1 = 2$ the equation $(2I - A)X = 0$ leads to the augmented matrix

$$\begin{bmatrix} -1 & -1 & 0 \\ -1 & -1 & 0 \end{bmatrix} \longrightarrow \begin{bmatrix} 1 & 1 & 0 \\ 0 & 0 & 0 \end{bmatrix}.$$

Thus

$$X_1 = \frac{1}{\sqrt{2}}\begin{bmatrix} 1 \\ -1 \end{bmatrix}$$

is a unit eigenvector for $\lambda_1 = 2$.

Similarly, we find that

$$X_2 = \frac{1}{\sqrt{2}}\begin{bmatrix} 1 \\ 1 \end{bmatrix}$$

is a unit eigenvector for $\lambda_2 = 4$. Thus

$$P = [X_1 \quad X_2] = \frac{1}{\sqrt{2}}\begin{bmatrix} 1 & 1 \\ -1 & 1 \end{bmatrix}$$

is an orthogonal matrix such that

$$P^T A P = \begin{bmatrix} \lambda_1 & 0 \\ 0 & \lambda_2 \end{bmatrix} = \begin{bmatrix} 2 & 0 \\ 0 & 4 \end{bmatrix}.$$

Note that $\det P = +1$ so P represents a rotation. Since $\cos\theta = 1/\sqrt{2}$ and $\sin\theta = -1/\sqrt{2}$, it follows that $\theta = -\pi/4$ and therefore P represents a rotation through an angle of -45 degrees. In Figure 7.11 we show the graph of the quadratic form and the two coordinate systems. Note that the x'-axis is in the direction of the eigenvector X_1 and that the y'-axis is in the direction of the eigenvector X_2. Relative to the $x'y'$ coordinate system the equation of the ellipse reduces to

$$2x'^2 + 4y'^2 = 16 \quad \text{or} \quad x'^2/8 + y'^2/4 = 1. \quad \blacksquare$$

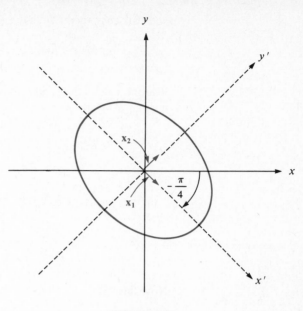

FIGURE 7.11

The discussion preceding the example generalizes directly to the case of n variables. The result is usually called the Principal Axis Theorem.

Theorem 7.4

PRINCIPAL AXIS THEOREM: Let A be an $n \times n$ symmetric matrix and let $Q(X) = X^T A X$ be the associated quadratic form. Let P be an orthogonal matrix, with $\det P = 1$, whose columns are eigenvectors of A that form an orthonormal basis for $\mathcal{R}^n$. Then the substitution $X = PY$ reduces Q to the equivalent form

$$Q(Y) = Y^T(P^T A P)Y = \lambda_1 y_1^2 + \lambda_2 y_2^2 + \ldots + \lambda_n y_n^2,$$

where the λ_i are the eigenvalues of A. The coordinate change $X = PY$ amounts to a rotation in $\mathcal{R}^n$.

Example 2

Identify and sketch the surface whose equation is

$$10x^2 + 8xy - 2y^2 - 20xz + 28yz + z^2 = 72.$$

The (symmetric) matrix of this quadratic form is

$$A = \begin{bmatrix} 10 & 4 & -10 \\ 4 & -2 & 14 \\ -10 & 14 & 1 \end{bmatrix},$$

and it is easy to verify that the characteristic polynomial of A is

$$\det(\lambda I - A) = \lambda^3 - 9\lambda^2 - 324\lambda + 2916$$

$$= (\lambda - 18)(\lambda + 18)(\lambda - 9).$$

Routine calculations, like those in Section 5.4, show that the eigenvectors of A are

$$X_1 = \frac{1}{3}\begin{bmatrix} -2 \\ 1 \\ 2 \end{bmatrix} \quad \text{for the eigenvalue } \lambda_1 = 18,$$

$$X_2 = \frac{1}{3}\begin{bmatrix} 1 \\ -2 \\ 2 \end{bmatrix} \quad \text{for } \lambda_2 = -18,$$

$$\text{and} \quad X_3 = \frac{1}{3}\begin{bmatrix} 2 \\ 2 \\ 1 \end{bmatrix} \quad \text{for } \lambda_3 = 9.$$

Now the matrix

$$P_1 = \frac{1}{3}\begin{bmatrix} -2 & 1 & 2 \\ 1 & -2 & 2 \\ 2 & 2 & 1 \end{bmatrix}$$

is an orthogonal matrix such that

$$P_1^T A P = Dg\{18, -18, 9\}.$$

Note that $\det(P_1) = 1$, so P_1 is the matrix of a rotation in $\mathcal{R}^3$. The substitution $X = PY$ reduces the quadratic form $Q = X^T A X$ to

$$Q = Y^T(P^T A P)Y = 18y_1^2 - 18y_2^2 + 9y_3^2.$$

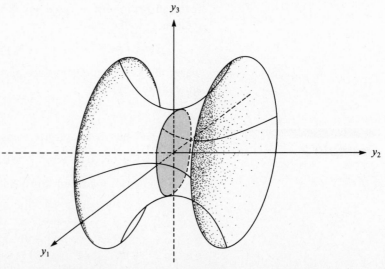

FIGURE 7.12

The y_1-coordinate axis is in the direction of $X_1 = \text{Col}_1(P)$, the y_2-coordinate axis is in the direction of $X_2 = \text{Col}_2(P)$, and the y_3-coordinate axis is in the direction of the eigenvector X_3. The surface is an elliptic hyperboloid of one sheet whose axis of symmetry is the y_2-axis; it is plotted in Figure 7.12 with respect to the y_i-coordinate axis and in Figure 7.13 with respect to the original coordinate system. ■

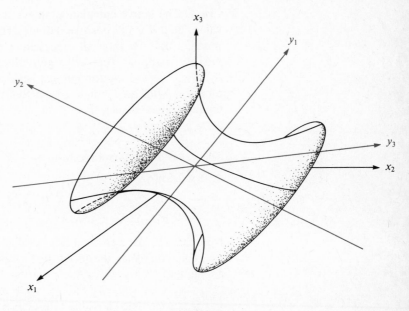

FIGURE 7.13

We now define a frequently cited property of quadratic forms and the associated symmetric matrices.

Definition 7.4

A symmetric matrix A, or the associated quadratic form $Q = X^T A X$, is said to be POSITIVE DEFINITE if $X^T A X > 0$ for every nonzero X in $\mathcal{R}^n$. Similarly, A is POSITIVE SEMIDEFINITE if $X^T A X \geq 0$ for all X, A is NEGATIVE DEFINITE if $X^T A X < 0$ for all $X \neq 0$, A is NEGATIVE SEMIDEFINITE if $X^T A X \leq 0$ for all X, and A is INDEFINITE otherwise.

If λ is an eigenvalue of A with associated eigenvector X, then $AX = \lambda X$ and the value of the quadratic form at the eigenvector X is

$$X^T A X = X^T(\lambda X) = \lambda X^T X.$$

This value has the same algebraic sign as λ since $X^T X$ is positive. Thus we see that the algebraic signs of the eigenvalues will be involved in deciding whether or not a given matrix is positive definite. The precise result is given in the next theorem.

Theorem 7.5

A symmetric matrix A, with eigenvalues $\lambda_1, \ldots, \lambda_n$, is

(1) positive definite if and only if all $\lambda_i > 0$;
(2) positive semidefinite if and only if all $\lambda_i \geq 0$;
(3) negative semidefinite if and only if all $\lambda_i \leq 0$;
(4) negative definite if and only if all $\lambda_i < 0$; and
(5) indefinite if A has both positive and negative eigenvalues.

Proof. The above calculation shows that if A has a negative eigenvalue, then A is not positive definite. The contrapositive is that if A is positive definite then all eigenvalues of A must be positive. Conversely, suppose that all eigenvalues of A are positive and let $X_1, X_2, \ldots, X_n$ be an orthonormal basis for $\mathcal{R}^n$ consisting of eigenvectors of A with $AX_i = \lambda_i X_i$. Then for any vector X in $\mathcal{R}^n$ there exist unique scalars a_i such that

$$X = a_1 X_1 + a_2 X_2 + \ldots + a_n X_n.$$

The value of the quadratic form $Q = X^T A X$ for this X is

$$
\begin{aligned}
X^T A X &= X^T (a_1 \lambda_1 X_1 + a_2 \lambda_2 X_2 + \ldots + a_n \lambda_n X_n) \\
&= (a_1 X_1^T + \ldots + a_n X_n^T)(a_1 \lambda_1 X_1 + a_2 \lambda_2 X_2 + \ldots + a_n \lambda_n X_n) \\
&= a_1^2 \lambda_1 X_1^T X_1 + a_2^2 \lambda_2 X_2^T X_2 + \ldots + a_n^2 \lambda_n X_n^T X > 0.
\end{aligned}
$$

Note that the last step used the fact that $X_i^T X_j = 0$ if $i \neq j$.

This completes the proof of the first assertion. The other proofs are similar. $\square$

It is sometimes possible to establish that a matrix is positive definite without finding all the eigenvalues. One such class of matrices for which this is possible is the class of *diagonally dominant* matrices. These matrices occur frequently in numerical analysis.

Definition 7.5

An $n \times n$ matrix A is DIAGONALLY DOMINANT if

$$|a_{ii}| \geq \sum_{j \neq i} |a_{ij}| \quad \text{for } i = 1, 2, \ldots, n$$

and STRICTLY DIAGONALLY DOMINANT if

$$|a_{ii}| > \sum_{j \neq i} |a_{ij}| \text{ for } i = 1, 2, \ldots, n.$$

The matrices

$$A = \begin{bmatrix} 5 & 2 & -2 \\ 2 & -4 & 1 \\ 0 & 0 & 1 \end{bmatrix} \quad \text{and} \quad B = \begin{bmatrix} 2 & 1 & 0 & 0 \\ 1 & 2 & 1 & 0 \\ 0 & 1 & 2 & 1 \\ 0 & 0 & 1 & 2 \end{bmatrix}$$

are both diagonally dominant, but only A is strictly diagonally dominant.

Theorem 7.6

A strictly diagonally dominant symmetric matrix, with positive diagonal elements, must be positive definite.

Proof. Let λ be an eigenvalue of A with $AX = \lambda X$ and suppose, without loss of generality, that X has been scaled so that all $|x_j| \leq 1$ and $x_i = 1$. Computing the ith component in the equation $AX = \lambda X$ yields

$$a_{i1}x_1 + \ldots + a_{ii}x_i + \ldots + a_{in}x_n = \lambda x_i \, .$$

Rearranging, using the fact that $x_i = 1$, and taking absolue values yields

$$\left|\lambda - a_{ii}\right| = \left|\sum_{j \neq i} a_{ij}x_j\right| \leq \sum_{j \neq i} \left|a_{ij}\right|\left|x_j\right| \, .$$

Since all $|x_j| \leq 1$ it follows that

$$\left|\lambda - a_{ii}\right| \leq \sum_{j \neq i} \left|a_{ij}\right| = r_i \, .$$

Since $a_{ii} > r_i$, by the assumed diagonal dominance of A, it follows that the eigenvalue λ must be positive. Thus A is positive definite by Theorem 7.5. $\square$

Exercise Set 7.3

Identify and sketch the graphs of the following curves.

1. $2x^2 + 6xy + 2y^2 = 10$
2. $2x^2 - 2xy + 2y^2 = 9$
3. $14x^2 + 16xy + 2y^2 = 36$
4. $16x^2 - 24xy + 9y^2 - 60x - 80y = 0$

Identify and sketch the graphs of the following surfaces.

5. $3x^2 - 2xy + 2xz + 3y^2 - 2yz + 3z^2 = 10$
6. $x^2 - 2y^2 + 4yz + z^2 = 12$
7. $3x^2 + 8xy + 4xz + 3y^2 + 4yz = 16$
8. $2x^2 + 2xy + 2xz + y^2 + z^2 = 10$
9. $-12xy - 12xz + 3y^2 - 3z^2 = 9$
10. Let $A = \begin{bmatrix} a & b \\ b & c \end{bmatrix}$ and $A' = \begin{bmatrix} a' & b' \\ b' & c' \end{bmatrix} = P^T A P$,

    where P is orthogonal.

(a) Show that $b^2 - ac = b'^2 - a'c'$; that is, the discriminant $(b^2 - ac)$ is invariant under rotation.

(b) Show how the result in part (a) can be used to identify the graph of

$$ax^2 + 2bxy + cy^2 + dx + ey + f = 0.$$

(*Hint:* Consider the case $b' = 0$.)

11. Let A be an $n \times n$ symmetric matrix and define $r_i = \sum_{j \neq i} |a_{ij}|$. Extend the discussion in the proof of Theorem 7.6 to show that all the eigenvalues of A lie in the union of the intervals $|x - a_{ii}| \leq r_i$.

12. Prove part 3 of Theorem 7.5.

13. Extend the result of Exercise 17 of Section 4.4 to show that every rotation in $\mathscr{R}^3$ is a product of plane rotations. Does this result generalize to $\mathscr{R}^n$?

7.4   *Linear Economic Models*

Requires: Chapter 1 with minor references to Sections 2.4 and 7.3

In this section we will take a brief look at the Leontief input-output model of an economy.* We begin with a simple example of a three-sector closed economy.

Example 1

Consider a cooperative arrangement between an electrician, a plumber, and a painter who each agree to contribute twelve hours per week working for themselves and the other two. After some negotiations they agree on the following work schedule.

		Supplier	
Consumer	*Electrician*	*Plumber*	*Painter*
Electrician	2	5	3
Plumber	4	2	6
Painter	6	5	3

---

*Wassily Leontief received the 1973 Nobel prize in economics for his work on mathematical modeling.

For tax reasons the three workers must establish a value for their services. They wish to do this in such a way that they all come out even. What hourly rates should each worker set for his services?

If we let the rate matrix be

$$X = \begin{bmatrix} x \\ y \\ z \end{bmatrix} = \begin{bmatrix} \text{electrician's hourly rate} \\ \text{plumber's hourly rate} \\ \text{painter's hourly rate} \end{bmatrix},$$

then the electrician requires services worth $2x + 5y + 3z$ and provides services worth $12x$. Thus the equilibrium (break-even) condition for the electrician is

$$2x + 5y + 3z = 12x$$

Similarly, the equilibrium conditions for the other two workers are $4x + 2y + 6z = 12y$ for the plumber, and $6x + 5y + 3z = 12z$ for the painter.

Thus, the unknown hourly wages must satisfy the linear system

$$\begin{bmatrix} 2 & 5 & 3 \\ 4 & 2 & 6 \\ 6 & 5 & 3 \end{bmatrix} \begin{bmatrix} x \\ y \\ z \end{bmatrix} = \begin{bmatrix} 12x \\ 12y \\ 12z \end{bmatrix}.$$

If we divide the last equation by 12 we obtain an equation of the form $EX = X$, where the coefficient matrix

$$E = \begin{bmatrix} 2/12 & 5/12 & 3/12 \\ 4/12 & 2/12 & 6/12 \\ 6/12 & 5/12 & 3/12 \end{bmatrix}$$

has the properties that all $e_{ij} \geq 0$ and that each column sum is 1.

The system $EX = X$ is equivalent to the homogeneous system $(I - E)X = 0$, for which the general solution can be obtained by Gaussian elimination. The solution is

$$X = \begin{bmatrix} x \\ y \\ z \end{bmatrix} = c \begin{bmatrix} 3/4 \\ 9/10 \\ 1 \end{bmatrix}, \quad \text{where } c \text{ is arbitrary.}$$

If we take $c = 20$, then $x = 15$, $y = 18$, and $z = 20$ are reasonable values for the desired hourly rates.   ■

More generally, suppose that we have n interdependent industries, each of which produces a single product and uses a part of the output of the other industries in its operation. Let us denote the industries by $S_1, S_2, \ldots, S_n$ and the unit price for their products by $x_1, x_2, \ldots, x_n$. Let e_{ij} be the amount of product j required in the production of one unit of product *i*. Let us assume that each industry produces one unit of its product and that the economy is CLOSED in the sense that no goods leave or enter the system. This means that the total consumption of each product must equal its total production.

It follows that the cost of producing one unit of product *i* is

$$e_{i1}x_1 + e_{i2}x_2 + \ldots + e_{in}x_n .$$

If we assume that each industry breaks even (income equals expenses), then we have the equations

$$e_{i1}x_1 + e_{i2}x_2 + \ldots + e_{in}x_n = x_i, \quad i = 1, 2, \ldots n .$$

In matrix form this equation can be written as

$$EX = X \quad \text{or} \quad (I - E)X = 0 ,$$

and the solution of this system, provided it is positive, would provide the equilibrium prices for the various products of the economy.

Note that all the elements of the matrix *E* are nonnegative and that, because the economy is closed, each of the columns of *E* has sum 1. Such matrices are called EXCHANGE MATRICES or COLUMN STOCHASTIC MATRICES. The fact that the column sums are all 1 can be interpreted to mean that each industry stores any of its product that is not used by the other industries.

For an exchange matrix *E* it is easy to see that the sum of the rows of *I* − *E* is the zero row. It follows that *I* − *E* is a singular matrix and, from Theorem 1.9, that $(I - E)X = 0$ and hence $EX = X$, has a nonzero solution. Note that this is equivalent to saying that $\lambda = 1$ is an eigenvalue of *E*.

It is more difficult to show that $EX = X$ has a nonnegative solution. The details are included in *The Theory of Linear Economic Models* by David Gale (New York: McGraw Hill, 1960).

The following theorem states this important property of exchange matrices. Some insight into why the solution is positive is provided in Exercise 10.

Theorem 7.7

If *E* is an $n \times n$ exchange matrix, then there exists a nonnegative vector *X* such that $EX = X$.

We will now discuss a model of an OPEN economy, where the industries produce more of their product than is required by the other industries. The remainder of their product is then available for consumer use. In this model we will assume that the prices of the products are fixed and will show how to compute the outputs from each industry needed to meet a given set of demands from the marketplace.

Suppose that we have, as above, *n* interdependent industries, each of which produces a single product. Let v_i represent the value of the output from industry i and let c_{ij} = value of product i needed to produce $1.00 worth of product j. For example, if $c_{32} = .37$, then $.37 worth of product 3 is required to make $1.00 worth of product 2.

Note that the first subscript represents input and the second subscript represents output. The $n \times n$ matrix C is called the CONSUMPTION MATRIX of the economy. A consumption matrix must have entries that satisfy $0 \leq c_{ij} \leq 1$.

For a consumption matrix C we observe that

(1) the sum of the elements in column j represents the total value of inputs needed to produce $1.00 worth of product j;
(2) industry j is profitable if the jth column sum is less than 1;
(3) $\text{row}_i(C)Y = c_{i1}y_1 + \ldots + c_{in}y_n$ is the total value of the output of industry i used by the other industries; and
(4) $y_i - \text{Row}_i(C)Y$ is the value of the output of industry i which is available to satisfy outside demand from consumers.

If D is a given demand vector, then the output vector Y must satisfy the linear system

$$Y - CY = D. \tag{7.7}$$

Example 2

Suppose that the following consumption matrix defines the demands of six industries on each other:

$$
\begin{array}{ccccccc}
 & A & M & P & Co & P & Ch \\
\textit{Automobile} & .10 & .05 & .10 & .05 & .05 & .10 \\
\textit{Metal} & .35 & .10 & .15 & .10 & .05 & .05 \\
\textit{Power} & .12 & .30 & .15 & .15 & .20 & .25 \\
\textit{Coal} & .10 & .20 & .30 & .05 & .15 & .10 \\
\textit{Plastics} & .15 & .01 & .05 & .10 & .10 & .10 \\
\textit{Chemical} & .05 & .10 & .05 & .05 & .20 & .15 \\
\end{array} = C.
$$

(a) What outputs are required to meet a demand of
$$D = [10 \quad 2 \quad 15 \quad 2 \quad 20 \quad 15]^T$$
million dollars?

(b) Suppose that the demand changes to
$$D1 = [12 \quad 2 \quad 15 \quad 2 \quad 20 \quad 15]^T$$
because of a good economic forecast. What output is required to meet demand $D1$?

(c) Suppose that the price of power is increased 10 percent while all other prices remain constant. What output is required to meet demand $D1$?

In order to answer the first question we need to solve the system
$$(I - C)Y = D.$$

The solution, obtained by Gaussian elimination, is

$$Y = [28.49 \quad 32.44 \quad 61.09 \quad 41.53 \quad 39.62 \quad 38.50]^T.*$$

Thus the total output of the automobile industry needs to be $28,490,000, while the total output of the power industry needs to be $61,090,000.

In order to answer the second question we need to solve the system

$$(I - C)Y1 = D1.$$

Note that the coefficient matrix is the same as for part (a). The solution, rounded to four significant figures, is

$$Y1 = [31.15 \quad 33.94 \quad 62.60 \quad 42.80 \quad 40.38 \quad 39.17]^T.$$

Note that all sectors of the economy are affected by the increased demand for only one product and that the increase in the demand for automobiles is greater than the increased consumer demand. In order to answer the third question we need to modify the consumption matrix C to reflect the change in power costs. Only the third row of C is changed; if the new matrix is called C', then

$$\text{Row}_3(C') = [.132 \quad .33 \quad .165 \quad .165 \quad .22 \quad .275].$$

The answer to part (c) is obtained by solving the linear system

$$(I - C')Y2 = D1,$$

for which the solution is

$$Y2 = [32.68 \quad 36.53 \quad 71.13 \quad 46.56 \quad 41.72 \quad 40.61]^T.$$

Note that the price change in only one industry affected the production plans of all the other industries.   ■

A consumption matrix C is said to be PRODUCTIVE if $(I - C)$ is nonsingular and all of the entries of $(I - C)^{-1}$ are nonnegative; that is, $(I - C)^{-1} \geq 0$. This of course implies that the solution of any system $Y - CY = D \geq 0$ will have a nonnegative (economically reasonable) solution.

The solution of (demand) Equation 7.7 can be written as $Y = (I - C)^{-1}D$. If one needs to make this calculation for many different values of D, not all of which are available at the same time, it might be worth the extra computational effort to find and store the inverse matrix $(I - C)^{-1}$.

---

*The calculations were done using the MATMAN program.

Note that

$$(I - C)(I + C^2 + C^3 + \ldots + C^m) = I - C^{m+1}$$

so that, if $\lim\limits_{m \to \infty} C^m = 0$, then

$$(I - C)^{-1} = I + C + C^2 + C^3 + \ldots \qquad (7.8)$$

can be expressed as an infinite geometric series. Approximations for $(I - C)^{-1}$ can be obtained by truncating the above series. Thus,

$$(I + C), (I + C + C^2), \quad \text{and} \quad (I + C^2 + C^3)$$

are all approximations to $(I - C)^{-1}$. Since $C^k \geq 0$ for all k, the latter approximations will be better than the earlier ones, but will require more computation.

It can be shown that $\lim_{m \to \infty} C^m = 0$ if and only if the eigenvalues of C satisfy $|\lambda| < 1$. Since, by Exercise 11 of Section 7.3,

$$|\lambda| \leq \max_j \left\{ \sum_{i=1}^{n} |c_{ij}| \right\},$$

it follows that C is productive if each column sum of C is less than 1, that is, if each industry is profitable. We summarize this discussion with Theorem 7.8.

Theorem 7.8

A consumption matrix C is productive if each industry is profitable.

Exercise Set 7.4

1. Show that each of the following is an exchange matrix and find a nonnegative price vector satisfying the equilibrium condition of Theorem 7.8.

   (a) $\begin{bmatrix} 1/6 & 1/4 & 1/3 \\ 2/3 & 1/4 & 2/3 \\ 1/6 & 1/2 & 0 \end{bmatrix}$ (b) $\begin{bmatrix} .25 & .40 & .50 \\ .50 & .20 & .50 \\ .25 & .40 & .00 \end{bmatrix}$

   (c) $\begin{bmatrix} 7/16 & 1/2 & 3/16 \\ 5/16 & 1/6 & 5/16 \\ 1/4 & 1/3 & 1/2 \end{bmatrix}$

2. Three doctors, an allergist, a dermatologist, and a surgeon, enter into a cooperative agreement of the type described in Example 1. They each agree to provide a total of six hours per month according to the following table.

$$\begin{array}{c} \\ A \\ D \\ S \end{array} \begin{array}{ccc} A & D & S \\ \begin{bmatrix} 1 & 4 & 3 \\ 2 & 1 & 3 \\ 3 & 1 & 0 \end{bmatrix} \end{array}$$

Determine reasonable rates for the three doctors so that each of the three doctors breaks even on the cooperative agreement.

3. Four neighbors, Tom, Pete, Bill, and Zeke, agree to share the produce from their backyard gardens. Tom will grow tomatoes, Pete will grow peas, Bill will grow beans, and Zeke will grow zucchini. The following table indicates how they will share the produce.

	Consumer			
Producer	*T*	*P*	*B*	*Z*
Tom	10	2	3	5
Pete	5	13	5	7
Bill	4	6	9	3
Zeke	5	3	7	9

What are reasonable prices for each of the four vegetables if each of the gardeners is to break even?

4. Suppose that the four gardeners in Exercise 3 decide to exchange their produce according to the following table.

	Consumer			
Producer	*T*	*P*	*B*	*Z*
Tom	10	2	3	5
Pete	5	3	5	7
Bill	4	1	2	4
Zeke	5	1	7	7

Is it still possible to assign prices to their products so that everyone breaks even?

5. The following consumption matrix defines the interdependence of four industries.

$$
\begin{array}{c}
\\ A \\ B \\ C \\ D
\end{array}
\begin{array}{cccc}
A & B & C & D \\
\left[\begin{array}{cccc}
.20 & .15 & .30 & .25 \\
.15 & .05 & .10 & .00 \\
.20 & .30 & .25 & .20 \\
.10 & .20 & .05 & .30
\end{array}\right]
\end{array}
$$

(a) What is the total value of the raw materials needed to make $1.00 worth of the product of industry C?

(b) What percentage of the output of industry C is used by the other industries?

(c) What percentage of the output of industry B is available to satisfy consumer demand?

6. **(a)** For the four-sector economy of Exercise 5 assume that prices are constant and find the production necessary to satisfy demands for $100,000 of product A, $75,000 of product B, $10,000 of product C, and $80,000 of product D.

   (b) Suppose that the demand for product C increases to $12,000 while the other demands remain constant. What production changes would be required to meet this new demand?

   (c) Suppose the cost of product C decreased 10 percent while the other prices remained constant. What production changes would be necessary to meet the demands of part (b)?

7. Read "The World Economy of the Year 2000" by Wassily Leontief in *Scientific American,* November 1980, page 207. How large is the matrix that describes the world economy?

8. Which of the following consumption matrices is productive?

   (a) $\begin{bmatrix} .10 & .10 & .20 \\ .20 & .15 & .30 \\ .40 & .30 & .15 \end{bmatrix}$ **(b)** $\begin{bmatrix} .35 & .25 & .00 \\ .15 & .55 & .35 \\ .45 & .30 & .60 \end{bmatrix}$ **(c)** $\begin{bmatrix} .2 & .4 & .1 & .4 \\ .3 & .2 & .2 & .1 \\ .0 & .4 & .5 & .3 \\ .5 & .0 & .2 & .2 \end{bmatrix}$

9. Let C be a consumption matrix. Compare the computational effort required to compute $(I - C)^{-1}$ using Gauss-Jordan elimination and Formula 7.8.

10. **(a)** Let E be an $n \times n$ exchange matrix. Show that, as a consequence of Exercise 11 of Section 7.3, that any eigenvalue of E must satisfy

$$|\lambda| \le \max\left\{ \sum_{j=1}^{n} |e_{ij}| \right\} = 1 .$$

   Thus $\lambda = 1$ is a dominant eigenvalue of E.

   (b) Let us use the power method of Section 5.5 to compute an eigenvector of E for $\lambda = 1$. If we use an initial vector with nonnegative entries, then the eigenvector will have nonnegative entries. (What if there is a complex eigenvalue of modulus one?)

Appendix: Linear Algebra Software

There has been a great expenditure of scientific effort aimed at producing high-quality software for handling the basic problems of linear algebra. Any scientific or academic computing facility is certain to have such programs in its program library. There are two mainframe packages that deserve special mention.

LINPACK is a collection of FORTRAN subroutines for handling all aspects of the linear equations problem. The LINPACK user must write a FORTRAN driver which dimensions the arrays, inputs the data, calls the appropriate subroutines, and outputs the desired results. For example, calling the two subroutines

$$SGECO(A, LDA, N, IPVT, RCOND, Z)$$

$$SGESL(A, LDA, N, IPVT, B, JOB)$$

will result in computing the solution of the linear system $AX = B$, where A is an $N \times N$ matrix with $N \leq LDA$. The solution is returned in B and the original matrix A is destroyed. LINPACK should be available at most scientific computing installations. On-line documentation is frequently inadequate. Full documentation is contained in the LINPACK Users Guide available from SIAM (The Society for Industrial and Applied Mathematics).

EISPACK is a collection of FORTRAN subroutines for handling all aspects of the algebraic eigenvalue problem. The EISPACK user must write a FORTRAN driver which dimensions the arrays, inputs the data, calls the proper subroutines, and outputs the desired results. For example, calling the three subroutines

$$ORTHES(NM, N, LOW, IGH, A, ORT)$$

$$ORTRAN(NM, N, LOW, IGH, A, ORT, Z)$$

$$HQR2(NM, N, LOW, IGH, A, WR, WI, Z, ERROR)$$

will result in computing all the eigenvalues of the $N \times N$ matrix A ($N \leq NM$). The real parts of the eigenvalues are returned in the vector WR and the complex parts in the vector WI. The eigenvectors are contained in the matrix Z. The matrix A is destroyed. EISPACK should be available at most scientific computing installations; usually with adequate on-line documentation. Full documentation is contained in Volume 6 of the Springer-Verlag Lecture Notes in Computer Science.

An increasing amount of linear algebra software is becoming available for personal computers. The program PC-Mathlab, although expensive, seems to be of high quality.

The software provided to users of this text (included with the *Instructor's Guide and Software Toolkit*) consists of two main programs, MATMAN and MATALG. These programs are designed primarily to help teach linear algebra; they are not designed as production codes for large-scale problems. The programs are menu driven and require no programming knowledge; all that is required is to know where to insert the program disk and how to turn on the machine. MATMAN (The MATrix MANipulator) is an interactive program to do row reductions; it can be used to solve linear systems, invert matrices, evaluate determinants, and solve linear programming problems using the tableaux implementation of the simplex algorithm. In the program's primary mode the user specifies the row operations to be performed and the machine does the arithmetic quickly and accurately. The arithmetic can be done using either rational or real numbers. An automatic switch can be unlocked after the student has progressed to the desired point. MATALG (The MATrix ALGebraist) is an interactive program for automating basic matrix operations. It can be used to check hand calculations, to save computation time with moderate-sized problems, to illustrate round-off error, and to provide experimental evidence in support of many of the theoretical observations in the text. These programs will run on most (all that we have tried) IBM-PC-compatibles; a version for the Apple Macintosh is available. These programs were written in FORTRAN; the disk contains only the compiled versions of the programs. Full documentation for these programs is contained both on the program disk and in the *Instructor's Guide and Software Toolkit*.

Answers to Selected Exercises

Chapter 1

Exercise Set 1.1   (page 10)

2. (a) $\left\{ \left(c, \dfrac{3c}{5} - 3 \right) \middle| c \text{ real} \right\}$ **(c)** $\left\{ \left(c_1, c_2, \dfrac{3c_1}{7} + \dfrac{5c_2}{7} - \dfrac{10}{7} \right) \middle| c_1, c_2 \text{ real} \right\}$

3. (a), (b)

4. (a) $x = 4, y = 0$

5. (a) $\begin{bmatrix} 2 & 0 & -3 \\ 4 & 5 & 1 \\ 0 & 4 & 12 \end{bmatrix}, \begin{bmatrix} 2 & 0 & -3 & 11 \\ 4 & 5 & 1 & 6 \\ 0 & 4 & 12 & 0 \end{bmatrix}$ **(e)** $\begin{bmatrix} 6 & -5 & 0 & 0 & 0 & 11 \\ 4 & 6 & -5 & 0 & 0 & 12 \\ 0 & 4 & 6 & -5 & 0 & 13 \\ 0 & 0 & 4 & 6 & -5 & 14 \\ 0 & 0 & 0 & 4 & 6 & 15 \end{bmatrix}$

6. (a) $x = -7/2, y = -6, z = 2$

7. (c)
$$\begin{aligned} 4x_1 + x_2 \qquad\qquad\quad &= 5 \\ x_1 + 4x_2 + x_3 \qquad\quad &= 0 \\ x_2 + 4x_3 + x_4 &= 0 \\ x_3 + 4x_4 &= 7 \end{aligned}$$

Systems **(b), (d), (e),** and **(f)** are easy to solve.

12. 40 cows and 20 chickens

13. 20 lbs. peanuts, 10 lbs. cashews, 10/3 lbs. almonds, and 66 2/3 lbs. sunflower seeds

15. $y = 2x^2 + 5x - 6$ **16.** $x^2 + y^2 - 2x - 4y = 19$ **17.** 14/13, 16/13, 8/13

Exercise Set 1.2   (page 20)

1. (a) $x = -193, y = 57, z = -11$

(b) $x_1 = -10 - 10c, x_2 = -2c, x_3 = c, x_4 = 5$ (c is arbitrary)

(d) $x_1 = 3 + 2c_1 - 5c_2, x_2 = c_1, x_3 = 2 - 3c_2 + c_3, x_4 = c_2, x_5 = c_3, x_6 = 3, x_7 = 0$

3. $x_1 = -1 - 3k/5, x_2 = 1 + k/5, x_3 = k$ (k is arbitrary)

5. $x_1 = -66, x_2 = 27, x_3 = 6, x_4 = 4$

7. inconsistent

9. $x_1 = c_2 - 1, x_2 = 2c_1, x_3 = c_1, x_4 = c_2$

12. $h = 5/2$

14. If $t = 4$ there are many solutions; if $t = -4$ there are no solutions. Otherwise the solution is unique.

15. (a) 17   **(c)** 17

18. $\begin{bmatrix} S \\ G \\ P \end{bmatrix} = \begin{bmatrix} 29 \\ 0 \\ 5 \end{bmatrix}$ or $\begin{bmatrix} 24 \\ 1 \\ 6 \end{bmatrix}$ or $\begin{bmatrix} 19 \\ 2 \\ 7 \end{bmatrix}$ or $\begin{bmatrix} 14 \\ 3 \\ 8 \end{bmatrix}$ or $\begin{bmatrix} 9 \\ 4 \\ 9 \end{bmatrix}$ or $\begin{bmatrix} 4 \\ 5 \\ 10 \end{bmatrix}$

19. $t_1 = 69.6, t_2 = 107.2, t_3 = 144.8, t_4 = 182.4$

Exercise Set 1.3   (page 33)

1. (a) $\begin{bmatrix} 1 & -3 & 0 & 0 & 0 & 1 \\ 0 & 1 & 1 & 1 & -1 & -1 \\ 0 & 0 & 1 & 7/6 & -4/3 & -5/6 \end{bmatrix}$ (not unique)

(c) $\begin{bmatrix} 1 & 1 & 2 & 6 \\ 0 & 1 & -2 & -3 \\ 0 & 0 & 1 & 2 \\ 0 & 0 & 0 & 0 \end{bmatrix}$ (not unique)

2. (b) $\begin{bmatrix} 1 & 0 & 0 & 0 & 14 & -9 \\ 0 & 1 & 0 & -1 & -14 & 5 \\ 0 & 0 & 1 & 1 & 5 & 0 \end{bmatrix}$

3. (a) $x_1 = 0, x_2 = 1, x_3 = 3$ **(c)** inconsistent   **5. (a)** $x = 0, y = -1, z = 1$
    (b) $x = 14, y = -14, z = 5$ **(c)** $x = -9, y = 5, z = 0$

7. $(R_2 \leftarrow 2R_1 + R_2), (R_3 \leftarrow 5R_1 + R_3), (R_1 \leftarrow 3R_3 + R_1), (R_2 \leftarrow -R_3 + R_2), (R_1 \leftarrow R_2 + R_1)$

$R = \begin{bmatrix} 1 & 0 & 0 & 41 \\ 0 & 1 & 0 & -11 \\ 0 & 0 & 1 & -6 \end{bmatrix}$

10. (a) $(n^2 - n)/2$ **(b)** $n^3/3 + n^2/2 + n/6$ **(c)** $n^3/3 + n^2 - n/3$

11. $x = 72, y = 108, z = 8$ **13.** $y = 2x^3 - x^2 - x + 1$

15. $C = 550$ for $x = 100, y = 50, u = 0, v = 150$

Exercise Set 1.4   (page 41)

1. (a) $x_1 = c_2 - 2c_1, x_2 = -4c_2 - 5c_1, x_3 = c_1, x_4 = c_2$ (c_1, c_2 arbitrary)
    (c) $x_1 = 46c, x_2 = 40c, x_3 = -35c, x_4 = 9c$ (c arbitrary)
    (e) $x_1 = -2c, x_2 = -2c, x_3 = c$ (c arbitrary)

3. (a) $4, 9$ **(c)** $1, 5$

4. (3) $x_1 = -3k/5, x_2 = k/5, x_3 = k$ (k arbitrary)   **(5)** $x_1 = x_2 = x_3 = x_4 = 0$ **(7)** $x_1 = x_2 = x_3 = x_4 = 0$
    (9) $x_1 = c_2, x_2 = 2c_1, x_3 = c_1, x_4 = c_2$

5. $B_1: x = 1, y = -1, z = -1$ $B_2: x = -1, y = 2, z = 1$ $B_3:$ inconsistent

7. $a_{11}a_{22} - a_{12}a_{21} = 0$ **8. (a)** $N_2 + 3H_2 = 2NH_3$ **(c)** $4C_3H_5(NO_3)_3 = 10H_2O + 12CO + 6N_2 + 7O_2$

9. (a) $15, 30, 45, 60$ **(c)** $4, 28, 52, 76$

Exercise Set 1.5   (page 52)

1. $AB = \begin{bmatrix} 5 & 18 \\ 14 & 39 \end{bmatrix}, BA = \begin{bmatrix} 1 & 2 & 3 \\ 14 & 19 & 24 \\ 16 & 20 & 24 \end{bmatrix}$

3. (a) $AB = \begin{bmatrix} 1 & 0 & 0 \\ 0 & 1 & 0 \\ 0 & 0 & 1 \end{bmatrix} = BA$ **(c)** $\begin{bmatrix} -9 & 17 & 22 \\ -5 & 12 & 6 \\ 5 & -7 & -1 \end{bmatrix}$

5. (a) $(AB)C = A(BC) = A$ **(b)** $A(B + C) = AB + AC = \begin{bmatrix} 6 & -8 & 14 \\ -5 & 16 & 28 \\ -28 & 54 & 64 \end{bmatrix}$

7. (a) $\begin{bmatrix} a & b \\ -b & a \end{bmatrix}$ **9.** $AB = BA = I$

11. (a) if $m = t$ and $r = s$; if $m = r = s = t$ **(b)** if A and B are square and of the same size

14. (a) $\begin{bmatrix} 3 & 5 & 8 \\ 4 & -2 & -4 \\ 5 & 3 & 12 \end{bmatrix}\begin{bmatrix} x \\ y \\ z \end{bmatrix} = \begin{bmatrix} 12 \\ 0 \\ 13 \end{bmatrix}$ **(b)** $\begin{bmatrix} 2 & 1 & 5 & 1 \\ 1 & 1 & -3 & -4 \\ 3 & 6 & -2 & 1 \\ 2 & 2 & 2 & -3 \end{bmatrix}\begin{bmatrix} x_1 \\ x_2 \\ x_3 \\ x_4 \end{bmatrix} = \begin{bmatrix} 5 \\ -1 \\ 8 \\ 2 \end{bmatrix}$

15. $\begin{bmatrix} 25 & 45 & 55 \\ 16 & 16 & -38 \\ 31 & 71 & 77 \end{bmatrix}\begin{bmatrix} u \\ v \\ w \end{bmatrix} = \begin{bmatrix} 12 \\ 0 \\ 13 \end{bmatrix}$

17. (a) $a_{ij} + (b_{ij} + c_{ij})$ **(e)** $\sum_k a_{ik}(b_{kj} + c_{kj})$
   (c) $(h + k)a_{ij}$ **(g)** $\sum_k \sum_s a_{ik}(b_{ks}c_{sj})$

19. $4x_2 - 2xy + 4xz + 2yz$

21. $mrn,\ n^3$

26. 1104.3, 1037.4, 858.3

Exercise Set 1.6   (page 61)

1. $(AB)^{-1} = \dfrac{1}{6}\begin{bmatrix} -3 & 9 & -3 \\ -3 & 7 & -3 \\ 3 & -7 & 9 \end{bmatrix}$, $A^{-2} = \begin{bmatrix} 1 & -1 & 2 \\ -3 & 8 & -2 \\ 3 & -7 & 3 \end{bmatrix}$, $(ABA)^{-1} = \dfrac{1}{6}\begin{bmatrix} 3 & -7 & 15 \\ -6 & 12 & -6 \\ 6 & -12 & 12 \end{bmatrix}$

3. $A^3 + A^2B + ABA + BA^2 + AB^2 + BAB + B^2A + B^3$; $A^3 + 3A^2B + 3AB^2 + B^3$

5. $B = \dfrac{1}{2}\begin{bmatrix} -6 & 4 \\ 5 & -3 \end{bmatrix}$

10. $J^{-1} = \begin{bmatrix} 1/3 & -1/9 & 0 \\ 0 & 1/3 & 0 \\ 0 & 0 & -1 \end{bmatrix}$ **16.** $\begin{bmatrix} 13 & -2 & -3 & 4 \\ 47 & 4 & -7 & 16 \\ 29 & -4 & -10 & 7 \\ 30 & 10 & -9 & 10 \end{bmatrix}$

17. $X = A^{-1}K$ **20.** $A^T = \begin{bmatrix} 1 & 2 & -1 \\ 2 & 0 & 1 \end{bmatrix}$, $B^T = \begin{bmatrix} 1 & 2 \\ -1 & 1 \\ 2 & 0 \end{bmatrix}$, $(AB)^T = B^T A^T = \begin{bmatrix} 5 & 2 & 1 \\ 1 & -2 & 2 \\ 2 & 4 & -2 \end{bmatrix}$,

$A^T B^T = \begin{bmatrix} -3 & 4 \\ 4 & 4 \end{bmatrix}$, and $A^T A = \begin{bmatrix} 6 & 1 \\ 1 & 5 \end{bmatrix}$.

22. $A^{-1} = \begin{bmatrix} 1 & -1 \\ -1 & 4/3 \end{bmatrix}$, B^{-1} does not exist, and $C^{-1} = \dfrac{1}{4}\begin{bmatrix} 3 & 1 & -1 \\ 1 & 3 & 1 \\ -1 & 1 & 3 \end{bmatrix}$.

Exercise Set 1.7 (page 72)

1. $X = \begin{bmatrix} -9 & -6 & -6 \\ 7 & 5 & 9/2 \end{bmatrix}$ 3. $X = \begin{bmatrix} 2-a & 1-b & 2-c \\ a & b & c \\ 0 & 1 & 1 \end{bmatrix}$

5. $A^{-1} = \begin{bmatrix} 2 & -1 & -1 \\ 1 & 0 & -1 \\ -2 & 1 & 2 \end{bmatrix}$ 7. $B^{-1} = \begin{bmatrix} 0 & 1 & 2 \\ -1 & 3 & 0 \\ 1 & -2 & 1 \end{bmatrix}$

10. $A^{-1} = \begin{bmatrix} 7 & -3 & -3 \\ -1 & 0 & 1 \\ -1 & 1 & 0 \end{bmatrix}$ 11. $B^{-1} = \dfrac{1}{18}\begin{bmatrix} 2 & 5 & -7 & 1 \\ 5 & -1 & 5 & -2 \\ -7 & 5 & 11 & 10 \\ 1 & -2 & 10 & 5 \end{bmatrix}$ 15. $L^{-1} = \begin{bmatrix} 1 & 0 & 0 & 0 \\ -2 & 1 & 0 & 0 \\ 13 & -5 & 1 & 0 \\ -39 & 13 & -3 & 1 \end{bmatrix}$

16. (7) $\begin{bmatrix} 0 & 0 & 1 \\ 0 & 1 & 0 \\ 1 & 0 & 0 \end{bmatrix}\begin{bmatrix} 1 & 0 & 0 \\ -1 & 1 & 0 \\ 0 & 0 & 1 \end{bmatrix}\begin{bmatrix} 1 & 0 & 0 \\ 0 & 1 & 0 \\ -3 & 0 & 1 \end{bmatrix}\begin{bmatrix} 1 & 0 & 0 \\ 0 & 1 & 0 \\ 0 & 2 & 1 \end{bmatrix}\begin{bmatrix} 1 & -1 & 0 \\ 0 & 1 & 0 \\ 0 & 0 & 1 \end{bmatrix}\begin{bmatrix} 1 & 0 & 0 \\ 0 & 1 & 1 \\ 0 & 0 & 1 \end{bmatrix}$ (not unique)

17. $P = \begin{bmatrix} 5 & -1 & -5 \\ 3 & -6/5 & -12/5 \\ -8 & 13/5 & 38/5 \end{bmatrix}$ 19. $[PA\,|\,P] \to [P_1PA\,|\,P_1P] = [I\,|\,A^{-1}]$

21. (d) $A^{-1} = \dfrac{1}{19}\begin{bmatrix} -1 & 4 & 4 \\ 5 & -1 & -1 \\ -7 & -10 & 9 \end{bmatrix}$

22. (a) Keep left arm straight. (c) Tack on the headers.

23. (a) 93, 113, 98, 36, 37, 41, 77, 96, 77, 126, 149, 141, 87, 87, 112

   (b) 73, 91, 74, 95, 95, 120, 61, 61, 81, 124, 138, 145, 94, 114, 99

Exercise Set 1.8 (page 86)

1. (a) $\dfrac{1}{7}\begin{bmatrix} 16 \\ 27 \\ 20 \end{bmatrix}$ (c) $\dfrac{1}{126}\begin{bmatrix} 120 \\ -63 \\ 3 \end{bmatrix}$

3. $A^{-1} = \dfrac{1}{44}\begin{bmatrix} 24 & -4 & -8 \\ 3 & 5 & -12 \\ 10 & 2 & 4 \end{bmatrix}$ $C^{-1} = \begin{bmatrix} .435 & -.135 & -.0156 & -.0451 \\ -1.97 & .310 & .160 & .444 \\ 1.90 & -.251 & -.193 & -.390 \\ .481 & -.0381 & -.00252 & -.113 \end{bmatrix}$ (correct to three significant figures.)

5. $\begin{bmatrix} 1 & 0 & 0 \\ 3 & 1 & 0 \\ 3 & 1/5 & 1 \end{bmatrix}\begin{bmatrix} 1 & 1 & 1 \\ 0 & -5 & -20 \\ 0 & 0 & 0 \end{bmatrix}$ 7. $\begin{bmatrix} 1 & 0 & 0 & 0 \\ 2 & 1 & 0 & 0 \\ 4 & 5 & 1 & 0 \\ 1 & 3 & 43 & 1 \end{bmatrix}\begin{bmatrix} 1 & -1 & 2 & 3 \\ 0 & 1 & -4 & -4 \\ 0 & 0 & 1 & 7 \\ 0 & 0 & 0 & -209 \end{bmatrix}$

9. $\begin{bmatrix} 1 & 0 & 0 & 0 \\ 0 & 0 & 1 & 0 \\ 0 & 1 & 0 & 0 \\ 0 & 0 & 0 & 1 \end{bmatrix} A = \begin{bmatrix} 1 & 0 & 0 & 0 \\ 1 & 1 & 0 & 0 \\ 3/2 & 0 & 1 & 0 \\ 2 & -3 & 10 & 1 \end{bmatrix}\begin{bmatrix} 2 & 4 & 3 & 2 \\ 0 & 1 & -1 & -5 \\ 0 & 0 & 1/2 & -1 \\ 0 & 0 & 0 & 5 \end{bmatrix}$

11. $\begin{bmatrix} 0 & 1 & 0 & 0 \\ 1 & 0 & 0 & 0 \\ 0 & 0 & 0 & 1 \\ 0 & 0 & 1 & 0 \end{bmatrix} A = \begin{bmatrix} 1 & 0 & 0 & 0 \\ 0 & 1 & 0 & 0 \\ 4 & 6 & 1 & 0 \\ -2 & 1 & 0 & 1 \end{bmatrix} \begin{bmatrix} 1 & -1 & 7 & -7 \\ 0 & 1 & 3 & 2 \\ 0 & 0 & -45 & 21 \\ 0 & 0 & 0 & -13 \end{bmatrix}$

13. $\begin{bmatrix} 0 & 0 & 1 \\ 0 & 1 & 0 \\ 1 & 0 & 0 \end{bmatrix} A = \begin{bmatrix} 1 & 0 & 0 \\ 2/3 & 1 & 0 \\ 1/3 & 1/2 & 1 \end{bmatrix} \begin{bmatrix} 3 & 5 & 6 \\ 0 & 2/3 & 1 \\ 0 & 0 & 1/2 \end{bmatrix}$

15. $\begin{bmatrix} 0 & 0 & 0 & 1 \\ 0 & 0 & 1 & 0 \\ 0 & 1 & 0 & 0 \\ 1 & 0 & 0 & 0 \end{bmatrix} A = \begin{bmatrix} 1 & 0 & 0 & 0 \\ 1/2 & 1 & 0 & 0 \\ 3/4 & 9/10 & 1 & 0 \\ 1/2 & 3/5 & 1 & 1 \end{bmatrix} \begin{bmatrix} 4 & 5 & 14 & 14 \\ 0 & 5/2 & -5 & -10 \\ 0 & 0 & -1 & 1/2 \\ 0 & 0 & 0 & 1/2 \end{bmatrix}$ **17.** $X = \dfrac{1}{209} \begin{bmatrix} 389 \\ 268 \\ 113 \\ -46 \end{bmatrix}$

Chapter 1 Review Exercises   (page 88)

1. I

2. Matrix multiplication is noncommutative and not every nonzero matrix has a multiplicative inverse.

3. $(-1/3)(B^2 - 7B + 7I)$

5. $\begin{bmatrix} 1 & -3 & 2 \\ -3 & 3 & -1 \\ 2 & -1 & 0 \end{bmatrix}$ **7.** $x = z, y = 2z, z$ arbitrary   **9.** $\begin{bmatrix} -31 & -17 & -20 \\ 19 & 10 & 12 \end{bmatrix}$ **11. (c)** $\begin{bmatrix} a & b \\ 0 & a \end{bmatrix}$

13. $\begin{bmatrix} 1 & -3 & 2 \\ -3 & 3 & -1 \\ 2 & -1 & 0 \end{bmatrix}$ **16.** $\begin{bmatrix} 2 \\ 1/5 \\ 0 \end{bmatrix}$ **18.** $\begin{bmatrix} 1 & 0 & 5/3 & -5/3 \\ 0 & 1 & 1/3 & 2/3 \\ 0 & 0 & 0 & 0 \end{bmatrix}$ **19.** $\lambda = 4, 9$ **22.** $\begin{bmatrix} -1 & 1 & 0 \\ 7 & -3 & -3 \\ -1 & 0 & 1 \end{bmatrix}$

Chapter 2

Exercise Set 2.1   (page 97)

1. $-2, 17, -34, -34, -34, -1/2, -2$

3. (a) 0   **(b)** $x^3 - 9x^2 + 8x + 5$ **(c)** $x^3 + 2x^2 - x - 2$

4. (a) 27   **(c)** -117 **9. (a)** $-\det(A)$ **(b)** $k \det(A)$ **(c)** $\det(A)$

11. (a) $\operatorname{tr}(A) = 9, \operatorname{tr}(B) = 2, \operatorname{tr}(AB) = \operatorname{tr}(BA) = 52$ **(b)** $\operatorname{tr}(A) = \Sigma a_{ii}$

13. 1.37×10^{53} years   **15.** $\begin{bmatrix} a & b & c \\ b & e & f \\ c & f & i \end{bmatrix}$ **17.** $X^T X = 29, XX^T = \begin{bmatrix} 4 & -6 & -8 \\ -6 & 9 & 12 \\ -8 & 12 & 16 \end{bmatrix}$

Exercise Set 2.2   (page 105)

1. $\det(A) = -12$ **3.** $\det(C) = -18$ **5.** $\det(E) = 132$ **7.** $\det(G) = 6$ **13.** $R_i \leftrightarrow R_j$

14. $R_i \leftrightarrow R_j$ **15.** $\det(A) = u_{11} u_{22} \ldots u_{nn}$ **17. (b)** plane through $(1, 2, 3), (4, 5, 6),$ and $(7, 8, 0)$

20. (a) -13 **(c)** no real solution   **21. (a)** 25.764462 (only three figures of which are significant)

Exercise Set 2.3   (page 112)

1. $A^{-1} = \dfrac{-1}{12}\begin{bmatrix} 4 & 4 & -4 \\ -8 & -14 & 5 \\ 0 & 6 & -3 \end{bmatrix}$ $D^{-1} = \dfrac{1}{117}\begin{bmatrix} 39 & 14 & 1 & 65 \\ 39 & -61 & 4 & 26 \\ 0 & 24 & -15 & -39 \\ 0 & 9 & 9 & 0 \end{bmatrix}$ $G^{-1} = \dfrac{1}{6}\begin{bmatrix} -2 & 2 & 6 & 2 \\ -1 & 1 & -6 & 2 \\ 5 & -5 & 6 & -4 \\ -1 & 7 & -2 & -2 \end{bmatrix}$

5. $z = 2$ 11. for A, 4×4;   for N, 3×3

12. (b) $(n + 1)(n^3 + 2n - 3)/3 + n$ (c) $n^2((n - 1)^3 + 2(n - 1) - 3)/3 + n + n^2$

14. $\lambda^4 + a_3\lambda^3 + a_2\lambda^2 + a_1\lambda + a_0$ 15. (a) $\lambda = 4, 9$ (b) $\lambda = 1, 5$

17. (a) circle through the points (x_1, y_1), (x_2, y_2), and (x_3, y_3) (c) cubic through (x_1, y_1), (x_2, y_2), (x_3, y_3), and (x_4, y_4)

Exercise Set 2.4   (page 122)

1. (a) X is an eigenvector for $\lambda = 2$, since $AX = 2X$.

$AY = \begin{bmatrix} -7 \\ 0 \end{bmatrix} \neq \lambda Y$, so Y is not an eigenvector of A.

   (b)   Both X and Y are eigenvectors for $\lambda = 1$.

3. $\lambda^2 - 25\lambda - 52$; $(25 \pm \sqrt{833})/2$; $X_1 = [28 \quad 7 - \sqrt{833}]^T$; $X_2 = [28 \quad 7 + \sqrt{833}]^T$

5. $\lambda^3 - 9\lambda^2 - 9\lambda + 81$; $\lambda_1 = 3, \lambda_2 = -3, \lambda_3 = 9$; $X_1 = [-2 \quad -2 \quad 1]^T, X_2 = [2 \quad -1 \quad 2]^T, X_3 = [-1 \quad 2 \quad 2]^T$

7. $\lambda^3 - 4\lambda^2 + 3\lambda$; $0, 3, 1$; $[1 \quad -2 \quad 4]^T$, $[1 \quad 1 \quad 1]^T$, $[1 \quad -1 \quad 1]^T$

9. $\lambda^3 + \lambda^2 - 12\lambda$; $-4, 3, 0$; $[1 \quad 0 \quad 1]^T$, $[1 \quad -1 \quad -1]^T$, $[-1 \quad -2 \quad 1]^T$

11. (8) $Dg\{5, -5, -4\}$;   (10) $Dg\{7, -4, -1\}$ 13. the diagonal elements

17. Eigenvalues of A^2 are the squares of the eigenvalues of A.   18. (e) $-4, 16, 0, \lambda^2 - 5\lambda + 4$

19. $(\lambda - 2)^2(\lambda^2 - 5\lambda + 6)$ 24. (a) $-6.653, 1.653$ (b) $-6.653, -3.414, 1.653, -.5858$

   (c) A: $5.425, -1.119, -2.306$; B: $3.732, 1.000, .2679$

Chapter 2 Review Exercises   (page 124)

1. (a) $\det A = -6$; $\det B = 6$ (b) 5 3. $\lambda^3 - 4\lambda^2 + 4\lambda - 1$

5. $x = 20/33, y = -4/33, z = 15/33$

7. T, F, F, T, T, T, F, F, T, F   10. $\begin{bmatrix} x_1 \\ x_2 \end{bmatrix} = \begin{bmatrix} \cos\theta & \sin\theta \\ -\sin\theta & \cos\theta \end{bmatrix}\begin{bmatrix} x \\ y \end{bmatrix}$ 12. (a) 0 (b) -20

13. (a) $\lambda^3 - 7\lambda^2 + 6\lambda$ (b) $[1 \quad 2 \quad 5]^T$

Chapter 3

Exercise Set 3.1   (page 137)

2. (a) $\mathbf{AB} = (-1, 3)$ (c) $\mathbf{AB} = (-5, 5, -4)$

3. (a) $\mathbf{v} = \mathbf{PQ}$ with $Q = (6, 5)$ is one possible solution.

   (b) $\mathbf{v} = \mathbf{PQ}$ with $Q = (4, -1, 2)$ is one possible solution.

5. (c) $(4, 0, 5)$ **(d)** $(11/2, -7/2)$ **7.** southwest $(217°)$, 25 mph

10. (a) $(23/25)\mathbf{v}$, $23/5$ **(b)** $(-17/169)\mathbf{v}$, $-17/13$ **11. (a)** $\mathbf{w} = 2\mathbf{u} - 3\mathbf{v}$ **(b)** $\mathbf{w} = 3\mathbf{u} + 5\mathbf{v}$

13. $z = -x + 2y$ **16. (b)** $\mathbf{u} \times \mathbf{v} = 19\mathbf{i} + 5\mathbf{j} - 7\mathbf{k} = (19, 5, -7) = -\mathbf{v} \times \mathbf{u}$

   (c) $\mathbf{u} \times \mathbf{v}$ is perpendicular to the plane of $\mathbf{u}$ and $\mathbf{v}$ **19.** $\mathbf{B, H, D, 2C, 0, H, 2F, B, 0}$

Exercise Set 3.2   (page 145)

1. (a) $\sqrt{14}, 9, \sqrt{6}, \sqrt{38}$ **(b)** $[1 \quad 0 \quad -3 \quad 0]^T$ (not unique)

4. (a) does not contain the zero vector   **(b)** not closed under multiplication by negative scalars

7. (a) and **(b)** only   **10. (a)** $(1/6)\mathbf{v}$;   **(b)** $3/\sqrt{5}$

Exercise Set 3.3   (page 150)

1. $\mathbf{w} = 3\mathbf{u} - \mathbf{v}$ **4. (a)** $\mathbf{x}_1 = [-2 \quad -5 \quad 1 \quad 0]^T$, $\mathbf{x}_2 = [1 \quad -4 \quad 0 \quad 1]^T$ **(b)** $\mathbf{0}$

5. $\mathbf{v}_2 = \mathbf{u}_1 + 2\mathbf{u}_2$, $\mathbf{v}_3 = 3\mathbf{u}_1 - \mathbf{u}_2$, $\mathbf{v}_1$ is not in SPAN$\{\mathbf{u}_1, \mathbf{u}_2\}$ (a single row reduction should suffice).

7. $\mathbf{u}_1 = 2\mathbf{v}_1 + \mathbf{v}_2$, $\mathbf{u}_2 = 3\mathbf{v}_1 + 2\mathbf{v}_2$ **9. (a)** $B = [1 \quad -2 \quad 1]$ **(c)** $\begin{bmatrix} 1 & -2 & 1 & 0 \\ 2 & -3 & 0 & 1 \end{bmatrix}$

11. (b) $\mathscr{CS}(A) \neq \mathscr{CS}\begin{bmatrix} 1 & 3 \\ 0 & 0 \end{bmatrix}$ since they represent different lines in $\mathscr{R}^2$.

Exercise Set 3.4   (page 157)

1. no (10 fails)   **3.** no (9 and 10 fail)   **5.** yes   **7.** yes   **9.** no (1 fails)   **11.** closed

13. not closed   **15.** not closed   **17.** closed

Exercise Set 3.5   (page 166)

1. (a), (b), (f)   **3. (a), (c)**   **5.** $\mathbf{d} = \mathbf{u}_1 + \mathbf{u}_2$ **10. (a)** $\begin{bmatrix} 1 & 0 & -1 \\ 0 & 1 & 2 \\ 0 & 0 & 0 \end{bmatrix}$ **(c)** $\begin{bmatrix} 1 & 0 & 0 & 1 \\ 0 & 1 & 0 & 0 \\ 0 & 0 & 1 & -1 \end{bmatrix}$

12. $A : \{[2 \quad 1 \quad 0 \quad 0]^T, [-4 \quad 0 \quad 3 \quad 1]^T\}$
   $B : \{[-5 \quad 2 \quad 0 \quad 0]^T, [-3 \quad 0 \quad 1 \quad 0]^T, [-7 \quad 0 \quad 0 \quad 2]^T\}$

Exercise Set 3.6   (page 175)

1. Does the system $\begin{bmatrix} 1 & 1 & 4 \\ -1 & 1 & 1 \\ 2 & 2 & 2 \\ 0 & 0 & 1 \end{bmatrix} \begin{bmatrix} a \\ b \\ c \end{bmatrix} = \begin{bmatrix} 0 \\ 0 \\ 0 \\ 0 \end{bmatrix}$ have a nonzero solution? The vectors are independent

2. independent

3. Does the system $\begin{bmatrix} 1 & 0 & 0 & 0 \\ 0 & 3 & 0 & 0 \\ 2 & 0 & 6 & 0 \\ 5 & 2 & 0 & 2 \end{bmatrix} \begin{bmatrix} a \\ b \\ c \\ d \end{bmatrix} = \begin{bmatrix} 0 \\ 0 \\ 0 \\ 0 \end{bmatrix}$ have a nonzero solution? The vectors are independent

5. independent   **7.** independent   **9.** independent   **11.** independent   **13.** dependent

21. $S_1 = \{[1 \quad 2 \quad 4 \quad 1], [2 \quad 3 \quad 7 \quad 1]\}$ **22.** $S_1 = \{\sin^2 x, \cos^2 x, e^x\}$

Exercise Set 3.7   (page 185)

1. (a), (d), (f)   **2.** (a), (d)   **3.** $[v_1]_{\mathfrak{A}} = [1 \quad 1 \quad 1 \quad 1]^T, [v_2]_{\mathfrak{A}} = [1 \quad 1 \quad 0 \quad 2]^T$

5. $[5/2 \quad -4 \quad 2 \quad 5/2]^T = [T]_{\mathfrak{A}}$ and $[I]_{\mathfrak{A}} = [1 \quad -4 \quad 1 \quad 0]^T$ **7.** $\dim(\mathcal{RS}(A)) = \dim(\mathcal{CS}(A)) = 2; \dim(\mathcal{NS}(A)) = 2$

9. $A = \begin{bmatrix} 0 & 1 & 3 & -2 \\ 1 & 2 & 6 & 0 \\ 2 & 3 & 9 & 2 \\ 1 & 1 & 3 & 2 \end{bmatrix} \mathcal{C} \begin{bmatrix} 1 & 0 & 0 & 0 \\ 0 & 1 & 0 & 0 \\ -1 & 2 & 0 & 0 \\ -1 & 1 & 0 & 0 \end{bmatrix}$; basis $= \{c_1, c_2\}$ **11.** $\{u_1, e_1, e_2\}$

14. rank$(A) = 2$, rank$(B) = 3$, rank$(C) = 4$ **15.** (a) $\mathfrak{B} = \{[-2/5 \quad 6/5 \quad 1 \quad 0 \quad 0]^T,$
$[-6/5 \quad 3/5 \quad 0 \quad 1 \quad 0]^T, [-1/5 \quad -3/5 \quad 0 \quad 0 \quad 1]^T\}$; dim $\mathcal{NS}(A) = 3$; dim$(\mathcal{RS}(A)) = 5 - 3 = 2$
    (b) $\mathfrak{B} = \{[3 \quad -4 \quad 1 \quad 0]^T\}$; dim$(\mathcal{NS}(A)) = 1$; dim$(\mathcal{RS}(A)) = 3$ **18.** $[v]_{\mathfrak{A}'} = [-3 \quad 1 \quad 6 \quad 2]^T$

Exercise Set 3.8   (page 196)

1. $u \cdot v = 9, \|u\| = \sqrt{31}, \|v\| = \sqrt{54}, \|u\| + \|v\| > \|u + v\| = \sqrt{103}$
    $a = 1/\sqrt{31}, h = -5/4, a = 34/13, b = -11/13, c = -6$

3. $X_1 = [-1 \quad -2 \quad 1]^T, X_2 = [-2 \quad 1 \quad 0]^T, X_3 = [1 \quad 2 \quad 5]^T; AX_1 = 0, AX_2 = X_2, AX_3 = 6X_3$

5. (a) $[1/2 \quad 1/2 \quad 1/2 \quad 1/2]^T$ (b) $[5 \quad 0 \quad -1 \quad -2]^T$ (c) $[7 \quad 2 \quad -2 \quad -9]^T$

7. The diagonals of a rectangle are equal.   **11.** $(1/3)u_1, (1/\sqrt{41})[6 \quad -1 \quad -2]^T$

13. (a) $c_1 = [1 \quad 1 \quad 1]^T/\sqrt{3}, c_2 = [2 \quad -1 \quad -1]^T/\sqrt{6}, c_3 = [0 \quad 1 \quad -1]^T/\sqrt{2}$
    (b) $c_1 = [3 \quad 0 \quad 4]^T/5, c_2 = [12 \quad -25 \quad -9]^T/5\sqrt{34}, c_3 = [4 \quad 3 \quad -3]^T/\sqrt{34}$

15. Applying the Gram-Schmidt process to the basis $\{u_1, u_2, e_3, e_4\}$ yields $v_1 = u_1/\|u_1\|, v_2 = u_2/\|u_2\|,$
    $v_3 = [2 \quad -3 \quad 2 \quad 2]^T/\sqrt{21}, v_4 = [1 \quad 0 \quad 0 \quad -1]^T/\sqrt{2}$

16. (a) $\{v_1 = [1 \quad 1 \quad 0]^T/\sqrt{2}, v_2 = [1 \quad -1 \quad -2]^T/\sqrt{6}\}; \{v_1, v_2, v_3 = [1 \quad -1 \quad 1]^T/\sqrt{3}\}$
    (c) $\{v_1 = [1 \quad 1 \quad 0]^T/\sqrt{2}, v_2 = [1 \quad -1 \quad 2]^T/\sqrt{6}, v_3 = [-1 \quad 1 \quad 1]^T/\sqrt{3}\}$

17. (a) $x_1 = [-5 \quad 2 \quad 1 \quad 0 \quad 0]^T/\sqrt{30}, x_2 = [-5 \quad -14 \quad 3 \quad 5 \quad 0]^T/\sqrt{255},$
    $x_3 = [-5 \quad -14 \quad 3 \quad -46 \quad 102]^T/\sqrt{12750}$
    (b) $v_1 = [3 \quad 4 \quad 5 \quad -1 \quad 0]^T/\sqrt{51}, v_2 = [192 \quad -118 \quad -20 \quad 4 \quad 51]^T/\sqrt{53805}$

19. (c) (a) $\{v_3 = [1 \quad -1 \quad 1]^T/\sqrt{3}\}$ (b) $\{v_4 = [0 \quad 1 \quad 0 \quad 1]^T/\sqrt{2}\}$ (c) $W^\perp = \{0\}$
    (d) $\{v_4 = [12 \quad -1 \quad -12 \quad 1]^T/\sqrt{270}\}$

20. (a) $H = \dfrac{1}{5} \begin{bmatrix} 3 & 0 & 4 & 0 \\ 0 & 5 & 0 & 0 \\ 4 & 0 & -3 & 0 \\ 0 & 0 & 0 & 5 \end{bmatrix}$ **(b)** $H = \begin{bmatrix} .3651 & -.5477 & .1826 & .7303 \\ -.5477 & .5274 & .1575 & .6301 \\ .1826 & .1575 & .9475 & -.2100 \\ .7303 & .6301 & -.2100 & .1600 \end{bmatrix}$

21. $v_1 = [1 \quad 0 \quad 1 \quad 0]/\sqrt{2}, v_2 = [1 \quad -1 \quad -1 \quad 1]/2, v_3 = [-1 \quad -1 \quad 1 \quad 1]/2$

23. $\dfrac{1}{169}\begin{bmatrix} 151 & -24 & -72 \\ -24 & 137 & -96 \\ -72 & -96 & -119 \end{bmatrix}$

27. (a) $A = \begin{bmatrix} 1/\sqrt{3} & 2/\sqrt{6} & 0 \\ 1/\sqrt{3} & -1/\sqrt{6} & 1/\sqrt{2} \\ 1/\sqrt{3} & -1/\sqrt{6} & -1/\sqrt{2} \end{bmatrix}\begin{bmatrix} \sqrt{3} & 2/\sqrt{3} & 1/\sqrt{3} \\ 0 & -2/\sqrt{6} & -1/\sqrt{6} \\ 0 & 0 & -1/\sqrt{2} \end{bmatrix}$

(b) $A = \begin{bmatrix} 3/5 & 12/5\sqrt{34} & 4/\sqrt{34} \\ 0 & -5/\sqrt{34} & 3/\sqrt{34} \\ 4/5 & -9/5\sqrt{34} & -3/\sqrt{34} \end{bmatrix}\begin{bmatrix} 5 & 37/15 & 19/5 \\ 0 & -\sqrt{34}/5 & 101/5\sqrt{34} \\ 0 & 0 & 11/\sqrt{34} \end{bmatrix}$

Exercise Set 3.9   (page 204)

1. (a) $P = \begin{bmatrix} -1 & -3 \\ 2 & 5 \end{bmatrix}$ **(b)** $\dfrac{1}{2}\begin{bmatrix} -1 & 5 & 7 \\ 1 & -9 & 5 \\ 0 & 0 & -24 \end{bmatrix}$ **(c)** $\dfrac{1}{6}\begin{bmatrix} 0 & 2 & -2 \\ -3 & -10 & 4 \\ 15 & 18 & 6 \end{bmatrix}$

(d) $\begin{bmatrix} 1 & 1 & 5 & -1 \\ 0 & 2 & -1 & 7 \\ 1 & 0 & 4 & 0 \\ 0 & 0 & 0 & 6 \end{bmatrix}$ **(e)** $\dfrac{1}{\sin 1}\begin{bmatrix} \sin 3 & -\cos 3 \\ -\sin 2 & \cos 2 \end{bmatrix}$ WRONG $\dfrac{1}{\cos(-1)}\begin{bmatrix} \cos 3 & \sin 3 \\ -\sin 2 & \cos 2 \end{bmatrix}$

3. $[7 \quad -3 \quad -16]^{T}/2$ **5.** $\mathscr{S}$ to $\mathscr{N}$: $\begin{bmatrix} \cos\theta & \sin\theta \\ -\sin\theta & \cos\theta \end{bmatrix}$; $\mathscr{N}$ to $\mathscr{S}$: $\begin{bmatrix} \cos\theta & -\sin\theta \\ \sin\theta & \cos\theta \end{bmatrix}$

7. $\mathscr{B}$ to $\mathscr{S}$: $M = [\mathbf{u_1} \quad \mathbf{u_2} \quad \ldots \quad \mathbf{u_n}]$; $\mathscr{S}$ to $\mathscr{B}$: $M^{-1} = M^{T}$ **9.** $\begin{bmatrix} 0 & 0 & 1 & 0 \\ 1 & 0 & 0 & 0 \\ 0 & 0 & 0 & 1 \\ 0 & 1 & 0 & 0 \end{bmatrix}$ ($\mathscr{B}$ to $\mathscr{B}'$)

Chapter 3 Review Exercises   (page 205)

1. $[-4 \quad -3 \quad 0 \quad 1]^{T}$ **3.** The zero vector is not in $\mathscr{W}$. $\mathscr{W}$ is not closed under addition.

5. (a) $\mathbf{u_1} \perp \mathbf{u_2}$ **(b)** $(1/\sqrt{17})\mathbf{u_1}$, $(1/\sqrt{6})\mathbf{u_2}$, $(1/\sqrt{21})[2 \quad -3 \quad 2 \quad 2]^{T}$

6. (a) Does the system $\begin{aligned} a + \quad 3b \quad\;\; &= \quad 1 \\ 5a + \quad\quad 5c &= \quad 5 \\ -11b + 7c &= -11 \end{aligned}$ have a solution?

(b) Does the system $\begin{aligned} x + 2y + 3z &= 0 \\ 5x - \;\, y + \;\, z &= 0 \\ 3x + 4y + 7z &= 0 \end{aligned}$ have a nonzero solution?

(c) Does the system $\begin{aligned} w + 5x + 9y + 3z &= a \\ 2w + 6x \quad\quad\; + 4z &= b \\ 3w + 7x + \;\, y + 5z &= c \\ 4w + 8x + 2y + 6z &= d \end{aligned}$ have a unique solution for all $\begin{bmatrix} a & b \\ c & d \end{bmatrix}$?

7. (a) $\begin{bmatrix} -2 \\ 1 \\ 0 \\ 0 \end{bmatrix}\begin{bmatrix} 1 \\ 0 \\ -1 \\ 1 \end{bmatrix}$ **(b)** rank$(A) = 4 - \dim(NS(A)) = 2$

9. (a) 13 **(b)** $\sqrt{35}$ **(c)** $(1/\sqrt{10})[1 \quad 0 \quad 3]^{T}$ **(d)** $13/\sqrt{29}$

11. (a) $\mathbf{u_1} \cdot \mathbf{u_2} = 0$ **(b)** $(1/2)\mathbf{u_1}$, $(1/6)\mathbf{u_2}$, $(1/\sqrt{18})[3 \quad -2 \quad -2 \quad 1]^{T}$.

Chapter 4

Exercise Set 4.1   (page 214)

1. (a) linear; domain is $\mathcal{R}^3$; $\mathcal{W} = \mathcal{R}^3$ (b) nonlinear; domain is $\mathcal{R}^3$; $\mathcal{W} = \mathcal{R}^3$ (c) nonlinear; domain is $\mathcal{R}^3$; $\mathcal{W} = \mathcal{R}^3$
(d) linear; domain is $\mathcal{P}_3$; $\mathcal{W} = \mathcal{P}_2$ (e) nonlinear; domain is $\mathscr{C}^2[a \quad b]$; $\mathcal{W} = \mathscr{C}^0[a \quad b]$
(f) linear; domain is $\mathcal{R}^{2 \times 2}$; $\mathcal{W} = \mathcal{R}^{2 \times 2}$ 3. $[11 \quad 34 \quad 22]^T$, $[5 \quad 14 \quad 9]^T$, $[8 \quad 32 \quad 21]^T$

5. $\mathbf{T(u)} = \begin{bmatrix} -1 \\ 12 \\ 6 \end{bmatrix}$, $\mathbf{T(T(u))} = \begin{bmatrix} -13 \\ 43 \\ 36 \end{bmatrix}$, $\mathbf{T(T(u)} + 3(\mathbf{u}) = \begin{bmatrix} -16 \\ 79 \\ 54 \end{bmatrix} = (A^2 + 3A)\mathbf{u}$

7. **S** is linear, one-to-one, and onto.   11. $[-45 \quad 90]^T$, $[10 \quad 15]^T$

15. (a) $[1 \quad 0 \quad 1/3]^T$ (b) $[-1/3 \quad -7/3 \quad 22/9]^T$

Exercise Set 4.2   (page 224)

1. (a) $\mathcal{NS}(\mathbf{T}) = \text{SPAN}\left\{ \begin{bmatrix} -2 \\ 1 \end{bmatrix} \right\}$, $\mathcal{R}(\mathbf{T}) = \text{SPAN}\left\{ \begin{bmatrix} 1 \\ 2 \end{bmatrix} \right\}$, rank = nullity = 1
(b) $\mathcal{NS}(\mathbf{D}^2) = \text{SPAN}\{1, x\} = \mathcal{R}(\mathbf{D}^2)$, rank = nullity = 2

2. (a) $\mathcal{NS}(\mathbf{T}) = \text{BASIS}\left\{ \begin{bmatrix} -1 \\ -1 \\ 1 \end{bmatrix} \right\}$, $\mathcal{R}(\mathbf{T}) = \text{BASIS}\left\{ \begin{bmatrix} 4 \\ 7 \\ 0 \end{bmatrix}, \begin{bmatrix} 5 \\ 0 \\ 14 \end{bmatrix} \right\}$ (b) $\mathcal{NS}(\mathbf{T}) = \text{BASIS}\left\{ \begin{bmatrix} 7 \\ -4 \\ 6 \\ 4 \end{bmatrix} \right\}$, $\mathcal{R}(\mathbf{T}) = \mathcal{R}^3$

(c) $\mathcal{NS}(\mathbf{T}) = \{\mathbf{0}\}$, $\mathcal{R}(\mathbf{T}) = \mathcal{R}^3$ (d) $\mathcal{NS}(\mathbf{T}) = \text{BASIS}\left\{ \begin{bmatrix} 1 \\ -2 \\ 1 \\ 0 \end{bmatrix} \begin{bmatrix} 1 \\ -1 \\ 0 \\ 1 \end{bmatrix} \right\}$, $\mathcal{R}(\mathbf{T}) = \text{BASIS}\left\{ \begin{bmatrix} 2 \\ 1 \\ 0 \end{bmatrix}, \begin{bmatrix} -1 \\ 0 \\ 1 \end{bmatrix} \right\}$

3. (a) $\{e^{2t}\}$ (b) $\{1, t\}$ (c) $\{\cos 2t, \sin 2t\}$ 5. $\dim(\mathcal{V}) < \dim(\mathcal{W})$, $\dim \mathcal{R}(\mathbf{T}) < \min\{\dim(\mathcal{V}), \dim(\mathcal{W})\}$

7. (a) 0 (b) 4 (c) nullity = 2, $\mathcal{R}(\mathbf{T}) = \mathcal{W}$ (d) rank(**T**) = 1, nullity = 4   9. $\mathcal{RS}(A) = \mathcal{R}(A^T)$

11. (a) $\begin{bmatrix} -1 \\ 0 \\ 1 \\ 2 \\ 0 \end{bmatrix} + \left(c_1 \begin{bmatrix} -5 \\ 1 \\ 0 \\ 0 \\ 0 \end{bmatrix} + c_2 \begin{bmatrix} -5 \\ 0 \\ -3 \\ -4 \\ 1 \end{bmatrix} \right)$ (b) $\begin{bmatrix} -29/3 \\ 2 \\ 0 \\ 5/3 \end{bmatrix} + c_1 \begin{bmatrix} 13/2 \\ -3/2 \\ 1 \\ 0 \end{bmatrix}$

Exercise Set 4.3   (page 229)

1. (a) $[2 \quad 1]$ (b) $[-24 \quad -5]$ (c) $[6 \quad 5]$

2. (a) $\begin{bmatrix} 5 & 6 \\ 4 & 15 \end{bmatrix}$ (b) $\begin{bmatrix} 1 & 2 \\ 3 & 4 \end{bmatrix}$ (c) $\begin{bmatrix} 4 & 0 \\ 0 & 6 \end{bmatrix}$ (d) $\begin{bmatrix} 3 & 0 \\ 0 & 7 \end{bmatrix}$ (e) $\begin{bmatrix} 4 & 2 \\ 3 & 11 \end{bmatrix}$

4. (a) $x^2 + x + 5, x^2 + x$ 5. (a) $\mathbf{D}^2 - 1$ (b) $\mathbf{D}^3 + 7\mathbf{D}^2 - \mathbf{D} - 22$
(c) $x\mathbf{D}^3 + (3x + 3)\mathbf{D}^2 + (9 + x)\mathbf{D} + 3$

6. $\mathcal{NS}(\mathbf{T}_2) \subseteq \mathcal{NS}(\mathbf{T}_1\mathbf{T}_2)$. If $\mathbf{T}_1\mathbf{T}_2 = \mathbf{T}_2\mathbf{T}_1$, then $\mathcal{NS}(\mathbf{T}_1) \subseteq \mathcal{NS}(\mathbf{T}_1\mathbf{T}_2)$ also.
9. A nonsingular

Exercise Set 4.4   (page 240)

1. (a) $\begin{bmatrix} 5/2 & -1/6 \\ -3/2 & 15/6 \end{bmatrix}$ (b) $\begin{bmatrix} -2 & 3 \\ -2 & 3 \end{bmatrix}$ 2. $\begin{bmatrix} 2 & -2 & 7 \\ 3 & -6 & 8 \\ 1 & 7 & -5 \end{bmatrix}$

5. $\begin{bmatrix} 1/2 & 1 \\ -5/4 & -3 \end{bmatrix}$, $T(1,1) = (3/2, -17/4)$, $T(1,0) = (1/2, -5/4)$, $T(0,1) = (1, -3)$

6. $\frac{\sqrt{2}}{2}\begin{bmatrix} -1 & 0 \\ -1 & 6 \end{bmatrix}$, $T(1,1) = (-\sqrt{2}/2, 5\sqrt{2}/2)$ 7. $\frac{1}{3}\begin{bmatrix} 2 & -5 \\ 1 & -3 \end{bmatrix}$, $T(1,1) = \frac{1}{3}(-3, -2)$

The answers to 9 and 11 are not unique; one answer is given.

9. (a) shear in the x-direction with factor -1 (b) shear in the y-direction with factor 3   (c) shear in the x-direction with factor 1

11. (a) shear in the x-direction with factor $-\tan\theta$ (b) expansion in the y-direction with factor $\sec\theta$ (c) shear in the y-direction with factor $\sin\theta$ (d) expansion in the x-direction with factor $\cos\theta$

13. $\begin{bmatrix} 0 & 0 & 0 \\ 1 & 0 & 0 \\ 3 & 0 & 2 \end{bmatrix}$

15. (a) shear in the x-direction of the xy-plane, factor -1 (b) expansion in the y-direction by a factor of 3   (c) reflection in the yz-plane   (d) shear in the x-direction of the xz-plane, factor 3   (e) shear in the y-direction of the yz-plane, factor 7   (f) shear in the z-direction of the yz-plane, factor 2   (g) shear in the z-direction of the xz-plane, factor -4 (h) shear in the y-direction of the xy-plane, factor -2

17. $\tan\theta = -c/a$ 23. $\frac{17}{30}\mathbf{v}_1 + \frac{2}{51}\mathbf{v}_2$

Exercise Set 4.5   (page 250)

1. (a) $\begin{bmatrix} 3 \\ 1 \\ -3 \end{bmatrix}, \begin{bmatrix} -2 \\ 6 \\ 0 \end{bmatrix}, \begin{bmatrix} 1 \\ 2 \\ 7 \end{bmatrix}, \begin{bmatrix} 0 \\ 1 \\ 1 \end{bmatrix}$ (b) $\begin{bmatrix} 11 \\ 5 \\ 22 \end{bmatrix}, \begin{bmatrix} -42 \\ 32 \\ -10 \end{bmatrix}, \begin{bmatrix} -56 \\ 87 \\ 17 \end{bmatrix}, \begin{bmatrix} -13 \\ 17 \\ 2 \end{bmatrix}$ (c) $\begin{bmatrix} -31 \\ 37 \\ 12 \end{bmatrix}$

3. $\begin{bmatrix} 2 & 2 & 1 \\ 0 & 0 & 0 \\ 2 & 1 & 2 \\ 0 & 0 & 0 \\ 2 & 1 & 1 \end{bmatrix}$ 5. $\begin{bmatrix} 6 & 1 & -7 & 12 \\ 8 & 1 & 9 & 9 \\ 6 & 1 & -7 & 9 \\ 0 & 0 & 0 & 0 \end{bmatrix}$ 8. $\frac{1}{9}\begin{bmatrix} 8 & -1 & -5 \\ 2 & 11 & 10 \\ 7 & 7 & 44 \end{bmatrix}$ 11. $\begin{bmatrix} 1 & 0 & 0 & 0 \\ 0 & 0 & 1 & 0 \\ 0 & 1 & 0 & 0 \\ 0 & 0 & 0 & 1 \end{bmatrix}$ 13. $\begin{bmatrix} 5 & 0 & 0 \\ 0 & 3 & 0 \\ 0 & 0 & 1 \end{bmatrix}$

Chapter 4 Review Exercises   (page 252)

1. (a) 2   (b) 8   (c) 7   (d) 10   (e) 1
3. not linear because $T(2A) = 4T(A)$
5. not linear because $T(0) \neq 0$ 7. $\mathcal{NS}(\mathbf{Q}) = \{a + dx^3\}$, $\mathcal{R}(\mathbf{Q}) = \{b + cx^2\}$, rank$(\mathbf{Q}) = 2$, nullity$(\mathbf{Q}) = 2$

9. $\begin{bmatrix} 2/3 & -5 \\ 1/3 & -3 \end{bmatrix} = \begin{bmatrix} 0 & 1 \\ 1 & 0 \end{bmatrix}\begin{bmatrix} 1 & 0 \\ 2 & 1 \end{bmatrix}\begin{bmatrix} 1 & -3 \\ 0 & 1 \end{bmatrix}\begin{bmatrix} 1/3 & 0 \\ 0 & 1 \end{bmatrix}$

10. $\frac{\sqrt{2}}{2}\begin{bmatrix} 1 & 3 \\ 3 & 3 \end{bmatrix} = \begin{bmatrix} \sqrt{2}/2 & -\sqrt{2}/2 \\ \sqrt{2}/2 & \sqrt{2}/2 \end{bmatrix}\begin{bmatrix} 0 & 1 \\ 1 & 0 \end{bmatrix}\begin{bmatrix} 1 & 0 \\ 2 & 1 \end{bmatrix}\begin{bmatrix} 1 & 0 \\ 0 & 3 \end{bmatrix}$

Chapter 5

Exercise Set 5.1   (page 261)

1. (a) $[T]_{\mathcal{R}} = \begin{bmatrix} 2 & 4 & 1 & 5 \\ -3 & 0 & 0 & 1 \\ 1 & 0 & 4 & 0 \\ 0 & -5 & 0 & -1 \end{bmatrix}$ (b) $P = \begin{bmatrix} 1 & 0 & 0 & 0 \\ -2 & 1 & 0 & 0 \\ 1 & 1 & 1 & 0 \\ 1 & -3 & 5 & 1 \end{bmatrix}$

(c) $[T]_{\mathcal{R}'} = \begin{bmatrix} 0 & -10 & 26 & 5 \\ -2 & -23 & 57 & 11 \\ 7 & 37 & -79 & -16 \\ -32 & -246 & 535 & 107 \end{bmatrix}$ (d) $T(v_1) = -2v_2 + 7v_3 - 32v_4$
$T(v_4) = 5v_1 + 11v_2 - 16v_3 + 107v_4$

3. (a) $\begin{bmatrix} 65 & 46 \\ -85 & -60 \end{bmatrix} = \begin{bmatrix} 4 & 3 \\ 3 & 2 \end{bmatrix}^{-1} \begin{bmatrix} 2 & -1 \\ 4 & 3 \end{bmatrix} \begin{bmatrix} 4 & 3 \\ 3 & 2 \end{bmatrix}$ 4. $[T]_{\mathcal{R}} = \begin{bmatrix} 1 & 0 \\ 0 & 2 \end{bmatrix}$ 5. $\begin{bmatrix} 8 & 4 \\ -9 & -4 \end{bmatrix}$

7. $[T]_{\mathcal{R}} = Dg\{5, 1, 1\}$, T is expansion by a factor of 5 in the u_1 direction.

9. $\begin{bmatrix} 3 & 0 & 0 & 0 \\ 0 & 2 & 1 & 0 \\ 0 & 0 & 2 & 0 \\ 0 & 0 & 0 & 1 \end{bmatrix}$ shear, by a factor of $1/2$ in the x_2 direction of the x_2, x_3 plane; expansions by a factor of 2 in the

x_2 and x_3 directions; expansion by a factor of 3 in the x_1 direction

17. $[T]_{\mathcal{C}'\mathcal{R}'} = [I]_{\mathcal{C}'\mathcal{C}}[T]_{\mathcal{C}\mathcal{R}}[I]_{\mathcal{R}\mathcal{R}'}$; that is, the two matrix representatives are equivalent matrices.

Exercise Set 5.2   (page 271)

In Exercises 1–15, the diagonalizing matrix P is not unique.

1. $\lambda^2 - 13\lambda + 36; 4, 9; P = \begin{bmatrix} 1 & 1 \\ -1 & 4 \end{bmatrix}$

3. $\lambda^2 - 6\lambda + 8; 2, 4; P = \begin{bmatrix} 3 & 1 \\ 1 & 1 \end{bmatrix}$

5. $\lambda^3 - 9\lambda^2 + 24\lambda - 20; 2, 2, 5; P = \begin{bmatrix} 1 & -1 & 1 \\ 1 & 0 & -1 \\ 0 & 1 & 1 \end{bmatrix}$ 6. $\lambda^3 - 9\lambda^2 + 26\lambda - 24; 2, 3, 4; P = \begin{bmatrix} 0 & 1 & 1 \\ 1 & -2 & -1 \\ 0 & -2 & -1 \end{bmatrix}$

7. $\lambda^3 - 9\lambda^2 + 24\lambda - 20; 2, 2, 5;$ not diagonalizable   9. $\lambda^3 + 4\lambda^2 - \lambda - 4; 1, -4, -1; P = \begin{bmatrix} 4 & 1 & 4 \\ 5 & 0 & -3 \\ 1 & -1 & -1 \end{bmatrix}$

11. $\lambda^4 - 13\lambda^2 + 36; 3, -3, 2, -2; P = \begin{bmatrix} 1 & 1 & 1 & 1 \\ 3 & -3 & 2 & -2 \\ 9 & 9 & 4 & 4 \\ 27 & -27 & 8 & -8 \end{bmatrix}$ 13. $(\lambda - 2)^4 - 3(\lambda - 2)^2 + 1; 2 \pm \sqrt{(3 + \sqrt{5})/2}$

15. $\lambda^3 - 3\lambda^3 + 4\lambda - 12; 3, \pm 2i; P = \begin{bmatrix} 1 & 1 & 1 \\ 3 & 2i & -2i \\ 9 & -4 & -4 \end{bmatrix}$

17. If the eigenvalues of A are $\lambda_1, \lambda_2, \ldots, \lambda_n$, then the eigenvalues of $p(A)$ are $p(\lambda_1), p(\lambda_2), \ldots, p(\lambda_n)$.

20. Define the characteristic polynomial of $\mathbf{T}$ to be the characteristic polynomial of any matrix representative of A. By Exercise 12 of Section 5.1, the result is independent of the choice of basis.

22. (a) $\lambda^3 - 2\lambda^2 - 5\lambda + 6$ **25.** $\dfrac{-3}{2}\begin{bmatrix} 3 \\ 1 \end{bmatrix} e^{2t} + \dfrac{11}{2}\begin{bmatrix} 1 \\ 1 \end{bmatrix} e^{4t}$

Exercise Set 5.3   (page 278)

1. $7 - 3i$, $-6 + 19i$, $7 + 7i$, $7 + 3i$, $22 + 7i$, $22 + 7i$, 13, 54

5. (a) $\lambda_1 = 2$, $\lambda_2 = 4$, $X_1 = \begin{bmatrix} 1 \\ -1 \end{bmatrix}$, $X_2 = \begin{bmatrix} 1 \\ 1 \end{bmatrix}$, $X_1 \cdot X_2 = 0$

7. $\dfrac{1}{5}\begin{bmatrix} 3 & -4 \\ 4 & 3 \end{bmatrix}$ **9.** $\dfrac{1}{3}\begin{bmatrix} 2 & 1 & 2 \\ -2 & 2 & 1 \\ -1 & -2 & 2 \end{bmatrix}$ **11.** $\dfrac{1}{\sqrt{5}}\begin{bmatrix} 0 & 0 & \sqrt{5} \\ 1 & -2 & 0 \\ 2 & 1 & 0 \end{bmatrix}$ **13.** $\begin{bmatrix} 1/\sqrt{2} & 1/\sqrt{2} & 0 & 0 \\ -1/\sqrt{2} & 1/\sqrt{2} & 0 & 0 \\ 0 & 0 & 3/5 & -4/5 \\ 0 & 0 & 4/5 & 3/5 \end{bmatrix}$

19. $A = (1/2)(A + A^T) + (1/2)(A - A^T)$

Exercise Set 5.4   (page 286)

1. $P = \dfrac{1}{\sqrt{17}}\begin{bmatrix} 1 & 4 \\ 4 & -1 \end{bmatrix}$, $P^T AP = \begin{bmatrix} 9 & 48/17 \\ 0 & 4 \end{bmatrix}$

2. not orthogonally similar to a triangular matrix because the eigenvalues are not real

3. $\det(\lambda I - A) = \lambda^3 - 3\lambda - 14$ has complex roots, so the given matrix is not orthogonally similar to a triangular matrix.

5. $\det(\lambda I - A) = (\lambda - 1)^2(\lambda - 5)$; $P = \dfrac{1}{\sqrt{6}}\begin{bmatrix} -\sqrt{3} & 1 & -\sqrt{2} \\ \sqrt{3} & 1 & -\sqrt{2} \\ 0 & -2 & -\sqrt{2} \end{bmatrix}$

13. (1) $A^2 - 13A + 36I = 0$ **(3)** $A^3 - 3A - 14I = 0$ **(5)** $A^{-1} = (A^2 - 7A + 11I)/5$

14. (1) $A^{-1} = (13I - A)/36$ **(3)** $A^{-1} = (A^2 - 3I)/14$

16. (1) $\begin{bmatrix} 9591 & 9089 \\ 36356 & 36858 \end{bmatrix}$ **(3)** $\begin{bmatrix} 96 & -44 & 27 \\ 176 & 35 & -24 \\ -36 & -8 & 55 \end{bmatrix}$ **(5)** $\begin{bmatrix} 463 & 468 & 468 \\ 936 & 931 & 936 \\ 468 & 468 & 463 \end{bmatrix}$

Exercise Set 5.5   (page 293)

1. $Y_0 = \begin{bmatrix} 1 \\ 1 \end{bmatrix} \longrightarrow \lambda_1 = -4.41421$, $X_1 = \begin{bmatrix} -.226540 \\ 1.0 \end{bmatrix}$

3. $Y_0 = \begin{bmatrix} 1 \\ 2 \\ 3 \end{bmatrix} \longrightarrow \lambda = 3$, $X = \begin{bmatrix} .5 \\ 1.0 \\ 1.0 \end{bmatrix}$ $Y_0 = \begin{bmatrix} 1 \\ 1 \\ 1 \end{bmatrix} \longrightarrow \lambda = 2$, $X = \begin{bmatrix} -1 \\ -1 \\ 0 \end{bmatrix}$

4. $Y_0 = \begin{bmatrix} 1 \\ 1 \\ 1 \end{bmatrix} \longrightarrow \lambda = 5, X = \begin{bmatrix} .6 \\ .6 \\ 1.0 \end{bmatrix}$

5. (a) $\lambda^2 - 4,\ \begin{bmatrix} 1 \\ 1 \end{bmatrix}, \begin{bmatrix} 1 \\ 2 \end{bmatrix}$ **(b)** $Y_1 = Y_3 = Y_5 = \begin{bmatrix} .8 \\ 1.0 \end{bmatrix},\ Y_2 = Y_4 = Y_6 = \begin{bmatrix} 1.0 \\ .5 \end{bmatrix}$ **(c)** Each sequence converges to

an eigenvector of A^2 for $\lambda = 4$.   **(e)** $\begin{bmatrix} 1 \\ 3 \\ 2 \end{bmatrix}$ and $\begin{bmatrix} 3 \\ 1 \\ 2 \end{bmatrix}$ are eigenvectors of A^2 for $\lambda = 1$.

7. (a) $\lambda^2 - 5\lambda + 22,\ (5 \pm i\sqrt{63})/2$
   (b) no apparent convergence   **(c)** $Z_8 - 5Z_7 + 22Y_6 = (A^2 - 5A + 22I)Y_6 = 0$

Chapter 5 Review Exercises   (page 294)

1. $\lambda^3 - 3\lambda^2 - 17\lambda + 3$ **3. (a)** $\begin{bmatrix} 1 \\ 1 \\ 0 \end{bmatrix}, \begin{bmatrix} -1 \\ 0 \\ 1 \end{bmatrix}$ **(b)** $\dfrac{1}{\sqrt{2}}\begin{bmatrix} 1 \\ 1 \\ 0 \end{bmatrix}, \dfrac{1}{\sqrt{6}}\begin{bmatrix} -1 \\ 1 \\ 2 \end{bmatrix}$

5. $P = \begin{bmatrix} 1 & -1 & -1 \\ 2 & 1 & 0 \\ 1 & 0 & 1 \end{bmatrix},\ P^{-1}AP = Dg\{5, 1, 1\}$ **6. (a)** $\begin{bmatrix} 2 & 9 & 11 \\ 19 & 66 & 83 \\ -12 & -49 & -61 \end{bmatrix}$ **(b)** $-13\mathbf{u}_1 - 98\mathbf{u}_2 + 23\mathbf{u}_3$

   (c) $P = \begin{bmatrix} 1 & 0 & 0 \\ 3 & 1 & 2 \\ 2 & 2 & 3 \end{bmatrix},\ [\mathbf{T}]_{\mathcal{G}} = A' = \begin{bmatrix} 3 & -5 & 7 \\ 0 & 3 & -4 \\ 1 & 1 & 1 \end{bmatrix}$

7. $\lambda^3 - 9\lambda^2 + 2\lambda + 48;\ 8, -2, 3$

9. (a) $[1\ \ 4\ \ 2\ \ 0]^T$ **(b)** $\mathbf{T(u_2)} = \begin{bmatrix} 11 & 12 \\ 5 & 3 \end{bmatrix}$ **(c)** $\begin{bmatrix} 41 & 28 \\ 18 & 7 \end{bmatrix}$

10. (a) $-4, -2, 8$ **(b)** $P = \begin{bmatrix} -1 & 1 & -1 \\ 2 & 0 & -1 \\ 1 & 1 & 1 \end{bmatrix}$

   (c) $Q = \begin{bmatrix} -1/\sqrt{6} & 1/\sqrt{2} & -1/\sqrt{3} \\ 2/\sqrt{6} & 0 & -1/\sqrt{3} \\ 1/\sqrt{6} & 1/\sqrt{2} & 1/\sqrt{3} \end{bmatrix}$

11. (a) $\begin{bmatrix} 1 & 0 & 2 \\ 0 & 2 & -1 \\ 5 & -5 & 1 \end{bmatrix}$ **(b)** $P = \begin{bmatrix} 1 & 0 & 2 \\ 1 & -1 & -1 \\ 1 & 1 & 4 \end{bmatrix}$

13. For $Q = \begin{bmatrix} -1/\sqrt{6} & 1/\sqrt{2} & -1/\sqrt{3} \\ 2/\sqrt{6} & 0 & -1/\sqrt{3} \\ 1/\sqrt{6} & 1/\sqrt{2} & 1/\sqrt{3} \end{bmatrix},\ Q^TAQ = Dg\{-2, 0, 10\}$

Chapter 6

Exercise Set 6.1   (page 306)

1.

3.

Unbounded feasible set

5.

7. max $z_1 = 750$ at $(15, 18)$,   min $z_1 = 250$ at $(0, 10)$,   max $z_2 = 370$ at $(20, 2)$,   min $z_2 = -300$ at $(0, 20)$

9. max $z = 34$ at $(3, 5)$ min $z = -\infty$; that is, z is not bounded below on this set

11.

$$3x + 2y + u \qquad\quad = 46$$
$$3x - 4y \qquad + v \qquad = 16$$
$$y \qquad\quad + w = 8$$

	1	2	3	4	5	6	7	8	9	10
x	0	0	0	0	46/3	16/3		12	10	16/3
y	0	23	−4	8	0	0	0	5	8	8
u	46	0	54	30	0	30		0	0	−18
v	16	108	0	48	−30	0		0	18	0
w	8	−15	12	0	8	8	0	3	0	0

There are ten possible basic solutions. System 7 is inconsistent, so there are only nine basic solutions. The basic feasible solutions are 1, 4, 6, 8, and 9 in the above table.

13. 40/3 acres of corn, 5/3 acres of beans; profit = $700

15. (a) $60,000 in government bonds and $540,000 in corporate bonds; $70,800

   (b) $360,000 in stocks, $180,000 in corporate bonds, and $60,000 in government bonds; $39,800

Exercise Set 6.2   (page 313)

1.
$$\begin{matrix} -2x - y \le -2 \\ x + y \le 8 \\ -x - y \le -3 \end{matrix} \quad \max[2 \quad 5]\begin{bmatrix} x \\ y \end{bmatrix} \text{ such that } \begin{bmatrix} -2 & -1 \\ 1 & 1 \\ -1 & -1 \end{bmatrix}\begin{bmatrix} x \\ y \end{bmatrix} \le \begin{bmatrix} -2 \\ 8 \\ -3 \end{bmatrix}, \begin{bmatrix} x \\ y \end{bmatrix} \ge 0$$

3. $\max[-3 \quad 4 \quad -4 \quad -6 \quad 6]X$ such that $\begin{bmatrix} 1 & 1 & -1 & 1 & -1 \\ -1 & -1 & 1 & -1 & 1 \\ 2 & 3 & -3 & 5 & -5 \\ 4 & 3 & -3 & 1 & -1 \end{bmatrix}\begin{bmatrix} x \\ y_1 \\ y_2 \\ z_1 \\ z_2 \end{bmatrix} \le \begin{bmatrix} 1 \\ -1 \\ 4 \\ 2 \end{bmatrix}, X \ge 0$

5. $\max[-1 \quad -3 \quad -2 \quad -3]\begin{bmatrix} x \\ y \\ u \\ v \end{bmatrix}$ such that $\begin{bmatrix} 1 & 1 & 0 & 0 \\ -1 & -1 & 0 & 0 \\ 0 & 0 & 1 & 1 \\ 0 & 0 & -1 & -1 \\ 1 & 0 & 1 & 0 \\ 0 & 1 & 0 & 1 \end{bmatrix}\begin{bmatrix} x \\ y \\ u \\ v \end{bmatrix} \le \begin{bmatrix} 150 \\ -150 \\ 150 \\ -150 \\ 200 \\ 250 \end{bmatrix}, \begin{bmatrix} x \\ y \\ u \\ v \end{bmatrix} \ge 0$

6. (1) $\max 2x + 5y$ such that $\begin{bmatrix} 2 & 1 & -1 & 0 & 0 \\ 1 & 1 & 0 & 1 & 0 \\ 1 & 1 & 0 & 0 & 1 \end{bmatrix}\begin{bmatrix} x \\ y \\ s_1 \\ s_2 \\ s_3 \end{bmatrix} = \begin{bmatrix} 2 \\ 8 \\ 3 \end{bmatrix}$

(3) $\max -3x + 4(y_1 - y_2) - 6(z_1 - z_2)$ such that $\begin{bmatrix} 1 & 1 & -1 & 1 & -1 & 0 & 1 \\ 2 & 3 & -3 & 5 & -5 & 1 & 0 \\ 4 & 3 & -3 & 1 & -1 & 0 & 1 \end{bmatrix}\begin{bmatrix} x \\ y_1 \\ y_2 \\ z_1 \\ z_2 \\ s_1 \\ s_2 \end{bmatrix} = \begin{bmatrix} 1 \\ 4 \\ 2 \end{bmatrix}$

(5) $\max C = -x - 3y - 2u - 3v$ such that $\begin{bmatrix} 1 & 1 & 0 & 0 & 0 & 0 \\ 0 & 0 & 1 & 1 & 0 & 0 \\ 1 & 0 & 1 & 0 & 1 & 0 \\ 0 & 1 & 0 & 1 & 0 & 1 \end{bmatrix}\begin{bmatrix} x \\ y \\ u \\ v \\ s_1 \\ s_2 \end{bmatrix} = \begin{bmatrix} 150 \\ 150 \\ 200 \\ 250 \end{bmatrix}$

7.

	1	2	3	4	5	6
x	0	0	0	3	2	4
y	0	6	2	0	0	-2
s_1	6	0	4	0	2	0
s_2	2	-4	0	-1	0	0

Solutions 1, 3, and 5 are feasible.

10. (a) i and iv   **11.** $\min .35x_1 + .5x_2$ such that $\begin{matrix} 2x_1 + 3x_2 \ge 90 \\ 4x_1 + 3x_2 \ge 120 \end{matrix}$

13. $\max P = 23PB + 22PA + 15RB + 14RA$ such that $\begin{bmatrix} 5 & 15 & 10 & 17 & 1 & 0 & 0 & 0 \\ 4 & 2 & 4 & 2 & 0 & 1 & 0 & 0 \\ 40 & 22 & 35 & 22 & 0 & 0 & 1 & 0 \\ 14 & 14 & 7 & 14 & 0 & 0 & 0 & 1 \end{bmatrix}\begin{bmatrix} PB \\ PA \\ RB \\ RA \\ S_1 \\ S_2 \\ S_3 \end{bmatrix} = \begin{bmatrix} 480 \\ 160 \\ 1190 \\ 280 \end{bmatrix}$

Exercise Set 6.3   (page 324)

1. basic variables: v and w; basic feasible solution: $[0 \quad 0 \quad 0 \quad 10 \quad 13]^T$; objective value: 0;

entering variable: u; departing variable: w; new tableau:

	x	y	u	v	w	z	
v	-11	-6	0	1	-2	0	4
u	7	4	1	0	1	0	3
	58	28	0	0	9	1	27

;

new basic feasible solution: $[0 \quad 0 \quad 3 \quad 4 \quad 0]^T$; new objective value: 27

3. basic variables: x_1, x_6, and x_4; basic feasible solution: $X = [12 \quad 0 \quad 0 \quad 10 \quad 0 \quad 6 \quad 10]^T$; objective value: 120; entering variable: x_2 (not uniquely determined); departing variable: x_1 (not unique);

new tableau:

	x_1	x_2	x_3	x_4	x_5	x_6	x_7	z	
x_2	1/4	1	1/4	0	1/4	0	3/4	0	3
x_6	$-1/2$	0	5/2	0	3/2	1	$-7/2$	0	0
x_4	5/4	0	25/4	1	17/4	0	31/4	0	25
	2	0	-6	0	6	0	0	1	144

new basic feasible solution: $X = [0 \quad 3 \quad 0 \quad 25 \quad 0 \quad 0 \quad 0]^T$; new objective value: 144

5. $x = 0$; $y = 160/3$; $z = 160$ **7.** no optimal solution since the feasible set is unbounded

9. $W = -2$ for $x_1 = 0$, $x_2 = 1$, and $x_3 = 2$.   **11.** $P = 620$ for $x_1 = 20$, $x_2 = 10$, and $x_3 = x_4 = 0$.   **13.** $P = 700$ for $C = 40/3$, $B = 5/3$, and $Z = 0$.   **15.** $R = 384$ at $[0 \quad 10 \quad 0]^T$.   **17.** $R_1 \leftarrow -R_2 + R_1$ and $R_3 \leftarrow 5R_2 + R_3$ yield

	x_1	x_2	x_3	x_4	z	
x_3	1	0	1	1	0	3
x_2	1	1	0	-1	0	6
	1	0	0	-5	1	30

$\longrightarrow$

	x_1	x_2	x_3	x_4	z	
x_4	1	0	1	1	0	3
x_2	2	1	1	0	0	9
	6	0	5	0	1	45

19. If the problem is degenerate, then B is a linear combination of $m - 1$ columns of A. These columns of A and B yield a singular submatrix.

Exercise Set 6.4   (page 331)

1. $x = 2$, $y = 1$, $z = 0$, $P = 4$.   **3.** no optimal solution since the feasible set is unbounded
5. $C = .45707$ for $E = 1.4146$ and $M = .5732$ **7.** $x = 150$, $u = 0$, $y = 0$, $v = 150$. Cost = 450

Chapter 7

Exercise Set 7.1   (page 340)

1. $M(G) = \begin{bmatrix} 0 & 1 & 0 & 2 & 0 \\ 1 & 0 & 1 & 1 & 1 \\ 0 & 1 & 0 & 1 & 2 \\ 2 & 1 & 1 & 0 & 0 \\ 0 & 1 & 0 & 2 & 0 \end{bmatrix}$ three paths of length 2 connecting T1 and T3; nine paths of length 3 connecting T4 and T3

2. (a) $\begin{bmatrix} 0 & 1 & 1 & 0 \\ 0 & 0 & 1 & 0 \\ 0 & 0 & 0 & 1 \\ 1 & 1 & 0 & 0 \end{bmatrix}$

3.

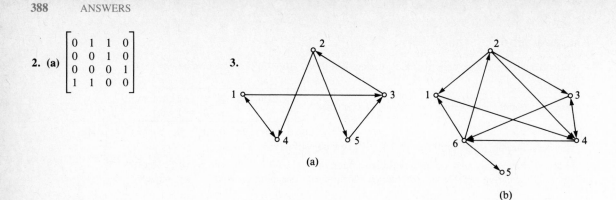

(a)

(b)

5. Two digraphs, G and G', are equivalent if there is a one-to-one mapping $\mathbf{T}$, of the vertices of G onto the vertices of G', which preserves edges in the sense that there is an edge connecting $\mathbf{T}(Vi)$ and $\mathbf{T}(Vj)$, in G' if and only if there is an edge connecting Vi and Vj in G. If A is the matrix of G and A' is the matrix of G', then there is a permutation matrix P such that $A' = PAP^T$.

7. $A(G) = \begin{matrix} & \begin{matrix} A & B & C & D & E & F \end{matrix} \\ \begin{bmatrix} 0 & 1 & 0 & 0 & 1 & 1 \\ 0 & 0 & 1 & 0 & 1 & 0 \\ 1 & 0 & 0 & 0 & 1 & 0 \\ 0 & 1 & 1 & 0 & 0 & 1 \\ 0 & 0 & 0 & 1 & 0 & 0 \\ 0 & 0 & 1 & 0 & 1 & 0 \end{bmatrix} & \begin{matrix} A \\ B \\ C \\ D \\ E \\ F \end{matrix} \end{matrix}$ Adam and Dan influence the most people. Ed is the most influenced. Dan is the best choice for leader.

8. This is a slight perturbation of the actual results of the 1985 season. Michigan is the graph-theory choice (35 to 32) to go to the Rose Bowl.

Exercise Set 7.2   (page 349)

1. (a) $y = -4/3 + 3x$; $\Sigma d^2 = .666$; for $y = 2x$, $\Sigma d^2 = 4$.

   (c) $y = 3.75 + .44x$; $\Sigma d^2 = .044$; for $y = 3.7 + .5x$, $\Sigma d^2 = .10$.

   (d) $y = 597 - 3.12x$; $\Sigma d^2 = 15.96$; for $y = 592 - 3x$, $\Sigma d^3 = 17$.

2. (a) $y = 9.55 - 15.175x + 1.3125x^2$ **3. (a)** $z = 2.3 - .9x - .6y$ **(c)** For $z = 7.4 + 3x - 5.5y$; $\Sigma d^2 = .7$.
   For $z = 3x - 5y + 6$, $\Sigma d^2 = 4$.   **5.** $(1, 1)$ and $(4, 0)$; the lines are nearly parallel.

6. 2.(a) $\begin{bmatrix} 1 & -2 & 4 \\ 1 & 0 & 0 \\ 1 & 1 & 2 \\ 1 & 4 & 16 \end{bmatrix} = \begin{bmatrix} .5 & .6708 & -.5345 \\ .5 & .2236 & .8018 \\ .5 & -.2236 & 0 \\ .5 & -.6708 & -.2673 \end{bmatrix} \begin{bmatrix} 2 & 2 & 12 \\ 0 & -4.472 & -8.944 \\ 0 & 0 & 8 \end{bmatrix}$ **7.** $a = 1.565, h = 5.565$

9. (d) $\mathbf{w} = \dfrac{1}{\sqrt{260}} \begin{bmatrix} 10 \\ -4 \\ -12 \end{bmatrix}$ **(f)** $A = QR$ where Q consists of the first n columns of $H_1 H_2 \ldots H_{n-1}$

Exercise Set 7.3   (page 359)

1. Hyperbola: $5x'^2 - y'^2 = 10$; $P = \dfrac{1}{\sqrt{2}} \begin{bmatrix} 1 & -1 \\ 1 & 1 \end{bmatrix}$ **3.** Hyperbola: $18x'^2 - 2y'^2 = 36$; $P = \dfrac{1}{\sqrt{5}} \begin{bmatrix} 2 & -1 \\ 1 & 2 \end{bmatrix}$

5. Ellipsoid: $2x'^2 + 5y'^2 + 2z'^2 = 10$; $P = \dfrac{1}{\sqrt{6}} \begin{bmatrix} -2 & -\sqrt{2} & 0 \\ -1 & \sqrt{2} & \sqrt{3} \\ 1 & -\sqrt{2} & \sqrt{3} \end{bmatrix}$

7. Hyperboloid of two sheets: $-x'^2 + 8y'^2 - z'^2 = 16$; $P = \begin{bmatrix} -5/\sqrt{45} & 2/3 & 0 \\ 4/\sqrt{45} & 2/3 & 1/\sqrt{5} \\ 2/\sqrt{45} & 1/3 & -2/\sqrt{5} \end{bmatrix}$

9. Hyperbolic cylinder along the z'-axis $x'^2 - y'^2 = 1$; $P = \begin{bmatrix} -2/3 & 2/3 & -1/\sqrt{6} \\ 2/3 & 1/3 & -2/\sqrt{6} \\ 1/3 & 2/3 & 1/\sqrt{6} \end{bmatrix}$

Exercise Set 7.4   (page 365)

1. (a) $c\begin{bmatrix} 30 \\ 56 \\ 33 \end{bmatrix}$ (b) $c\begin{bmatrix} 24 \\ 25 \\ 16 \end{bmatrix}$ (c) $c\begin{bmatrix} 4 \\ 3 \\ 4 \end{bmatrix}$ 3. $c[570 \quad 1058 \quad 723 \quad 739]^T$

5. (a) \$.70 (b) 95 percent (c) 70 percent   6. (a) $\begin{bmatrix} 298065 \\ 148016 \\ 209115 \\ 214096 \end{bmatrix}$ (b) $\begin{bmatrix} 299721 \\ 148650 \\ 212655 \\ 214766 \end{bmatrix}$ (c) $\begin{bmatrix} 284293 \\ 142745 \\ 179671 \\ 208517 \end{bmatrix}$

Index